W9-DBH-726

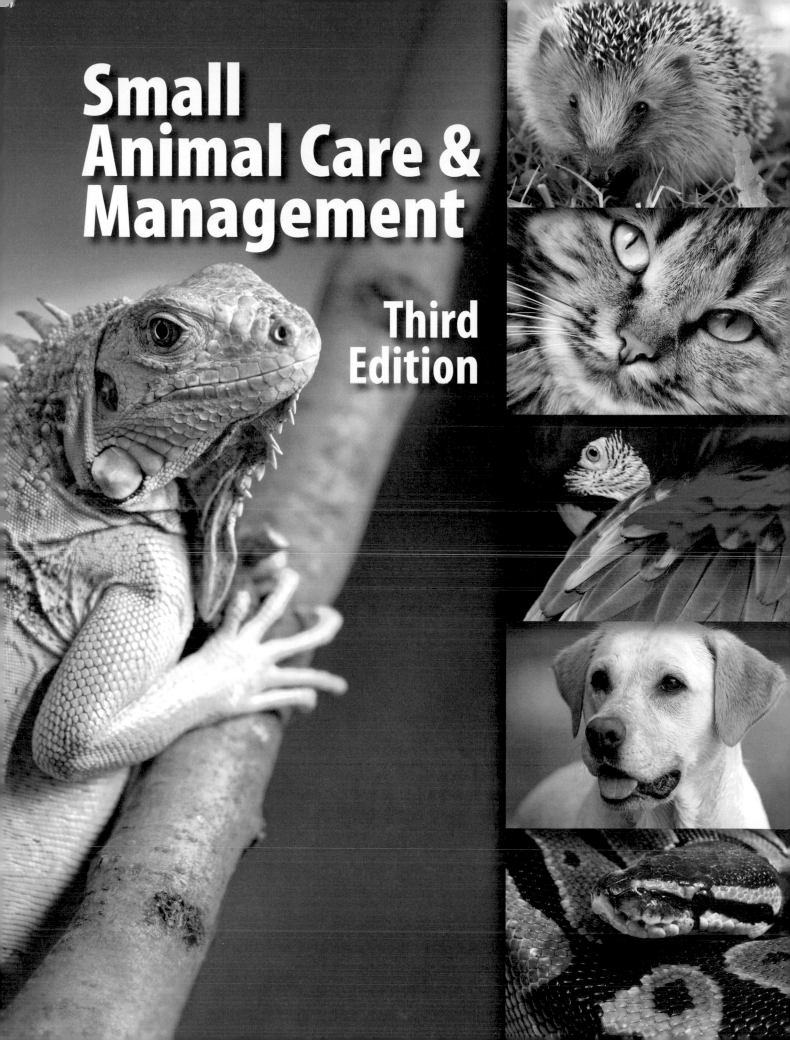

Small Animal Care & Management

Third Edition

Join us on the web at

agriculture.delmar.cengage.com

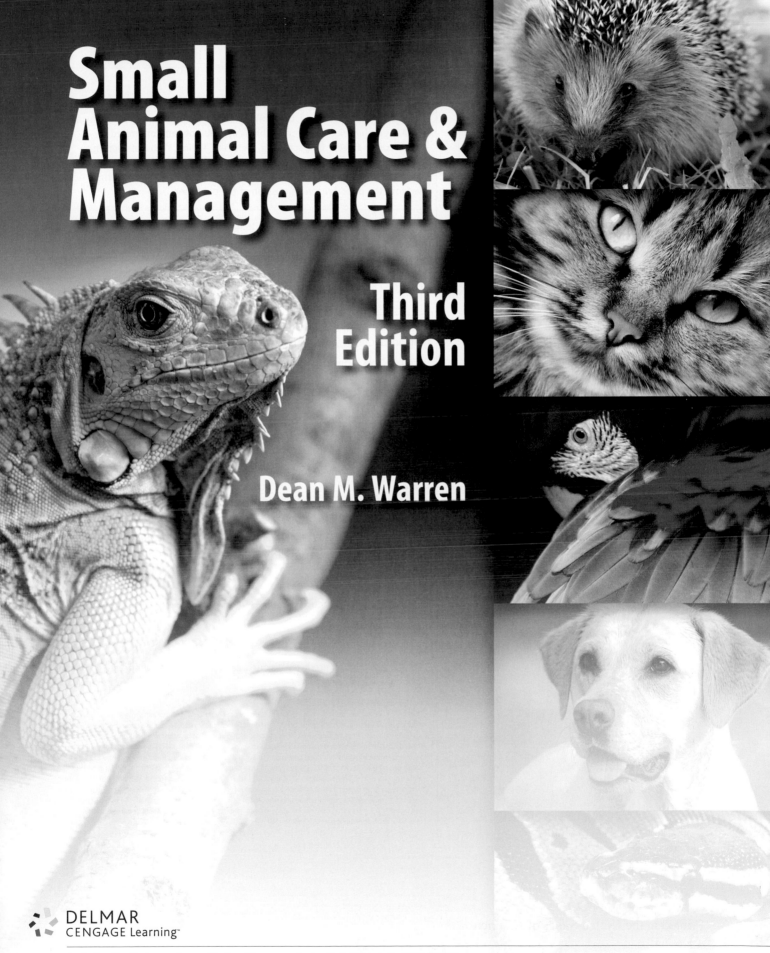

Small Animal Care & Management

Third Edition

Dean M. Warren

DELMAR
CENGAGE Learning™

Australia • Brazil • Japan • Korea • Mexico • Singapore • Spain • United Kingdom • United States

Small Animal Care and Management, Third Edition

Dean M. Warren

Vice President, Career and Professional Editorial: Dave Garza

Director of Learning Solutions: Matthew Kane

Acquisitions Editor: David Rosenbaum

Managing Editor: Marah Bellegarde

Product Manager: Christina Gifford

Editorial Assistant: Scott Royael

Vice President, Career and Professional Marketing: Jennifer McAvey

Marketing Director: Debbie Yarnell

Marketing Coordinator: Jonathan Sheehan

Production Director: Carolyn Miller

Production Manager: Andrew Crouth

Content Project Manager: Anne Sherman

Art Director: David Arsenault

Technology Project Manager: Tom Smith

Production Technology Analyst: Thomas Stover

© 2010, 2002, 1995 Delmar, Cengage Learning

ALL RIGHTS RESERVED. No part of this work covered by the copyright herein may be reproduced, transmitted, stored or used in any form or by any means graphic, electronic, or mechanical, including but not limited to photocopying, recording, scanning, digitizing, taping, Web distribution, information networks, or information storage and retrieval systems, except as permitted under Section 107 or 108 of the 1976 United States Copyright Act, without the prior written permission of the publisher.

For product information and technology assistance, contact us at **Cengage Learning Customer & Sales Support, 1-800-354-9706**

For permission to use material from this text or product, submit all requests online at **www.cengage.com/permissions**
Further permissions questions can be emailed to **permissionrequest@cengage.com**

Library of Congress Control Number: 2009921854

ISBN-13: 978-1-4180-4105-2

ISBN-10: 1-4180-4105-X

Delmar
Executive Woods
5 Maxwell Drive
Clifton Park, NY 12065
USA

Cengage Learning is a leading provider of customized learning solutions with office locations around the globe, including Singapore, the United Kingdom, Australia, Mexico, Brazil, and Japan. Locate your local office at **www.cengage.com/global**

Cengage Learning products are represented in Canada by Nelson Education, Ltd.

To learn more about Delmar, visit **www.cengage.com/delmar**

Purchase any of our products at your local bookstore or at our preferred online store **www.ichapters.com**

Notice to the Reader

Publisher does not warrant or guarantee any of the products described herein or perform any independent analysis in connection with any of the product information contained herein. Publisher does not assume, and expressly disclaims, any obligation to obtain and include information other than that provided to it by the manufacturer. The reader is expressly warned to consider and adopt all safety precautions that might be indicated by the activities described herein and to avoid all potential hazards. By following the instructions contained herein, the reader willingly assumes all risks in connection with such instructions. The publisher makes no representations or warranties of any kind, including but not limited to, the warranties of fitness for particular purpose or merchantability, nor are any such representations implied with respect to the material set forth herein, and the publisher takes no responsibility with respect to such material. The publisher shall not be liable for any special, consequential, or exemplary damages resulting, in whole or part, from the readers' use of, or reliance upon, this material.

Printed in the United States of America
4 5 6 7 12

CONTENTS

SECTION II

CHAPTER 21   Hedgehogs   **541**

CHAPTER 22   Sugar Gliders   **554**

PREFACE

When I started a class on small animal care several years ago, it soon became apparent that there weren't any textbooks available in the subject area. I purchased numerous books from pet shops and book stores to use in developing my course and spent many hours creating teaching materials, quizzes, and tests. Since some educators may be reluctant to start a course in small animal care because of their lack of knowledge and the lack of teaching materials in this subject area, I proceeded to write a text and an accompanying Instructor's Guide.

I aimed to write a book that would be flexible enough to be used by the general reader, pet enthusiast, and students with varying abilities at varying grade levels. If it is used as a text for an animal science class, the instructor can decide how difficult or easy to make the course. The instructor can also decide what chapters to cover and whether to cover all the material in a specific chapter. The reader of the text should realize that not all animals discussed in this text are used exclusively as pets. Some animals are used for food and for fur production. Handling and management practices may vary depending on what the animal is being produced for. Numerous scientific names are used; the instructor must decide if the student needs to learn them. This book also encourages the reader to research and explore subject areas. In a text of this type, it is impossible to thoroughly cover each area. Readers may wish to seek additional information, and for this reason, Internet sites were included in the second edition and this third edition. The author and Delmar made every attempt to ensure that all Web site URLs were accurate at the time of printing. However, due to the fluid nature of the Internet, we cannot guarantee they will remain current for the life of the edition.

The Instructor's Guide contains the same discussion questions that are found at the end of each chapter and the answers to the discussion questions. Also available through the publisher is a Classmaster CD-ROM, which contains the Teacher's Resource Guide, the Instructor's Guide, a PowerPoint slide presentation and a computerized test bank, as well as a Classroom Interactivity CD-ROM and Student Workbook. With these supplements, in addition to this text, I believe that an instructor will have the basic materials to meet local and/or state curriculum standards to teach small animal care courses.

The text contains a glossary of terms that might be unfamiliar to the reader. An index is also provided to assist one in quickly finding subject areas and topics. A strong point of the text is that it has many excellent full-color photographs and illustrations.

The pet and companion animal industry continues to be a rapidly growing segment of our economy. Employees with knowledge and skills to work with small animals are constantly needed.

It is the sincere hope of this author that this text meets the needs of those who wish to learn about the proper care and management of small animals.

ACKNOWLEDGMENTS

I wish to thank the many persons who have helped in putting this text together. I especially want to thank Mary Martha Scott, former English teacher and librarian at Columbia City High School, Columbia City, Indiana, for her assistance in checking the manuscript for grammatical errors and for assisting in locating reference materials.

I want to thank Isabelle Francais for her excellent photographs of many of the animals found in the book. Without her generous assistance, this book wouldn't have the exceptional visual element of illustrations that it contains. Aaron Norman provided the excellent photographs of fish. Chapter 20 is greatly enhanced because of his photographs.

I wish to thank the United States Department of Agriculture for providing photographs and resource material on diseases.

Dr. Jack L. Albright of Purdue University and William McVay, retired agriculture teacher at Whitko High School, South Whitley, Indiana, provided materials and input on animal rights and animal welfare. Cindy Raker, agriculture educator at Carroll High School, Fort Wayne, Indiana, provided me with her class outline and materials on dogs and cats. It was her material that I used in my early classes on small animal care.

I wish to thank my former students; the staff of Columbia City High School, Columbia City, Indiana; and the agriculture educators in Indiana for their support and encouragement throughout this project, especially during the first and second editions.

I want to thank Jack Caster, Computer Coordinator, for his assistance with word processing and other computer problems during the writing of the first and second editions.

A special thanks goes to Delmar for seeing a need for a text on small animal care and giving me an opportunity to provide one. And, also to Christina Gifford and all the staff at Delmar who had a hand in putting this text together. I truly appreciate their assistance and patience throughout this project.

The author and Delmar also wish to thank the following reviewers for their time and content expertise:

Sarah Hurley
Parkland College
Champaign, IL

John Borst
Millbrook High School
Winchester, VA

Vicki Judah
(ret) Jordan Applied Technology Center
Sandy, UT

Todd Kohlepp
Fall Creek High School
Fall Creek, WI

Samuel Shapiro
John Bowne High School
Flushing, NY

Shari Graffunder
River Valley High School
Spring Green, WI

Maxine Chipman
Wonewoc Center
Wonewoc, WI

SMALL ANIMAL CARE & MANGEMENT, 3E

ADDED FEATURES

- Pedagogical use of FULL COLOR to accelerate learning through highly illustrative material.
- Competency-based level of achievement needed for success is spelled out at the start of each chapter.
- FULL-COLOR species identification.

ENHANCED CONTENT

The updated and enhanced content addresses the evolving small animal curriculum:

- The physical restraint of animals and general guidelines for risk and safety management for small animals is discussed in depth.
- Expanded coverage of all species. This coverage provides the most up-to-date information available on small animal anatomy, handling, feeding, housing and equipment, history, breeds, uses, grooming, training, diseases and ailments, and reproduction.
- Extensive coverage of the skills and educational requirements needed for small animal students to move on to a specialized field of study.

EXTENSIVE TEACHING/ LEARNING PACKAGE

Instructor's Guide to Text

The Instructor's Guide provides answers to the end-of-chapter questions and the student workbook, as well as additional material to assist the instructor in the preparation of lesson plans.

Student Workbook

A new student workbook has been developed to provide students with fun activities to do. Word Scrambles, Crossword puzzles, and other self-directed activities are included to help reinforce the materials contained in the text.

Classmaster CD-ROM

This supplement provides the instructor with valuable resources to simplify the planning and implementation of the instructional program. It includes over 300 instructor slides in **PowerPoint** focusing on each chapter's key points to facilitate classroom presentations; a **Computerized Test Bank** of over 950 questions with answers and Internet testing capability, and a PDF of the **Instructor's Manual** containing answer keys to the end of chapter review questions, as well as the student workbook.

Classroom Interactivity CD-ROM

This technology provides the instructor with a tool to test students on their retention of the materials in a fun, interactive format. The Classroom Interactivity CD-ROM has 4 different software applications each resembling a popular game show format. The games are comprised of a variety of multiple choice, true/false, and fill-in-the-blank questions based on the content in the core textbook.

SECTION 1

Introduction to Small Animal Care

OBJECTIVES

After reading this chapter, one should be able to:

🐾 briefly describe the history of animals on the earth.

🐾 list the important time periods in the evolution of animals.

🐾 describe how animals probably became domesticated.

🐾 describe the importance of the small animal industry.

🐾 describe how organisms are classified.

TERMS TO KNOW

Animalia	Monera	Plantae
Aves	notochord	Protista
binomial nomenclature	Osteichthyes	Reptilia
	paleontologists	taxa
Chordata	pariah dogs	taxonomy
Fungi	pharyngeal	trinomial nomenclature
invertebrates	placental mammals	
Mammalia		vertebrate

A BRIEF HISTORY OF ANIMALS

The oldest direct traces of life on earth date back 3.4 to 3.5 billion years. In rocks that age in Australia and Southern Africa, geologists have found stromatolites, layered structures created through the activity of primitive algae and bacteria.

Other Australian rocks of similar age provide even more direct evidence of ancient life. Sections of these rocks, known as cherts, show fossilized remains of blue-green algae.

Rocks also reveal even more distant, indirect traces of life. Living things use particular isotopes (atomic forms) of the element carbon preferentially. The mix of carbon isotopes detected in rocks from Greenland more than 3.8 billion years old shows evidence of life on earth.[1]

Some 2.2 billion years ago, free oxygen was present in the atmosphere. Living things used this reactive substance in the biochemical functions of their cells. The free oxygen in the atmosphere also produced a layer of ozone, which filters out the ultraviolet light from the sun that is harmful to life. *See Figure 1–1.*

Era	Period	Epoch	Biologic events	Years before present (B.P.)	Geologic events Events refer especially to North America and do not reflect great worldwide variations
PRE-CAMBRIAN	Hadean Archeon Proterozoic		Fossil algae; other fossils extremely rare; evidence of sponges and worm burrows	4.5 billion 3.8 billion 2.5 billion	Volcanic activity; mountain building; glaciations; variable climate
PALEOZOIC	Cambrian		Origin of many invertebrate phyla and classes; trilobites dominant; marine algae	600 million	Three periods of land submergence; mild
PALEOZOIC	Ordovician		Ostracoderms (first vertebrates); abundant marine invertebrates; first land plants	500 million	Submergence of land; warm
PALEOZOIC	Silurian		First jawed fishes	425 million	Continental seas and reefs; mild
PALEOZOIC	Devonian		First amphibians; freshwater fishes abundant; bryozoans and corals	405 million	Inland seas; first forests; mild
PALEOZOIC	Carboniferous	Mississippian	Radiation of amphibians; abundant sharks; scale trees and seed ferns	345 million	Inland seas; warm to hot; swamplands
PALEOZOIC	Carboniferous	Pennsylvanian	First reptiles; giant insects; great conifer forests	310 million	Shallow inland seas; extensive coal deposits; warm and moist
PALEOZOIC	Permian		Radiation of reptiles; ; displacement of amphibians; extinction of many marine invertebrates	280 million	Continents elevated; building of Appalachians; cold and dry
MESOZOIC	Triassic		First dinosaurs; mammal-like reptiles; conifers dominate plants	230 million	Continents elevated; widespread deserts; cool and dry
MESOZOIC	Jurassic		First birds; first mammals; dinosaurs abundant	180 million	Continents with shallow seas, building of Sierra Nevada mountains; cool then mild
MESOZOIC	Cretaceous		Climax of giant land and marine reptiles, followed by extinction; flowering plants; decline of gymnosperms	135 million	Spread of inland seas and swamps; building of Andes, Himalayas, Rockies; mild to cool
CENOZOIC	Tertiary	Paleocene	First placental mammals	70 million	Mountain building; subtropical
CENOZOIC	Tertiary	Eocene	Radiation of placentals	60 million	Mountain erosion; rain and mild
CENOZOIC	Tertiary	Oligocene	Apes, monkeys, whales	40 million	Mountain erosion; mild
CENOZOIC	Tertiary	Miocene	Abundant grazing mammals	25 million	Plains and grasslands; moderate
CENOZOIC	Tertiary	Pliocene	Large carnivores	7 million	Continental elevation; cool
CENOZOIC	Quaternary	Pleistocene	Early man	0.5 to 3 million	Ice ages; then warmer
CENOZOIC	Quaternary	Recent	Modern man	11 thousand	

Figure 1–1 The history of the earth is divided into periods of time. This geologic time chart shows the geologic events that occurred in North America and the living organisms that were present during the geologic events.

It is believed that the first land plants were established about 420 million years ago during the Ordovician period. This is also the period of time when **invertebrates,** such as arthropods and worms, appeared on land. These first terrestrial invertebrates fed on decaying plant material.

Vertebrate animals, in the form of amphibians and reptiles, appeared during the Devonian period about 370 million years ago. The bony fish appeared during the Ordovician period and became abundant during the Devonian period. *See Figure 1–2.*

During the Triassic period, 180 million years ago, dinosaurs appeared along with mammal-like reptiles. The first birds and mammals appeared 135 million years ago during the Jurassic period, when dinosaurs became abundant. Most **paleontologists,** those who deal with prehistoric life forms through the study of fossils, agree that birds and dinosaurs are closely related. However, there is controversy among paleontologists about how they are related. One theory suggests that birds are a branch of the dinosaur family tree that survived the Cretaceous period and evolved into the forms we see today. A second theory suggests that birds and dinosaurs had a common ancestor that evolved into both groups. *See Figures 1–3 and 1–4.*

During the Cretaceous period 135 to 70 million years ago, dinosaurs and marine reptiles reached their period of greatest abundance and then disappeared. The reason for their demise is still debated by scientists.

With the extinction of the dinosaurs, an opportunity arose for smaller creatures to evolve and become abundant. During the Paleocene and Eocene epoch 70 to 40 million years ago, **placental mammals** evolved, dispersed, and adapted to new environments. Placental mammals have a placenta through which the embryo and fetus are nourished while in the uterus.

Many of the small animals discussed in this book are placental mammals and evolved during this period.

Figure 1–2   Coelacanth, a primitive, lobe-finned, bony fish that some believe may be the ancestor of land animals. Coelacanth was thought to be extinct until some living specimens were discovered off the East Coast of South Africa. Negative #325025 (Photo by Rota). *Courtesy Department of Library Services, American Museum of Natural History.*

Figure 1–3 Archaeopteryx is one of the ancestors of modern birds. It lived 150 million years ago during the Jurassic period. This fossil was discovered in a Bavarian stone quarry. Negative #34711. *Courtesy Department of Library Services, American Museum of Natural History.*

DOMESTICATION OF ANIMALS

Dogs are believed to be the first animals domesticated. There are two main theories about how canine domestication occurred. Dogs may have evolved from a canine ancestor similar to the pariah dogs that populate many underdeveloped countries today. **Pariah dogs** are semi-feral dogs that live in close approximation to towns, where they survive by scavenging on garbage. This theory holds that individual pariah-type dogs that allow humans to get fairly close to them before they flee obtain more food and therefore reproduce more successfully than those dogs that run more readily from people. If the greater tolerance to human presence is inherited, over time the canine ancestor could have evolved into the domestic dog. Another popular idea is that dogs are domesticated wolves. In this theory, wolf pups are hand-raised by early humans. These human-raised wolves would alert their owner to danger by barking or perhaps would have assisted in hunting. Currently, archeological evidence indicates that domestication of the dog occurred about 14,000 years ago.

The cat's association with humans dates back to about 3,500 years ago. Pictures of cats have been found in the Egyptian tombs. The domestic cat probably provided pleasure and companionship, as well as controlled rats and mice.

Figure 1–4 A reconstruction of Archaeopteryx from fossil remains. Illustrated by Ron Ervin.

THE SMALL ANIMAL INDUSTRY

Pet ownership is at an all-time high—59.5 percent of all American families, about 68.7 million households, include at least one companion animal. *See Figure 1–5.* Cats outnumber dogs as the most common household pet (*see Figure 1–6*); there are more than 81.7 million cats in the United States

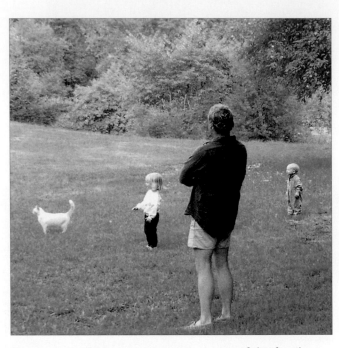

Figure 1–5 Pets are an important part of the family. *Courtesy of Carolyn Miller.*

Figure 1–6 Animals can teach children many things, such as responsibility, compassion, and trust. *Courtesy of Tracey A. Carter.*

compared with 72 million dogs. Dogs are found in 37.2 percent of the nation's households and cats in 32.4 percent (see Table 1–1). Most cat owners have an average of 2.0 cats. American households also contain 11.2 million birds, 6.1 million rabbits, 1.2 million hamsters, and 76 million fish.[2] See Table 1–2.

The pet supply industry is a large and growing one. The exact number of stores dealing in pet sales, pet equipment, and pet foods is hard to determine because of the great variation in size and type of stores (*see Figure 1–7*). Marshall Myers of the Pet Industry Joint Advisory Council gives his best estimate as follows:

1. Mass merchandisers, large stores that carry pet supplies and perhaps live fish, at 6,000 to 7,000 stores nationwide.

2. Mass merchandisers, mainly grocery and dairy stores, that carry some pet supplies.

3. Pet specialty, large-format, multi-chain stores at 1,800 to 1,900 stores.

4. Independents, which include a few large-format single stores or small chains, at 7,500 to 8,500, with the majority handling live animals and products.

5. Feed stores, garden centers, hardware centers, and so on that normally carry some pet products and perhaps a few live animals.

TABLE 1-1 United States Pet Ownership—The Percentage and Number of Pet-Owning Households and Pet Population Estimates, December 31, 1987, 1991, 1996, 2001, and 2006.

	Percentage of Households (%)					Number of Households (Millions)					Average (Mean) Number Owned Per Household					Total Population (Millions)				
	1987	1991	1996	2001	2006	1987	1991	1996	2001	2006	1987	1991	1996	2001	2006	1987	1991	1996	2001	2006
Dogs	38.2	36.5	31.6	36.1	37.2	34.7	34.6	31.2	37.8	43.0	1.51	1.52	1.69	1.6	1.7	52.4	52.5	52.9	61.6	72.0
Cats	30.5	30.9	27.3	31.6	32.4	27.7	29.2	27.0	33.2	37.5	2.04	1.95	2.19	2.1	2.2	54.6	57.0	59.1	70.8	81.7
Birds	5.7	5.7	4.6	4.6	3.9	5.2	5.4	4.6	4.8	4.5	2.48	2.16	2.74	2.1	2.5	12.9	11.0	12.6	10.1	11.2
Horses	2.8	2.0	1.5	1.7	1.8	2.6	1.9	1.5	1.5	2.1	2.63	2.54	2.67	2.9	3.5	6.6	4.9	4.0	5.1	7.3

Source: Statistics obtained from "U.S. Pet Ownership & Demographics Sourcebook," Center for Management, American Veterinary Medical Association, 1987, 1991, 1996, 2001, 2006.

TABLE 1-2 United States Pet Ownership—The Number of Specialty and Exotic Pet-Owning Households and Pet Population Estimates, December 31, 1991, 1996, 2001, and 2006.

Type of Pet	Number of Pets Per Household				Number of Households (1,000)				Population of Pets (1,000)			
	1991	1996	2001	2006	1991	1996	2001	2006	1991	1996	2001	2006
Fish*	9.05	8.92	7.70	8.4	2,652	6,228	6,396	9,036	23,997	55,554	49,251	75,898
Ferrets*	1.45	2.00	2.10	2.1	189	395	472	505	275	791	991	1,060
Rabbits*	3.22	2.63	2.70	3.3	1,420	1,878	1,783	1,870	4,574	4,940	4,813	6,171
Hamsters	1.39	1.86	1.20	1.5	947	1,008	734	826	1,316	1,876	881	1,239
Guinea Pigs	1.77	1.87	1.20	1.6	473	583	524	628	838	1,091	629	1.004
Gerbils	2.18	2.76	1.90	2.3	284	277	168	187	619	764	319	431
Other rodents	2.31	2.42	2.50	2.1	379	435	315	452	875	1,053	786	949
Turtles	1.87	1.78	1.70	1.8	379	534	629	1,106	708	950	1,070	1,991
Snakes	3.88	4.14	2.10	1.5	189	217	315	390	735	900	661	586
Lizards	1.66	1.55	1.30	1.5	189	455	419	719	314	705	545	1,078
Other reptiles	2.97	2.75	1.90	2.9	95	336	315	69	281	924	598	199
Other birds (pigeons & poultry)	13.78	13.16	9.19	10.7	379	336	315	464	5,220	4,423	2,894	4.966
Livestock	7.12	11.61	5.60	15.1	473	524	524	728	3,371	6,083	2,936	10,995
All others	3.37	3.26	2.40	3.1	189	376	839	1,182	638	1,225	2,013	3,664

*Fish, ferrets, and rabbits were specifically listed on the 1996 survey, but were not listed in 1991.

Source: Information obtained from "U.S. Pet Ownership & Demographics Sourcebook," Center for Information Management, American Veterinary Medical Association, 1997, 2001, and 2006.

Figure 1-7 Various pet foods are available on the market today. *Courtesy of Brian Yacur and Guilderland Animal Hospital.*

Figure 1–8 Animals are also used in research. *Courtesy of Agricultural Communications Department, University of Georgia.*

Altogether this would give us a number of between 15,000 and 18,000 stores that sell pets, pet equipment, pet foods, and other supplies.[3]

The American Pet Products Manufacturers Association (AAPMA) listed total U.S. pet industry expenditures at $38.5 billion in 2006. This is up from $28.5 billion in 2002, an increase of 35 percent. The breakdown of the $38.5 billion is as follows: food—$15.4 billion; veterinary care—$9.2 billion; supplies and over-the-counter medicines—$9.3 billion; live animal purchases—$1.9 billion; and pet services such as grooming and boarding—$2.7 billion. Estimates by the APPMA for 2007 are expenditures of $40.8 billion.[4]

Estimated veterinary expenditures in 2006 by the American Veterinary Medical Association (AVMA) in its publication, *U.S. Pet Ownership & Demographics Sourcebook,* were nearly $16.1 billion for dogs, $7.1 billion for cats, and $102.8 million for birds. These figures represent increases in veterinary expenses of 38.8 percent for dogs and 7.6 percent for cats since 2001.[5]

Figures from the United States Department of Agriculture in the Animal Welfare Enforcement Report for fiscal year 2004 showed a total of 1,101,958 animals used in laboratory experiments. Of this total, there were 64,932 dogs, 23,640 cats, 54,498 primates, 244,104 guinea pigs, 175,721 hamsters, and 261,573 rabbits. Since the use of birds, laboratory rats, and laboratory mice is currently excluded from Animal Welfare Act (AWA) regulation, no accurate figure is available, although the Animal Welfare Enforcement Report lists 171,312 other animals used in experiments. *See Figure 1–8.*

Registered with the AWA for fiscal year 2003 were 1,088 research facilities, 2,210 exhibitors, 4,179 dealers, 78 carriers, and 231 intermediate handlers. Research facilities include hospitals, colleges and universities, diagnostic laboratories, and many private firms in the pharmaceutical and biotechnology industries. Exhibitors operate animal acts, carnivals, circuses, public zoos, roadside zoos, and marine mammal displays. Carriers include airlines, motor freight lines, railroads, and other shipping lines. Intermediate handlers usually provide services for animals between consignor and carrier; they also care for animals delayed in transit. Dealers are persons selling laboratory animals for research and teaching, wild animals for exhibition, or pet animals at the wholesale level.

Animal care is a very large and rapidly growing industry. Pet care workers, zoo administrators, laboratory animal care workers, small animal breeders, animal trainers, animal groomers, veterinarians, veterinary technicians, and biologists will continue to be needed for this important industry.

CLASSIFICATION OF ORGANISMS

Taxonomy is the science concerned with the naming and classification of organisms. The Greek philosopher and biologist Aristotle is credited with the first attempts at classifying organisms based on their structural similarities. An English naturalist, John Ray (1627–1705), developed a more comprehensive system of classification, but Carolus Linnaeus (1707–1778), a Swedish botanist, is credited with developing our modern method of classification.

Linnaeus's system has been expanded since then and today is made up of seven categories, or **taxa.** These categories are kingdom, phylum, class, order, family, genus, and species. These categories can be subdivided into finer categories, such as subphylum or subclass. Today about 30 different taxa are recognized.

All organisms are divided into one of five kingdoms:

1. **Monera** are the bacterial organisms that lack a true nucleus in the cell and reproduce by fission.

2. **Protista** are single-celled or microscopic animals, including algae.

3. **Plantae** include multicellular photosynthesizing organisms, higher plants, and multicellular algae.

4. **Fungi** are molds, yeasts, and fungi.

5. **Animalia** is composed of the invertebrates and vertebrates.

All animals discussed in this book belong to the phylum **Chordata** in the kingdom Animalia. All members of this phylum have the following four distinctive characteristics:

1. an embryonic **notochord,** which is usually replaced by the spinal cord

2. a dorsal tubular nerve cord running down the back side

3. **pharyngeal** or throat area gill slits

4. a rear area tail

Although they possess these structures at some embryonic stage, these characteristics may be altered or disappear in later stages of life.

All animals discussed in this book belong to one of the following four classes:

1. **Mammalia** is the class of vertebrates that possess mammary glands, have a body that is more or less covered with hair, and possess a well-developed brain.

2. **Reptilia** is the class that includes snakes, lizards, and turtles. This class comprises cold-blooded vertebrates having lungs, an entirely bony skeleton, a body covered with scales or horny plates, and a heart with two atria and (usually) a single ventricle.

3. **Aves** is the class for birds. These are warm-blooded vertebrate animals with two legs, wings, and feathers and that lay eggs.

4. **Osteichthyes** are the bony fish. This is a large group of vertebrate animals that live in the water and have permanent gills for breathing, fins, and a body usually covered with scales.

Linnaeus's system of naming species is called **binomial nomenclature** and uses the Latin language. Each species has a name composed of two words. The first word is the genus and is written with the first letter capitalized. The second name is the species and is written with a lowercase first letter. The domesticated dog, for example, is named *Canis familiaria.*

Canis is the genus and *familiaria* is the species. In some cases, an organism may have a subspecies name and would then use **trinomial nomenclature.** An example of a complete classification for the domestic dog would be as follows:

1. Kingdom—Animalia—Animal kingdom

2. Phylum—Chordata—Animals with vertebrae

3. Class—Mammalia—Animals that suckle young

4. Order—Carnivora—Flesh or meat eaters

5. Family—Canidae—The dog family (includes wolves, fox, and coyotes)

6. Genus—*Canis*—dogs

7. Species—*familiaria*—domestic dog

SUMMARY

Most of the small animals used as pets or in laboratories evolved during the Paleocene and Eocene epochs of the Tertiary period 70 to 40 million years ago. Modern birds evolved earlier during the Cretaceous and early Tertiary periods.

The Cretaceous period, 135 to 70 million years ago, was the period of the dinosaurs. After the demise of the dinosaurs, smaller reptiles evolved. These small reptiles included snakes, lizards, and turtles, and they could compete with the large number of mammals that were also evolving.

The bony fish are the oldest vertebrate group. Some of them remain virtually unchanged since their evolution during the Devonian period 400 million years ago.

Domestication of animals probably began with the dog about 14,000 years ago. As humans moved from being hunters to farmers, other animals were domesticated. Cats are thought to have been domesticated some 3,500 years ago. Domesticated animals include food, draft, and companion animals.

Today, the pet industry is large and growing. Fully 59.5 percent of American households have a companion animal. A large industry exists to supply food, equipment, and health care for all of these animals. Millions of other small animals, birds, reptiles, and fish are used in exhibitions and laboratories. Together these industries provide jobs, careers, and opportunities for a significant segment of our population.

DISCUSSION QUESTIONS

1. When did life on earth begin according to scientific data? Please support your response.

2. When did plants first appear on earth? How do we know this?

3. Why weren't plants the first form of life to appear on earth?

4. During which time period were dinosaurs most abundant?

5. During which time period did the placental mammals appear on earth?

6. Which animals were first to be domesticated? Why?

7. What agency has responsibility for regulating the care and management of small animals?

SUGGESTED ACTIVITIES

1. Visit the library and use reference materials to trace the evolution of various classes of animals through time.

2. From references in the library, research the various theories as to why the dinosaurs became extinct.

3. Obtain a copy of the Animal Welfare Act. What are the regulations covered under the act?

4. Write out a complete classification for the following animals: domestic cat, domestic rat, Eastern box turtle, mynah bird.

5. Make a list of pet supplies you would need for a new puppy or kitten. Visit a pet supply store in person or on the Internet to find out the cost of these supplies and calculate the total cost.

6. Research the theories about how dogs were domesticated. What evidence supports each theory? What is your view? Explain.

ADDITIONAL RESOURCES

1. Dinosaurs and Birds
   http://www.enchantedlearning.com

2. Geologic Time Scale
   http://www.geosociety.org

3. Geologic Time—Enchanted Learning Software
   http://www.enchantedlearning.com

4. Geologic Time Chart
   http://www.littleexplorers.com/subjects/Geologictime.html

5. Guide to Internet Resources for Biological Taxonomy & Classification
   http://mclibrary.nhmccd.edu/taxonomy/taxonomy.html

6. All About Archaeopteryx
   http://www.talkorigins.org/faqs/archaeopteryx/info.html

7. Dog Breed Info Center
   http://www.dogbreedinfo.com

END NOTES

1. Excerpt from *Atlas of the Living World* by David Attenborough. Copyright © 1989 by Marshall Editions Ltd. Reprinted by permission of Marshall Editions Ltd. All rights reserved.

2. Excerpt from "U.S. Pet Ownership & Demographics Sourcebook," Center for Information Management, American Veterinary Medical Association, 2007.

3. Myers, Marshall, mmyers@pijac.org.org. RE: Retail pet stores in the U.S. 01 January 2007. Warren, Dean, Dean.warren@mchsi.com.

4. American Pet Products Manufacturers Association, Inc. (APPMA). "Pet Ownership" [Online]. 03 October 2007. http://www.appma.org

5. Excerpt from "U.S. Pet Ownership & Demographics Sourcebook," Center for Information Management, American Veterinary Medical Association, 2007.

CHAPTER 2

Safety

RISKS WITH SMALL ANIMALS

Owning and working with small animals can be very rewarding, but health risks are also involved. These risks are real, and everyone should take precautions.

Zoonoses are diseases that can be transmitted from animals to humans. Rabies is an example of a zoonosis that has been around for a long time. Ancient Greeks recognized that rabies can be transmitted by dog bites. The bubonic plague of the 1400s was caused by bacteria transmitted to humans by fleas carried by infected rats.

Ascarids (*Toxocara* species) and hookworms (*Ancylostoma* species and *Uncinaria stenocephala*) are common roundworms of dogs and cats. The larval stage of these parasites can infect humans. A **parasite** is an organism

Figure 2–1   Common roundworms of dogs, *Toxicana canis,* can be transmitted to humans. Transmission is usually through contact with feces or contaminated soil. *Courtesy of North Dakota Department of Education.*

that lives on or within another host organism and derives its **sustenance** (food or nourishment) from the host. Children are most at risk because they play in the soil where these animals have been. The parasites are transmitted to humans through contact with soil or other surfaces contaminated by the feces of infected animals. Play areas and sand boxes should be kept clean or covered and free of feces. To prevent becoming infected, it is important to wash your hands after handling animals or their feces. Symptoms of roundworms in humans are fever, headache, cough, and poor appetite. Hookworm larvae can penetrate the skin and leave an inflamed tract as they travel under the skin. *See Figure 2–1.*

Toxoplasmosis is a disease produced by infection with *Toxoplasma gondii,* a parasite that can be carried by several different animal species. It can be contracted when people touch their mouths after handling raw meat or contacting litter or other materials contaminated by cat feces. It can be carried by several different animal species, but it is usually spread to humans through cat feces or contaminated litter. Cats usually become infected by ingesting contaminated mice.

Most humans infected with *Toxoplasma* organisms will not develop symptoms because most people carry antibodies against the disease. People with immune system defects or those receiving immunosuppressive therapy can develop symptoms, however. Symptoms in humans include fever, headache, swollen lymph glands, cough, sore throat, nasal congestion, loss of appetite, and skin rash. Pregnant women should take extra precautions to prevent infection. Toxoplasmosis can cause miscarriage, premature births, and blindness in the unborn child. Pregnant women should avoid cleaning litter boxes. Daily cleaning of litter boxes is important because the organism in the feces of an infected cat becomes infective after 36 to 48 hours. Rubber disposable gloves should be worn when cleaning the litter box, and hands should be washed thoroughly afterward. Children's play areas and sand boxes should be prevented from being used as litter boxes. *See Figure 2–2.*

Figure 2–2   Toxoplasmosis is usually spread to humans through cat feces and cat litter or contaminated soil. Caution should be taken when disposing of cat litter and cleaning cat litter boxes.

Figure 2–3   Ringworm infection appears as round, scaly, or encrusted patches on the skin. The fungus that causes ringworm can be transmitted to humans. *Reprinted from Raising Rabbits, copyright 1977 by Ann Kanable. Permission granted by Rodale Press, Inc., Emmaus, PA 18098.*

Ringworm is a skin disease caused by a fungus; it is not caused by a parasitic worm as the name would imply. Dogs, cats, chinchillas, guinea pigs, rats, mice, and rabbits can be infected with ringworm and then transmit the infection to humans. Ringworm is spread by direct contact with a person or animal infected with the fungus. It can also be spread indirectly through contact with articles (such as combs or clothing) or surfaces that have been contaminated with the fungus. Symptoms of ringworm are the appearance of round, scaly, or encrusted lesions on the skin; hair is usually absent from these areas. The ringworm is infectious as long as the fungus remains present in the skin lesion. The fungus is no longer present when the lesion starts to shrink. Animals with rashes should be evaluated by a veterinarian. If the animal's rash is caused by a fungus, people should avoid contact with the animal until it heals. Caretakers should change clothing and wash themselves immediately after any contact. Topical and oral antifungal drugs are used to cure the problem in both humans and animals. *See Figure 2–3.*

Psittacosis (also known as ornithosis, parrot fever, and avian chlamydiosis) is caused by the bacterium *Chlamydophilia psittaci.* The disease can be transmitted to humans from birds. All birds are susceptible; however, pet birds such as parrots, parakeets, macaws, and cockatiels and poultry such as turkeys and ducks are usually involved in the transmission of the disease to humans.

Humans can be infected by contact with the feces of contaminated birds or from fecal dust. Symptoms in humans include coughing, chest pains, fever, chills, weakness, vomiting, and muscular pain.

Preventive measures include purchasing birds from reliable sources, testing, and quarantine of all new birds. When handling birds and cleaning cages, one should wear dust masks or protective shields. A blood test can be used to confirm psittacosis, and antibiotics may be used to treat the disease both in birds and humans.

Cat-scratch fever (*Bartonella henselae* infection) is associated with cat scratches or bites. The disease is caused by the bacteria *Bartonella henselae.*

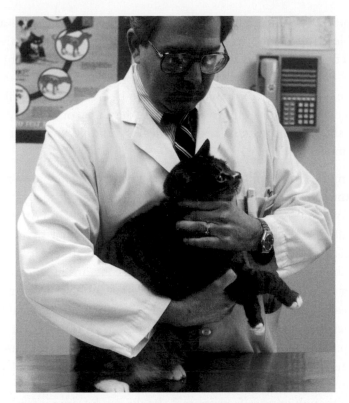

Figure 2–4 Cat scratches and injuries from cats can be avoided by properly handling the animals. The animal should be held close to the body and supported on the arm so that the animal feels secure. *Courtesy of Brian Yacur and Guilderland Animal Hospital.*

The affected area becomes swollen and is slow to heal, and the lymph nodes may swell and become tender and painful as well. The disease is usually not serious and can be treated with antibiotics. Learning to properly handle cats can help prevent scratches and bites. *See Figure 2–4.*

Salmonellosis is a disease caused by infection from *Salmonella* bacteria. The bacteria can be transmitted to humans and animals; children and the elderly are especially at risk. Inflammation of the stomach and intestines results in vomiting, diarrhea, and abdominal pain. Symptoms usually appear 12 to 72 hours after infection, and the illness usually lasts 4 to 7 days.

Salmonella bacteria can be carried by dogs, cats, birds, guinea pigs, hamsters, rats, mice, and rabbits. Pet turtles and other reptiles are a common source of infection in humans. Improper handling or consumption of raw meat or dairy products can also result in salmonellosis. Proper sanitation and husbandry practices will greatly reduce the risk for infection. *See Figure 2–5.*

Streptococcal bacteria infections cause sore throats, especially in children, and can be transmitted by dogs. Streptococcal bacteria infections can range from mild skin infections or sore throat to severe, life-threatening conditions. Approximately 10 millions cases of mild infections from streptococcal bacteria occur annually. Most infections are relatively mild illnesses such as *strep throat*, which causes soreness to the throat, or *impetigo*, a mild infection of the skin. Impetigo begins as tiny blisters that eventually burst and leave small wet patches of red skin that may seep fluid.

Figure 2–5 Proper sanitation is important in controlling salmonellosis. Children should be taught to wash their hands after handling or playing with animals.

Streptococcal bacteria can be transmitted by pets. Children are most susceptible. Proper hand washing and other sanitary measures are important to reduce the spread of the bacteria. Treatment with antibiotics for 24 hours or longer generally eliminates the infected person's ability to spread the bacteria. Infected animals can also be treated with antibiotics. If infected conditions don't improve, if redness develops, or if sores don't heal, a doctor should be consulted because these symptoms may be a sign of a more serious infection.

Injuries from animal attacks, especially animal bites, are a major community health problem in the United States. While many bite injuries are minor, a few are fatal. Approximately 4.7 million Americans are bitten by dogs each year. About 386,000 (129 people per 100,000 population) of these people require emergency treatment.[1]

Children 5 to 9 years old are the primary victims of animal bites. Better than 5 percent of that age-group report a bite every year; this is greater than the combined annual reports of measles, mumps, chicken-pox, and whooping cough. Although children in this age-group make up less than 9 percent of the population, they are the victims of nearly 30 percent of animal bites. Children 9 to 14 years old are the group with the next most bites.

Although injuries from animal attacks can become infected and fatalities do occur, the injury is more frequently psychological rather than surgical. Bite wounds can consist of lacerations, **evulsions** (a tear or a pulled-out wound), punctures, and scratches.

The potential for infection varies; although fewer than 5 percent of dog bites become infected, up to 50 percent of cat bites do. This increased rate is attributed to the difficulty in effectively irrigating the puncture wounds that typically occur from cat bites.

Generally, animal bites have not been shown to be a high risk for tetanus contaminations; very few cases are reported each year in the United States. However, this may be due in part to the high level of tetanus immunity in the general population.

The most important consideration in the treatment of the patient bitten or scratched is whether rabies treatment should be initiated. If rabies is a consideration and the animal cannot be identified with certainty, immunization is recommended.

Except in certain localities near the Mexican border and along the Atlantic Coast where rabies is especially prevalent, domestic dogs and cats in the United States have a low likelihood of rabies infection. Moreover, domestic animals can generally be identified and quarantined for observation or testing.

Bats and feral carnivores, particularly fox, coyotes, skunks, bobcats, ferrets, and raccoons, are frequently infected with rabies in the wild and are the most common source of human rabies in the United States.[2]

Successful vaccination programs that began in the 1940s caused a decline in rabies among domestic animals; however, as the number of cases of rabies in domestic animals decreased, the number in wild animals increased. *See Figure 2–7.*

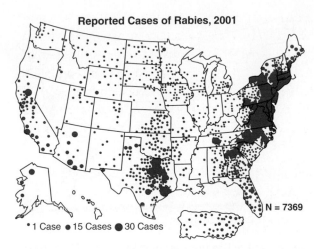

Figure 2–6 The number and location of reported cases of rabies in 2001.

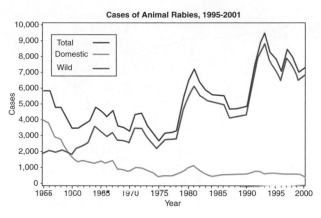

Figure 2–7 Cases of rabies increased from 1955 through 2001.

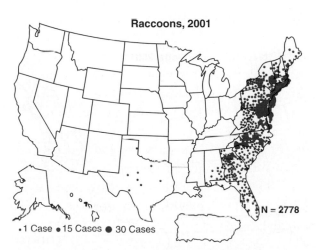

Figure 2–8 The number and location of reported cases of rabies caused by raccoons in 2001.

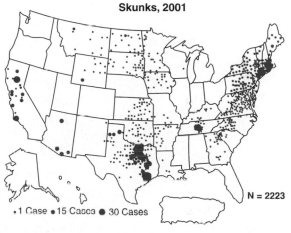

Figure 2–9 The number and location of reported cases of rabies caused by skunks in 2001.

In 1997, 49 states, the District of Columbia, and Puerto Rico reported 8,509 cases of rabies in animals and 4 cases in humans to the Centers for Disease Control and Prevention (CDC). Hawaii has never reported a case of rabies. In 2001, 7,437 cases of rabies in animals and no cases in humans were reported. That is a decrease of about 12.6 percent since 1997. *See Figures 2–6 and 2–7.*

In 2001, wild animals accounted for 93 percent of the reported rabies cases—37.2 percent raccoons, 30.7 percent skunks, 17.2 percent bats, 5.9 percent fox, and 0.7 percent other animals including rodents and rabbits. Reported cases in raccoons and fox decreased 0.4 percent and 3.5 percent, respectively, from the totals reported in 2000. Reported cases in skunks and bats increased 2.6 percent and 3.3 percent, respectively, from the totals reported in 2000. *See Figures 2–8 through 2–11.*

Domestic animals accounted for 6.8 percent of all rabid animals reported in 2001. Cases of rabies in cats increased 8.4 percent over cases reported in 2000. *See Figure 2–12.* Rabies cases in dogs decreased 21.9 percent. *See Figure 2–13.* Rabies cases appear to be most common during the summer months. *See Figure 2–14.* During the twentieth century, the number

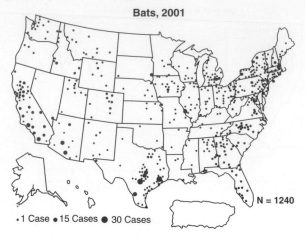

Figure 2–10 The number and location of reported cases of rabies caused by bats in 2001.

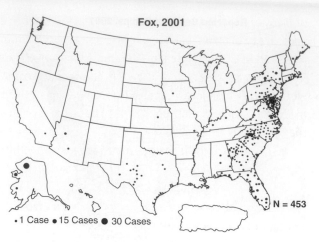

Figure 2–11 The number and location of reported cases of rabies caused by fox in 2001.

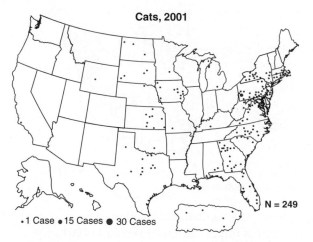

Figure 2–12 The number and location of reported cases of rabies caused by cats in 2001.

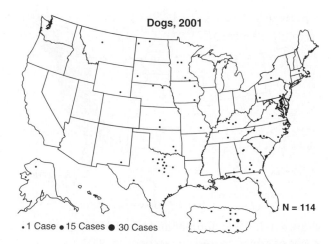

Figure 2–13 The number and location of reported cases of rabies caused by dogs in 2001.

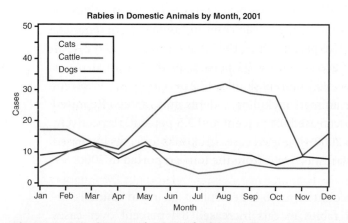

Figure 2–14 Reported cases of rabies in domestic animals by month, 2001.

of human deaths attributed to rabies declined from 100 or more each year to an average of 1 to 2 each year. One death in a human as a result of rabies was reported in 2001. (Refer to Chapter 7 for more information on rabies.)

If one is exposed to a potentially rabid animal, one should wash the wound with soap and water and seek medical attention immediately. Whether to provide rabies treatment for a patient bitten by an animal will depend on the following criteria:

- the species of the animal that caused the bite
- the frequency of rabies in the community
- the circumstances surrounding the bite (Was the animal provoked or was it an unprovoked attack?)
- the behavior of the biting animal
- whether the animal can be quarantined and observed
- whether the animal's head (in the case of a wild animal) can be sent in for laboratory examination of the brain[3]

Animal care workers should be alert to animals that are vicious or potentially hazardous. These workers can protect themselves against rabies. "Preexposure vaccinations can be given either intramuscularly or intradermally and consist of three injections, one injection per day on days 0, 7, and 21 or 28."[4] **Intramuscular** injections are given within the muscle and **intradermal** injections are given within the layers of skin. Workers who have received these injections would then need only two intramuscular boosters if they become exposed. Nonprotected workers will require a much more expensive series of five intramuscular injections of rabies vaccine and one rabies **immune gamma globulin** (antibodies) injection over a 28-day period. Current vaccines are relatively painless and are given in the arm.

In general, proper treatment of animal bites and scratches is very important in preventing infection and more serious complications. First-aid procedures for bites and scratches include immediately and thoroughly washing the area with soap and water. This should be followed by medical attention; most doctors will give a tetanus shot if none has been received within the past 5 years. Proper bandaging of the wound and appropriate antibiotics will also be given.

Proper handling of animals is also important to prevent injury to workers; all animal care workers should be instructed in proper techniques. To prevent animal bites, children should be taught to avoid unfamiliar animals and that any animal may bite if frightened, ill, or injured. Familiar animals may bite when startled or disturbed when sleeping or eating, or if they are handled roughly. Small children should always be supervised when handling or playing with animals.

The proper handling of animals is covered in each chapter; for information on a specific species of animals, refer to that chapter.

The following steps can be taken to protect one's self and family members from rabies:

1. Do not feed, touch, or adopt wild animals or stray dogs and cats. Children should be taught to stay away from stray animals, even though the animal may appear friendly or tame. One of the signs

of rabies is behavior changes. An animal may appear unusually aggressive or unusually tame. If a wild animal is on your property, leave it alone. If the animal exhibits unusual behavior, local animal control authorities should be contacted.

2. Be sure your pets have up-to-date rabies vaccinations. Check with your local veterinarian about when and how often to get the shots needed to protect your pets.

3. Family pets should be kept indoors at night. Pets should not be allowed to roam the neighborhood unattended.

4. Don't leave pet food outside that will attract stray and wild animals. Keep lids on garbage cans.

5. Make sure there are no openings to the attic, basement, porch, or garage. Make sure screens on doors and windows are in good repair so bats and other animals can't enter the house.

If one's pet has been bitten by another animal, the following steps should be taken:

1. Wear gloves when handling your pet. Rabies can be spread from the saliva of an infected animal. Saliva from an infected animal can enter a scratch or cut and can also be transferred to the eyes or mouth.

2. Isolate your pet from other animals and people.

3. Call your veterinarian. Vaccinated pets will need a booster shot.

4. Report the bite to local animal control authorities.

5. Unvaccinated pets will need to be isolated for a period of 10 to 14 days or longer for observation.

Ticks are frequent carriers of diseases that can cause illness in people and animals. Although an animal that has been infected through a tick bite will not pass the disease on to a person, the animal may carry disease-bearing ticks into households and yards, where the tick may later bite a human.

Rocky Mountain spotted fever is found in all areas of the country and is primarily transmitted by the American dog tick, although six other species of tick have also been found to be carriers.

The organism that causes the disease is *Rickettsia rickettsii*. The *Rickettsia* organism multiplies in the cells of the small peripheral blood vessels. Fever, headache, nausea, vomiting, and skin rash are symptoms of Rocky Mountain spotted fever. Early diagnosis and treatment with antibiotics is important. Anyone who experiences these symptoms within 2 weeks of a possible tick bite or exposure should see a doctor immediately. If not treated, Rocky Mountain spotted fever can cause death. Normal grooming of animals after they have been outside will help locate and eliminate ticks. Attached ticks should be removed carefully; one must avoid breaking off the mouthparts in the skin because they may cause infection. Forceps and gloves should be used to avoid contamination of the fingers, and the area should be treated with antiseptic.

Lyme disease is a tick-transmitted disease affecting both humans and animals. In 1969, a grouse hunter in Wisconsin developed a strange, red rash

Lyme Disease: A Growing Threat

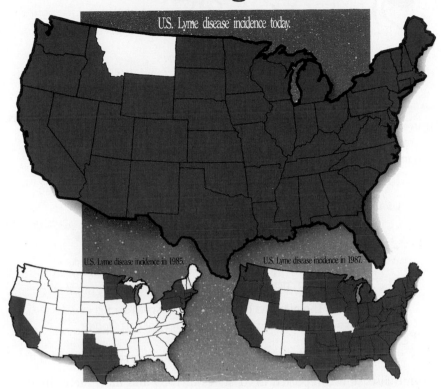

U.S. Lyme disease incidence today.

U.S. Lyme disease incidence in 1985.

U.S. Lyme disease incidence in 1987.

Figure 2–15 Lyme disease is a growing threat. These maps show the spread of the disease since 1985. *Courtesy of Fort Dodge Laboratories.*

Reported Cases of Lyme Disease – United States, 2004

1 dot placed randomly within county of residence for each reported case

Figure 2–16 The number and location of reported cases of Lyme disease in the United States in 2004.

around a tick bite; this is believed to be the first documented case of the rash that is often the first symptom of the disease.

In 1977, doctors observed arthritis-like symptoms in a group of children in the town of Lyme, Connecticut. These doctors are credited with giving the disease its present name. *See Figures 2–15 and 2–16.*

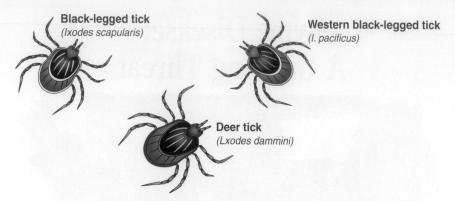

Black-legged tick
(Ixodes scapularis)

Western black-legged tick
(I. pacificus)

Deer tick
(Lxodes dammini)

Figure 2–17 These three ticks of the Ixodes species are associated with the transmission of the Lyme disease organism, *Borrelia burgdorferi,* to humans.

Lone star tick	Black-legged tick	Western black-legged tick
larvae nymph adult	larvae nymph adult	larvae nymph adult

Ticks shown actual size

Figure 2–18 The nymphs of ticks are very small and appear as specks on a human palm. *Courtesy of Fort Dodge Laboratories.*

Lyme disease was reported in 31 states in 1989 and in 47 states in 1992. In 1982, 492 cases were reported to the CDC, and more than 7,000 cases were reported in 1989. By 1998, 49 states and the District of Columbia had reported cases of Lyme disease. (See Appendix D for the reported cases of Lyme disease by each state in 2002.) Montana has never reported any cases. Almost 16,000 cases were reported in 1998. In 2002, nearly 154,000 cases were reported to the CDC. Based on reported cases, during the past 10 years, 90 percent of Lyme disease cases occurred in 10 states: New York, Connecticut, Pennsylvania, New Jersey, Wisconsin, Rhode Island, Maryland, Massachusetts, Minnesota, and Delaware.

Three species of ticks have now been recognized as carriers of the infectious bacteria *Borrelia burgdurferi* and can bite and infect humans. *See Figures 2–17* and *2–18*. In the Northeast and Midwest, the pinpoint-sized deer tick, *Ixodes dammini,* and in the Midwest and Southeast, the black-legged tick, *Ixodes scapularis,* are the recognized carriers. White-tailed deer serve as hosts for these two species of ticks. With no hosts present, they feed on other animals, including humans. *See Figure 2–19.*

On the East Coast, white-footed mice serve as a **reservoir** (an immune host) for the *Borrelia* organism. These ticks have a 2-year life cycle. They lay their eggs in the spring, and the larvae that emerge feed on white-footed mice. The larvae then remain dormant during the winter and develop into nymphs the following spring. The tick larvae pick up the bacteria from the white-footed mouse and transmit it to humans.

In the Pacific coastal states of California, Oregon, and Washington, the Western black-legged tick, *Ixodes pacificus,* is recognized as the carrier of the

LIFE CYCLE OF *IXODES DAMMINI* TICK

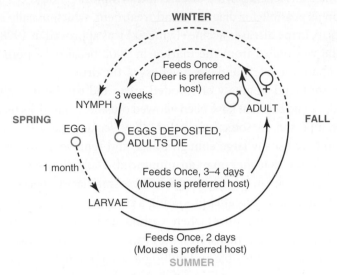

Figure 2–19 The life cycle of the *Ixodes dammini* tick.

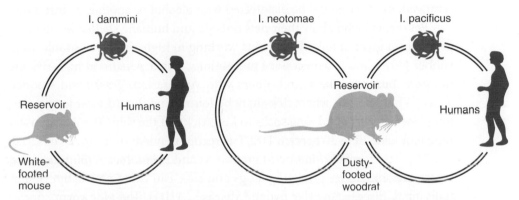

Figure 2–20 In the Northeast and Upper Midwest, the white-footed mouse serves as a reservoir for the Lyme disease bacteria. In Western areas, the dusty-footed woodrat serves as one of the reservoirs for the bacteria.

Borrelia organism. The dusty-footed woodrat, a common California rodent, is the reservoir for the *Borrelia* organism in the West. Another species of tick, *Ixodes neotomae,* can also be a carrier. This tick does not bite humans, but it carries the organism that infects the woodrat. *See Figure 2–20.*

The first symptom of Lyme disease is the distinctive skin lesion that appears in 3 to 32 days after the tick bite. The lesion begins as a small red spot that expands into a large irregular circular or oval-shaped area with a red outer border. Other lesions may appear as the bacteria spreads within the skin; these lesions may itch and burn. Variations of the distinctive symptoms may appear, including measles-like eruptions or hives; these lesions usually disappear in about 3 weeks. Flu-like symptoms, including aching muscles, stiff neck, fatigue, fever, chills, painful joints, nausea, coughing, sore throat, loss of appetite, swollen lymph glands, irritated eyes, and aversion to light, may also be present. These symptoms too may disappear within 3 weeks; however, the fatigue and muscle aches may last for several months. If treatment with antibiotics is not started, the bacteria can spread throughout the body and cause severe damage to the organs of the body.

Anyone experiencing Lyme disease-like symptoms should contact a doctor as soon as possible for diagnosis and treatment, which usually consists of antibiotics. A Lyme disease vaccine (Lymerix) was approved in 1998; however the vaccine was pulled from the market in 2002 because of poor sales and concerns about the safety and effectiveness of the drug.

Ticks are found in grassy or wooded areas and can be brought into the home by dogs and cats that have been allowed outside. Adult ticks usually crawl around on a person for some time before they decide to attach themselves to the skin and are usually large enough to see and remove; however, the small larvae are so tiny that they may go unnoticed. Daily showers and personal hygiene are important in helping eliminate these small larvae. Pet owners should wash their hands and arms after grooming a pet that has been outside, and pet care workers should wear gloves when washing and grooming animals.

If ticks have attached themselves, they can be removed by grasping the tick as close to the head as possible and pulling it straight out, being careful not to squeeze or twist it. One must be careful not to break off the head or to crush the tick, because this may serve as a means of infection. After the tick has been removed, the area should be disinfected with alcohol or another disinfectant.

Pet owners who allow their pets outside and humane shelter workers are probably at greatest risk. All persons working in high-risk areas should wear rubber gloves and take necessary precautions. Other persons at high risk are trappers, hunters, hikers, and others who venture into grassy and wooded areas. When in areas where ticks may be found, one should wear long pants, long-sleeved shirts, and long socks to keep ticks off the skin. The use of insect repellent with 20 to 30 percent DEET is recommended.

The tapeworms *Echinococcus granuosus* and *Echinococcus multilocularis*, which are sometimes carried by dogs and cats, can cause a rare, but potentially fatal, disease, alveolar hydatid disease (AHD). The tapeworm species *E. multilocularis* was originally confined to the Alaska coast, the cold tundras of Canada, and the eastern hemisphere; however, it has now been identified in the Dakotas, the Central Plains, and as far south as the Carolinas. Possible illegal shipments of animals have been linked to the spread of the disease. Fox, coyotes, and mice are the normal **intermediate hosts** (a host that the parasitic organism lives on or in during an immature stage) for this species of tapeworm, but the parasite is now appearing in domestic dogs and cats. The tapeworm species *E. granuosus* is found in California, Utah, Arizona, and New Mexico, where it normally uses sheep as the intermediate host. *See Figure 2–21.*

Infective *E. multilocularis* and *E. granuosus* eggs, accidentally ingested by humans, hatch in the small intestine, penetrate the intestinal wall, and migrate to the liver, where they produce parasitic tumors or cysts. *See Figures 2–22 and 2–23.* Clinical signs may not appear until the growth of the cyst has progressed for several years. Surgical removal of the cyst is the preferred treatment, although in many cases the cyst has affected other organs and may not be removable. AHD has a fatality rate of 50 to 75 percent with or without surgery.

E. multilocularis and *E. granuosus* are difficult to identify because the eggs are identical to those of the *Taenia* tapeworm species that is common to dogs and cats.

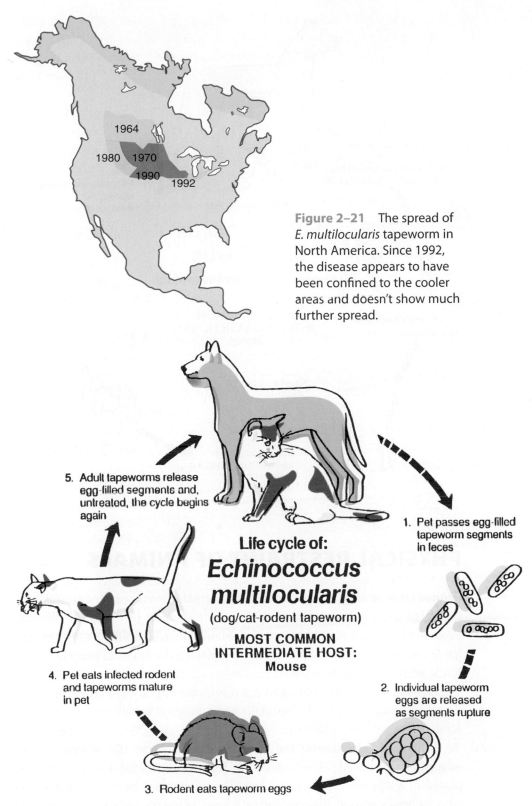

Figure 2–21 The spread of *E. multilocularis* tapeworm in North America. Since 1992, the disease appears to have been confined to the cooler areas and doesn't show much further spread.

5. Adult tapeworms release egg-filled segments and, untreated, the cycle begins again

Life cycle of:

Echinococcus multilocularis

(dog/cat-rodent tapeworm)

MOST COMMON INTERMEDIATE HOST: Mouse

1. Pet passes egg-filled tapeworm segments in feces

2. Individual tapeworm eggs are released as segments rupture

3. Rodent eats tapeworm eggs

4. Pet eats infected rodent and tapeworms mature in pet

Figure 2–22 Life cycle of *E. multilocularis* tapeworm. *Courtesy of Miles Inc.*

Preventive measures need to be taken to minimize the risk for human infection. Pet owners and pet care workers in rural areas are at higher risk and should take precautions through personal protection and hygiene to prevent hand-to-mouth transfer of eggs.

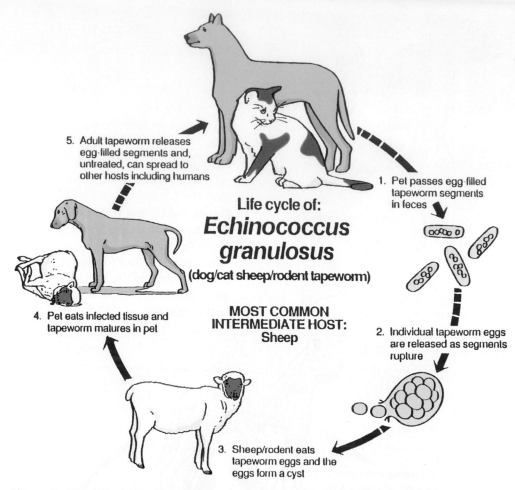

5. Adult tapeworm releases egg-filled segments and, untreated, can spread to other hosts including humans

Life cycle of:
Echinococcus granulosus
(dog/cat sheep/rodent tapeworm)

1. Pet passes egg-filled tapeworm segments in feces

MOST COMMON INTERMEDIATE HOST: Sheep

2. Individual tapeworm eggs are released as segments rupture

4. Pet eats infected tissue and tapeworm matures in pet

3. Sheep/rodent eats tapeworm eggs and the eggs form a cyst

Figure 2–23   Life cycle of *E. granulosus* tapeworm. *Courtesy of Miles Inc.*

PHYSICAL RESTRAINT OF ANIMALS

Animals may need to be restrained from normal movement for examination, collection of blood or other samples, administration of drugs, or therapy. Prolonged restraint should be avoided. Sick or injured animals may become frightened and confused. Special care may need to be taken to prevent injury to the animal and to the handler.

Cats are nervous animals, and when subjected to new situations or introduced to strange people, the normally good-natured family pet can bite and scratch.

When working around the head or neck of a cat, the animal can be wrapped in a blanket. This controls the legs of the animal. Cats can also be placed in a zippered-type canvas bag. The animal can be placed in the bag with its head exposed. To hold the cat's head, place the palm of the hand on the back of the head and grasp the head between the thumb and fingers.

Cats can be transported in small cages or cat carriers. Care may need to be taken when removing the cat from the carrier to prevent injury to the cat or the handler. If you need to take an unhappy cat out of a carrier, get a more experienced handler to show you how to do it safely. One may need to use gloves when reaching into the carrier. Some cats are able to bite through heavy leather gloves, and the gloves can inhibit your ability to get a good grip on the cat. The cat should be grabbed by the scruff of the neck with one hand and then grabbed by the rear legs with the other hand. The cat can then be lifted up and removed from the carrier.

Cats can be caught and restrained by a catch pole, which is a device that consists of a 5- or 6-foot pole with a rope and noose attached. The noose should be slipped over the animal's head and front legs and around the chest. This tool should be used only when all other methods of restraint have failed. Improper use of a catch pole can injure or even kill an animal. Cloth muzzles for cats can be used to keep a cat from biting during procedures. In addition to holding the mouth shut, these muzzles cover the eyes of the cat, helping to calm it.

Dogs can be restrained while the animal is in a standing or sitting position by placing one arm under the dog's neck with the forearm holding the head. The other arm is placed around the animal's body, and the animal is held close to the handler's body.

Dogs can be placed in a lying position by reaching over the back of the animal and grabbing the front legs with one hand and the rear legs with the other hand. The animal should be gently lifted and the animal's body allowed to slide down into a lying position. The handler's forearms can be lowered to rest against the animal's head and body to restrain the animal.

Muzzles can be used to prevent bites. Muzzles are placed over the dog's mouth. Different sizes are available, and care must be taken to ensure that the muzzle fits snugly and comfortably. A muzzle can be made from a narrow strip of gauze or cloth as follows:

1. Make a loop in the material and slip the loop over the dog's nose and mouth.

2. Then tighten the loop by pulling on the ends.

3. Then cross the ends under the dog's jaw and bring them up behind the ears and tie them in a bow (*see Figures 2–24 and 2–25*).

4. The bow can be untied quickly by pulling on the ends.

Figure 2–24 Applying muzzle material around dog's nose.

Figure 2–25 Securing muzzle behind the ears.

Figure 2–26 Elizabethan collar.

Elizabethan collars, or e-collars, are protective devices that are used to prevent an animal, usually a dog or cat, from licking or biting at its body or scratching at its head or neck while wounds or injuries heal. E-collars can be purchased from veterinarians or pet stores. The collars are usually made from a nonbrittle plastic. They are shaped like a cone and look similar to a lampshade. The collar slips over the animal's head and is fastened to the animal's usual collar. *See Figure 2–26.* The neck of the collar should be short enough to allow the animal to eat and drink. Some animals may find the collar very uncomfortable and may have trouble moving about and sleeping. Some animals may not eat or drink with the collar on, and it may need to be temporarily removed. Temporary collars can be made from cardboard, although they may not last very long. Collars can also be made from fast-food containers, plastic buckets, and plastic flower pots.

As noted, catch poles can also be used to restrain animals. The noose can be placed over the head of the animal and around the neck. The animal should be restrained in this manner only long enough to be tranquilized or subdued. Proper training in the use of a catch pole is essential; these devices can cause injury or death to the animal if not used carefully.

Rabbits can be picked up by grasping the scruff of the neck with one hand and lifting up while placing the other hand under the rump. Remember to support the hindquarters of the animal at all times. Rabbits seldom bite, but they can inflict injury with their hind legs. They can also injure their spine by hard kicking in an attempt to escape.

Rabbits should not be placed on a smooth surface. They may injure themselves while trying to hop or move. Rabbits should be placed on a piece of carpet or on a towel. Veterinarians and researchers sometimes use restraint boxes to hold rabbits during treatment.

Rats and mice that have not been handled may bite, and special care may be needed when handling them. Gloves can be used to place the animals on a piece of carpet or on a piece of screening. Once they are placed on the carpet or screening, they can be restrained with bare hands. The animal should be grasped gently by the tail. Don't grasp the end of the tail, but grasp as close to the body as possible. The animal will try to pull away, and while it is in a stretched-out position, use your other hand to grasp all the loose skin on the neck and shoulders that you can and lift the animal up. The animal should be restrained in this manner and unable to bite. Special restraining boxes are also available for rats and mice.

OTHER RISKS

Many of the disinfectants and some insecticides used in working with small animals are potentially dangerous. All workers should be instructed in the safe use, storage, and disposal of chemicals and containers. Chemicals such as pesticides can enter the body in several ways; the skin is the most common portal of entry into the body. To prevent absorption through the skin, the worker must wear protective clothing and equipment.

Gloves should be worn by workers when handling chemicals. Plastic or rubber unlined gloves that are liquid-proof are commonly used when handling chemicals. Canvas, cotton, or leather gloves should not be worn because they can be soaked with the chemicals and provide a continuous source of exposure. For handling animals that may bite or scratch, leather and Kevlar gloves are available.

Closed-toe shoes or boots are required for working safely around animals. Boots with no slip soles should be worn when working on wet floors. Shoes with steel toes may be beneficial to workers handling large cages and heavy equipment.

Workers' eyes should be protected at all times when working with chemicals. Tight-fitting goggles with anti-fog lenses and ventilation holes are recommended as the best protection; however, full-face shields also provide protection against chemicals splashed into the face. Workers in areas of high noise levels should wear ear protectors; damage to hearing can result from prolonged exposure to loud noises.

Uniforms, overalls, lab coats, or other personal protective equipment (PPE) should be worn when working with animals that may be sick or when working with potentially toxic substances. Protective clothing should be laundered daily and not worn outside the facility.

Respirators provide protection from inhaling toxic substances; workers should read labels to determine proper use of respirators.

An important component of workplace safety is material safety data sheets (MSDS). Material safety data sheets are forms that contain data regarding the properties of a particular substance. The sheets provide workers and emergency personnel with safe procedures for handling or working with a particular substance. Other information in an MSDS includes physical properties of the substance, toxicity, health effects, first-aid precautions, reactivity, storage requirements, proper disposal procedures, proper protective equipment to use when handling the substance, and safe handling procedures.

GENERAL GUIDELINES FOR SAFETY

No safety guidelines can cover all situations, so it is extremely important for workers and students to follow all instructions from supervisors and teachers and on all labels when handling chemicals. The proper procedures in treating and reporting accidents should be taught. The following are some general guidelines to follow for safety in the workplace or school laboratory.

1. Always wear protective clothing and equipment when the job requires it.

2. Always wash protective clothing and equipment after use to prevent contamination.

3. Wash hands and face after completing a job to make sure all chemical residue is removed.

4. If required, shower after completing a job so that chemical residue is completely removed from the body.

5. Wash hands frequently while working with animals, especially if working with different species and in different areas. This will prevent contamination of other animals and will prevent self-contamination.

6. Keep hands away from the mouth, eyes, and face when working with chemicals and animals to prevent self-contamination.

7. Do not consume food or drinks in areas where contamination could occur, and do not store these items in areas where contamination could occur.

8. Remove uniforms, lab coats, and coveralls when leaving an area that could be contaminated.

9. Never wash uniforms, lab coats, or coveralls with regular clothing.

10. Make sure all containers are correctly labeled to prevent the misuse of a chemical.

11. Dispose of all chemicals and their containers according to proper procedure or instructions on the label.

12. Students and small animal workers should be instructed in the proper methods of handling small animals.

13. First-aid kits should be kept in the work area or instructional area. Workers and students should be made aware of the location of first aid kits.

SUMMARY

Working with small animals is an interesting and rewarding career; however, there are risks that everyone should be aware of.

Several diseases can be transmitted from animals to humans. The potential for infection in caged, domestic animals is low, but when dealing with animals that have been exposed to the outdoors, the risks increase. Several of these diseases can be life-threatening, so it is important that workers and pet owners are knowledgeable about the prevention and treatment of them.

Several chemicals and pesticides may be used when working with small animals. Proper safety is important to prevent accidents.

Safe handling techniques for all small animals are important in preventing scratches and bites to handlers, as well as preventing injury to the animal.

Most injuries and accidents when dealing with small animals can be prevented by being aware of safety guidelines and proper procedures.

DISCUSSION QUESTIONS

1. What is a parasite?

2. What is a host? What is an intermediate host?

3. What is a zoonosis?

4. What is the life cycle of the deer tick and the black-legged tick?

5. What is the life cycle of the Western black-legged tick?

6. Why are children and the elderly more at risk from some of the diseases and injuries covered in this chapter?

7. Why shouldn't a pregnant woman handle a cat's litter box?

8. Why is early diagnosis and treatment important in cases of Rocky Mountain spotted fever?

9. What determines whether someone should be given immune gamma globulin injections for rabies?

10. What are some safety guidelines to follow when working with small animals?

SUGGESTED ACTIVITIES

1. Visit the school or community library and do some further research concerning Lyme disease.

   How can Lyme disease be prevented?

   Can the disease be eliminated? What are some difficulties in dealing with Lyme disease?

2. Check with local health officials and determine the incidence of animal bites in your area. What are the major types of bacterial infections associated with animal bites?

3. Research the tapeworm species covered in this chapter and the diseases caused by them. How can they be prevented or eliminated?

4. Bring in labels from pesticide containers. What is the proper mixture of the material? What safety practices are listed on the label? How should the container be disposed of?

ADDITIONAL RESOURCES

1. Occupational Safety and Health Administration (OSHA)
   http://www.osha.gov

2. McGill Laboratory Safety Manual, 3rd Edition, 1994
   http://www.mcgill.ca/eso/labman/1.htm

3. Bowling Green State University, Laboratory Safety and Chemical Hygiene Plan
   http://www.bgsu.edu/offices/envhs/labsafe.htm

4. Centers for Disease Control and Prevention
   http://www.cdc.gov

5. Division of Vector-Borne Infectious Diseases National Center for Infectious Diseases Centers for Disease Control and Prevention, P.O. Box 2087 Fort Collins, Colorado 80522, (970) 221-6400
   http://www.cdc.gov/ncidod/dvbid/lymeepi.htm

6. Guide for the Care and Use of Laboratory Animals
   http://www.nap.edu/readingroom/books/labrats/

END NOTES

1. http://www.cdc.gov/ncipc/duip/biteprevention.htm.

2. Excerpt from Wishon, P. M., and A. Huang. 1989. "Pet associated injuries: The trouble with children's best friends." *Children Today* 18 (May/June): 24–27.

3. Kreisle, M. D., Ph.D., "Dealing with Bites and Other Occupational Hazards," Seminar of the Societal Issues in Animal Management, Purdue University, May 19, 1991.

4. Centers for Disease Control and Prevention http://www.cdc.gov/rabies/veterinarians.html

CHAPTER 3

Small Animals as Pets

OBJECTIVES

After reading this chapter, one should be able to:

- 🐾 list the questions a person should ask before buying a pet.
- 🐾 compare the different options available for obtaining a pet.
- 🐾 describe what is meant by responsible pet ownership.
- 🐾 list the important lessons children can learn from their pets.
- 🐾 describe the benefits pets can have for the elderly.
- 🐾 define euthanasia.
- 🐾 describe the importance of euthanasia to a pet owner.

TERMS TO KNOW

allergies	euthanasia	pedigree
conformation	neutered	pocket pets
entropion	no-kill shelters	spayed

CHOOSING A PET

People have kept domesticated animals for more than 10,000 years; dogs were probably the first animals domesticated, followed by cats. Dogs have been called man's "best friend," and until recently, more dogs were kept as pets than any other animal; cats now outnumber dogs, however.

Before World War II, many pets were hunting, farm, or show animals. Space and poverty limited animal ownership in cities. Following World War II, more people were moving from the farms to the cities and people enjoyed increased incomes and prosperity. The demographics of the family changed. People were marrying and having families later in life and having fewer children. More people were electing to remain single. As children grew up and

Figure 3–1 Pets provide companionship in today's fast-paced society. *Courtesy of Tom Schin.*

left home, older couples were experiencing the loneliness of the "empty nest." Pets, therefore, provided companionship in these fast-paced, changing times. Pets offer nonjudgmental acceptance, love, and companionship, no matter what the age of a person. *See Figure 3–1.*

Unfortunately, many pets are abandoned each year because they don't fit in with the lifestyle of the family. Many pets lose their "cuteness" as they get older, cost too much to feed, or put too much time demand on the family. Millions of unwanted animals pass through animal shelters each year, a result of uncontrolled breeding and abandonment. Many of these animals are eventually put to sleep through **euthanasia.**

"Yet supporters say the United States is already well on its way to becoming a "no-kill nation." Euthanasia of animals is down 70 percent, from 17 million animals euthanized in the 1980s to fewer than 5 million today, notes Michael Mountain, president of the Best Friends Animal Society. The Best Friends Animal Society operates an animal sanctuary in Southern Utah. Michael Mountain is also creator of the nationwide "No More Homeless Pets" campaign. The number of **no-kill shelters** has grown from about 50 in the 1980s to about 250 today, notes Merritt Clifton, editor of *Animal People* magazine. That's still a small fraction of the country's nearly 5,000 shelters, foster care groups, and placement agencies."[1] The most widely accepted definition of a no-kill shelter is a place where all adoptable and treatable animals are saved and where only unadoptable or non-rehabilitative animals are euthanized.[2] Many more pets are abandoned to fend for themselves, and stray animals may become a problem in many areas.

No matter how cute an animal appears, careful thought should precede the purchase of a pet, because a hasty decision may lead to later disappointment.

When selecting a pet, a person should consider the following questions carefully before making a purchase:

1. Where should I obtain a pet?
2. How much space do I have for a pet?
3. What kind of animal does my lifestyle allow?
4. How much will the animal cost?
5. What will the future bring?
6. Does everyone in the family want a pet?
7. What kind of personality do I have?
8. Is this animal a fad or status symbol?
9. What am I going to use the animal for?

Where Should I Obtain a Pet?

The most obvious answer is a pet shop; however, other options are available. Pet shops are usually conveniently located centers where consumers can purchase pets and pet products. Pet shops serve as intermediaries between the breeder and consumer and make a profit on the difference between the buying price and the selling price of an animal. Animals in pet shops may come from anywhere in the country. Consumers have little knowledge of where the animals come from, do not know the conditions under which they were raised and shipped, or have any knowledge of pedigree. Dogs and cats may be registered, but were the parents healthy and bred by reputable breeders?

Purebred breeders usually are in business for the long haul. Their success depends on their reputation. If they produce animals that possess inherited defects or are ill-mannered, their business will suffer. Most breeders will guarantee satisfaction and are willing to work with the customer.

Purchasing direct from the breeder will save money by eliminating the middle-person. When visiting the breeder, the buyer can observe the parents and determine traits that can be passed on to the offspring.

Animal shelters are where many abandoned animals end up. Many of these animals would make excellent pets if given the opportunity. A person would not only get a wonderful pet, but would also save the animal's life. The cost of such an animal will be much less than one purchased from a pet shop or breeder. Shelters usually charge only enough to cover expenses. There may be a charge to have the animal **spayed** or **neutered,** and most shelters require this procedure. Information on the animal will not be available unless the previous owner can be contacted. *See Figure 3–2.*

Friends and neighbors offer one of the easiest and cheapest sources to obtain a pet. Dogs and cats running loose often result in unwanted pregnancies and animals; friends and neighbors may give these animals away for free. The sire of these litters may be unknown, but some of the traits of the dam can be observed to get an idea of what the animal will be like when grown.

Figure 3–2 Dogs do not have to be purebred to provide loving companionship. "Mutts" can make excellent family pets as well as companions. *Courtesy of Tracey A. Carter.*

How Much Space Do I Have for a Pet?

A small one-room apartment doesn't allow much space for active animals. Living in rural areas, on the other hand, may provide large acreage for an animal to roam. A person needs to be familiar with the activity requirements of an animal before acquiring it.

What Kind of Animal Does My Lifestyle Allow?

Animals require different amounts of space, time, and care. Dogs need daily exercise and grooming; this varies considerably among the different breeds. Long-haired dogs require more grooming, whereas active dogs, such as the greyhound, require more exercise. It is important that people be familiar with the breeds they plan to obtain and are sure the pets will fit into their lifestyle. Cat breeds are fairly uniform in their needs for exercise, attention, and other daily care. Like dogs, long-haired breeds will require extra time for grooming. Compared to dogs, however, cats require less effort for daily maintenance and training.

If a person's lifestyle doesn't allow for a dog or cat, then a hamster or gerbil may make a more desirable pet. When hamsters, gerbils and other similar small animals are provided with enough food and water, they can be left alone for several days and will fit well into small apartments.

How Much Will the Animal Cost?

Many people don't look past the initial purchase price of their pet. Puppies and kittens don't cost much to care for, but as they grow to adult size, food can easily cost from $13 to $30 per month for dogs and approximately $14 for cats. In addition to feeding, there will be expenses in the form of toys, grooming, and veterinary bills. Many cities and counties also have taxes on animals. Because of high veterinary costs, many people are opting for pet health insurance. Insurance plans vary but cost from $4 to $50 per month depending on the plan, its deductible, and the age of the pet.

The American Society for the Prevention of Cruelty to Animals (ASPCPA) lists the annual costs for caring for a dog at between $420 and $780 depending on the size of the dog. Annual costs for cats are estimated at $575, and $691 for a rabbit, $565 for a guinea pig, $305 for a small mammal, and $35 for a fish. Initial costs may include vaccines and spaying or neutering. Yearly costs include food, medical treatment, litter, toys and treats, license, and miscellaneous expenses. First-year costs will be higher because owners will need to purchase new equipment. See Table 3–1.[3]

What Will the Future Bring?

A person should plan ahead. Will that cute, cuddly animal still be wanted when it matures? Will the family continue to be able to afford a pet? If the animal was obtained for the children, will they be around or will Mom and

TABLE 3-1   Annual Costs of Caring for Various Species of Pets

Costs	Notes	Small Dog	Medium Dog	Large Dog	Cat	Rabbit	Guinea Pig	Small Mammal	Small Bird	Fish
Annual Costs										
Food	1	$160.00	$310.00	$350.00	$170.00	$126.00	$75.00	$50.00	$75.00	$20.00
Medical	2	$150.00	$175.00	$200.00	$150.00	$125.00	$50.00			
Litter	3				$175.00	$400.00	$400.00	$220.00		
Toys/Treats		$60.00	$75.00	$150.00	$50.00	$25.00	$25.00	$20.00	$30.00	
License		$15.00	$15.00	$15.00						
Misc.		$35.00	$45.00	$65.00	$30.00	$15.00	$15.00	$15.00	$15.00	$15.00
Annual total		$420.00	$620.00	$780.00	$575.00	$691.00	$565.00	$305.00	$120.00	$35.00
Capital Costs										
Spay/neuter		$75.00	$100.00	$125.00	$75.00	$75.00				
Collar/Leash		$25.00	$30.00	$35.00	$10.00					
Litter box					$25.00	$25.00				
Cage							$90.00	$80.00	$35.00	$75.00	
Carrier		$30.00	$50.00	$80.00	$30.00	$30.00				
Crate		$60.00	$90.00	$160.00						
Aquarium eqpt	4									$200.00
Capital total		$190.00	$270.00	$400.00	$140.00	$220.00	$80.00	$35.00	$75.00	$200.00
Special Costs										
Long hair groom		$200.00	$300.00	$400.00						
First year total		$810.00	$1,190.00	$1,580.00	$715.00	$911.00	$645.00	$340.00	$195.00	$235.00

Notes:

1. Premium brand dry kibble.

2. Exam, vaccinations, heartworm medication.

3. Scoopable litter for cats, scooped daily; hardwood shavings or recycled paper products for rabbits, guinea pigs and small mammals, changed at least weekly.

4. Basic 20-gallon set up with light/hood, outside filter, undergravel filters, air pump and gravel. Does not include fish.

Reprinted with permission from the American Society for the Prevention of Cruelty to Animals (ASPCA).

Dad be the ones to care for it? Can the son or daughter take the animal to college when the time comes?

Does Everyone in the Family Want a Pet?

A pet should not be brought into the family unless every member agrees to it and is willing to share in the responsibility for the animal's care. An animal brought into the family as a gift or surprise may become an unwanted and uncared for pet.

Another factor to consider is that of **allergies.** Allergies to certain animals and animal hair or dust are fairly common. It is important to determine whether any member of the family is allergic to a particular species of animal. Finding this out can prevent the disappointment of having to give up a pet.

What Kind of Personality Do I Have?

Animals have personalities just like people. It is important that the pet one decides to purchase complements his or her personality. Dogs vary greatly in their personalities; some are docile, whereas others can be active and playful. These personality traits are characteristics of the breeds; therefore, it is important to be familiar with the breed of animal you intend to obtain.

Cats, too, vary in their personality. Some cats are very human-centered. They like to be held and spend most of their time with family members. They may be upset if left alone for long periods of time. Other cats are less social and more independent. These cats may be friendly and enjoy short periods of petting, but will also seek to spend a lot of their time alone. When choosing a cat, it is important to decide what type of behavior you want and then take time to assess the personality of a potential pet. **Pocket pets** are small animals that are usually kept in aquariums or cages. These pets may be ideal for those who want greater control over when and how long to interact with their pet.

Is This Animal a Fad or Status Symbol?

Many times, animals or special breeds of animals become fads or status symbols. When the fad fades away, they are unwanted or abandoned animals. An example is the wrinkled-skinned Shar-Pei. *See Figure 3–3.* The following appeared in *Atlantic Monthly* magazine: "Twelve of these Chinese guard and fighting dogs were brought into the United States in the late 1960s. Offspring sold for as much as $10,000. Sloppily bred from a limited gene pool, these dogs have developed autoimmune deficiencies leading to severe skin disorders, a condition where teeth don't align properly that sometimes make it impossible for them to eat. Their hips are bad; to correct entropion their eye lids need to be surgically cut from over the eyes. Many display the foul temper of a punch-drunk fighter. Since the collapse of the market on Shar-Pei, many are appearing in animal shelters, abandoned by owners no longer interested in supporting a dog that has fallen from vogue."[4]

Figure 3–3 The Shar-Pei, once a valuable status symbol, has been appearing in shelters since the collapse of the market on these wrinkled-skin dogs. *Courtesy of Stacey Riggert.*

Shar-Pei can be born with **entropion,** which causes the eyelid to roll inward. The fur and eyelashes rub against the surface of the eye, causing irritation. Watery eyes, infection, and corneal ulcers can occur. Surgical correction is required, and animals with this condition should not be bred because the condition is believed to be hereditary.

Another example of a fad breed is the Dalmatian. The popularity of Dalmatians increased rapidly after Disney released the animated *101 Dalmatians* in 1984 and reached a peak in 1993 after the re-release of the animated film in 1991. The rank of registrations for Dalmatians rose from 35th in 1984 to 9th in 1993. Since 1993, the number of registrations has declined, and the Dalmatian now ranks 77th. Unfortunately, many of the animals purchased during the peak popularity of their breed found their way to animal shelters, were abandoned, or were euthanized once they grew up and did not fit in with their owners' lifestyles. Another recent fashion trend is to have a Chihuahua or other miniature sized dog peeking out of a designer handbag. Celebrities have been seen toting the small animals at upscale events. Chihuahuas have also seen a recent increase in popularity since the release of the Walt Disney movie "Beverly Hills Chihuahua".

Many organizations exist around the country to help rescue Shar-Pei, Dalmatians, and other breeds from animal shelters and to help find homes for unwanted animals.

People need to take some time and give serious thought to the decision to obtain a certain breed of dog. People should check out the different breeds, become familiar with their characteristics and needs, and ask the questions posed in this unit before obtaining an animal.

What Am I Going to Use the Animal For?

This consideration is especially important with dogs. A person may want their dog to assist them in a specific sport or hobby such as hunting or tracking competitions. Or they may want the dog to do a certain job such as herding or guarding livestock. The American Kennel Club (AKC) groups breeds according to their traditional use. For example, there is the working group and the sporting group. Choosing a dog of an appropriate breed is likely to make training easier. While it is probably not impossible to teach a short, heavy-bodied Bassett Hound to herd cattle, for example, one is more likely to be successful with an Australian Cattle Dog. Just picking the right breed, however, is only the first step, because there is much variation within each breed. It is also important to examine the puppy's **pedigree.** The pedigree is the family history of the pup and will list any titles earned by its ancestors. Titles earned by a pup's ancestors for competition in field trials for tracking or herding, for example, would be more relevant than titles in **conformation** (how closely a dog resembles the written breed standard for appearance and attitude) when predicting the abilities of a puppy. It is a good idea to obtain a puppy from a breeder who raises dogs for the specific purpose the purchaser has in mind.

CHILDREN AND PETS

Almost all children sooner or later will beg their parents for a pet; most will want a dog because of its playfulness. When contemplating bringing a new pet into the family, parents should consider the age and maturity of the children in the household. Many professionals believe that a child should be 3 years of age to be considered mature enough to act responsibly around a pet. Children younger than 3 would not be able to handle a pet gently without direct supervision. *See Figure 3–4.*

However, many children will quickly lose interest in a pet as it ages or once the novelty wears off. Others will not have developed the self-discipline required to take care of the pet's needs every day. Parents must either make sure the child cares for the pet or be willing to assume these duties themselves if the child doesn't follow through. A short-lived cage pet such as a hamster or gerbil may be a good choice for your child's first pet. If the child learns the basic care and behavior of this pet and demonstrates his or her ability to provide consistent care, he or she may be ready for a more demanding pet.

Children can learn things by caring for a pet, such as responsibility, social skills, respect and compassion, and coping with loss.

Responsibility

Responsibility can be learned from taking care of a pet. This in turn develops one's self-esteem and self-worth. By caring for an animal, children learn to

Figure 3–4 Children can learn many things from caring for a pet. *Courtesy of Carolyn Miller.*

be competent individuals. These characteristics can be reinforced by praising children for the way they care for their pets.

Social Skills

All children will experience the feeling of being unwanted or unloved. A pet will always accept children and provide them with a sense of being loved and wanted.

By learning to understand a pet's body language, emotions, and needs, children develop a stronger sense of responsibility and personal worth.

Children can develop skills that are important for when they become adults. Many of these same skills are important in caring for children, the elderly, and the disabled.

Respect and Compassion

Respect and compassion can be developed by the daily caring for a pet. A child will recognize changes in the animal's behavior, such as illness, depression, or annoyance at the child. By recognizing these changes and having consideration, a child can transfer these traits to others. The child also learns to appreciate the effect his or her actions have on others. Erika Friedman, a professor of health and nutrition sciences at Brooklyn College of the City University of New York, in an article for the *Courier,* states, "In early childhood pets provide a connection to the natural world and are instrumental in teaching respect for other living things. For slightly older children, the responsibility of a pet may be instrumental in developing self-esteem, self-confidence, and independence."[5]

Grief and Coping with Loss

Grief is the normal response to the loss of a pet, especially if the pet had become an important part of one's life. For many children, the loss of a pet may be the first time they experience death. The feelings and emotions that children experience should be discussed. Discussing the loss, and perhaps performing a ceremony or burial, may make children feel better.

In the grieving process, most people go through five stages. During stage one; there is shock, denial, and isolation. The loss may not be accepted, and one may block the loss out of one's mind or deny that the death has occurred. During the second stage, there is anger. The anger may be aimed at family members, friends, or the veterinarian who couldn't help the pet or performed the euthanasia. There may also be feelings of guilt or fear during this stage. During the third stage, there are feelings of helplessness. One may question whether he or she fed and cared for the pet correctly, whether the pet would have lived if taken to the veterinarian earlier, or what could have been done differently. During the fourth stage, there are feelings of depression, sadness, hopelessness, and regret. There may be worry about what to do with the pet's body and about the costs involved. The final stage is the acceptance of the loss.

Support of friends and family is important in dealing with a loss. Talking and discussing one's feelings are important. Discussions with the veterinarian may be helpful so that one understands the facts and procedures that were done to help the pet.

Everyone deals with loss differently, and the time involved in the grieving process varies with individuals. Children usually get over a loss more quickly than adults, but that does not mean they grieve any less. When dealing with the loss of a pet, children need to understand that most animals do not live as long as humans, and that they do not fear death.

Children need to focus on other activities to get their mind off the loss. After the child has accepted the loss, it may be time to get a new pet.

PETS AND PET THERAPY

Researchers are beginning to realize that pets are important in helping people overcome illness. Pets can help us live longer, happier lives by helping to lower our blood pressure. Pets help people to recover from heart attacks and other serious illnesses. "Heart rate, for example, is lower when people sit quietly or read aloud in the presence of a friendly animal, than when doing so alone," according to Erika Friedmann, associate professor of Health Sciences at Brooklyn College. A year after their heart surgery, Friedmann found higher survival rates for patients with pets in their homes than for those who had none, even after accounting for overall health differences among patients. Elderly people who own pets also make fewer visits to doctors than do those who are without animal companions, possibly because the animals mitigate loneliness."[6] *See Figure 3–5.*

When University of Pennsylvania researchers studied a group of seriously ill heart patients, they found that pet owners had much better survival records. In the year the study lasted, the death rate for patients who did not own pets

Figure 3–5 Pets can be valuable in helping the elderly lead longer, happier lives. *Courtesy of Deborah Rhodes (left) and Amy Tucker (right).*

was 28 percent, whereas pet owners had a death rate of 6 percent.[7] A study in the December 15, 1995, issue of the *American Journal of Cardiology* found that dog owners, in particular, were significantly less likely to die within 1 year of a heart attack, compared with heart attack survivors who did not own dogs.[8]

Other studies have documented that petting a dog or cat lowers blood pressure in patients with hypertension and that older people show improved alertness when caring for a pet. Pets often seem to be especially valuable for some people experiencing major life changes, such as unemployment, illness, or the loss of a loved one.[9] A study published in the April 1999 edition of the journal *AIDS Care* found that people with AIDS who have pets are less likely to suffer from depression than people with AIDS who do not own pets.[10] Several studies have also demonstrated that pet owners tend to have lower cholesterol and blood pressure levels than non–pet owners and are therefore at a reduced risk for cardiovascular disease.[11] A study published in the March 1999 *Journal of the American Geriatrics Society* showed that senior citizens who own pets are less likely to be depressed, are better able to tolerate social isolation, and are more active than those who do not own pets.[12]

H. Ellen Whiteley, DVM, in an article in the *Saturday Evening Post* states, "Pets help keep us young: they decrease loneliness, and they give us the opportunity to be needed. Pets also offer a healthier lifestyle by stimulating us to exercise while we take care of them."

Dr. Whiteley continues, "the elderly may benefit the most. At a time in their lives when they have returned to depending on others, they need to feel a sense of responsibility. Pets fulfill this need because they depend on their owners for care and attention. In return, the pets offer love and unqualified approval."[13]

Studies also indicate that pets can have a beneficial effect on the disabled. Dogs can be trained to perform numerous tasks, including opening and closing doors, turning switches on and off, pulling wheelchairs, carrying packages, and assisting individuals who are blind. *See Figure 3–6.* Not only can they perform these tasks, but they contribute to a person's self-esteem and psychological

Figure 3–6 Some companion animals, especially dogs, can be very valuable to people with disabilities. Dogs can be trained to provide assistance to people with blindness or deafness and to many who are confined to wheelchairs. *Courtesy of the Guide Dog Foundation for the Blind, Inc.*

well-being. Dogs and other animals have been used in programs in nursing homes to lift the spirits and encourage activity and involvement by patients. Severely withdrawn patients, especially children, have shown responses when pets have been introduced in their therapy. Pets are also beneficial for children coming home from school before their parents get home. The children have their pets to care for and enjoy until their parents get home.

A study in the April 3, 1996, issue of the *Journal of the American Medical Association* concluded that trained service dogs can be highly beneficial and potentially cost-effective components of independent living for people with physical disabilities.[14]

EUTHANASIA

Perhaps the kindest thing you can do for a pet that is severely sick or injured is to have your veterinarian induce death quickly and humanely; this is called euthanasia. The decision to have your pet euthanatized is a serious one and is never easy to make.

Your relationship with your pet is special. When you acquired your pet, you assumed responsibility for its health and welfare. Euthanasia may be the most difficult decision you will ever make regarding your pet. Your veterinarian, family, and friends can assist in the decision-making process. You should consider what is best for your pet, for you, and for your family. Quality of life is important for pets and people alike.

If your pet can no longer get around comfortably; if your pet isn't eating or willing to eat; if your pet is no longer aware of or appreciates your companionship; if your pet is apt to bite family members, strangers, or children; if your pet can no longer respond in normal ways; if there is more pain than pleasure in your pet's life; or if the care and treatment become too much financially, then euthanasia may be a necessary option.

Euthanasia is accomplished with an injection of a tranquilizer that stops nerve transmissions and causes muscle relaxation. The veterinarian may sometimes administer a tranquilizer to calm and relax your pet so that the euthanasia procedure can be carried out more humanely and peacefully. The euthanasia solution is usually injected into a vein in the foreleg. Six to twelve seconds after injection, the animal will lapse into a deep unconsciousness. Death will come quickly and painlessly.

OVERPOPULATION

Each year, approximately 27 million dogs and cats are born in the United States; 50 million are euthanized each year as unwanted or abandoned. This overpopulation is a result of unplanned matings, matings for commercial purposes, and parents who want their children to experience "the miracle of childbirth."

To solve the overpopulation and unwanted animal problem, the number of puppies and kittens born must be reduced. To accomplish this, three events need to occur:

1. People must be educated about the problem.

2. Animal control laws must be passed and existing laws enforced.

3. People must recognize the importance of spaying and neutering.

Many unwanted animals end up in animal shelters, where those that can't be placed in another home are euthanized; the shelters have taken the responsibility to eliminate the problem of unwanted animals. Adults and children need to be educated and take responsibility for controlling the breeding of their dogs and cats.

Licensing of animals, leash laws, and other animal control laws need to be enforced. Other laws and ordinances controlling the breeding of animals may also need to be passed.

Spaying and neutering are surgical procedures that prevent animals from reproducing. Spaying, in which the ovaries and uterus are removed, is performed on females. This prevents the female from coming in heat and from becoming pregnant. Neutering, in which the testicles are removed, is performed on males. This prevents the male from producing sperm and impregnating a female.

Spaying is generally recommended before the female's first estrus cycle. Many veterinarians now perform the spaying and neutering surgery on animals as young as 8 weeks of age. Your veterinarian can recommend the best time for your pet to be spayed or neutered.

During the surgical procedure, the animals are fully anesthetized, so they feel no pain. Post-surgical pain is usually minimal and can be treated with medication from the veterinarian.

Spaying eliminates potential problems and risks associated with pregnancy and birth. Females are also less susceptible to mammary cancers and uterine infections.

Neutering eliminates testicular tumors and reduces the risk for prostate disorders. Neutering males removes their motivation to roam the community in search of mates, so neutered males are less apt to be hit by cars or injured in fights. Neutering can reduce "mounting" behavior that can be embarrassing or annoying and urine marking behavior. These two effects are most pronounced if the surgery is done when the dog is young and these behaviors have not become well established.

SUMMARY

Since World War II, we have experienced a pet explosion in the United States. Pets provide companionship in today's fast-paced society. Unfortunately, many of these pets get abandoned each year. Millions of unwanted animals pass through animal shelters each year, a result of uncontrolled breeding and abandonment. A major reason owners relinquish their pets to a shelter is that they are unable to meet the pet's behavioral needs.

Pets can have an important influence on their owners, both young and old. Children can learn important lessons that can help them in later years, and pets can have therapeutic benefits for the elderly.

Pets have also been of great value to people with disabilities. The deaf, blind, and those confined to wheelchairs or bed can receive practical and psychological benefits.

People must be educated about the fact that they are responsible for their pets and for the controlled breeding of their pets.

DISCUSSION QUESTIONS

1. Are there unwanted animals in your area? What happens to them? Does a problem exist with unwanted animals? What should be done to correct the problem?

2. What are the questions one should ask before obtaining a pet?

3. What are four sources from which you can obtain pets? What are the advantages and disadvantages of each source?

4. What are some of the various lifestyles that people have? What pets might best fit those lifestyles?

5. What are some other costs of owning a pet besides the initial purchase price?

6. What lessons can children learn from having a pet? Are there other lessons not listed in the text?

7. Do animals have personalities? What are some of the personalities you have observed?

8. What are some animal fads or status animals of recent years, other than those mentioned in the text?

9. What do we mean by responsible pet ownership?

10. What do we mean by the terms *spay* and *neuter*? Should we have our pets spayed and/or neutered?

11. What are the benefits of pets to the elderly?

12. What are some of the therapeutic uses of pets?

13. What is euthanasia, and why should it be considered by a pet owner? How would you counsel someone who is dealing with the loss of a pet?

SUGGESTED ACTIVITIES

1. With the approval of the instructor, invite pet shop owners, various pet breeders, and animal shelter personnel to your class to make presentations.

2. Visit a veterinarian and discuss with him or her the process of euthanasia and how pet owners facing the loss of a pet are counseled. Report the information back to the class.

3. Choose one breed from each of the following AKC groups: Sporting, Toy, and Terrier. Make a chart that compares the following aspects of each breed: size, grooming needs, exercise

needs, health risks associated with the breed, training ease, and any purpose for which they may be used. Discuss which would NOT be a good fit for your family and why.

ADDITIONAL RESOURCES

1. American Veterinary Medical Association (AVMA)
   http://www.avma.org

2. Mayo Foundation for Medical Education and Research
   http://www.mayohealth.org

3. Dog-Play
   http://www.dog-play.com

4. Pet Life
   http://www.petlifeweb.com

5. Delta Society
   http://www.deltasociety.org/

6. American Kennel Club
   http://akc.org

END NOTES

1. Excerpt from Szabo, Liz. 2004. "Kinder, gentler animal shelters." *USA Today*, July 26, 2004.

2. "No kill now: Definition of no kill shelters." http://www.nokillnow.com (accessed January, 2007).

3. American Society for the Prevention of Cruelty to Animals (ASPCA), "Pet care costs." http://www.aspca.org/site/DocServer/PET_CARE_COSTS.pdf (accessed January 22, 2007).

4. Derr, M. 1990. "The politics of dogs." *The Atlantic Monthly* 265(3) (March): 70.

5. Friedmann, E. 1998. "People and pets." *Courier* 41 (February): 13.

6. Burke, S. 1991. "In the presence of animals." *U.S. News & World Report* 112 (February 24): 7, 64–65.

7. Reprinted with permission from *Mother Earth News,* copyright 1985, Sussex Publishers, Inc.

8. Friedmann, E., and S. A. Thomas. 1995. "Pet ownership, social support, and one year survival after acute myocardial infarction in the Cardiac Arrhythmia Suppression Trial (CAST)." *American Journal of Cardiology* 76(17) (December 15): 1213.

9. Mayo Foundation for Medical Education and Research. 1997. "Pet therapy, medical specialists on four legs." http://www.mayohealth.org/mayo/9701/ htm/pet_ther.htm (accessed July 13, 1999).

10. Siegel, J. M., F. J. Angulo, R. Detels, J. Wesch, and A. Mullen. 1999. "AIDS diagnosis and depression in the Multicenter AIDS Cohort Study: The ameliorating impact of pet ownership." *AIDS Care* 11(2): 157–170.

11. Mayo Foundation for Medical Education and Research. 1997. "Pet therapy, medical specialists on four legs." http://www.mayohealth.org/mayo/9701/htm/pet_ther.htm (assessed July 13, 1999).

12. Raina, P., D. Waltner-Toews, B. Bonnett, C. Woodward, and T. Abernathy. 1999. "Influence of companion animals on the physical and psychological health of older people: An analysis of a one-year longitudinal study." *Journal of the American Geriatrics Society* 47(3) (March): 323–329.

13. Whiteley, H. E., D.V.M. 1986. "The healing power of pets." *Saturday Evening Post* 258(7) (October): 22.

14. Allen, K., Ph.D., and J. Blascovich, Ph.D. 1996. "The value of service dogs for people with severe ambulatory disability." *Journal of the American Medical Association* 275(13): 967–1060. Available online at http://www.ama-assn.org/sci-pubs/journals/archive/jama/vol_275/no_13/abstract.htm (accessed April 31, 1996).

CHAPTER **4**

Animal Rights and Animal Welfare

After reading this chapter, one should be able to:

- define the terms *animal rights* and *animal welfare* and be able to compare and contrast them.

- identify important persons in the animal rights movement.

- identify important dates and acts of legislation associated with animal welfare.

- compare the issues concerning animal rights and animal welfare.

TERMS TO KNOW

animal rights	factory farming	unethical
animal welfare	humanize	vivisection
confinement systems	intensive operations	
ecoterrorism	specieism	

HISTORY

The terms *animal welfare* and *animal rights* are becoming more common today.

Many people believe the two terms to be synonymous, and they are often used interchangeably by the public and the media.

Animal welfare is the position that animals should be treated humanely. This includes proper housing, nutrition, disease prevention and treatment, responsible care, proper handling, and humane euthanasia or slaughter. Animal welfare advocates believe that animals can be used for human purposes but that they should be treated so that discomfort is kept to a minimum.

Animal rights is the position that animals should not be exploited. Animal rights advocates believe that animals should not be used for food, clothing, entertainment, medical research, or product testing. This also includes the use of animals in rodeos, zoos, circuses, and even as pets. They believe it is ethically, morally, and inherently wrong to use animals for human purposes under any condition. Animal rights advocates are against **vivisection** and believe that living animals should not be used for surgical operations and experiments in which the structure and function of organs are studied. They further believe that animals should not be used in experiments in which diseases and various therapy methods are studied.

The modern animal rights movement, which began in the early 1970s, draws heavily on philosophers and theologians; others who have been drawn into the movement have been activists in other areas such as human rights.[1] Other animal rights activists are drawn from vegetarians and vegans because of their belief that animals should not be used for food.

Public indignation over the welfare of animals in Great Britain resulted when a London housewife and vegetarian, Ruth Harrison, published a book titled *Animal Machines: The New Factory Farming Industry.* The main emphasis of the book centered on two areas: the use of antibiotics, hormones, and additives in animal feeds; and modern **factory farming,** in which chickens are kept in cages and veal calves are kept in small crates. This book had a major impact in England. An investigation was conducted, and as early as 1964, laws were passed concerning the care and treatment of chickens, turkeys, pigs, cattle, sheep, and rabbits.

The animal rights movement continued to gain momentum in 1975 with a publication by Peter Singer titled *Animal Liberation.* Peter Singer, an Oxford University-educated Australian philosopher and vegetarian, is often considered the founder of the modern animal rights movement. *Animal Liberation* condemned the use of animals produced for food and their use in research.

In 1990, animal rights groups staged a "March for the Animals" rally in Washington, D.C. Crowds estimated from 15,000 to 24,000 people marched down Pennsylvania Avenue. They carried banners and placards proclaiming "Animals Are Not for Wearing," "Fur Is Dead," and "Animals Have Rights, Too."

The United States has the distinction of having the first laws enacted to protect farm animals from cruel treatment. This was a legal code titled "The Body of Liberties" passed by the Massachusetts Bay Colony in 1641.

The first anticruelty law was passed by the New York legislature in 1828. The legislation claimed: "Every person who shall maliciously kill, maim, or wound any horse, ox, or other cattle, or sheep, belonging to another, or shall maliciously and cruelly beat or torture such animal, whether belonging to himself or another, shall upon conviction, be adjudged guilty of a misdemeanor."

In 1866, the American Society for the Prevention of Cruelty to Animals (ASPCA) was formed in New York, largely to look after the welfare of disabled horses and mules and save them from abandonment. This was America's first humane society and was founded by Henry Bergh.

Because no one had been prosecuted under the New York State Anti Cruelty Law, Henry Bergh drafted in 1867 "An Act for the more effectual prevention of cruelty to animals." This act had 10 sections and has served as the example for drafting many succeeding anticruelty laws. Forty-one states and the District of Columbia have present laws based on this act.

In 1906, the Animal Transportation Act was passed to protect animals traveling long distances by rail. This act was passed to provide humane care and treatment of animals destined for slaughter.

In 1958, the Humane Slaughter Act was passed and later amended in the 1970s to include the humane handling of animals before and during slaughter.[2]

In 1966, Congress enacted Public Law 89–544, known as the Laboratory Animal Welfare Act (AWA). This law regulates dealers who handle dogs and cats, as well as laboratories that use dogs, cats, hamsters, guinea pigs, rabbits, and primates in research.

The first amendment to the AWA was passed in 1970 (P.L. 91–579). This amendment authorized the regulation of other warm-blooded animals when used in research, exhibition, or wholesale pet trade. This amendment did not cover birds and rodents.

An amendment in 1976 (P.L. 94–279) prohibited animal fighting ventures and regulated the commercial transportation of animals.

An amendment in 1985 (P.L. 99–198), also called "The Improved Standards for Laboratory Animal Act," clarifies what is meant by "humane care" and mentions such items as sanitation, housing, and ventilation. It requires that standards for the exercise of dogs and for a physical environment adequate to promote the psychological well-being of non-human primates be set. It specifies that pain and distress must be minimized in experimental procedures and that alternatives to such procedures must be considered. It also defines practices that are considered to be painful. In addition, it specifies that no animal can be used in more than one major operative experiment with recovery. The amendment also called for the establishment of an information service at the national Agriculture Library. The Animal Welfare Information Center (AWIC) is to assist those regulated by the act in prevention of unintended duplication of research, employee training, searching for ways to reduce or replace animal use, and providing information on how to decrease pain and distress.[3] Additional requirements include the establishment of Institutional Animal Care and Use Committees (IACUC) at research facilities. These committees include at least one public member. The IACUC must review and approve all procedures involving animals before they are carried out. In order to get approval from the IACUC, a researcher must show that the use of animals is justified and that the research has established protocols to minimize pain and distress. The IACUC also investigates and responds to any complaints concerning animal care and use at the facility.

In 1990, injunctive relief and pet protection provisions were passed as part of the Food, Agriculture, Conservation, and Trade Act of 1990. The injunctive relief provision authorizes the Secretary of Agriculture to seek an injunction stopping a licensed entity from continuing to violate the AWA while charges are pending. The pet protection provision mandates the Secretary to issue additional regulations pertaining to random-source dogs and cats—animals that have been obtained from animal and humane shelters for use in laboratories. The act covers four categories of institutions: (1) dog and cat breeders; (2) zoos, circuses, and roadside menageries; (3) transportation of animals; and (4) research facilities.

The United States Department of Agriculture is charged with developing and implementing regulations to support the AWA. These regulations, Parts 1, 2, and 3, provide minimum standards for the care and handling, housing, feeding, sanitation, ventilation, shelter from extreme weather, veterinary care, and separation of species when necessary. Birds, laboratory rats, and laboratory mice are currently excluded from these regulations. The AWA regulations require the licensing of animal dealers, exhibitors, and operators of animal auction sales where animals regulated under the AWA are sold.[4]

The Horse Protection Act was passed in 1970 and amended in 1976. This act protects horses and regulates the horse show business. At issue is the showing or sale of horses whose gait is altered by inducing pain in their legs.[5]

In October 1986, an animal rights group known as the Animal Liberation Front (ALF) broke into laboratories on the Oregon State University campus and caused several thousand dollars' worth of damage. Six months later, ALF set fire to a laboratory on the campus of the University of California at Davis, causing nearly $5 million in damage. On Independence Day, 1989, ALF struck a research facility on the campus of Texas Tech, where Dr. John Orem was working with cats in an effort to understand the cause of Sudden Infant Death Syndrome (SIDS). The damages exceeded $55,000, but the break-in not only resulted in damage to the facility and equipment, but also caused a delay in the important research work of Dr. Orem. SIDS causes the death of more than 8,000 infants every year in the United States.

In 1989, the Farm Animal and Research Facilities Protection Act was introduced by Rep. Charles Stenholm (D-Texas). This act would have applied to facilities or premises where an animal is kept for food, agricultural research, testing, and education. It would make it a federal crime to disrupt the activities of the premises; steal or damage property or animals; enter a facility with the intent to commit a violation; or remain in a facility despite notice that entry was forbidden. Violators could be fined up to $10,000 and jailed for up to 3 years, or both. *See Figure 4–1.*

Figure 4–1 Animal rights activists have caused considerable damage to research facilities around the country. Fire was set to facilities on the campus of Michigan State University and sulfuric acid was doused throughout the laboratory. *Photo by Dave Weinstock, courtesy of The Farmer's Exchange.*

Figure 4–2 On the wall behind the clean-up crew are the words, "We will be back for the Otters!" Research at this laboratory was being conducted with mink. *Photo by Dave Weinstock, courtesy of The Farmer's Exchange.*

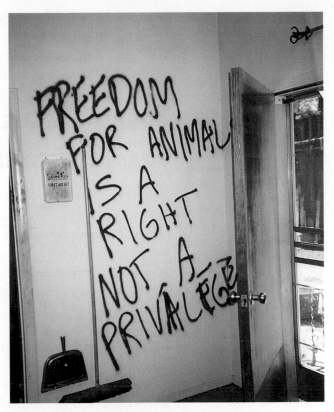

Figure 4–3 Animal rights activists vandalized a research facility at Oregon State University and spray-painted a message on the wall. *Photo by American Farm Bureau Federation staffer Julie Brown, courtesy of Farm Bureau News.*

Although the bill had 235 co-sponsors, it did not see floor action in the House before adjournment of the 101st Congress. Stenholm's office had released a list of 91 incidents between January 1981 and January 1990, ranging from bomb threats and trespassing to break-ins and arson, which were linked to groups associated with the animal rights movement.

On August 4, 1992, a bill introduced by Stenholm passed the House of Representatives and was merged with a similar proposal sponsored by Sen. Howell Heflin (D-Alabama), which had passed the Senate in the fall of 1991. On August 26, 1992, this bill, known as the Animal Enterprise Protection Act of 1992 (P.L. 102–346), was signed into law by President George Bush.

Facilities protected by this bill include farms, zoos, aquariums, circuses, rodeos, fairs, auctions, packing plants, and commercial or academic enterprises that use animals for food or fiber production, agricultural research, or testing. *See Figures 4–2 and 4–3.*

The law imposes up to 1 year in jail for causing damages of $10,000 or more on an animal enterprise; up to 10 years in jail if an intruder injures a person during an attack on an animal enterprise; or up to a life sentence in jail if an intruder kills another person during an attack on an animal enterprise.

These crimes are hard for law enforcement officers to solve and to prosecute. Although Section 43 of the Animal Enterprise Protection Act covers disruption of animal enterprises, many people are advocating an amendment to the Animal Enterprise Protection Act that would include adding to the list of protected persons loggers, miners, fishermen, farmers, trappers, ranchers, food outlets, processors, and all resources subject to **ecoterrorism,** terrorism or crimes committed under the disguise of saving nature.

CURRENT ISSUES

Do Animals Have Rights?

Animal rights activists believe that animals have the same rights as humans. They believe any use of animals by humans reflects a bias that humans are superior to animals. **Specieism** is the term they use for this belief. They believe that because animals have nervous systems and feel pain, it is wrong to use animals for food or experimentation. They contend that one cannot ignore the rights of animals while advocating the rights of humans, because humans are animals.

They contend that humans are supposed to care for and protect animals and not exploit them. To believe that animals have the same rights as humans is to **humanize** animals.

Animal welfare advocates believe that animals should be treated humanely and without cruelty. Animals should have proper housing, management, nutrition, disease prevention and treatment, responsible care, and handling. Proponents of this position believe that animals can be used in research as long as all measures are taken to ensure that the animals are cared for humanely.

Animals should be used only when no other alternatives exist to achieve the objectives of the research.

Should Animals Be Used for Food?

Animal rights activists argue that modern farming is inhumane and that eating meat is unhealthy. They contend that animals are constantly housed, kept in small cages with wire floors, and never allowed to see the light of day. It is their position that such "factory farming" results in severe physiological and behavioral problems, as well as causing a long list of diseases and ailments. They contend that modern farming is controlled by large corporations that care only about profits and not about the welfare of animals.

Farmers counter that they have led the way in animal welfare. It has always been in their best interest to care for animals humanely and to see that they are well fed and free of diseases and ailments. *See Figure 4–4.* Modern farmers argue that they could not stay in business long if their animals weren't healthy. Modern farming involves **intensive operations,** in which farmers try to increase their output through better breeding, feeding, and management. Many of these operations are **confinement systems,** where the animals are confined to cages or pens in partially enclosed or totally enclosed buildings. Farmers can get improved production through closer control of the environment. These buildings may have slotted floors, automatic feeding and watering systems, and forced air ventilation. *See Figure 4–5.*

Figure 4–4 Sows are usually placed in farrowing crates 1 week before they give birth. She will remain in the crate for 3 to 4 weeks. The use of a crate makes delivery easier, makes it easier for the farmer and veterinarian to work with her if necessary, and prevents the sow from rolling over and crushing her litter. *Courtesy of Agriscience.*

Figure 4–5 Laying hens are kept in cages under controlled environments that increase the efficiency of the laying operation. Each hen receives adequate feed and water. Eggs, a valuable part of the diet of many Americans, are available at affordable prices because of the efficiency of these operations. *Courtesy of Dr. Bill Muir, Purdue University.*

Figure 4–6 Veal calves are kept in individual stalls so that they can receive individual attention. Diseases are kept to a minimum and farmers are able to feed and care for the calves more efficiently. The economic welfare of farmers depends on the proper care and treatment of their animals. *Courtesy of Utah Agricultural Experiment Station.*

Most farm animals (except beef cattle) are kept in barns or suitable housing to protect the health and welfare of the animal. Beef cattle being raised for meat are usually fed outside in open-air feed lots. These facilities protect the animals from extremes in weather, diseases, and predators. Modern livestock facilities are well ventilated, are temperature controlled, and provide the animals with a constant supply of fresh water and feed. Beef cattle (cow-calf) operations are usually kept on pastures or open ranges, where they consume primarily nongrain feedstuffs such as grass, roughage, food-processing by-products, and crop residues. *See Figure 4–6.*

Farming in the United States is not controlled by large corporations. Of the 2.2 million farms in the United States, 97 percent are family owned and operated; only 7,000 are non-family-controlled corporations.[6]

Both the federal government and the American Heart Association contend that a diet containing meat, milk, and eggs is appropriate. No studies can substantiate that a vegetarian diet is healthier than a diet that includes meat, milk, and eggs. The American Dietetic Association, American Heart Association, and others recommend 5 to 7 ounces of lean, trimmed meat daily because it provides large amounts of essential nutrients.

Should Animals Be Used for Experimentation?

Animals are used in three areas of experimentation:

- research in biomedical and behavior sciences
- testing of products for their safety
- education in which animals are used for demonstrations and dissection

Animal rights activists believe that animal experimentation is **unethical** and unnecessary and that alternatives to animal experimentation should be used. Animal rights activists contend that the value of animal research is greatly exaggerated and that misleading animal tests can have a devastating effect on human health.

Much animal rights literature depicts maimed and tortured research animals. Those who believe that animals have the same rights as humans also believe that all forms of animal research should be eliminated. Animal rights activists also point out that biomedical research is a vast, lucrative industry, supported each year by $27 billion in taxes and charity, while killing millions of animals.

Several areas of animal experimentation have been attacked by animal rights activists:

1. Pound seizures, where dogs and cats from animal shelters have been sold to laboratory animal dealers and to research and educational facilities for use in animal experimentation.

2. The Draize Eye Test and the Skin Irritancy Test, which are used to measure the effects of products such as cosmetics on the unprotected eyes and skin of restrained rabbits. Rabbits are usually used for this test because they do not have tear ducts to wash the products from their eyes. In the skin tests, a patch of fur is shaved from the back and side of the animal. Products are applied to the area and then observed for any reactions. *See Figures 4–7 and 4–8.*

3. The dissection of animals in educational facilities. *See Figure 4–9.*

4. The LD50 Test, which is used to determine the dose of drugs. Test animals are exposed to increasing doses of the test substance until 50% of the test population dies.

5. Animals that are used in psychological research to discover the effects of pain, response to different stimuli, learning, and behavior.

Figure 4–7 These rabbits are being used in a Draize Eye Test. The animals are restrained and various products being tested are placed in their eyes. *Photo provided by People for the Ethical Treatment of Animals (PETA).*

Most critics of the animal rights activists' views on experimentation believe that biomedical research would be limited or stopped completely if animals were not available for research. Many medical advances would not exist if animals were not used in research. Kidney, heart, liver, lung, and endocrine organ transplants are now possible because of prior research on animals. New drugs, vaccines, and surgical techniques were perfected with animal research. Vaccines against diphtheria, hepatitis, tetanus, and polio were developed after research on animals. Insulin is a result of research conducted with dogs.

The number of animals used in testing is declining. Approximately 90 percent of the animals still used in testing are rats and mice. The Animal Welfare Act of 1966, which regulates the care and treatment of laboratory animals, does not include rats and mice. Many companies no longer use animals for testing their products or have reduced animal testing programs. Several companies have offered grants for the research of "alternative" (replacement methods for animal testing) methods of testing products.

Some of these alternative methods now being used or suggested include the following:

1. The Bovine Corneal Opacity (BCOP) Test is being offered as an alternative to the Draize Eye Test. Cow eyes obtained from slaughterhouses are used in this test. The cornea is removed, and the test substance is applied; the effects are then measured. This test does use the product of animals, but the eyes would have been discarded during the slaughter.

2. Computer videodisc simulations and three-dimensional computer programs are being used for teaching techniques without using animals.

3. Plastic anatomical models are being used for teaching in place of the actual dissection of animals.

4. "Cultured" skin and its use in testing irritation to cosmetics, household products, and other agents as well as the treatment of wounds

Figure 4–8 Skin Irritancy Tests are conducted on rabbits to test the effects of various products on their skin. A small patch of hair is shaved away, and the product being tested is applied to the skin. *Courtesy of the American Anti-Vivisection Society.*

Figure 4–9 Many small animals are used in educational facilities across the country for dissection. Many alternatives to dissection now exist and many people question the validity of dissection as an educational tool. *Photo provided by People for the Ethical Treatment of Animals (PETA).*

shows promise. Cultured skin is grown *in vitro* (meaning in glass or in laboratory containers) in the laboratory. The cultured skin contains live cells that respond to stimuli similarly to the way the skin of a live animal would.

5. The Fixed Dose Procedure (FDP) first described in 1984 is replacing the older KD50 test. The FDP uses less than half the number of animals and uses the dosage that produces signs of illness, rather than death, to estimate the LD50.

Primates make up about half of 1 percent of all animals used in experiments; however, their contributions have been significant. One example is the research into the cause and cure of poliomyelitis. The cause of poliomyelitis remained a mystery until 1908, when scientists were able to transmit the virus to monkeys. Years of research with primates followed until the early 1950s, when scientists were able to grow the virus in human cell cultures. This made the development of a vaccine possible. After a vaccine was developed, it was tested on monkeys to ensure that it was safe and effective. Kidney tissue from monkeys was also necessary to produce the vaccine. Today, the vaccine is produced from self-propagating cells.

Many people believe that chimpanzees and the pigtail macaque may be important in developing a cure for AIDS (acquired immune deficiency syndrome). The University of Washington reported in 1991 that it had successfully inoculated eight pigtail macaques with human AIDS and that the monkeys became critically ill with symptoms resembling those of human AIDS. Until then, the only other species successfully inoculated with the human AIDS virus was the chimpanzee, but no chimpanzee has yet developed clinical signs of the human AIDS.[7]

In mentioning AIDS research, it is also important to note that the United States Centers for Disease Control and Prevention (CDC) *Morbidity and Mortality Weekly Report* for September 11, 1992, announced that the blood of two laboratory workers working with monkeys had tested positive for simian AIDS. The two workers tested positive but had not become ill. This disease is highly fatal to monkeys. Laboratory workers need to be aware of guidelines published by the CDC in November 1988 to reduce the risk of transmission of "monkey AIDS."[8]

Animal agriculture and medical research have combined to make significant contributions to the health and well-being of people. From the adrenal glands of cattle, epinephrine is used to relieve some symptoms of asthma and some allergies. It is also used as a heart stimulant in some crisis situations and by dentists to prolong the effect of local anesthetics.

Thrombin from cattle blood helps blood clotting and is valuable in treating wounds in inaccessible parts of the body; it is also used in skin grafting. Liver extract is combined with folic acid and injected to treat various types of anemia.

Heart valves from hogs are used in heart valve replacement surgery in humans. They are in some cases superior to mechanical valves because they don't stick and do not need the same level of anticoagulant infusion. In the last 12 years, 250,000 lives have been saved through implantation of swine heart valves.

Because of its similarity to human skin, pig skin is used to treat massive burns and large accidental skin removal. Gelatin made from pig skin is used for capsules and pills.

Thyroid extracts from swine are used to regulate the rate of metabolism in humans. Another extract is used to treat low calcium and phosphate levels and to regulate heartbeat. Extract from swine pancreas is also a source of insulin. Even with synthetic insulin, an estimated 5 percent of all people with diabetes are allergic to all but insulin from hogs.[9]

Animal welfare proponents believe that it is important that all animals used for research be cared for humanely. Facilities, researchers, and laboratory technicians must learn how to correctly care for, handle, and meet the needs of their animals. Animals should be used only when no other alternatives exist to achieve the objectives of the research. Many people believe that animal experimentation and research hold the key to the successful cure of AIDS, cancer, and heart disease, as well as to improve the health of all humans.

Should Hunting and Trapping Be Allowed?

Many animal rights activists believe that regulated recreational hunting and trapping is inhumane and unnatural. They also believe that hunters and trappers kill merely to inflict suffering and for the joy of killing.

Animal rights activists would also like to see a stop to the use of hunting dogs. They contend that using dogs to hunt is unnatural and inhumane treatment.

Animal rights activists attack state game agencies, claiming that they manage wildlife numbers not to maintain the balance of nature, but to provide large numbers of animals so hunters will have continued hunting. They claim that hunting and trapping does not prevent overpopulation, that hunting for food is no longer necessary, and that hunting should be banned completely.

Hunting advocates note that they are the prime revenue source for preserving nonendangered wildlife. In 1996, hunters spent a total of more than $20.6 billion; $651 million was spent for licenses, stamps, tags, and permits. These figures compare to $12.3 billion in total expenses and $532 million spent for licenses, stamps, tags, and permits in 1991. In 2001, hunters spent $20.6 billion in total and $693 million for licenses, stamps, tags, and permits. The average expenditure per hunter in 2001 was $1,581. A portion of licenses, stamps, tags, permits, and excise taxes on ammunition, firearms, and equipment is used to finance game research and management and to help purchase habitat that benefits all creatures. There were 13.769 million hunters in 1996 compared to 13.771 million in 1991. There was a drop in total numbers of hunters between 1991 and 2001 of about 7 percent. In 2001 there were a total of 13.034 million hunters. This total makes up about 4 percent of our total population.[10]

Throughout history, hunting has been a tradition of humans. Until recent times, humans hunted through necessity for food. As civilization progressed, humans' need to hunt for survival was reduced; most hunting today is recreational. *See Figure 4–10.*

Hunters themselves have long been concerned for the welfare of wildlife. Through their actions, hunters have identified problems such as pollution and its effect on wildlife, habitat loss, and abuse. They helped focus public attention on the plight of rare and endangered species. Hunters have been largely responsible for initiating wildlife laws and their enforcement, wildlife research and management, and establishment of parks and wildlife preserves.[11]

Today, wildlife populations are controlled by game management practices, habitat improvement, and carefully regulated seasons and bag limits. Wildlife management programs have been successful in improving the populations of game species such as the whitetail deer, wild turkey, wood duck, and beaver, as well as populations of nongame species such as the trumpeter swan, egret, and heron.

At the turn of the century, the total population of whitetail deer in North America was estimated at about 500,000; in several states, whitetail deer had been completely eliminated. Nearly every state in the nation had closed its deer hunting seasons. After many years of closed hunting seasons, restocking programs, and management programs, the population had reached levels that were higher than ever before. By the 1960s, almost all states allowed some form of deer hunting. It is estimated that there are more than 30 million whitetail deer in the United States today. Many states have such large populations that the deer are causing destruction to the available vegetation and farmers' crops and are becoming hazards on our highways. *See Figure 4–11.*

The stories of the wild turkey and the wood duck are similar. By 1930, the wild turkey had been eliminated from almost all of our Northern states. A few states in the Southern United States still had small populations. Today, thanks to our nation's hunters, game agencies, and wildlife conservation organizations such as the National Wild Turkey Federation, there are more than 7 million wild turkeys roaming the continent in huntable populations in every state except Alaska, and they are even hunted in Ontario, Canada, and Mexico. Today, turkey hunting is one of the largest types of hunting in the United States, with close to 3 million sportsmen considering themselves turkey hunters.[12]

The wood duck, after being on the verge of extinction by 1915, is now one of our most common breeding waterfowl in the Eastern United States. Unregulated hunting and destruction of hollow tree nesting sites almost eliminated this beautiful bird. *See Figure 4–12a and b.*

Figure 4–10 Hunting is an opportunity to enjoy the companionship of other hunters, to enjoy the outdoors, and to enjoy the tradition of hunting. *Courtesy of Jeff Jackson.*

Figure 4–11 There are more whitetail deer in the United States now than at any time in history. *Courtesy of Brian Yacur.*

(a)

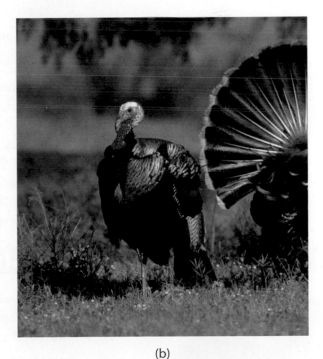

(b)

Figure 4–12 Successful restocking programs and careful management programs have brought the (a) wood duck and (b) wild turkey back from the verge of extinction. *Courtesy of PhotoDisc.*

Hunting programs are designed to harvest only the surplus animals. Wildlife cannot be stockpiled; many people do not realize the high natural mortality of wildlife. Nearly 75 to 80 percent of the bobwhite quail and dove populations will die annually from natural causes even if the birds are not hunted. Research shows that a good deer population can be maintained if hunters take 40 percent each year. Normally, hunters take only about 15 percent of the populations during annual hunting seasons; in many areas, hunters don't harvest enough deer. This leads to overpopulated areas where deer can severely damage the habitat.

SUMMARY

The terms *animal rights* and *animal welfare* are not synonymous. They represent two different viewpoints. Animal rights is the position that animals should have the same rights as humans and should not be exploited in any way. Activists for animal rights believe that animals should never be used for food, clothing, entertainment, medical research, or product testing. This also includes the use of animals in rodeos, zoos, and circuses, and even as pets. Animal rights activists believe that it is ethically, morally, and inherently wrong to use animals for human purposes under any condition.

Animal welfare is the position that animals should be treated humanely. Animal welfare advocates believe that animals can be used for human purposes but that they should be treated so that discomfort is minimized.

Among the issues being debated by animal rights and animal welfare groups are:

1. Do animals have rights?

2. Should animals be used for food?

3. Should animals be used for experimentation?

Before forming your own opinion on these issues, it is important to research the topics carefully. The issues involve more information than is presented here. Sources of information should be critically evaluated for accuracy and bias, taking care to separate opinion from fact.

More than 7,000 groups are involved in animal protection; 400 of these are considered animal rights groups. Many of the groups do not consider themselves as animal rights or animal welfare groups but are dedicated to the preservation of a specific species of animal and its habitat.

DISCUSSION QUESTIONS

1. What was the title and who was the author of the book that aroused public indignation over the welfare of animals in Great Britain?

2. What is a factory farm?

3. Who is considered the founder of the modern animal rights movement, and what was the name of his publication?

4. What do the terms *animal rights* and *animal welfare* mean?

5. Do animals have rights?

6. Should animals be used for food?

7. Should animals be used for experimentation?

8. Should animals be used for other purposes, such as, for hunting or trapping, for entertainment purposes in zoos and exhibits, for rodeos, or even for pets?

SUGGESTED ACTIVITIES

1. Visit the library and research the history of animal rights and animal welfare.

2. Research the various animal rights and animal welfare groups. Write letters to them requesting information, and then discuss the materials that were sent.

3. Research modern farming methods. Are they humane? Research and become familiar with the terms *factory farm, intensive farming,* and *confinement farming.*

4. Discuss the issue of using animals for dissection in classes. Divide into groups and debate the issue.

5. What is the ultimate goal of the animal rights movement? Is there one? Thoroughly study the animal rights movement, the groups, and the issues to answer these two questions.

SUGGESTED READING

1. *Animal Rights—Opposing Viewpoints,* David L. Bender and Bruno Leone, Series Editors, Janelle Rohr, Book Editor, Greenhaven Press, Inc., P.O. Box 28909, San Diego, CA 92198-0009, copyright 1989.

2. *The Animal Rights Crusade: The Growth of a Moral Protest,* by James M. Jasper and Dorothy Nelkin, 1992, The Free Press, a division of Macmillan, Inc. New York.

3. *The Animal Rights Handbook,* 1990, Living Planet Press, 558 Rose Ave., Venice, CA 90291.

4. *Animals Have Rights, Too,* by Michael W. Fox, 1991, The Continuum Publishing Company, 370 Lexington Avenue, New York, NY 10017.

5. *Animals In Society,* by Zoe Weil, 1991, ANIMALEARN, a division of the American Anti-Vivisection Society, Suite 204, Noble Plaza, 801 Old York Road, Jenkintown, PA 19046-1685.

6. "Beyond Cruelty," by Katie McCabe, 1990, *The Washingtonian Magazine,* Volume 25 (February), Number 5.

7. *Congressional Research Service Report for Congress, Humane Treatment of Farm Animals: Overview and Selected Issues,* by Geoffrey S. Becker, Specialist, Environmental and Natural Resources Policy Division, May 1, 1992.

8. *Guide for the Care and Use of Agricultural Animals in Agricultural Research and Teaching,* a booklet published by the Consortium for Developing a Guide for the Care and Use of Agricultural Animals in Agricultural Research and Teaching, March 1988, Editorial and Production Services, Association Headquarters, 309 West Clark Street, Champaign, IL 61820, (217) 356-3182.

9. *Animal Agriculture—Myths & Facts,* copyright 1988, Animal Industry Foundation, P.O. Box 9522, Arlington, VA 22209-0522, (703) 524-0810.

10. *Swine Care Handbook,* National Pork Producers Council, P.O. Box 10383, Des Moines, Iowa 50306, (515) 223-2600.

11. *Guide for the Care and Production of Veal Calves,* 3rd Edition, June 1, 1990, American Veal Association.

12. *Our Farmers Care,* Wisconsin Agribusiness Foundation, 2317 International Lane, Suite 109, Madison, Wisconsin 53704-3129.

ADDITIONAL RESOURCES

1. The National Fish and Wildlife Service
   http://www.fws.gov

2. The National Wild Turkey Federation
   http://www.nwtf.org

3. Ducks Unlimited
   http://www.ducks.org

4. U.S. Fish and Wildlife Service, Historic News Releases
   http://www.fws.gove/new/historian

5. *Environmental Health Perspectives.* Mary Weideman, "In Vitro Innovations," November, 1995, Vol. 103, No. 11, http://www.pubmedcentral.nih.gov.

END NOTES

1. Gillespie, J. R. 1997. *Modern Livestock and Poultry Production.* 5th ed. Albany, NY: Delmar.

2. Excerpts from *Animal Welfare and Animal Well-Being,* by Dr. Jack L. Albright, Purdue University. Presented to the Indiana High School Agriculture Teachers' Workshop, June 25, 1991.

3. Animal Welfare Information Center. "Animal Welfare Act and Regulations." http://www.nal.usda.gov/awic/legislat/usdalegl.htm#L5.

4. Excerpts from *Animal Welfare Enforcement,* Fiscal Year 1991, Animal and Plant Health Inspection Service, United States Department of Agriculture.

5. Excerpts from *Animal Welfare and Animal Well-Being,* by Dr. Jack L. Albright.

6. Excerpt from *Animal Agriculture: Myths and Facts.* 1988. Animal Industry Foundation, Arlington, Virginia.

7. Excerpts from International Primate Protection League. 1992. "The hit of the year." *News,* 19(3) (December).

8. Excerpt from International Primate Protection League. 1992. "Lab workers test positive for monkey AIDS," *News,* 19(3) (December).

9. Excerpts from *Animal Agriculture: Myths and Facts.*

10. National Fish and Wildlife Service, 2001 National Survey of Fishing, Hunting, and Wildlife-Associated Recreation, page 13, Table 4, Expenditures in the U.S. for Hunting: 2001. Available online at http://library.fws.gove/nat_survey2001.pdf surveys.html (accessed July 17,1999).

11. National Wild Turkey Foundation, "About the NWTF," available online at http://www.nwtf.org/about_us (accessed January, 2007).

12. Ibid.

CHAPTER 5

Careers in Small Animal Care

OBJECTIVES

After reading this chapter, one should be able to:

 compare opportunities in small animal care and management.

 describe the nature of the work, salaries, and requirements necessary for obtaining a job in the field of small animal care.

TERMS TO KNOW

anatomist

animal breeder

animal caretaker

animal curator

animal handler

animal husbandry technician

animal psychology

animal technician

animal trainer

aquarist

aquarium curator

aquarium director

aquatic biologist

biochemist

biological technician

biologist

biophysicist

biotechnician

botanist

cytologist

doctor of veterinary medicine

ecologist

embryologist

entomologist

gaming dog business

general curator

geneticist

histopathologist

ichthyologist

immunologist

laboratory animal technician

laboratory animal technologist

life scientist

limnologist

marine biologist

microbiologist

molecular biologist

mycologist

naturalist

nematologist

nutritionist

parasitologist

pet groomer

pharmacologist

physiologist

public-health microbiologist

toxicologist

veterinarian

veterinary technician

veterinary technologist

wildlife biologist

zoo curator

zoo director

zookeeper

zoologist

INTRODUCTION

The pet industry in the United States is a rapidly growing segment of today's business world. It is estimated that 37.2 percent of American households have dogs and 32.4 percent have cats. Cats outnumber dogs 81.7 million to 72.0 million, because most households have an average of two cats. Approximately 59.5 percent of U.S. households owned pets in 2006. Included are an estimated 76.0 million fish, 6.1 million rabbits, 1.2 million hamsters, 431,000 gerbils, and several million other household pets of various species.[1]

Americans spend in excess of $40.8 billion per year on their pets.[2] Veterinary expenses exceed $24.5 billion annually.[3] Pet food manufacturers achieve around $15.4 billion in yearly sales.[4] The pet care industry is just one area of small animal care. This chapter explores the jobs and careers that exist in the care and management of small animals.

ANIMAL BREEDERS AND TECHNICIANS

Animal breeders and technicians[5] help breed, raise, and market a variety of animals: cattle, sheep, pigs, horses, mules, and poultry for livestock; pets such as canaries, parrots, dogs, and cats; and other more exotic animals such as ostriches, alligators, minks, and many zoo animals. Technicians who are primarily involved with the breeding and feeding of animals are sometimes referred to as **animal husbandry technicians.** (*See Figure 5–1.*)

In general, animal breeders and technicians are concerned with the propagation, feeding, housing, health, production, and marketing of animals. These technicians work in many different settings and capacities. They may supervise unskilled farm workers; serve as field representatives assisting in the sales of animals to customers; work in kennels, stables, ranches, or zoos reproducing species and breeds for other clients or their own organization; or work on their own on a particular breed of interest. The diversity of

Figure 5–1 Animal breeders raise animals for sale to pet shops, laboratories, and the general public. Some may raise fur-bearing animals for their pelts, and others may raise rabbits for the fur and meat.

employment available for well-trained and well-qualified animal breeders and technicians makes this career extremely flexible. As science progresses, opportunities for these technicians should broaden.

Most animal breeders and technicians work as livestock production technicians with cattle, sheep, swine, or horses or as poultry production technicians with chickens, turkeys, geese, or ducks. Other animal breeders work with domesticated animals kept as pets, such as songbirds, parrots, and all dog and cat breeds. Even wildlife populations that are kept in reserves, ranches, zoos, or aquariums are bred with the guidance of a breeder or technician. Each category of animal and even some individual breeds within a category have technicians working on their reproduction if they are bred for livestock or domestic use. Within each of these categories the jobs may be specialized for one aspect of the animal's reproductive cycle.

One area of animal production technology that merits special mention because of the increasing focus on its use in animal husbandry is artificial breeding. Three kinds of technicians working in this specialized area of animal production are artificial-breeding technicians, artificial-breeding laboratory technicians, and artificial insemination technicians.

Artificial-breeding technicians collect and package semen for use in insemination. They examine the semen under a microscope to determine density and motility of sperm cells, and they dilute the semen according to standard formulas. They transfer the semen to shipping and storage containers with identifying data such as the source, date taken, and quality. They also keep records related to all of their activities. In some cases they may also be responsible for inseminating the females. Artificial insemination technicians do exactly what their name implies: they collect semen from the male species of an animal and artificially inseminate the female.

For all breeders, it is essential that they keep track of the lineage of the animals they breed. The genetic history for at least three previous generations is usually considered the minimum background required to ensure no inbreeding. For animals sold as pedigreed, these records are certified by some overseeing organizations.

Education and Training Requirements

High school students seeking to enter this field will find that the more agriculture and science courses they select in high school, the better prepared they will be. In addition, courses in mathematics, business, communications, chemistry, and mechanics are valuable.

Nine months to two years at a technical school or a college diploma are the usual minimum credentials for animal breeders and technicians. Many colleges now offer 2- and 4-year programs in animal science or animal husbandry, in which additional knowledge, skills, and specialized training may be acquired. Besides learning the scientific side of animal breeding, including instruction in genetics, animal physiology, and some veterinary science, students also take business classes that help them see the field from an economic point of view. With the increasing use of technology for breeding livestock and domesticated nonfarm animals, a bachelor's degree becomes more important for succeeding in the field. Master's and doctoral degrees are useful for the most specialized fields and the careers that require the most sophisticated genetic planning. Higher degrees are required for potential teachers in the field, and the current work being done in cloning is done exclusively by people with doctorates. Certification is not required, but nearly all major companies have certification programs that can enhance earnings and opportunities.

Animal breeders and technicians should have great love, empathy, and respect for animals. They must be patient and compassionate in addition to being very knowledgeable about the needs and habits of all the animals in their care. They must also have interest in reproductive science, genetics, and animal physiology. It is important to be able to communicate easily with agricultural scientists, farmers, and other animal owners and caretakers.

Exploring the Career

Organizations such as 4-H Clubs (http://www.4-H.org) and the National FFA Organization (formerly known as Future Farmers of America; http://www.ffa.org) offer good opportunities for hearing about, visiting, and participating in farm activities. Programs sponsored by 4-H allow students to learn firsthand about breeding small animals for show. Other organizations, such as the American Kennel Club, sponsor clubs dedicated to particular breeds, and these clubs usually provide educational programs on raising and breeding these animals.

Other opportunities might include volunteering at a breeding farm or ranch, kennel, or stable where animals are bred and sold. This will provide a chance to see the work required and begin to get experience in practical skills for the job. For at-home experience, raising pets is a good introduction to the skills needed in basic animal maintenance. Learning how to care for, feed, and house a pet provides some basic knowledge of working with animals. Other opportunities that provide animal maintenance experience include volunteering to work at animal shelters, veterinary offices, and pet breeders' businesses.

Advancement and Outlook

Many junior colleges participate in "learn-and-earn" programs, in which the college and prospective employer jointly provide the student's training, both in the classroom and through on-the-job work with livestock and other animals. Most technical programs offer placement services for graduates, and the demand for qualified people often exceeds the supply. Even when a good training or technical program is completed, the graduate often must begin work at a low level before advancing to positions with more responsibility. But the technical program graduate will advance much more rapidly to positions of major responsibility and greater financial reward than the untrained worker.

Continuing changes are expected in the next few years, in both the production and the marketing phases of the animal production industry. Because of the costs involved, it is almost impossible for a one-person operation to stay in business for farm animals. As a result, cooperatives of consultants and corporations will become more prevalent, with greater emphasis placed on specialization. This, in turn, will increase the demand for technical program graduates. Other factors, such as small profit margins, the demand for more uniform products, and an increasing foreign market, will result in a need for more specially trained personnel. This is a new era of specialization in the animal production industry; graduates of animal production technology programs have an interesting and rewarding future ahead of them.

Earnings

Salaries vary widely depending on employer, the technicians' educational and agricultural background, the kind of animal the technicians work with, and the geographical areas in which they work. In general, the salaries of breeders tend to be higher in areas with a heavy concentration and in the breeding of certain specialty animals. The U.S. Department of Labor reports that animal breeders had mean annual wages of $29,340 in 2002. The top 10 percent made $49,810 or more, whereas the bottom 10 percent made only $15,630 or less. Fringe benefits vary according to employer but can include paid vacation time, health insurance, and pension benefits.

Work Environment

Working conditions vary from operation to operation, but certain factors always exist. Much of the work is done inside in all types of facilities. Barns, pens, and stables are the most common facilities for farm animals; nonfarm animals may be bred in private homes or housing facilities. Both types of work often require long, irregular hours and work on Sundays and holidays. The work is also sometimes dangerous, especially when large animals such as stallions and bulls are involved.

Animal breeders and technicians are often their own bosses and make their own decisions. Although this can be an asset to those who value independence, prospective animal breeders and technicians must realize that self-discipline is the most valuable trait for success.

For More Information

For more information on becoming an animal breeder, contact
American Kennel Club
260 Madison Avenue
New York, NY 10016
Tel: 212-696-8200
Email: info@akc.org
http://www.akc.org

For information on careers and graduate programs, contact
American Society of Animal Science
1111 North Dunlap Avenue
Savoy, IL 61874
Tel: 217-356-9050

ANIMAL CARETAKERS

Animal caretakers,[6] as the name implies, take care of animals. The job ranges from the day-to-day normal activities for a healthy animal to caring for sick, injured, or aging animals. The daily animal routine usually involves feeding and providing drinking water for each animal and making sure that their enclosures are clean, safe, appropriately warm, and, if needed, stocked with materials to keep the animal active and engaged (*See Figure 5–2.*) Caretakers may be responsible for creating different enrichment materials so that the animal is challenged by new objects and activities. They may exercise or train the animals. They may assist veterinarians or other trained medical staff in working with animals that require treatment. Animal caretakers may also maintain the written records for each animal. These records can include weight, eating habits, behavior, medicines given, or treatment given. Animal caretakers and veterinary assistants hold about 151,000 jobs in the United States.

Animal caretakers also are referred to by several other names depending on their specialty. They perform the daily duties of animal care, which include feeding, grooming, cleaning, exercising, examining, and nurturing the individuals in their care. These caretakers have titles such as animal shelter workers, grooms, veterinary assistants, wildlife assistants, animal shelter attendants, **laboratory animal technicians, laboratory animal technologists,** and kennel technicians. (*See Figure 5–3.*)

Figure 5–2 Kennel attendants feed and care for animals and keep the facilities clean. *Courtesy of Brian Yacur and Guilderland Animal Hospital.*

Figure 5–3 Laboratory animal care workers take care of animals being used in research. They usually work closely with the scientists conducting the research and carry out duties that are assigned to them by the scientists. *Courtesy of PhotoDisc.*

Caretakers are responsible for some or all of the following tasks: selecting, mixing, and measuring out the appropriate food; providing water; cleaning the animal and the enclosure; changing bedding and groundcover if used; moving the animals from night facilities to day facilities or exercise spaces or different quarters; sterilizing facilities and equipment not in use; recording and filing statistics, medical reports, or laboratory reports on each animal; and providing general attention and affection to animals that need human contact.

The animal caretaker learns to recognize signs of illness, such as lack of appetite, fatigue, skin sores, and changed behavior. They check the animals they can physically approach or handle for lumps, sores, fat, texture of the skin, fur, or feathers, and condition of the mouth. The caretaker also maintains the animal's living quarters. For most animals in their care, this will be an enclosure of some type. The enclosure has to be safe and secure. The quarters need to be the right size for the animal. Caretakers set up and oversee enrichment activities that provide the animal with something to keep it engaged and occupied while in its home.

It is also an unpleasant side of the job that in almost every facility, the caretaker will have to deal with the death of an animal in his or her care. For veterinary offices, shelters, and wildlife facilities of any type, animal deaths are a part of everyone's experience. Shelters may choose to euthanize (kill) animals that are beyond medical treatment, deemed unadoptable, or unmaintainable because of their condition or the facilities' inability to house them. But even for places without a euthanasia policy, any center working with older, injured, sick, or rescued animals is going to lose the battle to save some of them.

As an animal caretaker gains experience working with the animals, the responsibilities may increase. Caretakers may begin to perform tasks that either senior caretakers or medical specialists were performing. This can include administering drugs; clipping nails, beaks, and wing feathers; banding wild animals with identification tags; and training the animal. Numerous clerical tasks may also be part of the animal caretakers' routine. Beyond the medical reports made on the animals, animal caretakers may be required to screen people looking to take an animal home and write status reports or care plans.

Education and Training Requirements

Students preparing for animal caretaker careers need a high school diploma. While in high school, classes in anatomy and physiology, science, and health are recommended. Students can obtain valuable information by taking animal science classes, where available. Any knowledge about animal breeds, behavior, and health is helpful. The basics of human nutrition, disease, reproduction, and aging help to give a background for learning about these topics for different species. A basic grasp of business and computer skills will help with the clerical tasks.

There are 2-year college programs in animal health that lead to an associate's degree. This type of program offers courses in anatomy and physiology, chemistry, mathematics, clinical pharmacology, pathology, radiology, animal care and handling, infectious diseases, biology, and current veterinary treatment. Students graduating from these programs go on to work in veterinary practices, shelters, zoos and aquariums, pharmaceutical companies, and laboratory research facilities. Students should look for programs accredited by the American Veterinary Medical Association. A bachelor's degree is required for many jobs, particularly in zoos and aquariums. Degrees in wildlife management, biology, zoology, animal physiology, or other related fields are most useful.

Animal caretakers should have great love, empathy, and respect for animals. They should have a strong interest in the environment. Patience, compassion, dependability, and the ability to work on repetitive, physically challenging, or unstimulating tasks without annoyance are essential characteristics for someone to be happy as an animal caretaker.

Exploring the Career

Volunteering is the most effective method of experiencing the tasks of an animal caretaker. Most shelters, rescue centers, and sanctuaries and some zoos, aquariums, and laboratories, rely on volunteers to fill their staff. Opportunities as a volunteer may include the ability to work directly with animals in some or all of the capacities of a paid animal caretaker.

Volunteering also provides a line on your resume that demonstrates that you bring experience to your first paying job. It gives you references that can vouch for your skills with animals, your reliability, and your dedication to the field.

Advancement and Outlook

Advancement depends on the job setting. There may be promotion opportunities to senior technician, supervisor, assistant facilities manager, or facilities manager. Some animal caretakers may open their own facilities or services.

Laboratory workers can move from assistant technician to technician to technologist with increased education and experience. But for most promotions, more education is usually required.

The animal care field is expected to grow faster than the average through 2012, according to the U.S. Department of Labor. More people have pets and are more concerned with their pets' care. Because most households have all the adults in full-time employment, animals are left home alone longer than in earlier times. Dog-walking services, pet sitting and in-house care, boarding facilities, kennels, and such that provide assistance with the daily care of an animal for the working or traveling owner are far more prevalent and successful than ever before.

There is a high turnover in the profession. This is due in part to the seasonal nature of some of the jobs, the low pay, and the lack of advancement opportunities in the field. Positions as animal caretakers in zoos, aquariums, and rehabilitation and rescue centers are the most sought after, partly because of the ability to work with exotic and wild species. Aspiring animal caretakers will find few openings in these facilities.

Graduates of veterinary technician programs have the best employment prospects. Laboratory animal technicians and technologists also have good opportunities. Increasing concern for animal rights and welfare means that these facilities are staffing more professionals to operate their laboratories.

Earnings

Salaried animal caretakers had average annual earnings of $18,750 in 2002, according to the *Occupational Outlook Handbook*. Wages for the middle half of people employed in the field were about $17,080. The top 10 percent earned more than $27,850, but the bottom 10 percent earned less than $12,750 a year.

Work Environment

Animals may be kept either indoors or outdoors, in any type of weather. Depending on the facilities, heavy lifting may be part of the job. You may have to lift crates, animals, food, equipment, or other items big enough to accommodate a large animal. The work can sometimes be hard, repetitive, and dirty. Cleaning enclosures and disinfecting spaces can involve hot or cold water and chemicals.

The work can also be dangerous, depending on the animals you work with. Although animals that are handled correctly and are treated with the proper respect and distance can be quite safe, situations can arise in which the animal is unpredictable or is frightened or cornered.

Many facilities require long workdays, long workweeks, odd hours, weekend work, holiday work, and intermittent schedules. Depending on the hours of the facility, the services provided, and the staffing, there may be several shifts, including a graveyard shift. Animal caretakers should be prepared to work a changing schedule. The needs of animals don't cease for weekends and holidays.

For More Information

For information about animal laboratory work and certification programs, contact

American Association for Laboratory Animal Science
9190 Crestwyn Hills Drive
Memphis, TN 38125
Tel: 901-754-8620
Email: info@aalas.org
http://www.aalas.org

For information on available training programs, such as the facility accreditation program, certification program for kennel operators, complete staff training program, and ethics program, contact

American Boarding Kennels Association
1702 East Pikes Peak Avenue
Colorado Springs, CO 80909
Tel: 719-667-1600
Email: info@abka.com
http://www.abka.com

For information on careers, schools, and resources, contact

American Veterinary Medical Association (AVMA)
1931 North Meacham Road, Suite 100
Schaumburg, IL 60173
Tel: 847-925-8070
Email: avmainfo@avma.org
http://www.avma.org

For information on the Student Career Experience Program, contact the U.S. Fish and Wildlife Service in your state or visit the following Web site:

U.S. Fish and Wildlife Service
http://www.fws.gov

ANIMAL HANDLERS

Anybody who works directly with animals, from the caretaker of your local park's petting zoo to the activist who reintroduces wild animals to national parks, is an **animal handler.**[7] Animal handlers care for, train, and study animals in such places as zoos, parks, research laboratories, animal-breeding facilities, rodeos, and museums. An animal handler's job involves feeding the animals, cleaning their living and sleeping areas, preparing medications, and other aspects of basic care. A handler may also be actively involved in an animal's training and in presenting animals to the public in shows and parks. Wrangling an iguana for a movie production; preparing the diet for a zoo's new albino alligator; comforting bison to keep them from committing "suicide"; training cats for an animal-assisted therapy program at a nursing home—all these responsibilities, strange as they may seem, actually exist for some animal trainers. Many western states have long telephone book listings of animal handlers who rent out trained iguanas, horses, cougars, cattle, and other animals for movie productions. Zoos and marine animal parks hire highly trained keepers to feed, shelter, and protect some of the most exotic animals in the world.

Whether taking on jobs like those just listed or working for a small park or large zoo, all animal handlers are called upon for the daily care and safety of animals. They may have special training in a particular animal or breed or work with a variety of animals.

With a wide knowledge of an animal's nutritional and exercise requirements, animal handlers make sure the animals in their care are well fed, well groomed, and healthy. They prepare food and formulas, which may include administering medications. Maintaining proper shelters for animals requires cleaning the area, ensuring good ventilation, and providing proper bedding. Animal handlers arrange for vaccinations, as well as look for diseases in their animals. They also prepare animals for transport, knowing how to use muzzles and kennels, and how to calm an animal. But an animal handler needs a rapport with the two-legged creatures as well; working with people is an important aspect of most animal care jobs because many of these animals are kept for presentation and performance.

Education and Training Requirements

Take biology, chemistry, and other science courses offered by your high school. The study of science will be important to any student of animals, as will the study of psychology and sociology. Knowing about animal nutrition, health, behavior, and biology will help you to understand the animals you care for, and how to best provide for them. And if you do choose to go on to college, most animal-related courses of study are science based.

Some may think of animal handlers as people who spend all their time separate from the rest of the community, communicating only with animals and having limited interaction with humans. However, nothing could be further from the truth—most animal handlers work actively with the public; they present the animals in zoos and public programs and may even perform with the animals. Join your speech and debate team, or your drama club, to prepare for speaking in front of groups of people.

Because so many animal programs, from petting zoos to animal-therapy programs, rely on community support, there are many volunteer opportunities for high school students looking to work with animals. Zoos, parks, and museums need volunteers, as do kennels, shelters, and local chapters of the Humane Society.

The value of a college degree depends on the work you do. Many animal handlers do not have degrees, but zoos often prefer to hire people with a post-secondary education. A degree can often determine promotions and pay raises among the workers of a zoo. Many universities offer degrees in animal sciences, zoology, and zoological sciences. There are also graduate degrees in zoology, which may require courses in physiology, animal behavior, and oceanography.

Some consider a job as an animal handler an internship in and of itself; after gaining experience in a petting zoo or teaching zoo, or working with a breeder or stable hand, some animal handlers pursue careers as zookeepers, veterinarians, and in animal research. Many unpaid internships are available for those willing to volunteer their time to researchers and other animal professionals. Check with your local university and zoo to find out about opportunities to study animals in the wild or to reintroduce animals to their native habitats.

Some animal handlers in very specialized situations, such as patrol dog trainers and laboratory animal technicians, are required to pursue certification. The American Association for Laboratory Animal Sciences offers certification for those working with laboratory animals. But for the majority of animal handlers, no certification program exists.

It is important for animal handlers to love the animals they care for. What might not be as apparent, however, is the need for animal handlers to enjoy working with people as well. Animal handlers are often required to present the animals to park and zoo visitors and to serve as tour guides; they also work as instructors in zoo and museum education programs. Some animal handlers even perform alongside their trained animals in theme parks and shows. Some shows, such as marine animal shows, can be particularly strenuous, calling for very athletic trainers.

Working with animals on a daily basis requires patience and calmness because animals faced with unfamiliar situations are easily frightened. Animal handlers must be very knowledgeable about the needs and habits of all the animals in their care.

Exploring the Career

Many zoos have programs in place to introduce young people to the duties and responsibilities of an animal handler. Many part-time jobs are available to high school students interested in working with animals. Pet shops, petting zoos, stables, and kennels are likely to have a few after-school positions. In larger cities, you may be able to start your own animal care business as a dog walker or pet sitter.

Advancement and Outlook

Depending on the area of animal care in which you want to work, you may be able to find many great opportunities. A high school job or internship is a good start; experience with animals is what is most important to employers hiring handlers. Any volunteering you've done will also look good to an employer because it shows that you have a personal dedication to the care of animals.

Jobs with zoos in major cities, or with animal shows, can be highly competitive. If you are hoping to work in a larger, more famous zoo, you should first pursue experience with a smaller zoo. Jobs working with marine mammals are also difficult to get; because there are few marine animal shows in the country (such as those performed at Sea World parks), you may first have to pursue experience with internships and college programs.

Most people who work with animals are not looking to climb any ladder of advancement. As a matter of fact, many people change from high-paying careers to lower-paying animal care jobs just to do something they love. Some who have gained experience handling and training animals may start their own businesses, perhaps building their own stables of trained "actors" to hire out for area movie shoots, stage shows, parades, and other performances. Some animal handlers may pursue higher education while working full or part time, taking courses toward veterinary sciences degrees or degrees in biology. With a degree, an animal handler may have a better chance at the higher-paying, supervisory zookeeper positions. After years with a particular zoo, an animal handler can take on more responsibility and make decisions that influence the direction of the zoo.

Earnings

The opportunity to work directly with a variety of different animals is often reward enough for animal handlers. Someone who owns a stable of well-trained animals used in performances may be able to negotiate for large contracts, or a successful dog breeder may make a comfortable living with an established business, but most animal handlers make do with small salaries and hourly wages. An experienced animal handler may draw an hourly wage of $16 to $20. Many full-time, salaried zoo positions include health benefits. The U.S. Department of Labor estimates that nonfarm animal handlers earned a median salary of $17,080 in 2002, with 10 percent earning less than $12,750 and 10 percent more than $27,850.

Work Environment

Depending on the lives of the animals for which they care, handlers usually work both indoors and outdoors. Be prepared for smelly, messy, and dusty environments; if you have allergies, they will be under constant assault. Sometimes institutions, particularly animal research laboratories, require handlers to have immunizations and physical examinations before working with the animals. In addition to allergies, there is some danger of diseases transferred from animals to humans. These risks can be lessened with protective clothing such as lab coats, gloves, and ventilated hoods.

The temperament of your animals will also affect your work environment. Handlers must be prepared for occasional scratches, bites, and kicks from animals with even the best dispositions.

With the popularity of cable channels such as the Discovery Channel and Animal Planet, as well as television specials and videos featuring animals, the public's interest in animals is only likely to increase.

Concerns about the treatment of animals will perhaps lead to more stringent laws and certification requirements. Some activists hope to end the capture of animals for display in zoos; some even object to filming animals in the wild. But zoos are likely to continue to operate and expand, with zoo professionals arguing that zoo animals are often safer and receive better care than they would in their natural habitats.

For More Information

For general information about zoos, aquariums, oceanariums, and wildlife parks, contact

American Zoo and Aquarium Association
8403 Colesville Road, Suite 710
Silver Spring, MD 20910-3314
Tel: 301-562-0777
http://www.aza.org

For information on the zoo animal technology program, and other career information, contact

Santa Fe Community College
3000 NW 83rd Street
Gainesville, FL 32606
Tel: 352-395-5604
http://www.inst.santafe.cc.fl.us/~zoo

ANIMAL TRAINERS

Animal trainers[8] teach animals to obey commands so the animals can be counted on to perform these tasks in given situations. The animals can be trained for up to several hundred commands, to compete in shows or races, to

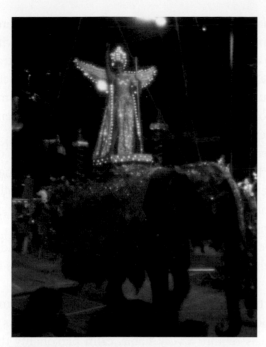

Figure 5–4 Animal trainers teach animals to obey commands and to perform tricks for audiences. *Courtesy of PhotoDisc.*

perform tricks to entertain audiences, to protect property, or to act as guides for the disabled. (*See Figure 5–4.*) Animal trainers may work with several types of animals or specialize with one type. Approximately 30,200 animal trainers are employed in the United States.

Many animals are capable of being trained. The techniques used to train them are basically the same, regardless of the type of animal. Animal trainers conduct programs consisting primarily of repetition and reward to teach animals to behave in a particular manner and to do it consistently.

First, trainers evaluate an animal's temperament, ability, and aptitude to determine its trainability. Animals vary in personality, just as people do. Some animals are more stubborn, willful, or easily distracted and would not do well with rigid training programs. All animals can be trained at some level, but certain animals are more receptive to training; these animals are chosen for programs that demand great skill.

One of the most familiar examples is the seeing-eye dog, now usually called a companion animal for the blind. These dogs are trained with several hundred verbal commands to assist their human and to recognize potentially dangerous situations. The dog must be able to, without any command, walk its companion around obstacles on the sidewalk. The companion dog must be able to read street lights and know to cross at the green, and only after traffic has cleared. The dog must also not be tempted to run to greet other dogs, grab food, or behave as most pet dogs do. Very few dogs make it through the rigorous training program. The successful dogs have proved to be such aids to the visually impaired that similar programs have been developed to train dogs for people who are confined to a wheelchair, or are hearing impaired, or are incapable of executing some aspect of a day-to-day routine with which a dog can assist.

Trainers usually specialize in one type of animal and are identified by this type of animal. Dog trainers, for example, may work with police dogs, training them to search for drugs or missing people. Some train guard dogs to protect private property; others train dogs for performance, in which the dog may learn numerous stunts or movements with hand commands so that the dog can perform on a stage or in a film without the audience hearing the commands spoken from offstage. (*See Figure 5–5.*)

Horse trainers specialize in training horses for riding or for harness. They talk to and handle a horse gently to accustom it to human contact, and then gradually get it to accept a harness, bridle, saddle, and other riding gear. Trainers teach horses to respond to commands that are either spoken or given by use of the reins and legs.

A highly specialized occupation in the horse-training field is that of race-horse trainers, who must create individualized training plans for every horse in their care.

Police horse trainers work with police horses to keep them from startling in crowds or responding to other animals in their presence. As with the police dogs, these animals require a very stable, calm personality that remains no matter what the situation in which the animal works.

Other animal trainers work with more exotic animals for performance or for health reasons. The dolphins and whales at the Shedd Aquarium in Chicago are trained to roll over, lift fins and tails, and open their mouths on command, so that much veterinary work can be done without anesthesia, which is always dangerous for animals.

Animal trainers work with hunting birds, training them to fly after an injury, or to hunt if the bird was found as a hatchling before a parent had trained it. Each species of animal is trained by using the instincts and reward systems that are appropriate to that species.

Figure 5–5 Dogs can be trained for many different jobs and skills depending on what they will be used for (i.e., herding, hunting and tracking, police work, or assisting people with disabilities). *Courtesy of PhotoDisc.*

Education and Training Requirements

For high school students interested in becoming an animal trainer, courses in anatomy, physiology, biology, and psychology will be helpful. Understanding how the body and mind work helps a trainer understand the best methods for training. Knowledge of psychology will help the trainer recognize behaviors in the animals they train, as well as in the people whom the animals are helping.

Although there are no formal education requirements to enter this field, some positions do have educational requirements that include a college degree. Animal trainers in circuses and the entertainment field may be required to have some education in animal psychology in addition to their caretaking experience. Zoo and aquarium animal trainers usually must have a bachelor's degree in a field related to animal management or animal physiology. Trainers of companion dogs prepare for their work in a 3-year course of study at schools that train dogs and instruct the disabled owner-companion.

Most trainers begin their careers as keepers and gain on-the-job experience in evaluating the disposition, intelligence, and "trainability" of the animals they look after. Although previous training experience may give job applicants an advantage in being hired, they still will be expected to spend time caring for the animals before advancing to a trainer position.

Establishments that hire trainers often require previous animal-keeping or equestrian experience, because proper care and feeding of animals is an essential part of a trainer's responsibilities. These positions serve as informal apprenticeships. The assistant may get to help an animal trainer on certain tasks but will be able to watch and learn from other tasks being performed around him or her.

Racehorse trainers must be licensed by the state in which they work. Otherwise, there are no special requirements for this occupation.

Prospective animal trainers should like and respect animals and have a genuine interest in working with them. With most of the career options for an animal trainer, there is an underlying desire to help people as well. Most trained animals work with people to accomplish a goal, so the relationship between the animal, the trainer, and the owner or companion is an important one.

Exploring the Career

Students wishing to enter this field would do well to learn as much as they can about animals, especially animal psychology, either through coursework or library study.

Volunteering offers an opportunity to begin training with animals and learning firsthand about the tasks and routines involved in managing animals, as well as training them. Part-time or volunteer work in animal shelters, pet-training programs, rescue centers, pet shops, or veterinary offices gives potential trainers a chance to discover whether they have the aptitude for working with animals. Experience can be acquired, too, in summer jobs as animal caretakers at zoos, aquariums, museums that feature live animal shows, amusement parks, and, for those with a special interest in horse racing, at stables.

Advancement and Outlook

People who wish to become animal trainers generally start out as animal keepers, stable workers, or caretakers and rise to the position of trainer only after acquiring experience within the ranks of an organization. You can enter the field by applying directly for a job as animal keeper, letting your employer or supervisor know of your ambition so you will eventually be considered for promotion. The same applies for volunteer positions. Learning as a volunteer is an excellent way to get hands-on experience, but you should be vocal in your interest in a paid position once you have gotten to know the staff and they have gotten to know you.

You should pay close attention to the training methods of any place at which you are considering working. No reputable organization, regardless of what it trains animals for, should use physical injury to train or discipline an animal.

The most coveted positions depend on the animals you want to work with. Sea mammals are a specialty of oceanariums and aquariums, and these positions are fiercely competitive. Dog-training programs are probably the most plentiful and offer the widest range of training philosophies and techniques. The Federal Emergency Management Agency (FEMA) works only with established dog and handler teams, who usually work within the emergency systems for the regional or local authorities in some capacity. These teams choose to also be trained within the FEMA guidelines.

Most establishments have very small staffs of animal trainers, which means that the opportunities for advancement are limited. The progression is from animal keeper to animal trainer. A trainer who directs or supervises others may be designated head animal trainer or senior animal trainer.

This field is expected to grow faster than the average through 2012, according to the U.S. Department of Labor. Although criticism of animals used for purely entertainment purposes has reduced the number of animals used for shows and performances, programs have expanded for companion animals and animals used in work settings. Also, a growing number of animal owners are seeking training services for their pets.

An increased number of trainers will be needed to train the increasing number of search-and-rescue and bomb-sniffing dog teams. The latter will be in demand to ensure the safety of airports, government buildings, corporations, amusement parks, sports facilities, and public utilities.

Earnings

Salaries of animal trainers can vary widely according to specialty and place of employment. Salaries ranged from $14,290 to $45,040 a year or more in 2002, according to the U.S. Department of Labor. The median salary for animal trainers was $22,950. Those who earn higher salaries are in upper management and spend more time running the business than working with animals.

In the field of racehorse training, however, trainers are paid an average fee of $35 to $50 a day for each horse, plus 10 percent of any money their horses win in races. Depending on the horse and the races it runs, this can exceed the average high-end earnings for a trainer. Show horse trainers may earn as much as $30,000 to $35,000 a year. Trainers in business for themselves set their own fees for teaching both horses and owners.

Work Environment

The working hours for animal trainers vary considerably, depending on the type of animal, performance schedule, and whether travel is involved. For some trainers, such as those who work with show horses, educational programs with hunting birds, or new animals being brought into zoos and aquariums, the hours can be long and quite irregular. Travel is common and will probably include responsibility for seeing to the animals' needs while on the road. This can include feeding, creative housing, and driving with the animal.

Much of the work is conducted outdoors. In winter, trainers may work indoors, but depending on the animal, they may continue outdoor training year-round. If the animal is expected to work or perform outdoors in winter, it has to be trained in winter as well.

Working with certain animals requires physical strength. Other aspects of the work may require lifting, bending, or extended periods of standing or swimming. Trainers of aquatic mammals, such as dolphins and seals, work in water and must feel comfortable in aquatic environments.

Patience is essential to the job as well. Just as people do, animals have bad days on which they won't work well and respond to commands. So even the best trainer encounters days of frustration during which nothing seems to go well. Trainers must spend long hours repeating routines and rewarding their pupils for performing well, while never getting angry with them or punishing them when they fail to do what is expected. Trainers must be able to exhibit the authority to keep animals under control without raising their voices or using physical force. Calmness under stress is particularly important when dealing with wild animals.

For More Information

For information on careers, contact

American Zoo and Aquarium Association
8403 Colesville Road, Suite 710
Silver Spring, MD 20910-3314
Tel: 301-562-0777
http://www.aza.org

Canine Companions for Independence is a nonprofit organization that assists people with disabilities by providing trained assistance dogs and ongoing support

Canine Companions for Independence
P.O. Box 446
Santa Rosa, CA 95402-0446

Tel: 800-572-2275

http://www.caninecompanions.org

The Delta Society currently provides certification programs in animal evaluation and in training animal handlers for animal-assisted therapy and companion animal training.

Delta Society

580 Naches Avenue, SW, Suite 101

Renton, WA 98055-2297

Tel: 425-226-7357

Email: info@deltasociety.org

http://www.deltasociety.org

AQUARISTS

Aquarists[9] (pronounced, like "aquarium," with the accent on the second syllable) work for aquariums, oceanariums, and marine research institutes. They are responsible for the maintenance of aquatic exhibits. Among other duties, they feed the fish, check water quality, clean the tanks, and collect and transport new specimens. Aquarists are not animal trainers and do not work in marine shows. They do, however, support the staff that does. Their work is generally technical and requires a strong science background.

Aquarists' job duties are similar to those of zookeepers. Aquarists feed fish, maintain exhibits, and conduct research. They work on breeding, conservation, and educational programs. Aquarists clean and take care of tanks every day. They make sure pumps are working, check water temperatures, clean glass, and sift sand.

Food preparation and feeding are important tasks for aquarists. Some animals eat live food and others eat cut-up food mixtures. Some animals need special diets prepared and may have to be individually fed.

Aquarists carefully observe all the animals in their care. They must understand their normal habits (including mating, feeding, sleeping, and moving) in order to be able to judge when something is wrong. Aquarists write daily reports and keep detailed records of animal behavior. Many aquarists are in charge of collecting and stocking plants and animals for exhibits. They may have to make several trips a year to gather live specimens.

Education and Training Requirements

If you want to become an aquarist, get your start in high school. Take as many science classes as you can; biology and zoology are especially important. Learn to pay attention to detail; marine science involves a good deal of careful record keeping.

Most aquariums, along with other institutions that hire aquarists, require that an applicant have a bachelor's degree in biological sciences, preferably with coursework in such areas as parasitology, ichthyology, or other aquatic sciences.

As an aquarist, you may be required to travel at different times throughout the year, to participate in research expeditions and trips for specimen collection.

On a more basic level, aquarists need to be in good physical shape, with good hearing and visual acuity. Some employers also require a certain strength level—say, the ability to regularly exert 100 pounds of force—because equipment, feed, and the animals themselves can be heavy and often unwieldy.

Exploring the Career

In addition to formal education, many aquariums, like other types of museums, look for a strong interest in the field before hiring an applicant. Most often, they look for a history of volunteering. That means you need to look for every avenue you can find to work around fish or other animals.

Advancement and Outlook

Full-time jobs for aquarists can be scarce, especially for those just starting in the field. Part-time or volunteer positions with zoos, aquariums, science institutes, nature centers, or even pet stores could provide valuable preliminary experience that may eventually lead to a full-time position.

The usual career path for an aquarist progresses from intern/volunteer through part-time work to full-fledged aquarist, senior aquarist, supervisor, and, finally, curator. Each step along the path requires additional experience and often additional education. Curators generally are expected to have a doctorate degree in a relevant marine science discipline.

Earnings

Aquariums often are nonprofit institutions, thus somewhat limiting the earnings ability in this job. In general, aquarists make between $23,000 and $32,000 a year, or between about $11 and $15 per hour. Aquariums do offer fairly extensive benefits, however, including health insurance, 401(k) plans, continuing education opportunities, tuition reimbursement, and reciprocal benefits with many other cultural institutions.

Work Environment

Aquarists may work indoors or outdoors, depending on the facility for which they work and the exhibit to which they are assigned. Aquarists spend a lot of time in the water. Their day will be filled with a variety of tasks, some repetitive, such as feeding, others unusual, such as working with rescued marine mammals, perhaps. In the beginning, aquarists work under the supervision of a senior aquarist or supervisor and may work as part of a team. Aquarists also can expect to travel as part of the job.

There is, in general, little change in the availability of positions for aquarists. While terrestrial zoos have begun to add aquarium complexes to their campuses in growing numbers, an actual boom in the construction of new aquariums is unlikely at this time. Many aquarists do advance to other positions, however, so openings do become available.

For More Information

For information on careers in aquatic and marine science, including job listings, contact

American Zoo and Aquarium Association
8403 Colesville Road, Suite 710
Silver Spring, MD 20910-3314
Tel: 301-562-0777
Email: generalinquiry@aza.org
http://www.aza.org

For information on guided tours and other opportunities, contact

Columbus Zoo and Aquarium
9990 Riverside Drive
Powell, OH 43065
Tel: 614-645-3550
http://www.colszoo.org

For information on educational programs, summer internships, and courses, contact the following organizations:

Marine Science Center
Northeastern University
430 Nahant Road
Nahant, MA 01908
Tel: 781-581-7370
http://www.dac.neu.edu/msc

Marine Science Institute
500 Discovery Parkway
Redwood City, CA 94063-4715
Tel: 650-364-2760
Email: info@sfbaymsi.org
http://www.sfbaymsi.org

BIOLOGISTS

Biologists[10] study the origin, development, anatomy, function, distribution, and other basic principles of living organisms. They are concerned with the nature of life itself in humans, microorganisms, plants, and animals, and with the relationship of each organism to its environment. (*See Figure 5–6.*) Biologists perform research in many specialties that advance the fields of medicine, agriculture, and industry. Approximately 75,000 biological scientists are employed in the United States. Biology can be divided into many specialties. The biologist, who studies a wide variety of living organisms, has interests that differ from those of the chemist, physicist, and geologist, who are concerned with nonliving matter. Biologists, or **life scientists,** may be identified by their specialties.

Anatomists study animal bodies from basic cell structure to complex tissues and organs. They determine the ability of body parts to regenerate and

Figure 5–6   The jobs and duties of biologists vary considerably depending on the specific field in which they are employed. *Courtesy of Utah Experimental Station.*

investigate the possibility of transplanting organs and skin. Their research is applied to human medicine. **Aquatic biologists** study animals and plants that live in water and how they are affected by their environmental conditions, such as the salt, acid, and oxygen content of the water and temperature, light, and other factors. **Biochemists** study the chemical composition of living organisms. They attempt to understand the complex reactions involved in reproduction, growth, metabolism, and heredity. **Biophysicists** apply physical principles to biological problems. They study the mechanics, heat, light, radiation, sound, electricity, and energetics of living cells and organisms and do research in the areas of vision, hearing, brain function, nerve conduction, muscle reflex, and damaged cells and tissues. **Biotechnicians,** or **biological technicians,** assist the cornucopia of biological scientists in their endeavors. **Botanists** study plant life. Some specialize in plant biochemistry, the structure and function of plant parts, and identification and classification, among other topics. **Cytologists,** sometimes called cell biologists, examine the cells of plants and animals, including those cells involved in reproduction. **Ecologists** examine such factors as pollutants, rainfall, altitude, temperature, and population size in order to study the distribution and abundance of organisms and their relation to their environment. **Entomologists** study insects and their relationship to other life forms. **Geneticists** study heredity in various forms of life. They are concerned with how biological traits such as color, size, and resistance to disease originate and are transmitted from one generation to another. **Histopathologists** investigate diseased tissue in humans and animals. **Immunologists** study the manner in which the human body resists disease. **Limnologists** study freshwater organisms and their environment. **Marine biologists** specialize in the study of marine species and their environment. **Microbiologists** study bacteria, viruses, molds, algae, yeasts, and other organisms of microscopic or submicroscopic size. **Molecular biologists** apply their research on animal and bacterial systems toward the goal of improving and better understanding human health. **Mycologists** study edible, poisonous, and parasitic fungi, such as mushrooms, molds, yeasts, and mildews, to determine which are useful to medicine, agriculture, and industry. **Nematologists** study nematodes (roundworms), which are parasitic in animals and plants. Nematodes transmit diseases, attack insects, or attack other nematodes that exist in soil or water. **Parasitologists** study animal parasites and their effects on humans and other animals. **Pharmacologists** may be employed as researchers by pharmaceutical companies. **Physiologists** are biologists who specialize in studying all the life stages of plants or animals. Some specialize in a particular body system or a particular function, such as respiration. **Wildlife biologists** study the habitats and the conditions necessary for the survival of birds and other wildlife. Their goal is to find ways to ensure the continuation of healthy wildlife populations, while lessening the impact and growth of civilization around them. **Zoologists** study all types of animals to learn their origin, interrelationships, classifications, life histories, habits, diseases, relation to the environment, growth, genetics, and distribution. Zoologists are usually identified by the animals they study: ichthyologists (fish), mammalogists (mammals), ornithologists (birds), and herpetologists (reptiles and amphibians). **Toxicologists** study the effects of toxic substances on humans, animals, and plants. **Public-health microbiologists** conduct experiments on water, foods, and the general environment of a community to detect the presence of harmful bacteria so pollution and contagious diseases can be controlled or eliminated. **Nutritionists** study how food is used and changed into energy.

Education and Training Requirements

High school students interested in a career in biology should take English, biology, physics, chemistry, Latin, geometry, and algebra. Prospective biologists should also obtain a broad undergraduate college training. In addition to courses in all phases of biology, useful related courses include organic and inorganic chemistry, physics, and mathematics. Modern languages, English, biometrics (the use of mathematics in biological measurements), and statistics are also useful. Courses in computers will be extremely beneficial. Students should take advantage of courses that require laboratory work, fieldwork, or specimen collection.

The best way to become a biologist is to earn a bachelor's degree in biology or one of its specialized fields, such as anatomy, bacteriology, botany, ecology, or microbiology. For the highest professional status, a doctorate is required. This is particularly true of top research positions and most higher-level college teaching openings.

A state license may be required for biologists who are employed as technicians in general service health organizations, such as hospitals or clinics.

Biologists must be systematic in their approach to solving the problems that they face. They should have probing, inquisitive minds and an aptitude for biology, chemistry, and mathematics. Patience and imagination are also required because they may spend much time in observation and analysis. Biologists must also have good communication skills in order to effectively gather and exchange data and solve problems that arise in their work.

Exploring the Career

Students can measure their aptitude and interest in the work of the biologist by taking courses in the field. Laboratory assignments, for example, provide information on techniques used by the working biologist. Many schools hire students as laboratory assistants to work directly under a teacher and help administer the laboratory sections of courses.

Part-time and summer positions in biology or related areas are particularly helpful. Students with some college courses in biology may find summer positions as laboratory assistants. Graduate students may find work on research projects conducted by their institutions. Beginning college and advanced high school students may find employment as laboratory aides or hospital orderlies or attendants.

Advancement and Outlook

Biologists who are interested in becoming teachers should consult their college placement offices. Public and private high schools and an increasing number of colleges hire teachers through the colleges at which they studied. Private employment agencies also place a significant number of teachers. Some teaching positions are filled through direct application. Biologists interested in private industry and nonprofit organizations may also apply directly for employment. Major organizations that employ biologists often interview college seniors on campus.

In a field as broad as biology, numerous opportunities for advancement exist. To a great extent, however, advancement depends on the individual's level of education. A doctorate is generally required for college teaching, independent research, and top-level administrative and management jobs. A master's degree is sufficient for some jobs in applied research, and a bachelor's degree may qualify one for some entry-level jobs.

The U.S. Department of Labor predicts employment for biological scientists to be as fast as average through 2012, although competition will be stiff for some positions. Private industry will need biologists to work in sales, marketing, and research management. Companies developing new drugs, modified crops, environmentally friendly products, and the like will need the expertise of biological scientists. Biologists with advanced degrees will be best qualified for the most lucrative and challenging jobs, although this varies by specialty, with genetic, cellular, and biochemical research showing the most promise.

Earnings

Earnings for biological scientists vary extensively based on the type and size of their employer, the individual's level of education and experience, and the area of biology in which the scientist specializes. The median salary for all biological scientists was $60,390 in 2002, as reported by the U.S. Department of Labor. In 2003 general biological scientists working for the federal government earned an average annual salary of $66,262. Those specializing in certain areas tended to have slightly higher annual incomes. Ecologists, for example, had an average salary of $65,207 per year, microbiologists reported an average of $73,513, geneticists averaged $78,652, and physiologists, $85,181.

According to the National Association of Colleges and Employers, those with bachelor's degrees in the biological sciences had beginning salaries averaging $23,000 to $35,000 per year in 2003. In general, the highest salaries were earned by biologists in business and industry, followed by those self-employed, working for nonprofit organizations, in military service, and working for the U.S. Public Health Service or other positions in the federal government. The lowest salaries were earned by teachers and by those working for various state and local governments.

Work Environment

The biologist's work environment varies greatly depending on the position and type of employer. One biologist may work outdoors or travel much of the time. Another wears a white smock and spends years working in a laboratory. Some work with toxic substances and disease cultures; strict safety measures must be observed.

Some biologists, including botanists, ecologists, and zoologists, may undertake strenuous, sometimes dangerous, fieldwork in primitive conditions. Wildlife biologists work in all types of weather and in all types of terrain and ecosystems. They may work alone or with a group in inhospitable surroundings in order to gather information.

For More Information

For information on careers in biology, contact

American Institute of Biological Sciences
1444 Eye Street, NW, Suite 200
Washington, D.C. 20005
Tel: 202-628-1500
Email: admin@aibs.org
http://www.aibs.org

For a career brochure, career-related articles, and a list of institutions that award academic degrees with a major in physiology, contact

American Physiological Society
9650 Rockville Pike
Bethesda, MD 20814-3991
Tel: 301-634-7164
http://www.the-aps.org

For information on careers, educational resources, and fellowships, contact

American Society for Microbiology
1752 N Street, NW
Washington, D.C. 20036
Tel: 202-737-3600
http://www.asm.org

For career information, including articles and books, contact

Biotechnology Industry Organization
1225 Eye Street, NW, Suite 400
Washington, D.C. 20005
Tel: 202-962-9200
http://www.bio.org

For information on careers in the marine sciences, contact

National Aquarium in Baltimore
501 East Pratt Street, Pier 3
Baltimore, MD 21202-3194
Tel: 410-576-3800
http://www.aqua.org

For information on specific careers in biology, contact

National Institutes of Health
9000 Rockville Pike
Bethesda, MD 20892

Tel: 301-435-1908
Email: orsinfo@mail.nih.gov
http://www.nih.gov

For information on specific careers, contact the FDA's job hotline

U.S. Food and Drug Administration (FDA)
5600 Fishers Lane
Rockville, MD 20857-0001
Tel: 888-463-6332
http://www.fda.gov

NATURALISTS

The primary role of **naturalists**[11] is to educate the public about the environment and maintain the natural environment on land specifically dedicated to wilderness populations. Their primary responsibilities are preserving, restoring, maintaining, and protecting a natural habitat. Among the related responsibilities in these jobs are teaching, public speaking, writing, giving scientific and ecological demonstrations, and handling public relations and administrative tasks. Naturalists may work in a variety of environments, including private nature centers; local, state, and national parks and forests; wildlife museums; and independent non-profit conservation and restoration associations. Some of the many job titles a naturalist might hold are wildlife manager, fish and game warden, fish and wildlife officer, land steward, wildlife biologist, and environmental interpreter. Natural resource managers, wildlife conservationists, and ecologists sometimes perform the work of naturalists. Because of the impact of human populations on the environment, virtually no area in the United States is truly wild. Land and the animal populations require human intervention to help battle against the human encroachment that is damaging or hindering wildlife. Naturalists work to help wildlife maintain or improve their hold in the world.

The work can be directly involved in maintaining individual populations of animals or plants, overseeing whole ecosystems, or promoting the work of those who are directly involved in the maintenance of the ecosystem. Fish and wildlife officers (or fish and game wardens) work to preserve and restore the animal populations, including migratory birds that may be part of the environment only temporarily. Wildlife managers and range conservationists oversee the combination of plants and animals in their territories. Wildlife managers, range managers, and conservationists work to maintain the plant and animal life in a given area. Wildlife managers can work in small local parks or enormous national parks. Range managers work on ranges that have a combination of domestic livestock and wild population. Wildlife managers also balance the needs of the humans using or traveling through the land they supervise and the animals that live in or travel through that same land. They keep track of the populations of animals and plants and provide food and water when it is lacking naturally.

Naturalists in all these positions often have administrative duties such as supervising staff members and volunteers, raising funds (particularly for independent nonprofit organizations), writing grant applications, taking and keeping records and statistics, and maintaining public relations.

Some naturalists, such as directors of nature centers or conservation organizations, have massive administrative responsibilities. They might recruit volunteers and supervise staff, organize long- and short-term program goals, and handle record keeping and the budget. To raise money, naturalists may need to speak publicly on a regular basis, write grant proposals, and organize and attend scheduled fundraising activities and community meetings. Naturalists also try to increase public awareness and support by writing press releases and organizing public workshops, conferences, seminars, meetings, and hearings. In general, naturalists must be available as resources for educating and advising the community.

Education and Training Requirements

If you are interested in this field, you should take basic science courses, including biology, chemistry, and earth science. Botany courses and clubs are helpful, because they provide direct experience in monitoring plant growth and health. Animal care experience, usually obtained through volunteer work, also is helpful. Take English courses in high school to improve your writing skills, which you will use when writing grant proposals and conducting research.

An undergraduate degree in environmental, physical, or natural sciences is generally the minimum educational requirement for becoming a naturalist. Common college majors are biology, forestry, wildlife management, natural resource and park management, natural resources, botany, zoology, chemistry, natural history, and environmental science. Graduate education is increasingly required for employment as a naturalist, particularly for upper-level positions.

If you are considering a career in this field, you should like working outdoors, because most naturalists spend the majority of their time outside in all kinds of weather. However, along with the desire to work in and with the natural world, you need to be capable of communicating with the human world. Excellent writing skills are helpful in preparing educational materials and grant proposals. Seemingly unrelated skills in this field, such as engine repair and basic carpentry, can be essential to managing a post.

Exploring the Career

One of the best ways to learn about the job of a naturalist is to volunteer at one of the many national and state parks or nature centers. These institutions often recruit volunteers for outdoor work. College students, for example, are sometimes hired to work as summer or part-time nature guides. You should also consider college internship programs. In addition, conservation programs and organizations throughout the country and the world offer opportunities for volunteer work in a wide variety of areas, including working with the public, giving lectures and guided tours, and working with others to build or maintain an ecosystem.

Advancement and Outlook

If you hope to become a park employee, the usual method of entry is through part-time or seasonal employment for the first several jobs, then a full-time position.

International experience is helpful with agencies that work beyond the U.S. borders. This can be through the Peace Corps or other volunteer organizations that work with local populations on land and habitat management or restoration. Other volunteer experience is available through local restoration programs on sites in your area.

In some settings, such as small nature centers, there may be little room for advancement. In larger organizations, experience and additional education can lead to increased responsibility and pay. Among the higher-level positions is that of director, handling supervisory, administrative, and public relations tasks.

Advancement into upper-level management and supervisory positions usually requires a graduate degree, although people with a graduate degree and no work experience will still have to start in nearly entry-level positions.

The outlook for naturalists is expected to be fair in the first decade of the twenty-first century. Although a growing public concern about environmental issues may cause an increased demand for naturalists, this trend could be offset by government cutbacks in funding for nature programs.

Earnings

The U.S. Fish and Wildlife Service reports that biologists working for this department have starting salaries at the GS-5 to GS-7 levels. In 2004, biologists at the GS-5 pay level earned annual salaries that ranged from $24,075 to $31,302, and those at the GS-7 level earned annual salaries that ranged from $29,821 to $38,767. The U.S. Fish and Wildlife Service further reports that biologists can expect to advance to GS-11 or GS-12 levels. In 2004, basic yearly pay at these levels was $44,136 and $52,899, respectively.

As with other fields, management positions are among the highest paying. For example, in 2002, the Idaho Department of Fish and Game advertised an opening for a wildlife game manager, offering a pay range of $22.78 to $35.59 per hour. These hourly wages translate into approximately $47,380 to $74,025 per year for full-time work.

For some positions, housing and vehicles may be provided. Other benefits, depending on employer, may include health insurance, vacation time, and retirement plans.

Work Environment

Field naturalists spend a majority of their working hours outdoors. Depending on the location, the naturalist must work in a wide variety of weather conditions: from frigid cold to sweltering heat to torrential rain. Remote sites are common, and long periods of working either in isolation or in small teams is not uncommon for field research and management. Heavy lifting, hauling, working with machinery and hand tools, digging, planting, harvesting, and tracking may fall to the naturalist working in the field. Indoor work includes scheduling, planning, and classroom teaching. Data gathering and maintaining logs and records are required for many jobs.

Those in supervisory positions, such as directors, are often so busy with administrative and organizational tasks that they may spend little of their workday outdoors. Work that includes guided tours and walks through nature areas is frequently seasonal and usually dependent on daily visitors.

Naturalists have special occupational hazards, such as working with helicopters, small airplanes, all-terrain vehicles, and other modes of transport through rugged landscapes and into remote regions. Adverse weather conditions and working in rough terrain make illness and injury more likely.

Naturalists also have a number of unique benefits. Most prominent is the chance to live and work in some of the most beautiful places in the world. For many individuals, the lower salaries are offset by the recreational and lifestyle opportunities afforded by living and working in such scenic areas.

For More Information

For information on environmental expeditions, contact

Earthwatch Institute
3 Clock Tower Place, Suite 100
P.O. Box 75
Maynard, MA 01754
Tel: 800-776-0188
Email: info@earthwatch.org
http://www.earthwatch.org

For information about internships, career conferences, and publications, contact

Environmental Careers Organization
30 Winter Street
Boston, MA 02108
Tel: 617-426-4375
http://www.eco.org

This group offers internships and fellowships for college and graduate students with an interest in environmental issues. For information, contact

Friends of the Earth
1717 Massachusetts Avenue, NW, 600
Washington, D.C. 20036-2002
Tel: 877-843-8687
Email: foe@foe.org
http://www.foe.org

For information on a variety of conservation programs, contact

National Wildlife Federation
11100 Wildlife Center Drive

Reston, VA 20190-5362
Tel: 800-822-9919
http://www.nwf.org

For information on volunteer opportunities, contact

Student Conservation Association
P.O. Box 550
Charlestown, NC 03603-0550
Tel: 603-543-1700
http://www.sca-inc.org

For information on careers, contact

U.S. Fish and Wildlife Service
U.S. Department of the Interior
Division of Human Resources
4401 North Fairfax Drive, Mailstop: 2000
Arlington, VA 22203
http://hr.fws.gov/HR/Careers_FWS.htm

For information on federal employment, contact

USAJOBS
Office of Personnel Management
http://www.usajobs.opm.gov

PET GROOMERS

Pet groomers[12] comb, cut, trim, and shape the fur of all types of dogs and cats. They comb out the animal's fur and trim the hair to the proper style for the size and breed. They also trim the animal's nails, bathe it, and dry its hair. (*See Figure 5–7.*) In the process, they check for flea or tick infestation and

Figure 5–7   Dog groomers specialize in grooming dogs. *Courtesy of Tracy Fera, PET SPAS of America.*

any visible health problems. In order to perform these grooming tasks, the pet groomer must be able to calm the animal down and gain its confidence.

Groomers are also called on to tend to more exotic pets such as ferrets, birds, and reptiles. New developments in animal grooming include high-performance clippers and cutting tools and more humane restraining devices. Current trends toward specialized services include perfuming, powdering, styling, and even massage, aromatherapy, and tattooing for pets!

Education and Training Requirements

A high school diploma generally is not required for people working as pet groomers. A diploma or GED certificate, however, can be a great asset to people who would like to advance within their present company or move to other careers in animal care that require more training, such as veterinary technicians. Useful courses include English, business math, general science, anatomy and physiology, health, zoology, psychology, bookkeeping, office management, typing, art, and first aid.

A person interested in pet grooming can be trained for the field in one of three ways: enrolling in a pet grooming school, working in a pet shop or kennel and learning on the job, or reading one of the many books on pet grooming and practicing on his or her own pet.

To enroll in most pet grooming schools, a person must be at least 17 years old and fond of animals. Previous experience in pet grooming can sometimes be applied for course credits. Students study a wide range of topics, including the basics of bathing, brushing, and clipping; the care of ears and nails; coat and skin conditions; animal anatomy terminology; and sanitation. They also study customer relations, which is very important for those who plan to operate their own shops.

Students can also learn pet grooming while working for a grooming shop, kennel, animal hospital, or veterinarian's office. They usually begin with tasks such as shampooing dogs and cats and trimming their nails, then gradually work their way up to brushing and basic haircuts.

The essentials of pet grooming can also be learned from any of several good books available on grooming. These books contain all the information a person needs to know to start his or her own pet grooming business, including the basic cuts, bathing and handling techniques, and type of equipment needed. Still, many of the finer points of grooming, such as the more complicated cuts and various safety precautions, are best learned while working under an experienced groomer. There still is no substitute for on-the-job training and experience.

Presently, state licensing or certification is not required and there are no established labor unions for pet groomers. To start a grooming salon or other business, a license is needed from the city or town in which a person plans to practice. The National Dog Groomers Association of America offers certification to groomers who pass written and practical examinations.

The primary qualification for a person who wants to work with pets is a love of animals. Animals can sense when someone does not like them or is afraid of them. A person needs certain skills in order to work with nervous, aggressive, or fidgety animals. They must be patient with the animals, able to gain their respect, and enjoy giving the animals a lot of love and attention. Persistence and endurance are also helpful because grooming one animal can take several hours of strenuous work. Groomers should enjoy working with their hands and have good eyesight and manual dexterity to accurately cut a clipping pattern.

Exploring the Career

To find out if you are suited for a job in pet grooming, you should familiarize yourself with animals as much as possible. This can be done in many ways, starting with the proper care of your dog, cat, or other family pet. Youth organizations such as the Boy Scouts, Girl Scouts, and 4-H Clubs sponsor projects that give members the chance to raise and care for animals. You can also volunteer or work part-time caring for animals at an animal hospital, kennel, pet shop, animal shelter, nature center, or zoo.

Advancement and Outlook

Graduates from dog grooming schools can take advantage of the schools' job placement services. Generally, there are more job openings than qualified groomers to fill them, so new graduates may have several job offers to consider.

The National Dog Groomers Association of America (NDGAA) promotes professional identification through membership and certification testing throughout the United States and Canada. The NDGAA offers continuing education, accredited workshops, certification testing, seminars, insurance programs, a job placement program, membership directory, and other services and products.

Other sources of job information include the classified ads of daily newspapers and listings in dog and pet magazines. Job leads may be available from private or state employment agencies or from referrals of salon or kennel owners. People looking for work should phone or send letters to prospective employers, inform them of their qualifications, and, if invited, visit their establishments.

Pet groomers who work for other people may advance to a more responsible position such as office manager or dog trainer. If dog groomers start their own shops, they may become successful enough to expand or to open branch offices or area franchises.

The pet industry is booming, so there are many avenues of advancement for groomers who like to work with pets. With more education, a groomer may get a job as a veterinary technician or assistant at a shelter or animal hospital. Those who like to train dogs may open obedience schools, train guide dogs, work with field and hunting dogs, or even train stunt and movie dogs. People can also open their own kennels, breeding and pedigree services, **gaming dog businesses,** or pet supply distribution firms.

The demand for skilled dog groomers has grown faster than average, and is expected to continue to grow at this rate through 2012, according to the U.S. Department of Labor. Every year more people are keeping dogs and cats as pets. They are spending more money to pamper their animals, but often don't have enough free time or the inclination to groom their pets themselves.

Earnings

Groomers charge either by the job or the hour. If they are on the staff of a salon or work for another groomer, they get to keep 50 to 60 percent of the fees they charge.

The U.S. Department of Labor reports that median hourly earnings of nonfarm animal caretakers (the category in which pet groomers are classified) were $8.26 (or $17,181 annually) in 2003. Salaries ranged from less than $6.11 (or $12,709 annually) to more than $13.40 per hour (or $27,872 annually). Those who own and operate their own grooming services can earn significantly more, depending on how hard they work, the clientele they service, and the economy of the area in which they work.

Groomers generally buy their own clipping equipment, including barber's shears, brushes, and clippers. A new set of equipment costs around $325; used sets cost less. Groomers employed full-time at salons, grooming schools, pet shops, animal hospitals, and kennels often get a full range of benefits, including paid vacations and holidays, medical and dental insurance, and retirement pensions.

Work Environment

Salons, kennels, and pet shops, as well as gaming and breeding services, should be clean and well-lighted, with modern equipment and clean surroundings. Groomers who are self-employed may work out of their homes. Some groomers buy vans and convert them into grooming shops. They drive them to the homes of the pets they work on, which many owners find very convenient.

Groomers usually work a 40-hour week and may have to work evenings or weekends. Those who own their own shops or work out of their homes, like other self-employed people, work very long hours and may have irregular schedules. Groomers are on their feet much of the day, and their work can get very tiring when they have to lift and restrain large animals. They must wear comfortable clothing that allows for freedom of movement, but they should also be presentable enough to deal with pet owners and other clients.

When working with any sort of animal, groomers may encounter bites, scratches, strong odors, fleas, and other insects. They may have to deal with sick or bad-tempered animals. The groomer must regard every animal as a unique individual and treat it with respect. Groomers need to be careful while on the job, especially when handling flea and tick killers, which are toxic to humans as well as the pests.

For More Information

For more Information about grooming and related professions, contact

Intergroom
76 Carol Drive
Dedham, MA 02026
Tel: 781-326-3376
http://www.intergroom.com

For information on shows, new grooming products and techniques, and workshop and certification test sites and dates, contact

National Dog Groomers Association of America Inc.
P.O. Box 101
Clark, PA 16113
Tel: 724-962-2711
Email: ndga@nationaldoggroomers.com
http://www.nationaldoggroomers.com

For information on pet grooming schools, contact the following institutions:

California School of Dog Grooming
655 South Rancho Santa Fe Road
San Marcos, CA 92069
Tel: 800-949-3746
Email: gtaurasi@csdg.net
http://www.csdg.net

Nash Academy of Animal Arts (Kentucky)
857 Lane Allen Plaza
Lexington, KY 40504
Tel: 859-276-5301
Email: info@nashacademy.com
http://www.nashacademy.com

Nash Academy of Animal Arts (New Jersey)
228 Anderson Avenue
Fairview, NJ 07022
Tel: 201-945-2710
Email: nashdog@prodigy.net
http://www.nashacademy.com

New York School of Dog Grooming
455 Second Avenue
New York, NY 10010
Tel: 212-685-3776
http://www.nysdg.com

For links to additional pet grooming career information, check out the following Web site:

PetGroomer.com
http://www.petgroomer.com

VETERINARIANS

The **veterinarian,** or **doctor of veterinary medicine,**[13] diagnoses and controls animal diseases, treats sick and injured animals medically and surgically, prevents transmission of animal diseases, and advises owners on proper care of pets and livestock. Veterinarians are dedicated to the protection of the health and welfare of all animals and to society as a whole. There are about 58,000 veterinarians in the United States. (*See Figure 5–8.*)

Figure 5–8 One-third of the veterinarians in the United States work exclusively with small animals. *Courtesy of Brian Yacur and Guilderland Animal Hospital.*

Veterinarians ensure a safe food supply by maintaining the health of food animals. They also protect the public from residues of herbicides, pesticides, and antibiotics in food. Veterinarians may be involved in wildlife preservation and conservation and use their knowledge to increase food production through genetics, animal feed production, and preventative medicine.

In North America, about 70 percent of veterinarians are employees of established veterinary practices. Although some veterinarians treat all kinds of animals, more than half limit their practice to companion animals such as dogs, cats, and birds. A smaller number of veterinarians work mainly with horses, cattle, pigs, sheep, goats, and poultry. Today, a veterinarian may be treating llamas, catfish, or ostriches as well. Others are employed by wildlife management groups, zoos, aquariums, ranches, feed lots, fish farms, and animal shelters.

Some veterinarians work in public and corporate sectors. Many are employed by city, county, state, provincial, or federal government agencies that investigate, test for, and control diseases in companion animals, livestock, and poultry that affect both animal and human health. Veterinarians also play an important public health role.

Pharmaceutical and biomedical research firms hire veterinarians to develop, test, and supervise the production of drugs, chemicals, and biological products such as antibiotics and vaccines that are designed for human and animal use. Some veterinarians are employed in management, technical sales and services, and marketing in agribusiness, pet food companies, and pharmaceutical companies. Still other veterinarians are engaged in research and teaching at veterinary and human medical schools, working with racetracks or animal-related enterprises, or working within the military, public health corps, and space agencies.

The U.S. Department of Agriculture has opportunities for veterinarians in the food safety inspection service and the animal and plant health inspection service, notably in the areas of food hygiene and safety, animal welfare, animal disease control, and research.

Veterinarians are often assisted by veterinary technicians, who may conduct basic tests, record an animal's medical history for the veterinarian's review, and assist the veterinarian in surgical procedures.

Education and Training Requirements

For the high school student who is interested in admission to a school of veterinary medicine, a college preparatory course is a wise choice. A strong emphasis on science classes such as biology, chemistry, and anatomy is highly recommended.

The doctor of veterinary medicine (DVM) degree requires a minimum of 4 years of study at an accredited college of veterinary medicine. Although many of these colleges do not require a bachelor's degree for admission, most require applicants to have completed 45 to 90 hours of undergraduate study. It is possible to obtain preveterinary training in a junior college, but because admission to colleges of veterinary medicine is an extremely competitive process, most students receive degrees from 4-year colleges before applying.

There are 28 colleges of veterinary medicine in the United States that are accredited by the Council of Veterinary Medicine of the American Veterinary Medical Association. Each college of veterinary medicine has its own preveterinary requirements, which typically include basic language arts, social sciences, humanities, mathematics, chemistry, and biological and physical sciences. (See Appendix A.)

All states and the District of Columbia require that veterinarians be licensed to practice private clinical medicine. To obtain a license, applicants must have a DVM degree from an accredited or approved college of veterinary medicine. They must also pass one or more national examinations and an examination in the state in which they plan to practice.

Individuals who are interested in veterinary medicine should have an inquiring mind and keen powers of observation. Aptitude and interest in the biological sciences are important. Veterinarians need a lifelong interest in scientific learning as well as a liking and understanding of animals. Veterinarians should be able to meet, talk, and work well with a variety of people. An ability to communicate with the animal owner is as important in a veterinarian as diagnostic skills.

Veterinarians may have to euthanize (that is, humanely kill) an animal that is very sick or severely injured and cannot get well. When a beloved pet dies, the veterinarian must deal with the owner's grief and loss.

Exploring the Career

High school students interested in becoming veterinarians may find part-time or volunteer work on farms, in small animal clinics, or in pet shops, animal shelters, or research laboratories. A good way to learn about the care of animals is by participation in extracurricular activities such as 4-H.

Advancement and Outlook

The only way to become a veterinarian is through the prescribed degree program, and veterinarian schools are set up to assist their graduates in finding employment.

New graduate veterinarians may enter private clinical practice, usually as employees in an established practice, or become employees of the U.S. government as meat and poultry inspectors, disease control workers, and commissioned officers in the U.S. Public Health Service or the military. New graduates may also enter internships and residencies at veterinary colleges and large private and public veterinary practices or become employed by industrial firms.

The veterinarian who is employed by a government agency may advance in grade and salary after accumulating time and experience on the job. For the veterinarian in private clinical practice, advancement usually consists of an expanding practice and the higher income that will result from it or becoming an owner of several practices. Those who teach or do research may obtain a doctorate and move from the rank of instructor to that of full professor, or they may advance to an administrative position.

Employment of veterinarians is expected to grow faster than the average through 2012. The number of pets is expected to increase slightly because of rising incomes and an increase in the number of people aged 34 to 59, among whom pet ownership has been highest in the past.

The outlook is good for veterinarians with specialty training. Demand for specialists in toxicology, laboratory animal medicine, and pathology is expected to increase. Most jobs for specialists will be in metropolitan areas. (See Appendix C.)

Despite the availability of additional jobs, competition among veterinarians is likely to be stiff. First-year enrollments in veterinary schools have increased slightly, and the number of students in graduate-degree and board-certification programs has risen dramatically.

Earnings

The U.S. Department of Labor reports that the median annual earnings of veterinarians was $64,750 in 2003. Salaries ranged from less than $38,400 to more than $123,040. The average annual salary for veterinarians working for the federal government was $72,208 in 2003.

According to a survey by the American Veterinary Medical Association, the average starting salary for veterinary medical college graduates who worked exclusively with small animals was $48,178 in 2002. Those who worked exclusively with large animals earned an average of $48,303. Equine veterinarians earned an average of $34,273 to start.

Work Environment

Veterinarians usually treat companion and food animals in hospitals and clinics. Those in large animal practice also work out of well-equipped trucks or cars and may drive considerable distances to farms and ranches. They may work outdoors in all kinds of weather.

Most veterinarians work long hours, often 50 or more hours a week. Although those in private clinical practice may work nights and weekends, the increased number of emergency clinics has reduced the amount of time private practitioners have to be on call. Large animal practitioners tend to work more irregular hours than those in small animal practice, industry, or government. Veterinarians who are just starting a practice tend to work longer hours.

For More Information

For more information on careers, schools, and resources, contact

American Veterinary Medical Association
1931 North Meacham Road, Suite 100
Schaumburg, IL 60173-4360
Attn: Education and Research Division
Tel: 847-925-8070
Email: avmainfo@avma.org
http://www.avma.org

For information on veterinary opportunities in the federal government, contact

Animal and Plant Health Inspection Service
U.S. Department of Agriculture
12th and Independence Avenue, SW
Washington, D.C. 20250
Email: APHIS.Web@aphis.usda.gov
http://www.aphis.usda.gov

The following Web site offers links to educational and career resources for veterinarians:

NetVet
http://netvet.wustl.edu/vet.htm

VETERINARY TECHNICIANS

Veterinary technicians[14] provide support and assistance to veterinarians. They work in a variety of environments, including zoos, animal hospitals, clinics, private practices, kennels, and laboratories. Work may involve large or small animals or both. Although most veterinary technicians work with domestic animals, some professional settings may require treating exotic or endangered species. There are approximately 53,700 veterinary technicians employed in the United States.

A veterinary technician is the person who performs much of the laboratory testing procedures commonly associated with veterinary care. In fact, approximately 50 percent of a veterinary technician's duties involve laboratory testing. Laboratory assignments usually include taking and developing X-rays films, performing parasitology tests, and examining various samples taken from the animal's body, such as blood and stool. A veterinary technician may also assist the veterinarian with necropsies in an effort to determine the cause of an animal's death.

In a clinic or private practice, a veterinary technician assists the veterinarian with surgical procedures. This generally entails preparing the animal for surgery by shaving the incision area and applying a topical antibacterial agent. Surgical anesthesia is administered and controlled by the veterinary technician. Throughout the surgical process, the technician tracks the surgical instruments and monitors the animal's vital signs. If an animal is very ill and has no chance for survival, or an overcrowded animal shelter is unable to find a home for a donated or stray animal, the veterinary technician may be required to assist in euthanizing it.

During routine examinations and checkups, veterinary technicians will help restrain the animals. They may perform ear cleaning and nail clipping procedures as part of regular animal care. Outside the examination and surgery rooms, veterinary technicians perform additional duties. In most settings, they record, replenish, and maintain pharmaceutical equipment and other supplies.

Education and Training Requirements

Veterinary technicians must have a high school diploma. High school students who excel at math and science have a strong foundation on which to build. Those who have had pets or who simply love animals and would like to work with them also fit the profile of a veterinary technician.

The main requirement is the completion of a 2- to 4-year college-based accredited program. Upon graduation, the student receives an associate's or bachelor's degree. Currently, there are more than 90 accredited programs in the United States. (See Appendix B.)

In addition to classroom study, accredited programs offer practical courses. Hands-on education and training are commonly achieved through a clinical practicum, or internship, in which the student has the opportunity to work in a clinical veterinary setting. During this period, a student is continuously evaluated by the participating veterinarian and encouraged to apply the knowledge and skills learned.

Although the AVMA determines the majority of the national codes for veterinary technicians, state codes and laws vary. Most states offer registration or certification, and the majority of these states require graduation from an AVMA-accredited program as a prerequisite for taking the examination.

As a veterinarian technician, one should be able to meet, talk, and work well with a variety of people. An ability to communicate with the animal owner is as important as diagnostic skills.

Exploring the Career

High school students can acquire exposure to the veterinary field by working with animals in related settings. For example, a high school student may be able to work as a part-time animal attendant or receptionist in a private veterinary practice. (*See Figure 5–9.*) Paid or volunteer positions may be available at kennels, animal shelters, and training schools.

Figure 5–9 High school students can get employment as animal hospital workers. Responsibilities usually include cleaning and disinfecting cages and equipment, feeding animals, and performing other duties assigned by the veterinarians. *Courtesy of USDA.*

Advancement and Outlook

Veterinary technicians who complete an accredited program and become certified or registered by the state in which they plan to practice are often able to receive assistance in finding a job through their college's placement offices. Students who have completed internships may receive job offers from the place where they interned.

Where a career as a veterinary technician leads is entirely up to the individual. Opportunities are unlimited. With continued education, veterinary technicians can move into allied fields such as veterinary medicine, nursing, medical technology, radiology, and pharmacology. By completing 2 more years of college and receiving a bachelor's degree, a veterinary technician can become a **veterinary technologist.**

Employment for veterinary technicians will grow much faster than the average for all other occupations through 2012, according to the U.S. Department of Labor.

Earnings

Earnings are generally low for veterinary technicians in private practices and clinics, but pay scales are steadily climbing because of the increasing demand. Most veterinary technicians are employed in private or clinical practice and research. The U.S. Department of Labor reports that the median annual salary for veterinary technicians and technologists was $23,340 in 2003. The lowest paid 10 percent made less than $16,410 annually, and the highest paid 10 percent made more than $34,380 annually.

Figure 5–10 Veterinary technicians assist veterinarians. *Courtesy of Brian Yacur and Guilderland Animal Hospital.*

Work Environment

Veterinary technicians generally work 40-hour weeks, which may include a few long weekdays and alternated or rotated Saturdays. Hours may fluctuate because veterinary technicians may need to have their schedules adjusted to accommodate emergency work.

Veterinary technicians must be prepared for emergencies. In field or farm work, they often have to overcome weather conditions in treating animals. Injured animals can be very dangerous, and veterinary technicians have to exercise extreme caution when caring for them. A veterinary technician also handles animals that are diseased or infested with parasites. Some of these conditions, such as ringworm, are contagious, so the veterinary technician must understand how these conditions are transferred to humans and take precautions to prevent the spread of diseases. (*See Figure 5–10.*)

People who become veterinary technicians care about animals. For this reason, maintaining an animal's well-being or helping to cure an ill animal is very rewarding work.

For More Information

For more information on careers, schools, and resources, contact the following organizations:

American Veterinary Medical Association
1931 North Meacham Road, Suite 100
Schaumburg, IL 60173
Tel: 847-925-8070
Email: avmainfo@avma.org
http://www.avma.org

Association of Zoo Veterinary Technicians
http://www.azvt.org

National Association of Veterinary Technicians in America

P.O. Box 224

Battle Ground, IN 47920

Tel: 765-742-2216

Email: navta@navta.net

http://www.navta.net

For information on veterinary careers in Canada, contact

Canadian Veterinary Medical Association

339 Booth Street

Ottawa, ON K1R 7K1 Canada

Tel: 613-236-1162

Email: info@canadianveterinarians.net

http://www.canadianveterinarians.net

ZOO AND AQUARIUM CURATORS AND DIRECTORS[15]

Zoos are wild kingdoms, and aquariums are underwater worlds. The word *zoo* comes from the Greek for "living being" and is a shortened term for *zoological garden* or *zoological park*. Although this may imply that zoos are created just for beauty and recreation, the main functions of modern zoos are education, conservation, and the study of animals. The term *aquarium* comes from the Latin for "source of water"; in such places, living aquatic plants and animals are studied and exhibited. These land and water gardens are tended by people with an affinity for animals.

Zoo and aquarium directors, or chief executive officers, are administrators who coordinate the business affairs of these centers. Directors execute the institution's policies, usually under the direction of a governing authority. They are responsible for the institution's operations and plans for future development and for such tasks as fundraising and public relations. They also serve as representatives of, and advocates for, their institutions and their entire industry. **Zoo and aquarium curators** are the chief employees responsible for the day-to-day care of the creatures. They oversee the various sections of the animal collections, such as birds, mammals, and fishes.

General curators of zoos and aquariums oversee the management of an institution's entire animal collection and animal management staff. **Animal curators** are responsible for the day-to-day management of a specific portion of a zoo's or aquarium's animal collection (as defined taxonomically, such as mammals or birds, or ecogeographically, such as the Forest Edge or the Arizona Trail); the people charged with caring for that collection; supervising assistant curators, zookeepers, administrative staff such as secretaries, as well as researchers, students, and volunteers; and the associated facilities and equipment.

Education and Training Requirements

High school students who want to prepare for careers in upper management in zoos and aquariums should take classes in the sciences, especially biology, microbiology, chemistry, and physics, as well as in mathematics, computer science, language, and speech.

Extracurricular activities for students interested in becoming zoo and aquarium curators and directors should focus on developing leadership and communication skills—these include student body associations, service clubs, debate teams, and school newspapers.

The minimum formal education requirement for curators is a bachelor's degree in one of the biological sciences, such as zoology, ecology, biology, mammalogy, and ornithology. Typically, an advanced degree is required for curators employed at larger institutions; many curators are required to have a doctoral degree. But advanced academic training alone is insufficient; it takes years of on-the-job experience to master the practical aspects of exotic animal husbandry. Also required are management skills, supervisory experience, writing ability, research experience, and sometimes the flexibility to travel.

A director's education and experience must be rather broad, with a solid foundation in animal management skills. Therefore, a good balance between science and business is the key to finding a position in this field. Directors need courses in zoology or biology as well as business courses, such as economics, accounting, and general business, and humanities, such as sociology.

Today, most directors have a master's degree; many at larger institutions have doctoral degrees.

Curators who work for zoos and aquariums must have a fondness and compassion for animals. But as managers of people, strong interpersonal skills are extremely important, including conflict management and negotiating. Curators also need excellent oral and written communication skills. They must be effective and articulate public speakers. They need to be good at problem solving. Curators should have an in-depth knowledge of every species and exhibit in their collections and how they interact.

Zoo and aquarium directors are leaders and communicators. Inspiring others and promoting their institution are among their most important tasks. Their most important traits include leadership ability, personal charisma, people skills, and public speaking ability.

Exploring the Career

Reading about animals or surfing the Internet, taking classes at local zoos and aquariums, or joining clubs such as 4-H or Audubon can help you learn about animals.

Volunteering at zoos or aquariums, animal shelters, wildlife rehabilitation facilities, stables, or veterinary hospitals demonstrates a serious commitment to animals and provides firsthand experience with them.

Advancement and Outlook

Neither the position of zoo and aquarium curator nor the position of director is an entry-level job. Although there are exceptions, most curators start their careers as zookeepers or aquarists and move up through the animal-management ranks.

Working on a part-time or volunteer basis at an animal facility could provide an excellent opportunity to improve your eligibility for higher-level jobs in later years. Moving up from supervisory keeper positions to curator and director positions usually involves moving to another institution, often in another city and state.

Today's zoo and aquarium directors often are people who began their careers in education, marketing, business, research, and academia, as well as animal management. Curatorial positions are often the top rung of the career ladder for many zoo and aquarium professionals. Curators do not necessarily wish to become zoo or aquarium directors, although the next step for specialized curators is to advance to the position of general curator. Curators and directors who work for a small facility may aspire to a position at a larger zoo or aquarium, with greater responsibilities and a commensurate increase in pay.

There are fewer than 200 professionally operated zoos, aquariums, wildlife parks, and oceanariums in North America. Considering the number of people interested in animal careers, this is not a large number. Therefore, it is expected that competition for jobs as curators (as well as for most zoo and aquarium jobs) will continue to be very strong.

The employment outlook for zoo curators is not favorable. Because of the slow growth in new zoos and in their capacity to care for animals, job openings are not expected to grow rapidly. The prospects for aquarium curators are somewhat better because of planned construction of several new aquariums.

Earnings

Yearly salaries for curators range from as low as $20,000 to as high as $79,000 for general curators in major metropolitan areas. According to the U.S. Department of Labor, median annual earnings for curators were $35,270 in 2002. Directors tend to be the highest-paid employees at zoos and aquariums; the range of their salary is also broad, from $28,000 to more than $100,000 per year, with some directors at major institutions earning considerably more than that. Given the scope of their responsibilities, salaries are not very high.

Most zoos and aquariums provide benefits packages including medical insurance, paid vacation and sick leave, and generous retirement benefits. As salaried employees, curators are not eligible for overtime pay, but they may get compensatory time for extra hours worked. Larger institutions may also offer coverage for prescription drugs, dental and vision insurance, mental health plans, and retirement savings plans. Private corporate zoos may offer better benefits, including profit sharing.

Work Environment

The work atmosphere for curators and directors of animal facilities will always center on the zoo or aquarium in which they work. Curators spend most of their time indoors at their desks, reading email, talking on the phone, writing reports, meeting deadlines for budgets, planning exhibits, and so forth. Particularly at large institutions, the majority of their time is spent on administrative duties rather than hands-on interaction with animals. Like other zoo and aquarium employees, curators often work long hours tending to the varied duties to which they are assigned.

When the unexpected happens, curators get their share of animal emergencies. In difficult situations, they may find themselves working late into the night with keepers and veterinarians to help care for sick animals or those that are giving birth.

Directors tend to spend a great deal of time in their offices conducting business affairs. They attend a lot of meetings.

For More Information

Visit the AZA Web site to read the following online pamphlets: **Zoo and Aquarium Careers and Careers in Aquatic and Marine Science.**

American Zoo and Aquarium Association
8403 Colesville Road, Suite 710
Silver Spring, MD 20910-3314
Tel: 301-562-0777
Email: generalinquiry@aza.org
http://www.aza.org

For information on training, contact the following educational institutions:

Friends University of Wichita
2100 West University Street
Wichita, KS 67213
Tel: 800-794-6945
Email: science@friends.edu
http://www.friends.edu

Santa Fe Community College Zoo Studies Programs
3000 Northwest 83rd Street
Gainesville, FL 32606
Tel: 352-395-5000
Email: information@santafe.cc.fl.us
http://www.santafe.cc.fl.us

ZOOKEEPERS

Zookeepers[16] provide the day-to-day care for animals in zoological parks. They prepare the diets, clean and maintain the exhibits and holding areas, and monitor the behavior of animals that range from the exotic and endangered to the more common and domesticated. (*See Figure 5–11.*) Zookeepers interact with visitors and conduct formal and informal educational presentations; they sometimes assist in research studies and, depending on the species, may also train animals.

Zookeepers are responsible for providing the basic care required to maintain the health of the animals in their care. Daily tasks include preparing food by chopping or grinding meat, fish, vegetables, or fruit; mixing prepared

Figure 5-11 Zoo keepers are responsible for the daily care and feeding of animals. *Courtesy of Getty Images.*

commercial feeds; and unbaling forage grasses. Administering vitamins or medications may be necessary as well. In addition, zookeepers fill water containers in the cages. They clean animal quarters by hosing, scrubbing, raking, and disinfecting.

Zookeepers must become experts on the species—and the individuals—in their care. They must observe and understand all types of animal behaviors, including courtship, mating, feeding, aggression, sociality, sleeping, moving, and even urination and defecation.

The professional zookeeper works closely with zoo staff on research, conservation, and animal reproduction. Many keepers conduct research projects, presenting their findings in papers or professional journals or at workshops or conferences.

Zookeepers must be able to interact with zoo visitors and answer questions in a friendly, professional manner. Keepers may participate in formal presentations for the general public or for special groups.

Taking care of animals is hard work. About 85 percent of the job involves custodial and maintenance tasks, which can be physically demanding and dirty. These tasks must be done both indoors and outdoors, in all types of weather. In addition, there is the risk or an animal-inflicted injury or disease. Although direct contact with animals is limited and strictly managed, the possibility for injury exists when a person is working with large, powerful animals or even small animals that possess sharp teeth and claws.

Because animals require care every day, keepers must work weekends and holidays. They also may be called on to work special events outside their normal working hours.

Education and Training Requirements

If you are planning a career in zookeeping, take as many science classes while in high school as possible. A broad-based science education including courses in biology, ecology, chemistry, physics, and botany, coupled with mathematics and computer science, will be helpful. Courses in English and speech develop vocabulary and hone public speaking skills.

Although practical experience may sometimes be substituted for formal education, most entry-level positions require a 4-year college degree. Animal management has become a highly technical and specialized field. Degrees in animal science, zoology, marine biology, conservation biology, wildlife management, and animal behavior are preferred.

Smaller zoos may hire keeper trainees, who receive on-the-job training to learn the responsibilities of the zookeeper. Several major zoos offer formal keeper training courses, as well as on-the-job training programs, to students who are studying areas related to animal science and care; contact the American Zoo and Aquarium Association (AZA) for further information about which schools and animal facilities are involved in internship programs. Such programs could lead to full-time positions.

Many institutions offer unpaid internships for high school and college students interested in investigating a career in animal care. Internships may involve food preparation, hands-on experience with the animal collection, interpretive services for the public, exhibit design and construction, or the collection and analysis of data. The length of the internships varies. The minimum age for most of these programs is 18.

Some zoos require written aptitude tests or oral examinations. Applicants must pass a physical examination because keepers must be physically able to do such demanding work as lifting heavy sacks of feed or moving sick or injured animals.

Zookeepers must first and foremost have a fondness and empathy for animals. The work of the zookeeper is not glamorous. It takes a special kind of dedication to provide care to captive animals that require attention 24 hours a day, 365 days a year.

Exploring the Career

Many zoos offer volunteer opportunities for teens, such as Explorers or Junior Zookeeper programs, which are similar to programs for adult volunteers but with closer supervision. Most volunteer programs require a specific time commitment. Prospective zookeepers can volunteer or work part-time at animal shelters, boarding kennels, wildlife rehabilitation facilities, stables, or animal hospitals.

Advancement and Outlook

Despite the low pay and challenging working conditions, competition for jobs at zoos is intense. There are many more candidates than available positions. Most zookeepers enjoy their work, and turnover is low. Part-time work, summer jobs, or volunteering at a zoo increases an applicant's chances of getting full-time employment. Many zoos fill new positions by promoting current employees.

Job advancement in zoos is possible, but the career path is more limited than in some other professions requiring a college degree. The possibility for advancement varies according to a zoo's size and operating policies and an employee's qualifications. Continuing professional education is a must to keep current on progress in husbandry, veterinary care, and technology, and in order to advance.

Most zoos have different levels of animal management staff. The most common avenue for job promotion is from keeper to senior keeper to head keeper, then possibly to area supervisor or assistant curator and then curator.

Zoos hire more animal keepers than any other classification. But this is still a very small field. For this reason, employment should grow at a slower rate than other occupations through 2012. According to the American Zoo and Aquarium Association, of the more than 32,000 full-time professionals working in zoos and aquariums, fewer than 5,000 are zookeepers. Each year, there are many more applicants than positions available. Competition for jobs is stiff in the fewer than 200 professionally operated zoological parks, aquariums, and wildlife parks in North America.

Earnings

Most people who choose a career as a zookeeper do not do so for the money, but because they feel compassion for and enjoy being around animals. Salaries vary widely among zoological parks and depend on the size and location of the institution, whether it is publicly or privately owned, the size of its endowments and budget, and whether the zookeepers belong to a union.

The zookeeper's salary can range from slightly above minimum wage to more than $40,000 a year, depending on the keeper's background, grade, and tenure and the zoo's location.

According to a 2000 salary survey by the American Association of Zoo Keepers, the average salary for zookeepers was $24,925. Those with a college degree fared better, earning an average of $26,715; however, salaries as low as $15,000 and as high as $50,000 to $75,000 were also reported.

Work Environment

Cleaning, feeding, and providing general care to the animals are a necessity 7 days a week, sometimes outdoors and in adverse weather conditions. The zookeeper must be prepared for a varied schedule that may include working weekends and holidays. Sick animals may need round-the-clock care. A large portion of the job involves routine chores for animals that will not express appreciation for the keeper's efforts.

The zookeeper may be exposed to bites, kicks, diseases, and possible fatal injury from the animals he or she attends. He or she must practice constant caution because working with animals presents the potential for danger.

Many keepers would agree that the disadvantages of the job are outweighed by the advantages. A chief advantage is the personal gratification of successfully maintaining wild animals, especially rare or endangered species. A healthy, well-adjusted animal collection provides a keeper with a deep sense of satisfaction.

For More Information

Visit the AAZK Web site to read the online pamphlet Zoo Keeping as a Career.

American Association of Zoo Keepers Inc.
3601 29th Street, SW, Suite 133
Topeka, KS 66614
http://www.aazk.org

Visit the AZA Web site to read the following online pamphlets: Zoo and Aquarium Careers and Careers in Aquatic and Marine Science.

American Zoo and Aquarium Association
8403 Colesville Road, Suite 710
Silver Spring, MD 20910-3314
Tel: 301-562-0777
Email: generalinquiry@aza.org
http://www.aza.org

ZOOLOGISTS

Zoologists[17] are biologists who study animals. They often select a particular type of animal to study, and they may study an entire animal, one part or aspect of an animal, or a whole animal society. There are many areas of specialization from which a zoologist can choose, such as origins, genetics, characteristics, classifications, behaviors, life processes, and distribution of animals.

Although zoology is a single specialty within the field of biology, it is a vast specialty that includes many major subspecialties. Some zoologists study a single animal or a category of animals, whereas others may specialize in a particular part of an animal's anatomy or study a process that takes place in many kinds of animals.

Many zoologists are classified according to the animals they study. For example, entomologists are experts on insects, **ichthyologists** study fish, herpetologists specialize in the study of reptiles and amphibians, mammalogists focus on mammals, and ornithologists study birds. **Embryologists,** however, are classified according to the process they study. They examine the ways in which animal embryos form and develop from conception to birth.

Some zoologists are primarily teachers, whereas others spend most of their time performing original research. Teaching jobs in universities and other facilities are probably the most secure positions available, but zoologists who wish to do extensive research may find such positions restrictive.

Education and Training Requirements

To prepare for a career in zoology, make sure to get a well-rounded high school education. Although a solid grounding in biology and chemistry is an absolute necessity, you should remember that facility in English will also be invaluable. Computer skills are also essential because most zoologists not only use the computer for writing, communication, and research, but they also use various software packages to perform statistical analyses.

A bachelor's degree is the minimum requirement to work as a zoologist; advanced degrees are needed for research or administrative work. Academic training, practical experience, and the ability to work effectively with others are the most important prerequisites for a career in zoology.

Success in zoology requires tremendous effort. It would be unwise for a person who wants to work an 8-hour day to become a zoologist, because hard work and long hours (sometimes 60 to 80 hours per week) are the norm. Also, although some top scientists are paid extremely well, the field does not provide a rapid route to riches. A successful zoologist finds satisfaction in work, not in a paycheck.

Exploring the Career

One of the best ways to find out if you are suited for a career as a zoologist is to talk to zoologists and find out exactly what they do. Try to become an intern or a volunteer at an organization that is involved in an area that you find interesting. Most organizations have internships, and if you look with determination for an internship, you are likely to find one.

Advancement and Outlook

You will be ahead of the game if you have made contacts as an intern or as a member of a professional organization. It is an excellent idea to attend the meetings of professional organizations, which generally welcome students.

Though it is possible to find work with a bachelor's degree, it is likely that you will need to continue your education to advance further in the field. Competition for higher-paying, high-level jobs among those with doctoral degrees is fierce; as a result, it is often easier to break into the field with a master's degree than it is with a doctorate Higher education and publishing are two of the most important means of advancing in the field of zoology. The holder of a doctorate will make more money and have a higher status than the holder of a bachelor's or master's degree. The publication of articles and books is important for both research scientists and professors of zoology.

According to the *Occupational Outlook Handbook,* job opportunities for biological scientists should grow at an average rate through 2012. Competition for good positions—especially research positions that require a doctorate—is high. Those with a bachelor's or master's degree will face less competition because of a larger number of available positions, especially in nonresearch areas.

Earnings

A study conducted by the National Association of Colleges and Employers determined that in 2003, beginning salaries averaged $29,456 for holders of bachelor's degrees in biological science (including zoologists), $33,600 for those with master's degrees, and $42,244 for holders of doctoral degrees.

According to the U.S. Department of Labor, the median wage for biological scientists in 2003 was $49,330. Salaries ranged from less than $30,000 a year to more than $75,000 a year, depending on the zoologist's education and experience.

The benefits that zoologists receive as part of their employment vary widely. Employees of the federal government or top universities tend to have extensive benefit packages, but the benefits offered by private industry cover a wide range, from extremely generous to almost nonexistent.

Work Environment

There is much variation in the conditions under which zoologists work. Professors of zoology may teach exclusively during the school year or may both teach and conduct research.

Zoologists who do fieldwork may have to deal with difficult conditions. Zoologists spend much of their time corresponding with others in their field, studying the latest literature, reviewing articles written by their peers, and making and returning phone calls. No zoologist works in a vacuum. Even those who spend much time in the field have to keep up with developments within their specialty.

For More Information

For information on careers, schools, internships, and job opportunities, check out the following Web sites or contact

American Association of Zoo Keepers
3601 29th Street, SW, Suite 133
Topeka, KS 66614
http://www.aazk.org

American Institute of Biological Sciences
1444 Eye Street, NW, Suite 200
Washington, D.C. 20005
Tel: 202-628-1500

American Zoo and Aquarium Association
8403 Colesville Road, Suite 710
Silver Spring, MD 20910-3314
Tel: 301-562-0777
Email: generalinquiry@aza.org
http://www.aza.org

The following society publishes the bimonthly journal Integrative and Comparative Biology, *and it is a good source of information about all areas and aspects of zoology. For more information, contact*

Society for Integrative and Comparative Biology
1313 Dolley Madison Boulevard, Suite 402
McLean, VA 22101
Tel: 800-955-1236
http://www.sicb.org

SUMMARY

There are few get-rich jobs working with animals. Salaries can be comfortable and increase with education and experience. Most people who choose careers in animal care do so because of their love of animals.

Persons involved in animal care professions must realize that cleaning, feeding, and general care is a constant job. Overtime hours are common, especially when caring for sick or pregnant animals.

Employment opportunities look good for the future because 37.2 percent of American households have a dog, 32.4 percent have one or more cats, 1.6 percent have rabbits, and 0.7 percent have hamsters. Americans spend in excess of $40.8 billion a year on their pets; pet food manufacturers accrue some $15.4 billion in yearly sales, and veterinary care easily exceeds $24.5 billion annually.

The pet industry is just one area of small animal care. Numerous jobs exist in laboratory animal care, zoo animal care, animal health, animal training, and the many areas of the biological sciences.

DISCUSSION QUESTIONS

1. How large is the animal care industry? How important is it in your area?

2. What are the duties of the workers in the various small animal careers?

3. What are some of the different working conditions one might encounter?

4. What are the job requirements for the various animal care careers?

5. What are some background requirements that would be helpful when applying for a job in small animal care?

6. What requirements are necessary for advancement in the various careers?

7. What are some high school courses that would be helpful for students to take that might help prepare them for a career in small animal care?

SUGGESTED ACTIVITIES

1. Make a list of small animal career workers in your area.

2. Visit some of the small animal career workers in your area and interview them to find out how they started in their profession, what education they needed, and what other requirements were necessary.

3. With the approval of the teacher, invite animal care workers to your classroom to present programs on their careers.

4. Make a list of desired characteristics, other than general education requirements, that are helpful for a successful career in small animal care.

5. Do research and make a list of the various colleges and institutions that offer career preparation in the different areas of small animal care.

6. Write letters to the different organizations that are listed in this chapter to receive more information.

ADDITIONAL RESOURCES

1. 2006–2007 *Occupational Outlook Handbook*
   http://www.bls.gov/oco

2. Career Consulting Corner
   http://www.careercc.com

END NOTES

1. Excerpt from *U.S. Pet Ownership and Demographics Sourcebook*, Center for Information Management, American Veterinary Medical Association, 2007.

2. American Pet Products Manufacturers Association, Inc. (APPMA). "Pet Ownership" [Online]. 03 October 2007. http://www.appma.org

3. Excerpt from U.S. Pet Ownership and Demographics Sourcebook, Center for Information Management, American Veterinary Medical Association, 2007.

4. American Pet Products Manufacturers Association, Inc. (APPMA). "Pet Ownership" [Online]. 03 October 2007. http://www.appma.org

5. "Animal breeders and technicians," *Encyclopedia of Careers and Vocational Guidance* (13th ed., vol. 2, pp. 124–129). Ferguson, an Imprint of Facts on File, Inc., 2005.

6. "Animal caretakers," *Encyclopedia of Careers and Vocational Guidance* (13th ed., vol. 2, pp. 130–135). Ferguson, an Imprint of Facts on File, Inc., 2005.

7. "Animal handlers," *Encyclopedia of Careers and Vocational Guidance* (13th ed., vol. 2, pp. 135–139). Ferguson, an Imprint of Facts on File, Inc., 2005.

8. "Animal trainers," *Encyclopedia of Careers and Vocational Guidance* (13th ed., vol. 2, pp. 139–144). Ferguson, an Imprint of Facts on File, Inc., 2005.

9. "Aquarists," *Encyclopedia of Careers and Vocational Guidance* (13th ed., vol. 2, pp. 169–171). Ferguson, an Imprint of Facts on File, Inc., 2005.

10. "Biologists," *Encyclopedia of Careers and Vocational Guidance* (13th ed., vol. 2, pp. 335–341). Ferguson, an Imprint of Facts on File, Inc., 2005.

11. "Naturalists," *Encyclopedia of Careers and Vocational Guidance* (13th ed., vol. 2, pp. 405–411). Ferguson, an Imprint of Facts on File, Inc., 2005.

12. "Pet groomers," *Encyclopedia of Careers and Vocational Guidance* (13th ed., vol. 2, pp. 642–647). Ferguson, an Imprint of Facts on File, Inc., 2005.

13. "Veterinarians," *Encyclopedia of Careers and Vocational Guidance* (13th ed., vol. 2, pp. 677–680). Ferguson, an Imprint of Facts on File, Inc., 2005.

14. "Veterinary technicians," *Encyclopedia of Careers and Vocational Guidance* (13th ed., vol. 2, pp. 680–684). Ferguson, an Imprint of Facts on File, Inc., 2005.

15. "Zoo and aquarium curators and directors," *Encyclopedia of Careers and Vocational Guidance* (13th ed., vol. 2, pp. 733–738). Ferguson, an Imprint of Facts on File, Inc., 2005.

16. "Zookeepers," *Encyclopedia of Careers and Vocational Guidance* (13th ed., vol. 2, pp. 738–744). Ferguson, an Imprint of Facts on File, Inc., 2005.

17. "Zoologists," *Encyclopedia of Careers and Vocational Guidance* (13th ed., vol. 2, pp. 744–748). Ferguson, an Imprint of Facts on File, Inc., 2005.

CHAPTER 6

Nutrition and Digestive Systems

OBJECTIVES

After reading this chapter, one should be able to:

- define the terms *nutrition* and *nutrient*.
- list the six basic nutrient groups.
- discuss the difference between the ruminant and nonruminant digestive system.
- discuss the importance of the various nutrients in the diets of animals.

TERMS TO KNOW

absorption	digestion	nutrient
amino acid	DNA	nutrition
antibodies	enzymes	pH
assimilation	gizzard	pellagra
biochemical reaction	grit	proventriculus
black tongue factor	hemoglobin	respiration
black tongue disease	hormones	ruminant animals
cation	macrominerals	solubility
ceca (cecum)	microminerals	ventriculus
coprophagy	nitrogen-free extract	
cytochromes	nonruminant animals	

NUTRITION

Adequate nutrition is important for pets. **Nutrition** refers to the animal receiving a proper and balanced food and water ration so that it can grow, maintain its body, reproduce, and supply or produce the things we expect from it. Animals can provide us with work, egg production, milk production, meat production, production of offspring for sale, fur or pelt production, and companionship. *See Figure 6–1.*

The term **nutrient** refers to a single food or group of foods of the same general chemical composition that supports animal life.

In discussing nutrients, we cover six basic nutrient groups or classes: water, proteins, carbohydrates, fats, vitamins, and minerals.

Water

Water is in every cell of the animal and is necessary for the following:

1. supporting **biochemical reactions** in the body, including **respiration, digestion,** and **assimilation** (see Glossary)

2. transporting other nutrients

3. helping maintain body temperature

4. helping give the body its form

5. carrying waste from the body

Water is more important than any other nutrient group. An animal can go several days without food; it may become hungry and lose body weight, but it will not suffer significantly from the lack of food. However, without

Figure 6–1 All animals must receive proper nutrition if they are to grow well and maintain their health. *Courtesy of Catherine Wein.*

drinking water, there will be a loss of water from the blood, which results in a failure of proper circulation and decrease in the oxygen-carrying capacity. The tissues become dehydrated and the body becomes overheated. The cells of the body are starved for oxygen, and there is a slowdown in normal body functions. There is also a lowering of the body's resistance to disease.

Water is quickly absorbed into the body through the walls of the stomach. It is important that the animal receive a supply of good clean, fresh water. Any disease organisms (bacteria and parasites) in the water are also quickly absorbed into the animal's body, which creates a possibility of infection. Water is also a good carrier for medications for disease control.

Water makes up about 55 to 65 percent of an animal's body. A beef steer has about 55 percent of its body composition as water, whereas a mouse has about 65 percent. The composition of an animal's blood is 90 to 95 percent water, muscle is 72 to 78 percent water, and bone is 30 to 40 percent water.

Proteins

Proteins are complex nutrients composed of carbon, hydrogen, oxygen, and nitrogen. Proteins are needed in an animal's body for the following:

1. developing and repairing body organs and tissues, such as muscles, nerves, skin, hair, hooves, and feathers
2. producing milk, wool, and eggs
3. developing the fetus
4. building material for **enzymes** and **hormones**
5. developing **antibodies**
6. transmitting **DNA**

Proteins found in feed materials are broken down into **amino acids** during the digestive process. These amino acids are then carried by the bloodstream to various parts of the body.

Twenty-five amino acids are found in animal feeds. Of these, 10 or 11 are considered essential and the others are nonessential. Essential amino acids are those that cannot be produced by the animal's body and therefore must be supplied in the ration. The nonessential amino acids are those that can be synthesized by the animal's body.

Ruminant (see Glossary) animals are capable of manufacturing all the amino acids they require if any one or more amino acids are given in large enough quantity. In this process, nitrogen combines with carbohydrates to produce the various amino acids.

The essential amino acids are arginine, isoleucine, histidine, leucine, methionine, threonine, phenylalanine, lysine, tryptophan, and valine. Poultry also require the amino acid glycine. Cats need the amino acid taurine added to their diet. The cat's body does synthesize taurine, but not in sufficient quantity. A cat that lacks sufficient taurine in its diet will develop eye lesions, which can lead to blindness, and weakened heart walls, which can lead to heart failure. In animal diets, cystine can be used to replace insufficient amounts of methionine and thus is considered by some to be an essential amino acid.

The amount of the essential amino acids needed varies depending on the animal's function and stage of growth.

Carbohydrates

Carbohydrates contain the chemical elements carbon, hydrogen, and oxygen. Animals convert carbohydrates into energy that is needed for the following:

1. supporting bodily functions, such as breathing, digestion, and exercising
2. producing heat to keep the body warm
3. storing fat

Carbohydrates are made up of the group of chemicals called sugars, starches, and crude fiber. These groups are usually classified as **nitrogen-free extract** (NFE) and crude fiber on most feed bags. Nitrogen-free extracts are the more easily and completely digested sugars and starches. Crude fiber is composed mostly of nondigestible bulk or roughage. Some animals, such as the ruminants, can handle larger amounts of crude fiber than the single-stomached animals.

Fats

Fats are made up of the same chemical elements as carbohydrates, but in different combinations. They contain 2.25 times as much energy as an equivalent amount of carbohydrates and proteins. Fats are essential in the diet for the following:

1. providing energy
2. aiding in the absorption of fat-soluble vitamins
3. providing fatty acids (Certain fatty acids such as linoleic, linolenic, and arachidonic acids are essential in the diet of animals. Most animal requirements for fat are less than 3 percent.)

Vitamins

Vitamins are organic substances required in very small amounts. Vitamins do not build body tissues but are necessary for specific biochemical reactions such as the following:

1. regulating digestion, absorption, and metabolism
2. developing normal vision, bone, and external body coverings such as hair and feathers
3. regulating body glands
4. forming new cells
5. fighting disease and strengthening the immune system
6. developing and maintaining the nervous system

TABLE 6–1 Fat-Soluble and Water-Soluble Vitamins

Vitamin Name	Chemical Name	Solubility
A	retinol	Fat
D (two forms)	D_2-ergocalciferol	Fat
	D_3-cholecalciferol	
E	Tocopherols	Fat
K (three forms)	phylioquinone	
	menaquinone	Fat
	menadione	
C	ascorbic acid	Water
B_1	thiamine	Water
B_2	riboflavin	Water
B_3	niacin	Water
B_5	pantothenic acid	Water
B_6 (three forms)	pyridoxine	
	pyridoxal	Water
	pyridoxamine	
B_7	biotin	Water
B_9	folic acid or folacin	Water
B_{12}	cyanocobalamin	Water
	choline	Water

Vitamins are classified on the basis of their **solubility**; they are fat soluble or water soluble (see Table 6–1). Fat-soluble vitamins can be stored and accumulated in the liver, fatty tissues, and other parts of the body, whereas only limited amounts of the water-soluble vitamins are stored. Fat-soluble vitamins include A, D, E, and K. Water-soluble vitamins include C and the B-complex vitamins; the B-complex group includes thiamine, riboflavin, pyridoxine, cyanocolalamin, pantothenic acid, niacin (nicotinic acid), folic acid (folacin), biotin, and choline.

Vitamin A is required to prevent poor vision, respiratory ailments, digestive problems, and reproductive difficulties. Green leafy plants contain carotene (provitamin A), which animals convert to vitamin A in the body.

Vitamin D is made in the animal's body after exposure to ultraviolent (UV) rays from the sun. Vitamin D occurs in two forms: vitamin D_2 (ergocalciferol) and D_3 (cholecalciferol). Vitamin D is associated with calcium and phosphorus in the body. Animals deficient in vitamin D will develop a condition called *rickets* (metabolic bone disease), or soft and deformed bones. Animals exposed to sunlight have plenty of vitamin D; animals raised in confinement may need supplemental feeding of vitamin D. Good-quality, sun-cured hay is a good source of vitamin D.

Vitamin E is required in female rats for normal gestation and in the male rat to prevent sterility; it is also considered important in other animals for successful reproduction. In ruminants, a lack of vitamin E is associated with a form of degenerative muscle disease referred to as *white muscle disease* (and as *stiff lamb disease* in lambs). A lack of vitamin E in rabbits can cause a lack of interest in breeding, poor fertility, abortions, and miscarriages. Vitamin E has also been associated with muscle development. Most animals being fed proper rations are usually receiving sufficient amounts of vitamin E. *See Figure 6–2.*

Figure 6–2 Vitamin E is important for successful reproduction. *Courtesy of Catherine Wein.*

Figure 6–3 Adequate nutrition is one of the most important factors in keeping and maintaining pets.

Vitamin K is necessary for maintenance of normal blood coagulation. Green forages, seeds, and good hay should provide sufficient amounts of vitamin K. Supplementation of vitamin K is rarely necessary.

Vitamin C, or ascorbic acid, is synthesized by most animals and is not a consideration in feeding rations. However, guinea pigs and primates must receive dietary vitamin C to prevent scurvy, which is characterized by swollen, painful joints and bleeding gums. Diarrhea, rough hair, and hair loss are also signs of vitamin C deficiency. Vitamin C is unstable and, when combined in feeds, begins to break down and be of little value. Bags of feed used for guinea pigs must be used within 3 months of opening or the vitamin C dissipates. *See Figure 6–3.*

Vitamin B_1, or thiamine, was one of the first vitamins discovered; a deficiency of this vitamin was probably among the first known nutritional deficiency diseases. Symptoms of thiamine deficiency are decreased appetite, muscular weakness, and a characteristic paralysis known as polyneuritis or beri-beri. Thiamine is also required for normal metabolism of carbohydrates. Animals receiving raw, whole grains or their products do not usually develop thiamine deficiency. Freshwater fish contain an enzyme that destroys thiamine in the ration. Mink and fox fed raw, fresh fish develop thiamine deficiency; this condition can be overcome by cooking the fish.

Vitamin B_2 is also referred to as riboflavin. Vitamin B_2 deficiency causes curled toe paralysis in chicks; reduces egg production and hatchability in poultry; causes lesions, reduced growth, reproductive failures, and high neonatal mortality in pigs; and causes eye problems in rabbits and other animals. Riboflavin functions in several enzyme systems that are important in the metabolism of amino acids and carbohydrates. Whole or skim milk, green forages, and good-quality hay are sources of this vitamin. Riboflavin is available at low cost for the fortification and supplementation of breeding and growing rations.

Niacin was first recognized and designated as the vitamin preventing **pellagra,** or **black tongue factor.** Pellagra in humans shows up as a reddening of the skin and development of sores in the mouth and the intestinal tract, resulting in bloody diarrhea. In other animals, a lack of niacin causes digestive problems and retarded growth. Niacin deficiency causes **black tongue disease** in dogs. Ducks and pheasants appear to have a high niacin requirement. Niacin is a white powder that is available in adequate quantities at rather low cost. Most feeds contain some niacin, but usually not in sufficient amounts.

Pantothenic acid or panacid deficiency appears as abnormal skin condition around the beak and retarded growth in chicks and turkeys. In swine, a deficiency results in the development of a peculiar gait of the hind legs referred to as *goose stepping.* This condition is caused by permanent damage to the sciatic nerve. Corn contains a low amount of pantothenic acid, but other feed materials have adequate amounts. Pantothenic acid is available from chemical sources at reasonable prices.

Vitamin B_{12}, or cyanocobalamin, is essential in several biochemical reactions. It is essential for normal growth, reproduction, and blood formation. This vitamin was originally referred to as the animal protein factor (APF). The most economical and reliable source of vitamin B_{12} is the chemically pure form. Animal protein feeds such as fish meal, fish solubles, liver meal, and dried milk products are also good sources of vitamin B_{12}.

Choline functions in the transportation and metabolism of fatty acids. Deficiencies of choline show up as slipped tendons or perosis in chicks and turkeys, and as kidney and liver damage in other animals. Animals lacking choline develop what is referred to as *fatty liver.* Most feeds are good sources of choline; supplements of choline chloride ensure an adequate supply.

Folic acid, or folacin, is required for normal cell development and is essential in certain biochemical reactions; poor growth and various blood disorders are common symptoms of deficiency. Most feeds are fair to good sources of folic acid, and supplementary rations are usually not required.

Biotin is important in carbohydrate and fat metabolism. Dermatitis, loss of hair, and retarded growth are common deficiency symptoms. Biotin usually is supplied in adequate amounts by common food materials and is usually not supplemented. When a diet contains raw egg white, avidin in the egg white combines with the biotin in the intestinal tract to make biotin unavailable to the animal.

Vitamin B_6, or pyridoxine, helps build proteins and is essential in the metabolism of carbohydrates. It is needed in the biosynthesis of unsaturated fatty acids. Deficiency symptoms may appear as poor growth, anemia, muscle contractions, disorders of the head and neck muscles, and convulsions. Most feed materials contain adequate amounts of vitamin B_6.

Minerals

Minerals are inorganic elements that are essential to life's processes. Sources disagree on exactly the number of minerals that play a role in the body processes, but most agree there are seven that are required in the largest amounts and these are called **macrominerals.** They are calcium, phosphorus, potassium, sodium, sulfur, chlorine, and magnesium. Eight are needed in lesser or trace amounts and are called **microminerals:** iron, iodine, copper, cobalt, manganese, zinc, molybdenum, and selenium. There are seven other minerals that have been recognized as needed in very small amounts: chromium, silicon, nickel, arsenic, vanadium, tin, and boron. This discussion will only include the macrominerals and the micromenerals. (See Table 6–2.)

TABLE 6–2 Minerals in Animal Nutrition	
Macrominerals	**Microminerals**
Calcium (Ca)	Iron (Fe)
Phosphorus (P)	Iodine (I)
Potassium (K)	Copper (Cu)
Sodium (Na)	Cobalt (Co)
Sulfur (S)	Manganese (Mn)
Chlorine (Cl)	Zinc (Zn)
Magnesium (Mg)	Molybdenum (Mo)
	Selenium (Se)

As noted, the macrominerals are calcium, phosphorus, potassium, sodium, sulfur, chlorine, and magnesium, and the microminerals are iron, iodine, copper, cobalt, fluorine, manganese, zinc, molybdenum, and selenium.

Calcium is required by vertebrates in larger amounts than any other mineral. Ninety-nine percent of the calcium found in the body is deposited in the bones and teeth. Calcium is essential for bone, teeth, and eggshell formation; for normal blood coagulation; and for milk production.

A lack of calcium causes retarded growth and deformed bones in the young. A prolonged and extreme deficiency of calcium causes a condition known as rickets, which is characterized by soft, flexible bones. In older animals, lack of calcium may lead to two common conditions: osteoporosis, a condition in which bones lose their density and become porous and brittle, and osteomalacia, a softening of the bones. Egg-producing birds and animals lay soft-shelled eggs when fed rations low in calcium. The most economical sources of calcium are limestone, oyster shell, bone meal, and defluorinated phosphates; crushed eggshells and cuttlebone are also good sources.

Phosphorus is closely associated with calcium in animal nutrition. Seventy-five percent of the phosphorus found in the body is deposited in the bones and teeth. Phosphorus is essential for the formation of bones, teeth, and body fluids and is required for metabolism, cell respiration, enzyme-based reactions, and normal reproduction. Symptoms of phosphorus deficiency are similar to those of calcium deficiency in growing animals.

The most economical sources of phosphorus are low-fluorine phosphates and bone meal. Other good sources are wheat bran and middlings, high-protein feeds, bone-containing feeds, whole or skim milk, and most grains.

Potassium is required for many body functions, such as maintaining osmotic pressure and acid-base balance, activating enzymes, regulating neuromuscular activity, and helping to regulate heartbeat and digestion. Symptoms of a potassium deficiency may appear as decreased feed consumption, slow growth, stiffness, and a loss of weight. Grains and most feeds are sufficient in potassium; supplements are usually not required.

Most of the sodium of the animal body is present in the soft tissues and body fluids. Like potassium, sodium is concerned with the acid-base balance and osmotic regulation of the body fluids. Sodium is the chief **cation** of blood plasma and other extracellular fluids of the body. The sodium concentration within the cells is relatively low, the element being replaced largely by potassium and magnesium. Sodium also plays a role in the transmission of nerve impulses and in the absorption of sugars and amino acids from the digestive tract.

Much of the sodium is ingested in the form of sodium chloride (common salt), and it is also mainly in the form that the element is excreted from the body. There is evidence that sodium rather than chlorine is the chief limiting factor in salt-deficient diets of sheep and cows.

Chlorine is associated with sodium and potassium in acid-base relationships and osmotic regulation. Chlorine also plays an important part in the gastric secretion, where it occurs as hydrochloric acid as well as chloride salts. Chlorine is excreted from the body in the urine and is also lost from the body, along with sodium and potassium in perspiration.[1]

The main source of this element is common salt.

Under most conditions, there are no deficiency symptoms. The only signs may be unthrifty-appearing animals and impaired performance. In heavy-working or heavy-producing animals, a lack of these two minerals may cause sudden death. In birds, nervousness, poor feathering, and feather plucking are signs of deficiency.

Supplemental salt given for animals normally provides them with a source of sodium and chlorine. Salt can be provided in the form of blocks, loose or free salt, spools that attach to cages, or mineral mix.

Most of the sulphur in the animal body occurs in proteins containing the amino acids, cystine, cysteine, and methionine. The two vitamins biotin and thiamin, the hormone insulin, and the important metabolite coenzyme A also contain sulphur. The structural compound chondroitin sulphate is a component of cartilage, bone, tendons, and the walls of the blood vessels. Sulphur-containing compounds are also important in elements of the respiratory process from haemoglobin through **cytochromes.** Only a small amount of sulphur is present in the body in inorganic form, though sulphates are known to occur in the blood in small quantities. Wool is rich in cystine and contains about 4 per cent sulfur.[2]

Most feeds contain more-than-adequate supplies of sulfur; thus, deficiencies are usually not a concern. If a deficiency does appear, it expresses itself as a protein deficiency, which is a general unthrifty condition and causes poor performance.

Magnesium is necessary in the activation of many enzyme systems, particularly those involved with carbohydrate and lipid metabolism and for proper functioning of the nervous system. Approximately 70 percent of the magnesium in the body is found in the bones and the remaining 30 percent in the soft tissues and fluids. Most common feeds contain enough magnesium so that a deficiency usually does not appear; however, if feed rations are low in magnesium, problems may occur. The most common problems are with cattle and sheep being fed low-quality forages, especially in late winter and early spring. Animals will develop a hypomagnesemia associated with severe toxic symptoms and frequently death. The condition is also referred to as *grass tetany* or *grass staggers*.

Iron is necessary in the formation of **hemoglobin.** Deficiencies of iron usually show early in an animal's life because milk is low in iron content; thus, nursing animals are more susceptible. However, newborn animals normally have sufficient iron reserves in the liver and spleen to carry them through the early nursing period.

Anemia is a condition associated with iron deficiency. Animals with anemia show a loss of appetite, poor growth, and listlessness and become weak. Sow's milk is not very high in iron content, and a baby pig's system contains only enough iron for about 3 days. Around 3 days of age, piglets begin to show signs of anemia unless iron shots are administered. Additional symptoms in baby pigs are rough hair coat, wrinkled skin, and paleness of mucous membranes. In addition, the animals will have trouble breathing and breathe with a characteristic dull "thumping" sound. Young calves and lambs on pasture usually do not show signs of anemia. However, those raised on only mother's milk may develop anemia. Anemia may occur in laying hens. Anemia can occur in other animals if not fed properly formulated diets.

Anemia can be caused by many other factors besides iron deficiency. Sudden and severe blood loss can lead to shock and even death in cases in which large amounts of blood are lost. Hemolytic anemia can occur when red blood cells are destroyed by toxins, trauma, infections, and immune system defects. *See Figure 6–4.*

Copper is necessary for proper iron absorption, for hemoglobin formation, in various enzyme systems, and in the synthesis of keratin for hair and

Figure 6–4 Newborn animals should be watched carefully for signs of iron deficiency. *Courtesy of Laurette Richen.*

wool growth. Most feeds contain sufficient supplies of copper; however, if the soil is lacking in copper, foods grown in those areas may contain low amounts of copper. Copper is closely associated with iron, and a deficiency normally shows up as an anemic condition.

Iodine is important in the production of thyroxin by the thyroid gland. Historically, most feeds and water found in the United States contained sufficient natural amounts of iodine.

In some areas of the U.S., including mountain regions, the Mississippi River Valley, the Ohio River Valley, and the Great Lakes region, the soil has always had a very low iodine content. But even in other areas of once iodine-rich soil, overfarming has frequently depleted the iodine content. Hence, we no longer get adequate iodine via the plants we consume. To compensate for this, iodine was added to salt, bread, and milk. Today, iodine is no longer added to bread or to milk, and the amount of iodine added to salt has steadily declined over the years. All of these factors contribute to the current prevalence of iodine deficiency in the United States.[3]

Common symptoms at birth or soon afterward are goiter (a swelling of the thyroid glands in the neck area), dead or weak animals, hairlessness, and impaired brain development. Suppressed estrus in females and infertility in males may also be caused by a deficiency of iodine.

Cobalt is an important component of the vitamin B_{12} molecule and is necessary in ruminant animals for proper synthesis of vitamin B_{12}. The microorganisms in the rumen require cobalt for growth. Symptoms of cobalt deficiency include poor appetite, unthriftiness, weakness, anemia, decreased fertility, slow growth, and decreased milk and wool production. Areas that have cobalt-deficient soils need to supplement this mineral in the diet.

Manganese is involved with the enzyme systems that influence estrus, ovulation, fetal development, udder development, milk production, growth, and skeletal development. A deficiency of manganese takes the form of delayed estrus, reduced ovulation, reduced fertility, deformed young, poor growth, lowered serum alkaline phosphates, abortions, resorptions, and lowered tissue manganese. Most feeds contain adequate amounts of manganese.

Zinc is needed for the body's immune system, for the manufacture of proteins and genetic material, for the activity of more than 200 enzymes, for stimulating hair growth, for the sense of smell and taste, and for healing wounds. Deficiencies will appear as poor appetite, slow growth, wounds that do not heal, loss of hair, frequent infections, and trouble with sense of smell and taste. In swine, a condition called *parakeratosis* is associated with zinc deficiency; too much calcium in the diet may render zinc unavailable or inadequate. Most feeds are adequate in zinc. Yeast is a rich source. Zinc is found in the bran and germ of cereals. Fish meal is also a good source of zinc.

Molybdenum is a component of the enzyme xanthine oxidase, which is important in poultry for uric acid formation. This mineral is also important in stimulating action of rumen organisms. Molybdenum is usually adequate in most feeds.

Selenium is necessary for the **absorption** and utilization of vitamin E. Symptoms of selenium deficiency are similar to those of vitamin E deficiency and appear in the form of heart failure, paralysis, poor growth, low fertility, liver necrosis, and muscular dystrophy in cattle and lambs. Many areas of the country produce forages that are deficient in selenium, and supplementing diets is very important.

A lot of research has been conducted in recent years on pet nutrition. Proper feeding is easy today because of the variety of commercial pet foods available. Usually, a brand-name pet food contains a complete and balanced diet for pets, and recommendations for proper amounts to feed are included in the labeling.

Deficiencies usually do not occur among animals fed properly formulated diets; most problems occur from feeding animals table scraps or homemade diets. It is difficult to feed a homemade diet that contains all the nutrients required by animals.

Supplemental feeding of vitamins and minerals is made easy with the use of trace mineralized salt, liquid vitamins, and minerals that are easily added to water. The problem of oversupplementation with vitamins and minerals rarely occurs in animals being fed commercial diets. Oversupplementation of vitamins and minerals is usually seen when owners try to formulate their own diets and do not get the balance correct or when they use unnecessary vitamins and minerals supplements in addition to a commercial diet. Some problems that may occur in oversupplementing of vitamins and minerals are listed in Table 6–3.

TABLE 6-3 Toxicity of Vitamins and Minerals			
Vitamin	**Toxicity**	**Mineral**	**Toxicity**
A	hair loss, blurred vision	Calcium	interferes with iron absorption and kidney function, reduces copper absorption, may cause magnesium deficiency
D	nausea, weakness, irritability, brain or liver damage, jaundice, destruction of red blood cells (usually not supplemented)	Iron	alimentary disturbances, reduced growth, phosphorus deficiency, reduces copper absorption
E	interferes with absorption of other fat-soluble vitamins	Copper	accumulates in the liver leading to necrosis of the liver walls, jaundice, loss of appetite
pyridoxine	permanent nerve damages	Iodine	reduction in weight gain, reduction in feed intake, reduced egg production in chickens
niacin	itching, skin flushing, gastrointestinal distress	Manganese	depressed appetite, retarded growth
		Zinc	depressed food consumption, induced copper deficiency
		Selenium	dullness, stiffness of joints, reduced food intake, loss of hair, acute poisoning—results in death
		Phosphorus	calcium deficiency, bone malformation, fractures
		Sodium Chloride	nervousness, weakness, seizure, and death usually associated with low intake of water

ANIMAL DIGESTION

Digestion in animals is the process of breaking down food materials into their various nutrient forms that can be absorbed into the animal's bloodstream.

Digestive systems in animals are basically of two types: ruminant and nonruminant. To understand the nutritional requirements of various animals, we must understand how they digest their foods. Ruminant animals are often referred to as *forage-consuming* or *multistomached* animals, and include farm animals such as cattle, sheep, and goats and many animals commonly found in the zoo, such as elk, deer, giraffes, buffalo, camel, and antelope. The nonruminant animals are often referred to as *single-stomached* or *monogastric* animals, and they include all of the small animals discussed in this book. Although the digestive systems of birds, rabbits, and horses differ somewhat, they are considered as nonruminant animals because they have a single-compartmented stomach.

Ruminant animals have four compartments to their stomach. These compartments are the rumen, reticulum, omasum, and abomasum. *See Figure 6–5.*

The rumen is the largest compartment and makes up about 80 percent of the total capacity of the stomach. A mature dairy cow may have a rumen capacity of 50 to 60 gallons. A ruminant animal chews food just enough to aid in swallowing. When the food material reaches the rumen, it is worked on by millions of bacteria and microorganisms. These bacteria and microorganisms transform low-quality protein and some nitrogen compounds into essential amino acids. They also manufacture many needed vitamins, including the B-complex group. Bacteria and microorganisms utilize some of these nutrients in food and, when they die, are digested and utilized by the animal.

Food material not fully digested in the rumen can be regurgitated in the form of cud. The animal chews on this cud and then swallows it back down into the rumen for further digestion.

Closely associated with the rumen is the reticulum. The reticulum works with the rumen in the formation of the cud for regurgitation. Foreign bodies such as nails or pieces of wire can be held in the reticulum for long periods without causing a serious injury. The reticulum is referred to as the *hardware*

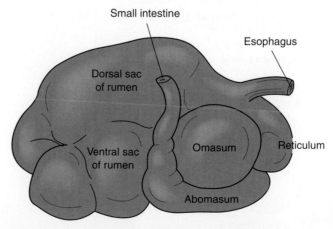

Figure 6–5 A ruminant digestive system. Note the four compartments of the ruminant stomach.

stomach for this reason. If the walls of the reticulum are punctured, the metal pieces can puncture the heart and cause death or bacteria can enter the abdominal cavity and cause infections and liver abscesses. These infections are referred as *hardware disease.* To prevent hardware disease, a permanent magnet can be placed in the reticulum, which will collect these dangerous pieces of metal and prevent them from puncturing the walls of the reticulum. Because of its beehive interior appearance, it is also referred to as the *honeycomb.*

The third compartment is the omasum, which is made up of many layers of strong muscular tissue referred to as the *many plies.* The function of this third compartment is to remove large amounts of water from food as it moves from the rumen to the fourth compartment.

The fourth compartment, or abomasum, is referred to as the *true stomach* because it functions very similarly to the stomach found in single-stomached animals.

The first step in the digestive process is the breaking, tearing, cutting, and grinding that takes place in chewing. Saliva is mixed with the food material in the mouth; saliva aids in the chewing process, helps the animal swallow, and aids in controlling the **pH** of the stomach. Saliva contains the first digestive enzymes in the digestive process.

In **nonruminant animals,** food passes from the mouth to the stomach through the esophagus. Food is moved through the esophagus by a series of involuntary muscle contractions. *See Figures 6–6, 6–7, and 6–8.*

The functions of the stomach are to break down food material by muscular movement and to secrete digestive juices. These digestive juices begin to break down proteins and to break down fat into fatty acids and glycerol.

Food passes from the stomach to the small intestine, which is the primary site for the digestion and absorption of carbohydrates, fats, and proteins. The small intestine is divided into three sections: the duodenum, the jejunum, and the ileum. Pancreatic juices containing enzymes from the pancreas enter the duodenum and convert fats into fatty acids and glycerol. Bile from the

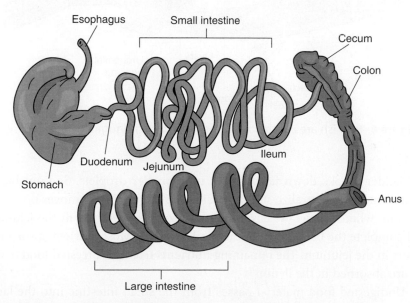

Figure 6–6 A single-stomach or monogastric digestive system.

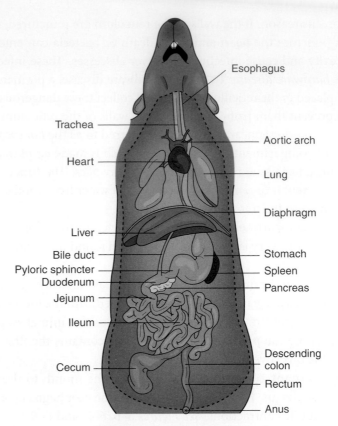

Figure 6–7 The rat is a monogastric animal.

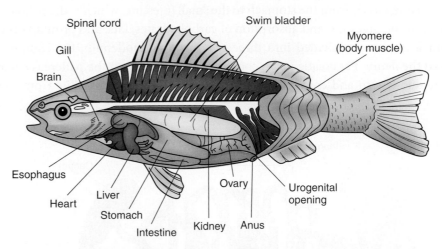

Figure 6–8 Fish are another representative of a monogastric animal.

gallbladder breaks down fat particles into smaller droplets. Sodium bicarbonate enters to neutralize stomach acid. The material then moves on to the jejunum, where pancreatic enzymes and enzymes produced in the jejunum wall complete the breakdown of the food material. Most nutrient absorption occurs in the jejunum. The remaining nutrients from the digested food material are absorbed in the ileum.

Undigested food material passes from the small intestine into the large intestine. The large intestine is also divided into three sections: the cecum, the colon, and the rectum. The primary function of the large intestine is to absorb the water from the undigested material; to add lubricating mucus, which aids in the passage of the material through the large intestine; and to shape the material for easier passage. The undigested material moves through the rectum, the last part of the large intestine, and out the anus, the opening at the lower end of the alimentary canal.

The horse can consume large amounts of forages but is not considered a ruminant animal. It has a small, single stomach, but unlike other single-stomached animals, it has a large **cecum** and colon located between the small and large intestines. Bacterial action takes place in the cecum and allows the horse to digest roughage material, but not as efficiently as a true ruminant animal. Horses do not have a gallbladder. Bile is secreted into the duodenum directly from the liver. In feeding horses, one needs to reduce the amounts of concentrates and increase roughages, which contain higher-quality proteins. *See Figure 6–9.*

The rabbit's digestive system is similar to that of the horse. It too has a large cecum that allows for the utilization of high-quality roughage material; bacteria are present in the cecum and help break down roughages. The maintenance of proper levels of bacteria in the cecum is very important in maintaining the health of rabbits and horses.

Many animals, including rabbits and rodents, eat their feces. This is referred to as **coprophagy.** In rabbits this practice is usually done late at night or early in the morning. Rodents will often be seen consuming feces directly from the anus. Feces are eaten because the food material is not held for a long enough time and the products of the digestive process are not fully absorbed. Feces eaten by the rabbit are usually light green, are soft, and have not been completely digested. Consumption of nutrients, primarily proteins and vitamins, in the undigested feces allows the rabbit's digestive system to make full use of the bacterial action of the cecum. *See Figure 6–10.*

Birds, although considered single-stomached animals, have several different organs in their digestive systems. Birds do not have teeth, so no chewing or breaking down of food material takes place in the mouth, although saliva is added here to aid in swallowing. *See Figure 6–11.*

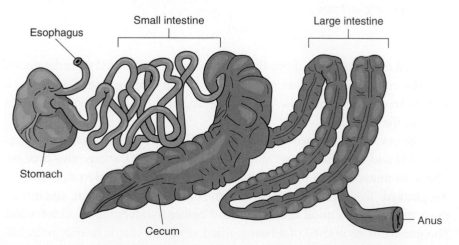

Figure 6–9 The digestive system of a horse. Note the large cecum that enables the animal to consume large amounts of roughage.

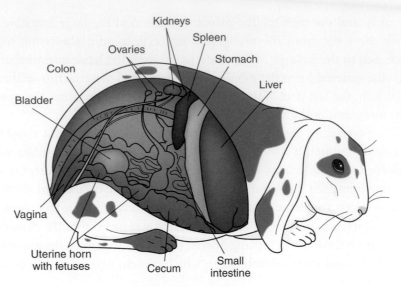

Figure 6–10 The digestive system of the rabbit is similar to that of the horse, because rabbits also have a large cecum. The rabbit consumes roughage materials that are broken down in the cecum.

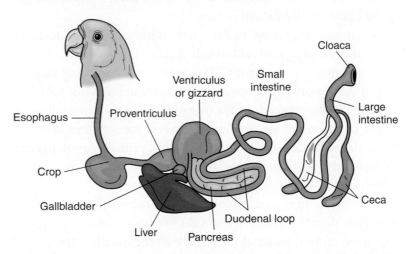

Figure 6–11 Birds are considered monogastric animals; however, their digestive system is somewhat different from that of other monogastric animals.

Food material passes down the esophagus, often referred to as the gullet, into the crop. The crop is an enlargement of the esophagus and serves as a storage area for food material. Some softening of the food material also takes place in the crop.

The next step is the **proventriculus,** which is the *true stomach,* where gastric acids and enzymes are secreted to begin the chemical digestive process. The food material then passes to the **ventriculus,** commonly referred to as the **gizzard.** This is the largest organ of the bird's digestive system, and its primary purpose is to grind and crush food before it enters the small intestine. The gizzard is composed of a horny, lined structure that is heavily muscled. Involuntary muscular action serves to break up the food material much like teeth action in other animals. Some birds, those that consume seeds and whole grains, are often fed **"grit"** in the form of crushed granite, oyster shell, or other insoluble material that aids in the grinding and breaking down of coarse food material. The presence of grit is not essential but does increase the breakdown of grain. Birds that feed on nectar, fruits, and other soft foods do not need grit in their gizzard.

Food material then passes on to the small intestine. The first part of the small intestine is the duodenum, where enzymes are secreted from the pancreas to help break down the proteins, starches, and fats. Bile is secreted from the gallbladder to aid in the breakdown of fats. Most of the absorption takes place in the lower section of the small intestine.

Birds have two pouches, called **ceca,** located where the small intestine connects into the large intestine. Although some bacteria are present in these pouches, very little digestion of the fiber is believed to occur here. Food material passes from the small intestine into the large intestine and then into the cloaca. The cloaca serves as a common junction for the bird's digestive, urinary, and reproductive systems.

SUMMARY

Nutrition refers to an animal receiving a proper and balanced food and water ration for optimum health. The term *nutrient* refers to a single component of food that supports animal life. There are six basic nutrient groups: water, proteins, carbohydrates, fats, vitamins, and minerals.

Digestion in animals is the process of breaking down food material into various nutrient forms that can be absorbed into the animal's bloodstream. Digestive systems in animals are basically of two types: ruminant and nonruminant. Nonruminant animals, such as swine, have a single digestive stomach. Ruminant animals, such as cows and sheep, have four compartments to their stomach. Nonruminant herbivores utilize a large cecum to break down complex carbohydrates from consumed plant material. Birds, although considered a single-stomached animal, have several different organs in their digestive system.

DISCUSSION QUESTIONS

1. Define the terms *nutrient* and *nutrition.*
2. Define the term *biochemical reaction.* What are the biochemical reactions that take place in the animal's body?
3. What are the nutrient groups, and what is their importance in the animal's diet?

4. List the various vitamins. What is their importance in the animal's body?
5. List the various minerals necessary in the diet of an animal. What is the important role each plays in the diet of the animal?
6. What is the difference between the digestive systems of a ruminant and a nonruminant animal?

7. List the four compartments of the ruminant stomach and explain the function of each.

8. How does the digestive system of a horse, rabbit, and chicken differ from a ruminant digestive system? Although they are considered nonruminants, how does the digestive system of a horse, rabbit, and chicken differ from those of most other nonruminants?

SUGGESTED ACTIVITIES

1. Visit a local feed processing plant and see how feeds are mixed, formulated, and produced.

2. Obtain ingredient tags from several different feeds. Compare the tags. How are the feed formulations different?

ADDITIONAL RESOURCES

1. Purina Pet Care Center
   http://www.purina.com/

2. Hill's Health Center
   http://www.hillspet.com/public/health_center/index.html

3. DC Nutrition.com
   http://www.dcnutrition.com

ENDNOTES

1. McDonald, P., R. A. Edwards, J. F. D. Greenhalgh, and C. A. Morgan. 2002. Animal Nutrition (6th ed., pp. 122–123). Pearson Educational Limited.

2. Ibid.

3. Greene, Winston, W. B. A., B. S., D. C. "Iodine (I)—general discussion," http://www.dcnutrition.com (accessed January 11, 2008).

SECTION 2

CHAPTER 7

Dogs

CLASSIFICATION

Kingdom — Animalia
Phylum — Chordata
Class — Mammalia
Order — Carnivora
Family — Canidae
Genus — *Canis*
Species — *C. familiaria*

OBJECTIVES

After reading this chapter, one should be able to:

- discuss the history of the dog.
- compare the seven major groups of dogs.
- list the breeds that make up the seven major groups of dogs.
- discuss the proper methods of feeding and exercising dogs.
- discuss methods of training dogs.
- discuss the grooming and health care of dogs.
- discuss the various diseases of dogs.

TERMS TO KNOW

anemia	estrus	placental membrane
colostrum	gestation	proestrus
conformation	heat period	rodenticide
congenital	pedigree	styptic

HISTORY

The dog is probably the first animal domesticated by humans. The domestic dog originated about 12,000 to 14,000 years ago in Europe and Asia and was domesticated about 10,000 years ago. Many of our modern dogs probably descended directly from the wolf. These animals roamed in packs and probably gradually found it was easier to find food discarded by humans than to hunt for it. They probably became less wary of humans and gradually found their way into human encampments. Humans found that they could depend on the dog to warn of danger, and the dog depended more on humans for food and shelter. The modern dog evolved as a result of selective breeding for specific purposes and as a result of the environment in which they lived.

The ancestor of the entire dog family is believed to have been a civet-like animal, *Miacis*, that lived 40 or 50 million years ago. *See Figure 7–1. Miacis* was

Figure 7–1 *Miacis,* a small carnivorous mammal that lived in trees, is believed to be an early ancestor of several mammal species. *Reprinted with permission of T.F.H. Publications, Inc.*

a small animal with an elongated body and was probably arboreal, or at least seemed to have spent considerable time in the lush forests of its era. Next was the appearance, about 35 million years ago, of two apparently direct descendants of *Miacis.* These were *Daphaenus,* a large, heavy-boned animal with a long tail, and *Hesperocyon,* a small, slender animal who was the forerunner of the long-bodied, coyote-like "bear dogs" that eventually evolved into modern bears. *Hesperocyon* can be considered as the "grandfather" of the dog family. *Hesperocyon* retained the long body and short legs of the primitive carnivores, but unlike *Miacis,* *Hesperocyon* spent little time in the trees and began to hunt on the ground. Its claws were retractile, enabling it to walk on the ground and to climb trees.

From *Hesperocyon* there evolved two distinct dog types. The first, *Temnocyon,* was an important link in the evolutionary chain that led to the modern hunting dog of Africa, the Cape hunting dog. The second, *Cynodesmus,* is regarded as the ancestor of a large and diversified group of dogs that includes the modern Eurasian wolf and the American dogs, fox, and wolves.

The animal considered to be the "father" of modern dogs, *Tomarctus,* was directly descended from *Hesperocyon. Tomarctus* had a body built for speed and endurance as well as for leaping and differed little in appearance from the modern dog. This was a hunter, an animal geared for the chase, that brought down prey with slashing teeth. The modern dog still retains much of *Tomarctus'* anatomical features. *See Figure 7–2.*

Figure 7–2 *Tomarctus* is considered the ancestor of all canine-type mammals. *Reprinted with permission of T.F.H. Publications, Inc.*

As the evolutionary progress of the Canidae family continued, the progeny of *Tomarctus* developed into the modern dogs, wolves, fox, coyotes, fennecs, and jackals. At the same time, descendants of *Temnocyon* gradually emerged as the Cape hunting dog of today. *See Figure 7–3.*

Figure 7–3 This family tree traces the domestic dog's ancestors back about 40 million years to the tree-climbing mammal called *Miacis*. This tree also shows that domestic cats descended from this same tree-climbing mammal.

Descending in a direct line from *Tomarctus* are four major lines of dogs: the herd dogs, the hounds and terriers, the Northern and toy dogs, and the guard dogs. It is from these four lines, or groups, that modern dogs are descended.[1] Today, there are seven major groups, and more than 400 breeds that have been developed in the past 100 to 250 years.

In 2006 the American Kennel Club recognized 155 different breeds in seven groups. The seven groups are sporting dogs, hounds, working dogs, terriers, toys, herding dogs, and the nonsporting dogs. In addition to the seven groups, the American Kennel Club has established a miscellaneous class for breeds that have considerable interest but are not presently admitted to one of the regular groups.

GROUPS AND BREEDS

The Sporting Group

This group of dogs was developed to assist the hunter in the pursuit of game. There are 24 breeds recognized in this group.

The Pointer

The Pointer is a breed that gets its name from the stance taken in the presence of game. The pointer was developed about 200 years ago in England from pointing breeds imported from Spain and Portugal. This is a powerful and agile animal with speed and endurance developed for hunting. Its short hair makes it neat and clean, and it requires less attention than some of the other hunting breeds. The Pointer possesses an even temperament and is very affectionate, intelligent, and patient with children and is a congenial companion. *See Figure 7–4.*

Figure 7–4   Pointer. *Photo by Isabelle Francais.*

The German Shorthaired Pointer

The German Shorthaired Pointer originated in Germany around 1600 from crossing the German Bird Dog, the Spanish Pointer, the English Pointer, and several other track and trail dogs. This versatile dog can be used for hunting almost all types of game. It is a rustic-looking dog with outstanding temperament and extraordinary hunting capabilities in all types of weather and terrain. The tail is usually docked to 40 percent of its length to prevent injury during the hunt. *See Figure 7–5.*

Figure 7–5   German Shorthair. *Photo by Isabelle Francais.*

The German Wirehaired Pointer

The German Wirehaired Pointer was developed from crossing the German Pointer, the Wirehaired Griffon, the Pointer, the Bloodhound, and the Airedale. The hair is hard and bristly like steel wool and is weather-resistant. The breed is noted for its keen nose, intelligence, staunchness at point, and tough constitution. The dog responds well to its master but can be somewhat unfriendly to strangers. *See Figure 7–6.*

Figure 7–6 German Wirehair. *Photo by Isabelle Francais.*

The Labrador Retriever

The Labrador Retriever is native to Newfoundland. It is a short-coupled, strongly built, very active breed. It is an active water dog bred to assist the hunter in retrieving downed waterfowl, and it will readily plunge into the water in almost any weather. Its short, thick hair is water-resistant. The Labrador is noted for its excellent sense of smell and as being alert, friendly, and good-natured. Its excellent capabilities, fine temperament, and dependability have established it as one of the prime breeds for service as a guide dog for the blind or for search and rescue work. *See Figure 7–7.*

Figure 7–7 Labrador Retriever. *Photo by Isabelle Francais.*

The English Setter

The English Setter is believed to have been produced from crossing the Spanish Pointer, the large Water Spaniel, and the Springer Spaniel. The breed has been trained as a bird dog in England for more than 400 years. It has gained popularity in the United States because of its usefulness and beauty. The breed is noted for having a mild, sweet disposition and aristocratic appearance. Because of its temperament and lovable disposition, it makes an excellent companion. *See Figure 7–8.*

Figure 7–8 English Setter. *Photo by Isabelle Francais.*

The Irish Setter

The Irish Setter is a solid red breed. It is probably older than the English Setter, but they may have common ancestry. Many believe that the breed is a result of crossing the Spanish Pointer, the English Setter, and perhaps the Gordon Setter. The big red breed can be used for almost all kinds of game. They are noted for their speed and toughness in all types of terrain. Other qualities include being independent and bold. The breed can also be gentle, lovable, and loyal. They may learn a little slower, but once trained, they will retain their abilities for a long time. The beauty, gaiety, courage, and personality of this breed have made it an ideal show dog. When it is bred for show, its field ability has been sacrificed. *See Figure 7–9.*

Figure 7–9 Irish Setter. *Photo by Isabelle Francais.*

The Brittany

Until 1982, the Brittany was registered by the American Kennel Club as Brittany Spaniel. Although called a spaniel, its method of working game is more like a setter. The breed is named for the French province in which it originated, and it is believed to be a cross of native spaniels of Brittany and English pointing dogs. The Brittany has steadily gained in popularity because of its merits as a shooting dog. It is smaller and works closer in. It is noted for its excellent sense of smell and its desire to please. It is adapted to all types of terrain and is always active, enthusiastic, and untiring. Because of its jovial character, it makes an excellent companion dog. *See Figure 7–10.*

Figure 7–10   Brittany. *Photo by Isabelle Francais.*

The English Springer Spaniel

The Springer Spaniel is the foundation breed for all of the English hunting spaniels. During the 1600s, it was an ideal companion for a hunter in Europe. The breed is compact and strong and is noted for being cheerful, friendly, eager to please, courageous, and thorough. It is affectionate, good-natured, and sincere. It can be used in all types of terrain, especially one that is thick with brambles. The breed is hardy, long-legged, and more powerful and faster than other spaniels. *See Figure 7–11.*

Figure 7–11   Springer Spaniel. *Photo by Isabelle Francais.*

The American Cocker Spaniel

The American Cocker Spaniel dates back to 1368. The Cocker Spaniel has retained its early classification as a sporting dog and is recognized as the smallest member of the sporting dog family. The breed gets its name from its proficiency at hunting woodcock. The breed is a cheerful, hardy, sporting dog with an elegant appearance. The Cocker Spaniel makes an excellent companion dog and is a lover of the home and family. This is a soft-hearted, sensitive dog that must be trained with a gentle hand. *See Figure 7–12.*

Figure 7–12   American Cocker Spaniel. *Photo by Isabelle Francais.*

Other breeds of sporting dogs include the following:

American Water Spaniel	Irish Water Spaniel
Chesapeake Bay Retriever	Nova Scotia Duck Tolling Retriever
Clumber Spaniel	Spinone Italino
Curly-Coated Retriever	Sussex Spaniel
English Cocker Spaniel	Vizsla
Field Spaniel	Weimaraner
Flat-Coated Retriever	Welsh Springer Spaniel
Golden Retriever	Wirehaired Pointing Griffon
Gordon Setter	

		Average Weight in Pounds		Average Height at Shoulder in Inches	
Breed	**Color**	**Females**	**Males**	**Females**	**Males**
Pointer	Liver, lemon, black, orange; either in combination with white or solid-colored.	45–65	55–75	23–26	25–28
German Shorthaired Pointer	Solid liver or any other combination of liver and white, such as liver and white ticked, liver spotted and white ticked, or liver roan.	45–60	55–70	21–23	23–25
German Wirehaired Pointer	Liver and white, usually either liver and white spotted, liver roan, liver and white spotted with ticking and roaning, or sometimes solid liver.	45–60*	50–65*	22–24	24–26
Labrador Retriever	Black, yellow, or chocolate.	55–70	60–75	21½–23½	22½–24½
English Setter	Black, white, and tan; black and white; blue belton; lemon and white; lemon belton; orange and white; orange belton; liver and white; liver belton; and solid white.	50–60*	55–70*	24	25
Irish Setter	Mahogany or rich chestnut red with no trace of black.	60	70	25	27
The Brittany	Orange and white or liver and white in either clear or roan patterns. Some ticking desirable.	30–40	30–40	17½–20½	17½–20½
English Springer Spaniel	Black or liver with white markings or predominantly white with black or liver markings; tricolor: black and white or liver and white with tan markings.	49–55	49–55	19	20
American Cocker Spaniel	Many colors exist in solid and parti-colors.	26–32	28–34	15–16	16–17

TABLE 7–1 Color, Size, and Weight Comparison of the Sporting Breeds

*These figures are not given in the Official Standards and are thus approximate.

Information obtained from *"The Complete Dog Book,"* 17th edition, the official publication of the American Kennel Club, Howell Book House, 1986, Macmillan Publishing Company.

See Table 7–1 for a color, size, and weight comparison of the sporting breeds.

The Hound Group

This group of dogs is of two basic types: one type hunts by scent and the other by sight.

The Greyhounds, Afghans, Borzoi, Saluki, and Whippet are among the approximately 26 breeds of sight hounds; all others described or listed in this book hunt by scent.

The Afghan Hound

The Afghan Hound is a very old breed, and little is known about its origin and history. It is believed that the Afghan Hound existed thousands of years ago in Egypt and is native to the Sinai. The breed was discovered by the Western world in Afghanistan during the early 1800s. Originally, the breed was used for hunting. They would hunt in packs, pursuing game by sight and followed by huntsman on horseback. As a hunting hound, they were noted for their speed and their ability to travel rough terrain swiftly and sure-footedly. Today, the breed is popular as a show dog and companion. Because of their long, silky hair coat, Afghans require regular care and grooming. They are sensitive, intelligent, somewhat aloof, and dignified. Afghans respond to kindness and attention. *See Figure 7–13.*

Figure 7–13   Afghan Hound. *Photo by Isabelle Francais.*

The Basset Hound

The Basset Hound is an old breed developed in France. It is thought that the friars of the French Abbey of St. Hubert were instrumental in the development of this breed. Because of its excellent sense of smell, it is used for trailing rabbits, deer, raccoon, fox, and other game. It is a short-legged dog, heavier in bone (considering its size) than any other breed of dog. The Basset is gentle, affectionate, good with children, and devoted. In the field, it is capable of great endurance and determination. It may be stubborn at times and difficult to house-break. *See Figure 7–14.*

Figure 7–14   Basset Hound. *Photo by Isabelle Francais.*

The Beagle

The Beagle is one of the oldest breeds, but its actual origin is unknown. The modern-day Beagle was developed in the 1700s. The Beagle is noted as a rabbit hound and hunts in packs or singly. This little hound is cheerful, friendly, lively, and merry and makes an excellent family pet and companion. *See Figure 7–15.*

Figure 7–15   Beagle. *Photo by Isabelle Francais.*

The Black and Tan Coonhound

The Black and Tan Coonhound descended from the Talbot hound known in England during the eleventh century and was further developed through the bloodhound and the foxhound. The breed was developed in the United States and was officially recognized in 1945. It was specifically developed for hunting raccoons, but it can be used for hunting deer, bear, mountain lion, fox, and opossum. The breed is a passionate hunter and will work a scent with consummate skill and determination. The breed is intelligent, alert, friendly, eager, and aggressive. It responds well to its owner. *See Figure 7–16.*

Figure 7–16   Black and Tan Coonhound. *Photo by Isabelle Francais.*

The Bloodhound

The Bloodhound dates back to the third century AD. They were developed by the monks of St. Hubert in Belgium and were later brought to England by the Normans. The purebred Bloodhound is one of the most docile of all breeds. Its sense of smell is highly developed, and they have been used for tracking people and finding lost children, buried miners, and earthquake victims. Bloodhounds have been used by the police to track criminals, but unlike other breeds of police dogs, bloodhounds will not attack the criminal they are tracking. They are capable of following a scent that is several hours old. The Bloodhound is extremely affectionate, somewhat shy, and equally sensitive to kindness and correction by its owner. They are very good with children. *See Figure 7–17.*

Figure 7–17 Bloodhound. *Photo by Isabelle Francais.*

The Dachshund

The Dachshund is one of the older breeds of dog and dates back to the fifteenth century. In early Germany, it was known as the Teckel. The breed was perfected during the 1800s in Germany, where it was used for hunting badger. The dogs would trail the badger and follow it into its den. *Dachs* in German means *badger,* and *hund* in German means *dog,* thus the badgerdog or Dachshund. There are three varieties of the Dachshund: the shorthaired, the wirehaired, and the longhaired. Within the three varieties there are three sizes: normal, miniature, and toy. This is a short-legged, long-bodied, and muscular dog. It is considered clever, lively, courageous, bold, and confident. The Dachshund is good with children and is primarily used as a companion dog. It is considered a "barker" and makes a good watchdog. *See Figure 7–18.*

Figure 7–18 Shorthaired Dachshund. *Photo by Isabelle Francais.*

The American Foxhound

The American Foxhound descended from English hounds brought to America in 1650 by Robert Brooke. They were bred a century later to hounds imported from England in 1742 by Thomas Walker and to French hounds sent as a gift by Lafayette to George Washington. Although it is similar to the English hounds, the American Foxhound is smaller, possesses a keener sense of smell, and will work harder and faster. During the 1600s, these dogs were used to track Native Americans and later were used for hunting all animals. The breed is bold and aggressive in the field. They have been used primarily for hunting in packs and prefer the company of other dogs. They have not been used very much as house pets, although they are considered amiable and affectionate when not in the field. *See Figure 7–19.*

Figure 7–19 American Foxhound. *Photo by Isabelle Francais.*

The Greyhound

The Greyhound is a very ancient breed that traces its history back to the Tomb of Amten in the Valley of the Nile, which dates between 2900 and 2751 BC. The Greyhound has always had a cultural and aristocratic background and was a favorite of royalty in Egypt. The breed has been used for hunting practically all kinds of game, including deer, fox, wild boar, and rabbit. They hunted in packs and would catch up to their quarry and pull them down. Its greatest gift is speed, and it is primarily used today for racing, where it can reach speeds up to 40 miles per hour. The animal is also valued as a show dog. The Greyhound is very intelligent, sensitive, courageous, and loyal. Although they are great runners, they are calm and generally do well even in small homes. This breed can have a high drive for prey and may kill small animals such as cats, geese, ducks, and chickens. *See Figure 7–20.*

Figure 7–20   Greyhound. *Photo by Isabelle Francais.*

The Norwegian Elkhound

The Norwegian Elkhound traces its history back to the early Vikings in 4000 to 5000 BC, when it was used for hunting elk. Over its history, it has been used to defend flocks from wolves and bears, to protect man from wild animals, and to hunt lynx, mountain lion, and raccoon. The animal is short and compact, and is noted for its stamina during the hunt and its eagerness to please. It is an exceedingly versatile breed also noted for its loyalty, affection, and intelligence. *See Figure 7–21.*

Figure 7–21   Norwegian Elkhound. *Photo by Isabelle Francais.*

Other breeds in the hound group include the following:

Basenji	Saluki
Petit Basset Griffon Vendeen	Ibizan Hound
Borzoi	Scottish Deerhound
Pharaoh Hound	Irish Wolfhound
English Foxhound	Whippet
Rhodesian Ridgeback	Otterhound
Harrier	

See Table 7–2 for a color, size, and weight comparison of the hound group.

The Terrier Group

This group is divided into two subgroups: the long-legged, larger breeds such as the Fox Terriers, and the short-legged, small breeds such as the Dandie Dinmont.

		Average Weight in Pounds		Average Height at Shoulder in Inches	
Breed	**Color**	**Females**	**Males**	**Females**	**Males**
Afghan Hound	All colors.	50	60	24–26	26–28
Basset Hound	Usually white with chestnut or sand-colored markings.	40–51	40–51	Not over 14	14
Beagle	White, black, orange, or typical hound tricolor.	18–30	18–30	13–15	13–15
Black and Tan Coonhound	Black coat with tan markings on the muzzle, limbs, and chest.	60–70*	65–80*	23–25	25–27
Bloodhound	Black and tan, red and tan, and tawny.	80–100	90–100	23–25	25–27
Dachshund					
(normal)	Solid colors of red, tan, red-yellow, and yellow. Bi-colors of black, chocolate, gray, and white. There are also speckled, streaked, and harlequin.	20	20	5–9 depending on the variety	
(miniature)		9	9		
(toy)		8	8		
American Foxhound	Any color.	35–40*	40–45*	21–24	22–25
Greyhound	All colors.	60–65*	65–70*	20–25*	23–27*
Norwegian Elkhound	Gray with darker gray saddle. The muzzle, ears, and tail tip are black.	48	55	19½	20½

TABLE 7–2 Color, Size, and Weight Comparison of the Hound Group

*These figures are not given in the Official Standards and are thus approximate.

Information obtained from *"The Complete Dog Book,"* 17th edition, the official publication of the American Kennel Club, Howell Book House, 1986, Macmillan Publishing Company.

The Airedale Terrier

The Airedale Terrier was developed in the county of York in England and takes its name from the Aire River located there. It descended from crossing the Old English Terrier or Broken Hair Terriers with the Otterhound. The breed was originally used for hunting bears, wolves, fox, badgers, weasels, and other game. The breed is very docile, obedient, and lively, and it gets along well with children. It needs to feel loved and respected. During wartime, the dog was used to carry messages and was used as a police dog and bodyguard. *See Figure 7–22.*

Figure 7–22 Airedale Terrier. *Photo by Isabelle Francais.*

The Bedlington Terrier

The Bedlington Terrier takes its name from the small mining village of Bedlington, in the county of Northumberland in England, where it served as a watchdog. *See Figure 7–23.*

Figure 7–23 Bedlington Terrier. *Photo by Isabelle Francais.*

The Border Terrier

The Border Terrier was developed by the farmers and shepherds in the border area between Scotland and England for the purpose of helping drive out the fox, which played havoc on sheep flocks. They needed a breed of dog with sufficient leg to follow a horse, but short enough to allow it to follow the fox into the den. The dogs also needed to be active, strong, and tireless and to have weather-resistant coats to withstand the prolonged exposure to drenching rains and mist in the area. The Border Terrier was thus developed and met all the necessary criteria. Because of its temperament and adaptability, it makes an excellent companion dog and is especially fond of children. *See Figure 7–24.*

Figure 7–24   Border Terrier. *Photo by Isabelle Francais.*

The Bull Terrier

The Bull Terrier dates back to the 1830s, when dogfights and bullfights were permitted and well attended. It is a cross of the Bulldog, the Old English Terrier, and the Spanish Pointer. This breed is noted for its great strength, agility, and courage. It can be exceedingly friendly and thrives on affection, but if provoked, it will readily fight to protect itself and its master. Over the years, the breed has been used as a guardian of livestock, a mouse and rat hunter, a companion dog, a guard dog, and a watchdog. There are two varieties of the breed: white and colored. *See Figure 7–25.*

Figure 7–25   Bull Terrier. *Photo by Isabelle Francais.*

The Dandie Dinmont Terrier

The Dandie Dinmont Terrier breed dates back to 1700 and is a cross of the Scottish and Skye Terriers. The Dandie Dinmonts were developed to hunt otters, badgers, weasels, martens, and skunks. Sir Walter Scott made the breed famous in his novel *Guy Mannering*, published in 1814. A character in the book, a farmer named Dandie Dinmont, owned six of the small dogs, and since the time of the novel, the breed has been known as "Dandie Dinmont Terriers." The breed is noted for its excellent hunting qualities and has retained those qualities. The Dandie Dinmont is intelligent, fond of children, and makes an excellent watchdog and companion. *See Figure 7–26.*

Figure 7–26   Dandie Dinmont Terrier. *Photo by Isabelle Francais.*

The Fox Terrier

The Fox Terrier has developed over many years into two distinct varieties: the Smooth and the Wire. Wires were crossed with the Smooths in the early days, but this practice has been almost universally discontinued. In 1984, the American Kennel Club approved separate breed standards for the Smooth Fox Terrier and the Wire Fox Terrier. It is believed that the Wire was developed from the old rough-coated black and tan working terriers and that the ancestors of the Smooth were the smooth-coated black and tans (the Bull Terriers, the Greyhound, and the Beagle). Fox Terriers are one of the best known and most widely distributed of purebred dogs. The breed is English in origin and dates back to the 1700s. Like many other terriers, the breed was developed to hunt den animals and was noted for its courage. Today, they are considered affectionate and trainable and make excellent companion dogs. *See Figure 7–27.*

Figure 7–27   Smooth-Coated Fox Terrier. *Photo by Isabelle Francais.*

The Miniature Schnauzer

The Miniature Schnauzer is of German origin and dates back to the fifteenth century. The Miniature Schnauzer was developed from the Standard Schnauzer through crossings with the Affenpinschers and Poodles. The Miniature Schnauzer was recognized as a separate breed in 1899. As with the other breeds in the terrier group, it was developed to hunt badgers, fox, weasels, and other vermin. The breed is characterized by its stocky build, wiry coat, and abundant whiskers and leg furnishings. The breed is hardy, healthy, and intelligent and is fond of children. It makes a charming and attractive companion. It is not as large or active as its larger cousin (the Giant Schnauzer) and makes a good apartment dog and family pet. *See Figure 7–28.*

Figure 7–28   Miniature Schnauzer. *Photo by Isabelle Francais.*

The Skye Terrier

The Skye Terrier traces its origin back to a Spanish shipwreck off the island of Skye in the Scottish Hebrides. Among the survivors were Maltese dogs, which mated with the native terriers and produced this small breed of terrier. Originally used for hunting in dens, the breed became a favorite at kennel shows because of its long, flowing, beautiful hair coat. The breed is characterized as being considerate, well mannered, and affectionate. It is friendly and lively with those it knows, but it is shy and cautious of strangers. *See Figure 7–29.*

Figure 7–29   Skye Terrier. *Photo by Isabelle Francais.*

The Welsh Terrier

The Welsh Terrier is a very old breed and in olden times was more commonly known as the Old English Terrier, or Black and Tan Wirehaired Terrier. The breed was recognized as Welsh Terriers in 1884–1885 and was developed to hunt in dens and work in packs with hounds. Today, it is considered an apartment dog and watchdog. It is friendly and outgoing with people and other dogs, showing spirit and courage. *See Figure 7–30.*

Figure 7–30   Welsh Terrier. *Photo by Isabelle Francais.*

		Average Weight in Pounds		Average Height at Shoulder in Inches	
Breed	**Color**	**Females**	**Males**	**Females**	**Males**
Airedale Terrier	The head and ears are tan, with the ears being a darker shade than the rest. The sides and upper parts of the body should be black or dark grizzle.	42	44	22	23
Bedlington Terrier	Two distinct colors, liver and blue.	15–21*	17–23	15½	16½
Border Terrier	Red, grizzle and tan, blue and tan, or wheaten.	11–14*	13–15	11–12*	12–13*
Bull Terrier (two color types)	The White: all white, or white with black markings on the head. Colored: any color but white, with brindle the preferred color.	17–20*	18–21*	30–50*	40–60*
Dandie Dinmont Terrier	Two distinct colors: Pepper: blue-gray to light silver with light tan or silver points and a very light gray or white topknot.	18–24	18–24	8–11	8–11
	Mustard: dark ocher color to cream with white points and topknot.				
Fox Terrier	White should predominate; brindle, red, or liver are objectionable.	16	18	14 ½	15½
Schnauzer	Salt and pepper, black and silver, and solid black.	13–16*	15–19*	12–14	12–14
Skye Terrier	Coat is one overall color at the skin, but may be of varying shades of the same color in the full coat, which may be black, blue, dark or light gray, silver platinum, fawn, or cream.	23	25	9½	10
Welsh Terrier	The jacket is black spreading up onto the neck, down onto the tail, and into the upper thighs. The legs, quarters, and head are clear tan. The tan is deep reddish brown.	20	20	14	15

TABLE 7–3 Color, Size, and Weight Comparison of the Terrier Group

*These figures are not in the Official Standards and are thus approximate.

Information obtained from "*The Complete Dog Book,*" 17th edition, the official publication of the American Kennel Club, Howell Book House, 1986, Macmillan Publishing Company.

Other breeds in the terrier group include the following:

American Staffordshire Terrier	Norfolk Terrier
Australian Terrier	Norwich Terrier
Cairn Terrier	Parson Russell Terrier
Glenn of Imaal Terrier	Scottish Terrier
Irish Terrier	Sealyham Terrier
Kerry Blue Terrier	Soft-Coated Wheaten Terrier
Lakeland Terrier	Staffordshire Bull Terrier
Manchester Terrier	West Highland White Terrier
Miniature Bull Terrier	

See Table 7–3 for a color, size, and weight comparison of the terrier group.

The Working Dog Group

This group of dogs has been developed to labor or work for humans. They may serve as guard dogs, sled dogs, police dogs, rescue dogs, and messenger dogs.

The Alaskan Malamute

The Alaskan Malamute is one of the oldest Artic sled dogs and was named for the native Innuit tribe called Mahlemuts (now spelled *Malamute*), who settled along the shores of Kolzbue Sound in the upper western part of Alaska. The origin of these dogs has never been ascertained, but we know they have been in Alaska for generations. When Alaska became settled by non-natives, this Artic breed was bred with outside strains; with the increase in the popularity of sled racing as a sport, there was renewed interest in developing a pure strain of the Alaskan Malamute. The clean, odorless Malamutes are loyal, intelligent, and affectionate toward their owner, although they are fighters among dogs. Therefore, firm handling and training are necessary. The Malamute needs a lot of physical exercise and mental stimulation. Without it they become bored and restless, which can result in destructive behavior. The Malamute can be loud and rowdy. *See Figure 7–31.*

Figure 7–31   Alaskan Malamute. *Photo by Isabelle Francais.*

The Boxer

The Boxer is one of the many descendants of the old fighting dogs of the high valleys of Tibet. The breed showed up in many parts of Europe during the sixteenth and seventeenth centuries and reached its perfection in Germany within the past 100 years. The Boxer carries Bulldog, Terrier, and Bullenbeisser Mastiff blood. The breed was developed for two purposes: to fight bulls and to hunt bear. The breed is considered good-natured, loyal, playful, tolerant, and somewhat suspicious of strangers. It is a short-lived dog and often does not reach 10 years of age. It is subject to rheumatism and must be exercised regularly. The Boxer is a watchdog, bodyguard, and guide dog for the blind and is used in police work. *See Figure 7–32.*

Figure 7–32   Boxer. *Photo by Isabelle Francais.*

The Doberman Pinscher

The Doberman Pinscher breed originated in Apolda, in Thuringia, Germany, around 1890 and takes its name from Louis Dobermann. The breed was developed by crossings among Great Danes, German Shepherds, Rottweilers, and Pinschers, with possibly some blood of the Beauceron and the English greyhound. At first, the Doberman was used as a guard dog and watchdog. Because of its intelligence and excellent nose, it has been used as a police dog, war dog, and hunting dog. The males and females have different temperaments. The females are calm, responsive, and affectionate with the family, but wary of strangers. The males are extremely intelligent, but impetuous, often aggressive, and must be trained by a dominant owner who is a good disciplinarian. The life expectancy of the Doberman is 8 to 12 years. *See Figure 7–33.*

Figure 7–33   Doberman Pinscher. *Photo by Isabelle Francais.*

The Great Dane

Drawings of dogs much like the Great Dane date back to Egyptian monuments of 3000 BC; early Chinese literature of 1121 BC describes a dog resembling the Great Dane. It is believed that the Mastiff breeds of Asia and the Irish Wolfhounds are the principal ancestors of this breed. Its development into the modern type began in Germany during the 1800s. This giant dog was used for hunting wild boar and bear. During its history, it has seen service as a battle dog, a hunting dog, a cart dog, a watchdog, and a bodyguard. The Great Dane is spirited, courageous, friendly, and dependable. Today, it is used primarily as a watchdog and companion dog. *See Figure 7–34.*

Figure 7–34   Great Dane. *Photo by Isabelle Francais.*

The Great Pyrenees

The remains of this breed have been found in the fossil deposits of the Bronze Age, which dates between 1800 and 1000 BC in Europe. It is believed to originate from Central Asia or Siberia. The Great Pyrenees were developed in the Pyrenees Mountains, where they served as guardians to the shepherd's flocks. Their excellent sense of smell and sight proved invaluable to the shepherd, and because of their size and strength, they were almost unbeatable as guardian of the flock. Also known as the Great Dog of the Mountains, the breed is docile and makes a faithful, loving, and affectionate house pet. It does require considerable space, exercise, and grooming. Besides a guardian of sheep, the breed has been used as a guard dog for people and property, a guide and pack dog in ski areas, a rescue dog, and a messenger dog during World War I. *See Figure 7–35.*

Figure 7–35   Great Pyrenees. *Photo by Isabelle Francais.*

The Standard Schnauzer

The Standard Schnauzer is a German breed that dates back to the fifteenth and sixteenth centuries. It was developed from crossings of a black German poodle and a gray wolf spitz with wirehaired Pinscher stock. The dogs were originally used as rat catchers, guard dogs, and yard dogs. Before World War I, they were used in Germany to guard the carts of farm produce in the marketplace while farmers rested themselves and their teams at the inns. Today, the breed is used as a watchdog and a bodyguard as well as a family dog. It is feisty, intelligent, bold, and courageous. The breed is noted for being long-lived, clean, and almost odor-free. *See Figure 7–36.*

Figure 7–36   Standard Schnauzer. *Photo by Isabelle Francais.*

The Rottweiler

The Rottweiler is a descendant of the Drover dogs of Rome, which were a Mastiff type. Rottweilers were bred and developed in the German town of Rottweil, which is how they got their name. Drover dogs were brought into the area by Roman armies around 74 AD. The breed was used as a herding dog to drive cattle to and from the markets. This massive, powerful animal is considered peaceful, obedient, courageous, and intelligent. They are known to be very protective of their owners. Today, the breed is used as a herd dog, police dog, watchdog, and bodyguard. They also make excellent companion dogs. *See Figure 7–37.*

Figure 7–37   Rottweiler. *Photo by Isabelle Francais.*

The Saint Bernard

The Saint Bernard probably descended from large Mastiff dogs that were brought to the area that is now Switzerland by the Roman armies during the first two centuries AD. These dogs were crossed with native dogs that existed in this region and were used for a variety of herding, guarding, and drafting duties. The breed appeared around 1660–1670 in the Hospice of St. Bernard de Menthon near the Great St. Bernard Pass. The monks of the Hospice used the dogs for companionship and took them on trips of mercy, where they discovered the dogs were excellent pathfinders in the drifting snow, and the dogs' excellent sense of smell made them invaluable in locating helpless persons overcome during storms. In 1830, the breed was crossed with the Newfoundland, producing the first long-haired Saint Bernards. The long-haired variety proved unsuited for the cold weather condition because snow clung to their fur, obstructing the dog's movements. The Saint Bernard is obedient, peaceful, quiet, extremely loyal, and very gentle and loves the company of humans. Some lines may show aggressiveness, stubbornness, and dominance problems. Saint Bernards have a short life expectancy because of numerous health problems. *See Figure 7–38.*

Figure 7–38   Saint Bernard. *Photo by Isabelle Francais.*

The Samoyed Breed

The Samoyed breed was developed by and takes its name from the Samoyed tribe found in Siberia. The breed was used to herd reindeer and, as sled dogs, is noted for being able to pull heavy loads over long distances. It is a strong, dynamic, active dog that has proved itself as an excellent companion and watchdog. It is very gentle, kind, unusually intelligent, and affectionate and loves children. The Samoyed is considered by many the most beautiful breed in existence. Its only fault would be its frequent barking. *See Figure 7–39.*

Figure 7–39   Samoyed. *Photo by Isabelle Francais.*

The Siberian Husky

The Siberian Husky breed was originated by the Chukchi people of Northeastern Asia as an endurance sled dog. The breed was brought to Alaska in 1909 to compete in the All Alaska Sweepstakes Race. A team of Siberian Huskies won the race in 1910 and most of the racing titles in Alaska over the next decade as well. The breed is naturally friendly and gentle in temperament. The Husky possesses an independent nature at times, and although very alert, it lacks the aggressive or protective tendencies of a watchdog. It may be stubborn, easily bored, cranky, or ill-humored around children. *See Figure 7–40.*

Figure 7–40   Siberian Husky. *Photo by Isabelle Francais.*

The Komondor

The Komondor is a heavily coated dog that probably descended from the Aftscharka, which the Huns found on the southern steppes of Russia and brought with them into the area that is now Hungary. The breed is characterized by its great size, strength, courageous demeanor, and unusual, heavy white coat. The breed makes an excellent guard dog and will readily fight wolves, bears, and other larger predators that would attack the flock. It is obedient, loyal, and devoted to its owner. *See Figure 7–41.*

Figure 7–41   Komondor. *Photo by Isabelle Francais.*

Other breeds in the working dog group include the following:

Akita	Great Swiss Mountain Dog
Anatolian Shepherd Dog	Kuvasz
Bernese Mountain Dog	Mastiff
Black Russian Terrier	Neopolitan Mastiff
Bullmastiff	Newfoundland
German Pinscher	Portuguese Water Dog
Giant Schnauzer	

See Table 7–4 for a color, size, and weight comparison of the working dog group.

The Herding Dog Group

This group was developed to aid the livestock herder with various species of livestock.

TABLE 7–4   Color, Size, and Weight Comparison of the Working Dog Group

Breed	Color	Average Weight in Pounds		Average Height at Shoulder in Inches	
		Females	Males	Females	Males
Alaskan Malamute	The usual colors range from light gray through the intermediate shadings to black, always with white underbodies, parts of the legs, feet, and part of mask markings.	75	85	23	25
Boxer	Fawn and brindle. Fawn in various shades of light tan to stag red or mahogany. White markings may appear on the chest and on the face.	50–65*	60–70*	21–23½	22½–25
Doberman Pinscher	Black, red, blue, and fawn. Rust markings may appear above each eye and on the muzzle, throat and forechest; on all legs and feet; and below the tail. The nose will be solid black on black dogs, dark brown on red ones, dark gray on blue ones, and dark tan on fawns.	60–70*	65–80*	24–26	26–28
Great Dane	Five basic types: brindle, fawn, blue, black, and harlequin.	110–115*	120–150*	28+	30+
Great Pyrenees	White.	90–115	100–125	25–29	27–32
Standard Schnauzer	Pepper and salt or pure black.	60–70*	70–80*	17½–18½	18½–19½
Rottweiler	Always black with rust to mahogany markings.	80–90*	90–110*	22–25	24–27
Saint Bernard	Red with white markings or white with red markings.	90–120*	110–160*	27½+	27½+
Samoyed	Pure white.	50–60*	55–65*	19–21	21–23½
Siberian Husky	All colors from black to pure white. A variety of markings on the head are common.	35–50	45–60	20–22	21–23½
Komondor	White.	80–110*	90–125*	23½+	25½+

*These figures are not given in the Official Standards and are thus approximate.

Information obtained from *"The Complete Dog Book,"* 17th edition, the official publication of the American Kennel Club, Howell Book House, 1986, Macmillan Publishing Company.

The Collie

The Collie comes in two varieties: the more familiar rough-coated or long-haired Collie, and the smooth-coated Collie. The exact origin of the Collie is not known, but both varieties existed long ago as herding dogs of Scotland and Northern England. Because sheepherding is one of the world's oldest occupations, the Collies date far back in history. Being no longer in demand as a herder, today's Collie serves as a devoted family dog. The Collie is kind, sensitive, intelligent, and loyal. It is very protective of its owner and family, but because of its sensitive nature, careful, gentle training is required. The Collie is wary of strangers and may also be lazy or stubborn on occasion. The breed is very versatile, being used as rescue dogs, as guide dogs for the blind, and in police work. Keeping burrs and knots out of the beautiful coat of the longhaired Collie requires frequent brushing. *See Figure 7–42.*

Figure 7–42   Collie. *Photo by Isabelle Francais.*

The German Shepherd

Several theories exist regarding the origin of this breed. Some believe that it is a result of crosses between various herding dogs and farm dogs existing in Germany, whereas others believe it is the result of matings between herding dogs, farm dogs, and wolves. The German Shepherd is courageous, obedient, steady, and loyal. It is affectionate with its owner and family, but wary of strangers. Originally used for sheepherding, the German Shepherd has, over its history, been used in time of war to carry messages, and as a rescue dog, police dog, guard dog, and guide dog for the blind. *See Figure 7–43.*

Figure 7–43   German Shepherd. *Photo by Isabelle Francais.*

The Old English Sheepdog

The Old English Sheepdog originated early in the nineteenth century in the counties of Devon, Somerset, and the Duchy of Cornwall in England. Its exact ancestry is not known, but some believe its ancestors were the Scotch Bearded Collie, whereas others believe it descended from a long-haired Russian breed called the Owtchar. It is a steady, muscular dog with a characteristic gait much like that of a bear. It is an outstanding herd dog and has been used as a guard dog, sled dog, and retriever. Its friendly, affectionate disposition makes it an excellent companion dog and family pet. Grooming and coat care is important to keep the coat in good condition. Matting can become a problem if the animal does not receive regular grooming. Some Old English Sheepdog owners have reported stubbornness and aggressive behavior. Firm handling and training is needed. *See Figure 7–44.*

Figure 7–44   Old English Sheepdog. *Photo by Isabelle Francais.*

The Shetland Sheepdog

The Shetland Sheepdog is a breed that probably descended from the Collies that were brought to the Scottish island of Shetland. These Collies developed in miniature because of the environment and from being crossed with small Border Collies and other small breeds native to the islands. The breed resembles the larger, rough Collie in miniature. It is noted for its devoted, docile nature and its keen and seemingly human intelligence and understanding. It is considered very lively, perky, and affectionate. The Shetland Sheepdog makes an excellent family pet and companion. *See Figure 7–45.*

Figure 7–45   Shetland Sheepdog. *Photo by Isabelle Francais.*

The Welsh Corgi

There are two types of the Welsh Corgi: the Cardigan and the Pembroke. The ancestors of the Cardigan came into the hill country of Wales that is now known as Cardiganshire, in the British Isles around 1200 BC. The dog is a member of the same family that produced the Dachshund. The ancestors of the Pembroke came across the English Channel with Flemish weavers to Pembrokeshire in Wales in 1107. This dog sprang from the same family that includes the Keeshond, the Pomeranian, the Samoyed, the Chow Chow, the Norwegian Elkhound, and the Finnish Spitz. It has little or nothing of the Dachshund characteristics. *See Figure 7–46.*

Figure 7–46   Welsh Corgi, Cardigan. *Photo by Isabelle Francais.*

In comparing the two types, the Pembroke is shorter of body; the legs are straighter and lighter boned, and the coat is of finer texture. Cardigan ears are rounded, whereas the Pembroke's are pointed at the tip and stand erect. The Cardigan has a long tail, and the Pembroke a short one. The Pembroke is more restless and more easily excited. Both types were used to herd cattle, and they possessed unusual speed for having such short legs, as well as agility, balance, and endurance. The Corgi is noted for its intelligence and for being alert, lively, affectionate, friendly, and easy to train. *See Figure 7–47.*

Figure 7–47   Welsh Corgi, Pembroke. *Photo by Isabelle Francais.*

Other breeds in the herding dog group include the following:

Australian Cattle Dog	Border Collie
Australian Shepherd	Bouvier des Flandres
Bearded Collie	Briard
Belgian Malinois	Canaan Dog
Belgian Sheepdog	Polish Lowland Sheepdog
Belgian Tervuren	Puli

See Table 7–5 for a color, size, and weight comparison of the herding dog group.

The Toy Dog Group

Certain dogs are called toy dogs because of their size; some are as small as 1.5 pounds. These dogs are very alert and popular as house pets and companions.

TABLE 7–5 Color, Size, and Weight Comparison of the Herding Dog Group

Breed	Color	Average Weight in Pounds		Average Height at Shoulder in Inches	
		Females	Males	Females	Males
Collie	Four recognized colors: sable and white, tricolored, blue merle, and white.	50–65	60–75	22–24	24–26
German Shepherd	Several colors exist. Black iron gray, ash gray, either uniform or regular shading of brown, yellow, or light gray. Nose is black.	60–75*	70–85*	22–24	24–26
Old English Sheepdog	Any shade of gray, grizzle, blue, or blue merle with or without white markings or in reverse.	70–100*	80–110*	20+	22+
Shetland Sheepdog	Black, blue merle, and sable; marked with varying amounts of white and/or tan.	18–25*	21–28*	13–16	13–16
Welsh Corgi	Cardigan: All shades of red, sable, and brindle. Blue merle with or without tan or brindle points. White flashings are usual on the neck, chest, legs, muzzle, underparts, tip of the tail, and as a blaze on the head.	25–34	30–38	10½–12½	10½–12½
	Pembroke: Outer coat colors of red, sable, fawn, black, and tan, with or without white markings.	28	30	10–12	10–12

*These figures are not given in the Official Standards and thus are approximate.

Information obtained from *"The Complete Dog Book,"* 17th edition, the official publication of the American Kennel Club, Howell Book House, 1986, Macmillan Publishing Company.

The Chihuahua

The Chihuahua is recognized as the smallest breed in the world. The Toltecs, a group of natives that date back to the ninth century AD in what is now Mexico, had a breed of dog called Techichi. This small breed indigenous to Central America is believed to be one of the ancestors of the Chihuahua. It is believed that the Techichi was crossed with a small hairless dog brought from Asia or Alaska over the land bridge where the Bering Strait now is. This hairless dog, similar to the one found in China, was responsible for the reduction in size. The name comes from the state of Chihuahua in Mexico. Because it dates back to the ninth century AD, it is considered the oldest breed of dog on the American continent. There are two varieties: a short-haired variety and a wavy-haired variety. The Chihuahua is a companion dog and is noted for being courageous, energetic, lively, alert, proud, affectionate, loyal, and intelligent. It is clannish, recognizing and preferring its own kind, and as a rule not liking dogs of other breeds. *See Figure 7–48.*

Figure 7–48 Long-haired Chihuahua. *Photo by Isabelle Francais.*

The Italian Greyhound

The Italian Greyhound is a very old breed, descending from the Egyptian Greyhound of over 6,000 years ago. This is the smallest of the family of dogs that hunt by sight. It is believed to have been brought into Europe by the Phoenician traders and developed in Italy, thus the Italian Greyhound name. There is some disagreement as to whether the breed was developed for hunting or for a pet and companion. The breed is very affectionate and thrives best when this affection is returned. The coat is short and smooth and requires very little grooming. The Italian Greyhound is odorless and sheds very little. The breed is sensitive, alert, intelligent, and playful. It may be timid, meek, and wary of strangers. *See Figure 7–49.*

Figure 7–49 Italian Greyhound. *Photo by Isabelle Francais.*

The Manchester Terrier

The Manchester Terrier breed was developed in the Manchester district of England during the nineteenth century by crossings between the Black and Tan Terrier, the Whippet, the Greyhound, and the Italian Greyhound. Its nickname is the "rat terrier" because it has long been recognized for its ability to hunt rats and other rodents. It is considered intelligent, active, and alert. Today, it is primarily a small companion dog. In 1959, two varieties were recognized: the toy and the standard. *See Figure 7–50.*

Figure 7–50 Manchester Terrier. *Photo by Isabelle Francais.*

The Pekingese

In ancient times, the Pekingese was held sacred in China, the land of its origin. The exact date of origin is not known, but its existence traces back to the Tang dynasty of the eighth century. The very old strains (held only by the imperial family) were kept pure, and the theft of one of these sacred dogs was punishable by death. Introduction of the Pekingese into the Western world occurred as a result of the looting of the Imperial Palace at Peking by the British in 1860. The breed is sensitive, extremely affectionate with its owner, loyal, somewhat aloof, reserved, and quiet. Although Pekingese are not aggressive, they are courageous and will stand their ground against larger adversaries. They are primarily an apartment lapdog, but they make good watchdogs as well. The long, fine coat needs frequent care. *See Figure 7–51.*

Figure 7–51 Pekingese. *Photo by Isabelle Francais.*

The Pug

The Pug is a very old breed that dates back to 400 BC. It is of Oriental origin, first appearing in the Buddhist monasteries of Tibet. From China it made its way to Japan and then to Europe. The Pug has a compact body and well-developed musculature. It has a round, massive head; a short, square muzzle; and a deep-wrinkled forehead. Its eyes are prominent with a sweet expression. The Pug is affectionate and loving, likes lots of attention, and loves children. Pugs are intelligent and easily trained. *See Figure 7–52.*

Figure 7–52   Pug. *Photo by Isabelle Francais.*

The Yorkshire Terrier

The Yorkshire Terrier was developed in the 1800s by the working class in the English county of Yorkshire for the purpose of catching rats and other vermin. Breeds used in the development of the Yorkshire Terrier were the Waterside Terrier, the Black and Tan English Terrier, the Paisley Terrier, and the Clydesdale Terrier.

Because of its long, silky coat, it has become a favorite lapdog and companion. The Yorkshire Terrier is lively, alert, affectionate with its owner, and wary of strangers. The Yorkshire Terrier's long, silky coat needs constant care and attention. They are best suited to adults who have plenty of time to devote to their required care. They are demanding, dependent, and need lots of attention. Yorkshire Terriers can be difficult to housebreak. *See Figure 7–53.*

Figure 7–53   Yorkshire Terrier. *Photo by Isabelle Francais.*

The Shih Tzu

The Shih Tzu breed dates back to 624 AD, when a pair was given to the Chinese court. They were said to have come from Fuh, which was believed to be in the Byzantine Empire. These dogs were bred in the Forbidden City of Peking. *Shih Tzu* means lion, and these small dogs resembled lions with their long, flowing hair. They were brought to England in 1930. This is a typical drawing-room dog. It is very lively and alert, with a distinctly arrogant carriage. Its long, silky hair needs daily attention to prevent matting and knotting. *See Figure 7–54.*

Figure 7–54   Shih Tzu. *Photo by Isabelle Francais.*

TABLE 7–6 Color, Size, and Weight Comparison of the Toy Dog Group

Breed	Color	Average Weight in Pounds		Average Height at Shoulder in Inches	
		Females	Males	Females	Males
Chihuahua	Any color—solid, marked, or splashed.	1–6*	1–6*	6–9	6–9
Italian Greyhound	All colors and markings recognized.	11	11	13–15	13–15
Manchester Terrier					
(standard)	Jet black with rich mahogany tan around the face and lower legs.	12–22	12–22	9–13*	9–13*
(toy variety)		<12	<12	<9*	<9*
Pekingnese	Red, fawn, black, black and tan, sable, brindle, white, and parti-colored.	<14	<14	8–9*	8–10*
Pug	Silver or apricot fawn with black mask.	14–18	14–18	10–12*	11–13*
Yorkshire Terrier	Puppies are born black and tan. Adults are usually dark steel-blue or tan.	>7	>7	8–9*	8–10*
Shih Tzu	All colors; nose and eye rims black.	12–15	12–15	9–10½	9–10½

*These figures are not given in the Official Standards and are thus approximate.

Information obtained from *"The Complete Dog Book,"* 17th edition, the official publication of the American Kennel Club, Howell Book House, 1986, Macmillan Publishing Company.

Other breeds in the toy dog group include the following:

Affenpinscher	Maltese
Brussels Griffon	Miniature Pinscher
Cavalier King Charles Spaniel	Papillon
Chinese Crested	Pomeranian
English Toy Spaniel	Silky Terrier
Havanese	Toy Fox Terrier
Japanese Chin	Toy Poodle

See Table 7–6 for a color, size, and weight comparison of the toy dog group.

Nonsporting Dogs

This group consists of miscellaneous breeds with a wide variety of sizes and characteristics. Basically, they are used as companion dogs.

The Boston Terrier

The Boston Terrier is an American breed that dates back to 1870 and was developed in the Boston area. It originated from a crossing of an English Bulldog and a white English Terrier; later there was considerable inbreeding. The breed was recognized by the American Kennel Club in 1893. This is a clean-cut dog with a short head, snow-white markings, soft black eyes, and a body conformation of the terrier rather than the Bulldog. It is an excellent companion and house dog, noted for its gentle disposition and affection for its owner. The Boston Terrier is intelligent, alert, well mannered, and even-tempered, and it gets along well with children. *See Figure 7–55.*

Figure 7–55 Boston Terrier. *Photo by Isabelle Francais.*

The Bulldog

The Bulldog originated in England and was used for the sport of bull baiting. Its ancestry is probably that of the Asiatic Mastiff and other fighting dogs. The original Bulldogs were vicious and courageous. In 1835, dog fighting and bull baiting became illegal in England. The Bulldog no longer had a use, but fanciers of the breed wanted to preserve it. Over the next generations, the ferocity was bred out of the dog and other desirable qualities were retained. The Bulldog we know today is good natured, reserved, dignified, loyal, clean, and aristocratic. Over its history, it has been used as a war dog, guard dog, and police dog. Problems may occur at whelping because of the large size of the puppy's head. A veterinarian should be kept advised, and cesarean delivery may be necessary. *See Figure 7–56.*

Figure 7–56 Bulldog. *Photo by Isabelle Francais.*

The Dalmatian

The Dalmatian's origin, according to historians, is the center of much debate and disagreement. Models, engravings, paintings, and writings place the dog in ancient Egypt, Asia, and Africa. The dogs were frequently found in bands of Romanies, and like their gypsy masters, Dalmatians have been well known but not located long in any one place. Authoritative writers place them first as a positive entry in Dalmatia, a region in west Yugoslavia. During their history, they have been used as a dog of war, a draft dog, and a herd dog. Dalmatians are excellent for hunting rats and vermin. As sporting dogs, they have been used as bird dogs, trail hounds, and retrievers, as well as in packs for boar and bear hunting. But most important has been their use as a carriage or coach dog. The breed is noted for its intelligence and adaptability. The instinct for coaching is inbred and also trained. Dalmatians will follow their master whether on foot, on horseback, or in a carriage. Dalmatians were originally noted for being calm, quiet, sensitive and loyal; they thrived on human companionship. They loved children and made natural guardians. However, after the fad period following the release of the movie *101 Dalmatians* (See page 39), many Dalmatians were bred without regard to genetics and heredity. This resulted in many with strong-willed, neurotic, stubborn, and aggressive behaviors. It has been estimated that as many as one-third of Dalmatians are deaf in one or both ears. *See Figure 7–57.*

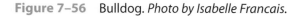

Figure 7–57 Dalmatian. *Photo by Isabelle Francais.*

The Lhasa Apso

The Lhasa Apso is a small dog with a heavy coat of hair. The breed originated in Tibet, where it was used as a guard dog because of its intelligence, quick hearing, and an instinct for distinguishing intimates and strangers. This breed is easily trained and responsive to kindness. The dogs are most obedient to anyone they trust. They can be very affectionate to their owners and are more suited to adults who will pay lots of attention to them. *See Figure 7–58.*

Figure 7–58 Lhasa Apso. *Photo by Isabelle Francais.*

The Poodle

The Poodle breed probably originated in Germany, where it was known as the Pudel. However, for years, it has been known as the national dog of France, where it was commonly used as a retriever and as a circus trick dog. In France, it was known as Canichi, which is derived from *chien canard*, or duck dog. The English word *poodle* comes from the German *pudel* or *pudelin*, meaning to splash in the water. There are three sizes: the standard, miniature, and the toy. All are of one breed, governed by the same standard of perfection. The breed is considered good-natured and very intelligent and responds well to their owners. Poodles are no longer used primarily as water dogs or retrievers, but they have gained fame as a show dog and companion. *See Figures 7–59 and 7–60.*

Figure 7–59   Standard Poodle. *Photo by Isabelle Francais.*

Other breeds in the nonsporting dog group include the following:

American Eskimo Dog	Keeshond
Bichon Frise	Lowchen
Chinese Shar-Pei	Schipperke
Chow Chow	Shiba Inu
Finnish Spitz	Tibetan Spaniel
French Bulldog	Tibetan Terrier

See Table 7–7 for a color, size, and weight comparison of the nonsporting dog group.

Figure 7–60   Toy Poodle. *Photo by Isabelle Francais.*

TABLE 7–7   Color, Size, and Weight Comparison of the Nonsporting Dog Group

Breed	Color	Average Weight in Pounds		Average Height at Shoulder in Inches	
		Females	Males	Females	Males
Boston Terrier	Brindle with white markings and black with white markings.				
	(lightweight)	<15	<15	11–15	15–17
	(middleweight)	15–20	15–20	depending on weight	
	(heavyweight)	20–25	20–25	depending on weight	
Bulldog	Red brindle, all other brindles, solid white, solid red, fawn or fallow, and piebald.	40	50	13–15	14–16
Dalmatian	Two varieties: white with black spots and white with liver brown spots.	35–45*	40–50*	19–24	19–24
Lhasa Apso	All colors.	10–13*	13–16*	9–10	10–11
Poodle	All colors with the coat being an even and solid color at the skin.				
	(standard)	35–45*	40–50*	>15	>15
	(miniature)	16–35*	18–30*	10–15	10–15
	(toy)	14–16*	15–18*	<10	<10

*These figures are not given in the Official Standards and are thus approximate.

Information obtained from *"The Complete Dog Book,"* 17th edition, the official publication of the American Kennel Club, Howell Book House, 1986, Macmillan Publishing Company.

Miscellaneous Class

Currently, the American Kennel Club recognizes five breeds in the miscellaneous class. These are breeds for which interest exists. Breeds in these classes may compete in obedience trials, earn obedience titles, and compete in conformation shows, but they are limited to the miscellaneous class and may not compete for championship points. If interest and growth of the breed continues, they may be admitted to one of the regular classes.

The Tibetan Mastiff is one of the breeds currently listed in the miscellaneous class. Written accounts tell of a large dog dating back to around 1100 BC in China. Little was known of the dogs outside of China and the Himalayan region until the 1800s. The large dogs were used to protect flocks of sheep, goats, and yaks. The dogs were also expected to defend women, children, and their masters against predators and intruders.

The breed is considered highly intelligent, independent, strong-willed, and a rather reserved dog. They are aloof with strangers and highly protective of their charges and their property. The coat colors are black, brown, and blue-grey, all with or without tan markings. The breed has a fairly long, thick, coarse guard hair with a heavy soft undercoat. The dogs typically weigh between 100 and 160 pounds. *See Figure 7–61.*

Figure 7–61 Tibetan Mastiff. *Photo by Isabelle Francais.*

Other breeds in the miscellaneous class include the following:

Beauceron	Redbone Coonhound
Plott	Swedish Vallhund

Hybrid Breeds

The hybrid breeds are also referred to as designer breeds. They are becoming very popular. Hybrid breeds are produced by crossing purebred dogs of two different breeds. The parents of the dog are known. The difference between a hybrid dog and a "mutt" is that the ancestry of the "mutt" is unknown. The hybrid has documented ancestry. Breeders of hybrid dogs believe that the crossbreeding eliminates many of the problems that purebreds encounter. They believe the dogs are heartier, healthier, and some believe the dogs are smarter. Some of the more common hybrids are:

Schnoodle—a Schnauzer and Poodle cross
Goldendoodle—a Golden Retriever and Poodle cross
Labradoodle—a Labrador Retriever and standard Poodle cross
Puggle—a female Beagle and a male Pug
Poochin—a Poodle and Japanese Chin cross

ANATOMY

The anatomy and skeletal structure of the dog are identified in *Figures 7–62* and *7–63*.

A Healthy Dog

Figure 7–62   The anatomy of the dog.

Figure 7–63   The skeletal structure of a dog.

CHOOSING A DOG

In selecting a dog, several major decisions need to be made:

1. what breed to buy
2. what source to buy from
3. what individual animal to buy

Within these decisions are many other questions that should be answered before making a final selection.

Bringing a puppy or dog home into the family can be an enjoyable experience for everyone, but unfortunately, many dogs end up in animal shelters, abandoned to run through the streets, dumped along the roadside, or destroyed. Time should be given to studying, researching, and fully considering the decision before bringing a dog home.

Selecting a Breed

In selecting a specific breed, the following questions should be considered:

1. Does one want a large breed or small breed of dog?
2. Is an active or quiet breed of dog wanted?
3. What type of hair coat should the dog have?
4. What is the purpose of the dog?
5. What price will have to be paid?

Dogs are available in all sizes, from the small Chihuahua that weighs less than 2 pounds, to the large Tibetan Mastiff that may reach a height of 31 inches and weigh up to 220 pounds. The decision will also be influenced by where one lives and the environment in which the dog will be raised. The Chihuahua and many toy dogs are fragile and are strictly apartment dogs. They require very little space and are more comfortable in the apartment environment. On the other hand, many of the large breeds require more space and outdoor exercise.

Along with the size of the dog, consideration needs to be given to how active or quiet the breed is. Even active breeds can be kept successfully in small homes or apartments if the owner is an active person who is willing to take them out for vigorous daily exercise. Small active breeds, such as many of the toy breeds, may be able to get plenty of exercise running around inside the home. Some of the largest breeds are much less active than many smaller dogs and may be a better match for an inactive owner with a modest dwelling. For example, the Border Collie weighs about 35 pounds, but is extremely active and will often develop destructive behaviors if not exercised (preferably involving running and playing with the owner) daily. Mastiffs and Bullmastiffs, on the other hand, weigh over 100 pounds, but have low activity levels and exercise needs that can be met with a walk on a lead or a short romp in the yard.

Hair coats on dogs are mainly of four types: short haired, long haired, wire haired, and curly haired. This decision should probably be based on the amount of time one has. Dogs with short-hair coats require very little grooming, although frequent brushing will remove old, dead hair and distribute the skin's oils. Long-haired dogs need regular brushing and combings, with extra attention given for mats and tangles. Some breeds, such as the Yorkshire Terrier, because of their long, silky hair, may require daily care. Dogs with wirehair and curly hair require regular plucking and special attention to remove dead hair and debris that may accumulate.

In making a selection of breed, one needs to consider how the dog will be used. Over the years, dog breeds have been selected and bred for several purposes. Some dogs were used as hunting dogs, herd dogs, guard dogs, show dogs, companion dogs, or simply as pets. Some dogs were needed for killing rats, mice, and other "varmints." More specific uses would be as guide dogs for the blind, as service dogs, as rescue dogs, for smelling out and detecting drugs or explosives, and (in the case of the Greyhound) for racing. Service dogs are used as hearing or signal dogs to assist the deaf, as mobility dogs to pull a person's wheelchair, and to carry a backpack, pick things up, or to open and close doors. Service dogs can also be used as walker dogs to help a handler walk by providing balance; as seizure alert dogs to respond to a person's seizure either by staying with the person or going for help; and as psychiatric service dogs to help a person with a mental disability go out in public or to help a person with autism to stay focused.

Finally, one needs to consider the price. The common rules of supply and demand control the price of the various breeds. The price will be lower if the breed is common and readily available. The price will be higher if the breed is rare. There are exceptions to this rule, but this is usually the case. According to the number of registrations in 2006 with the American Kennel Club, the most popular breeds today are as follows: Labrador Retrievers, Yorkshire Terriers, German Shepherds, Golden Retrievers, Beagles, Dachshunds, Boxers, Poodles, Shih Tzu, and Miniature Schnauzers.[2]

Choosing a Source

After determining the breed to buy, the next step is to consider where and from what source to buy a dog. Several sources are available. These sources include your friends and neighbors, from newspaper and magazine advertisements, from purebred breeders, from pet shops, and from animal shelters.

If one is just interested in obtaining a pet or companion dog and is not interested in a particular breed, dogs can be obtained from friends and neighbors. These "mutts" make lovable, friendly companions. Newspapers and magazines run advertisements for purebred as well as mixed-breed animals. A purebred breeder should be contacted for purebred dogs. In addition to newspaper and magazine advertisements, breeder names can be obtained from telephone directories; from local, state, or national kennel clubs; and from breed associations. Breeders of purebred animals are especially careful to maintain their reputations and will go to great lengths to see that future owners get a dog that meets their needs, satisfies them, and is healthy, genetically sound, and free of any hereditary imperfections. Finally, dogs are available from pet shops. Pet shops obtain animals from local breeders; the dog may be sound and have registration papers, but many times they obtain their puppies from various "puppy mills" or "puppy factories." These places exist only to supply puppies to pet shops, and very little care is taken to promote sound breeding practices, sanitation, and good general care. Many of these "mills" and "factories" have been the subject of magazine articles and television shows. One should resist the impulse to buy an adorable puppy from a pet store and spend a little extra time in locating an animal that is right for one's family.

Animals are also available at animal shelters. Shelters are located throughout the country and offer temporary housing for animals while new, permanent homes are sought. Most shelters have purebred and mixed breed dogs awaiting adoption. They have been relinquished or abandoned by their owner; a few may have been abused and are seeking a caring home. Approximately 20 percent of family pets come from animal shelters. Adoption fees are charged and vary depending on the organization, type of animal, and services provided. Fees range from about $30 to $75 for cats and from $45 to $125 for dogs. The fees will cover such services as spay/neuter, first series of shots, flea and tick treatment, general worming, collar and tags, and, sometimes, microchipping.

Selecting an Individual Animal

Once the decision has been made as to what breed is wanted and from what source the animal will be obtained, the prospective owner is ready to select an individual animal. Again, certain questions need to be answered:

1. What is the purpose for which the dog is being obtained?
2. How much does it cost?
3. What is its pedigree?
4. Is there a history of show winners or field trial champions in its ancestry?
5. Does its conformation meet standards?
6. Is it a male or a female?
7. How old is it?

The first four questions are closely related. Individual animals in a breed may be used for different purposes. The Irish Setter, for example, because of its beautiful, long, flowing hair coat, has long been admired as a show dog. Breeders who are primarily interested in show animals will breed for show quality and will pay very little attention to hunting instinct. As a result, it is possible to purchase an animal that has retained very little of its hunting instincts and would be of little use to a hunter. An animal that is of superior show quality or that has superior hunting instincts would be far more valuable than an ordinary animal sold as a pet or companion. **Pedigree,** or the ancestry of an animal, is very important in selection for show or for hunting. Many times animals are purchased by mail or phone, sight unseen, strictly because of the breeding that is used to produce the dog. The reputation of the breeder is very important in a purchase of this type because the purchaser is relying on the breeder to select an animal that is best suited to the purchaser. If the ancestry of an animal has a history of winning shows or field trials, the value of that animal is greatly increased. It comes from proven stock, and chances are it will be of superior show quality or a superior hunter.

Conformation

Conformation is the general structure, look, and makeup of the animal. In selecting animals for show, it is important that the animal meet the standards for the breed as set by the breed association. The official standard for each breed and any revision of the standard originates with the parent club, which is the national organization for each respective breed. Again, animals that come from proven show stock usually meet these standards. Personal preference usually enters the decision when selecting specific color or markings of an individual.

Most puppies should not be taken from their mother until they are at least 8 weeks old. It is often hard to determine at an early age the show potential of an individual, and one may want to wait until an animal is 6 months or older. When selecting animals, most people prefer young puppies because they are more adorable, usually are easier to train, have not been subjected to any other training, have not been subjected to long periods of abuse, and usually adapt easier to a family. However, one of the major problems with bringing young puppies into the home is the fact that they grow up to be dogs that are not as cute, require more time to feed and care for, and require more expense for food. When one buys an older animal, he or she knows exactly what the dog is like. Many older dogs that have been taken to shelters make excellent pets and companions for people interested in providing a good home for an animal.

FEEDING AND EXERCISE

To keep an animal healthy and in good condition, an adequate diet and clean water must be supplied. Beginning at about 3 weeks of age, fresh water should be available to puppies at all times. Even if they are still nursing, they may need additional liquid.

Types of Food

Several types of commercial foods are on the market for dog owners to choose from. Although one can mix his or her own rations, these commercial foods save time and are very convenient to use.

Three main types of commercial foods are available today: dry, semi-moist, and canned. These differ primarily in their ingredients and the amount of moisture they contain. The main ingredients in dry foods are corn, soybean meal, wheat millings, meat, and bone meal. Ingredients in semi-moist foods are corn, meat by-products, soybean meal, and corn syrup. Canned foods are of two types: (1) a ration-type that contains barley, meat by-products, wheat grain, and soy flour, and (2) a meat-type that contains meat by-products, meat, poultry, and soy flour. Dry foods contain about 10 percent moisture, semi-moist contain 30 percent, and canned foods 75 percent moisture. *See Figure 7–64.*

Their nutrient composition also differs. Dry foods contain approximately 23 percent protein, 9 percent fat, and 6 percent fiber. Semi-moist foods have about 25 percent protein, 9 percent fat, and 4 percent fiber. The ration-type canned foods contain about 30 percent protein, 16 percent fat, and 8 percent fiber, whereas the meat-type canned foods contain 44 percent protein, 32 percent fat, and 4 percent fiber. Most foods contain about the same amount

```
┌─────────────────────────────────────────────────────────────────────────┐
│                          GUARANTEED ANALYSIS                              │
│  Crude Protein (Minimum) ............21.00%    Crude Fiber (Maximum) ........5.00%  │
│  Crude Fat (Minimum) .................12.00%    Moisture (Maximum) .........10.00%  │
│  Linoleic Acid (Minimum) .............3.50%     Vitamin E (Minimum) ......30 IU/kg  │
├─────────────────────────────────────────────────────────────────────────┤
│                              INGREDIENTS                                  │
│  Lamb Meal, Ground Rice, Rice Bran, Rice Flour, Sunflower Oil (Preserved with mixed Tocopherols, a  │
│  Source of Natural Vitamin E and Ascorbic Acid), Rice Gluten, Dried Whole Egg, Natural Flavors,    │
│  Monosodium Phosphate, Dried Kelp (Source of Iodine), Choline Chloride, Vitamin E Supplement,      │
│  Zinc Sulfate, Ascorbic Acid (Source of Vitamin C), Niacin, Ferrous Sulfate, Calcium Pantothenate,│
│  Vitamin A Supplement, Manganous Oxide, Thiamine Mononitrate (Source of Vitamin B₁), Vitamin D₃   │
│  Supplement, Riboflavin Supplement (Source of Vitamin B₂), Vitamin B₁₂ Supplement, Pyridoxine     │
│  Hydrochloride (Source of Vitamin B₆), Inositol, Folic Acid, Cobalt Carbonate, Biotin.            │
└─────────────────────────────────────────────────────────────────────────┘
```

GUARANTEED ANALYSIS

Crude Protein (Minimum) 21.00%	Crude Fiber (Maximum) 5.00%
Crude Fat (Minimum) 12.00%	Moisture (Maximum) 10.00%
Linoleic Acid (Minimum) 3.50%	Vitamin E (Minimum) ....30 IU/kg

INGREDIENTS

Lamb Meal, Ground Rice, Rice Bran, Rice Flour, Sunflower Oil (Preserved with mixed Tocopherols, a Source of Natural Vitamin E and Ascorbic Acid), Rice Gluten, Dried Whole Egg, Natural Flavors, Monosodium Phosphate, Dried Kelp (Source of Iodine), Choline Chloride, Vitamin E Supplement, Zinc Sulfate, Ascorbic Acid (Source of Vitamin C), Niacin, Ferrous Sulfate, Calcium Pantothenate, Vitamin A Supplement, Manganous Oxide, Thiamine Mononitrate (Source of Vitamin B_1), Vitamin D_3 Supplement, Riboflavin Supplement (Source of Vitamin B_2), Vitamin B_{12} Supplement, Pyridoxine Hydrochloride (Source of Vitamin B_6), Inositol, Folic Acid, Cobalt Carbonate, Biotin.

Figure 7–64 Always check the label when buying pet food.

TABLE 7-8 Daily Rations in Ounces Needed by Dogs of Various Weights*

Dog's Weight (lb)	Kcal Needed/lb Body Weight	Dry Type	Semi-moist	Canned
2	65	1.3	1.7	3.4
5	52	2.5	3.2	7.00
10	44	4.3	5.5	10.75
15	39	6.1	7.8	15.25
20	37	7.5	9.7	18.75
30	33	10.0	12.8	25.0
45	30	14.0	18.0	35.0
75	26	20.4	26.2	51.0
110	24	25.0	32.2	62.5

*Caloric requirements—maintenance—and recommended daily food intakes of average adult dogs of various body weights. Individuals may require one-quarter more or less than these averages.

Source: Reprinted with permission of the American Kennel Club, Inc. *The Complete Dog Book,* 17th edition, copyright 1986, Howell Book House, Inc., New York, NY.

of energy. These are approximate figures and will vary with different brands and with different methods of preparation. Table 7–8 shows the daily ration necessary for dogs of various weights.

Dry foods have the advantage of being cheaper to purchase, are convenient to use, do not spoil quickly, and help keep the dog's teeth clean.

The AAFCO (Association of American Feed Control Officials) regulates the commercial pet food industry and ensures that foods are uniformly labeled and marketed. Nationally marketed foods can be of two types: (1) "Formulated" foods, which means the product has been produced to meet guidelines set by the AAFCO, but has not actually been fed to dogs. (2) "Tested" foods, which means the product was produced to meet guidelines and then fed to dogs prior to sale to ensure it meets important criteria related to the growth, maintenance, and reproduction of the dogs.

Not all foods are alike. Some bags or cans of food contain exactly what the outside label says. These are *fixed-formula foods.* Fixed-formula ingredients do not change from bag to bag or from can to can. Others may be *variable* or *open-formula foods* and may not contain what the label says, because the manufacturer is allowed to change the formula before the label is changed to reflect what's inside the bag or can. Premium dog foods are available only through veterinarians, pet supply stores, and feed stores. The ingredients in the premium foods do not change in relation to the price or availability of the ingredients. One should look on the label that shows that the nutritional value of the product was confirmed by feeding trials. A veterinarian should be consulted with any questions about a pet's diet and the foods that are available.

Feeding Pregnant and Lactating Females

The amount of ration given to the pregnant female should increase as her weight increases. Her weight normally starts to increase about the 4th week of pregnancy, and just before giving birth, she may be consuming 35 to 50 percent more food than before. She should be fed three or four evenly spaced meals to avoid the discomfort that one large meal may cause. Modern commercial food provides an adequate and balanced ration during pregnancy, and supplemental feeding should not be needed.

Following whelping, the female needs two to three times as much food as before. This should be given in three or four meals per day. The ration should be the same as has been provided throughout the pregnancy.

Beginning at about 3 weeks of age, the young puppies will begin to lap food from their mother's food bowl, and the owner will need to consider weaning the pups. This can be done gradually; as the puppies begin to consume more solid food. Remove the female from the puppies during the day, and increase the amount of solid food given to the puppies. Weaning should be completed when the puppies are about 6 weeks of age. The female can then be returned to her regular maintenance diet.

Feeding Puppies

As mentioned, puppies begin to lap solid food at about 3 weeks of age. Puppies can be started on solid food by giving them canned puppy food. Special puppy formulas are high in protein. Dry puppy food can also be used. One may want to mix some of the puppy replacement milk formulas that are available with the dry puppy food or simply soak the food with water for an hour to soften it prior to feeding. Read the label on the packaging and follow the directions. One should consult with a veterinarian if there are any questions.

The amount to feed a puppy varies depending on its breed. Usually a puppy should be fed three or four meals a day; its digestive system is not yet developed to handle one large feeding per day. A young puppy must receive enough food to provide for its energy requirements and for adequate growth. Puppies will experience their most rapid growth during the first 6 months of life. One must be careful not to overfeed because the puppy may become overly fat or grow too rapidly. Overfeeding can lead to obesity and may worsen the symptoms of hip dysplasia and other inherited bone diseases. The bones of a young, overweight puppy may not be developed to adequately carry excess weight, which could cause permanent injury to the bones and joints. A thin layer of fat should be felt when you place your fingers over the ribs and gently move them forward and backward. If you cannot feel the ribs or the edges of the ribs, the animal has an excess of fat.

Feeding Older Dogs

As dogs get older, their metabolism slows down and their need for calories decreases. Owners must feed a ration containing adequate protein and nutrients to maintain body condition. One should guard against overfeeding, which will lead to obesity and cut down on the expected life of the animal. Older dogs should receive smaller meals, which are easier to digest, several times a day.

Other Feeding Suggestions

Many people believe that feeding raw eggs to dogs improves the hair coat and produces more shine. Although the fat in egg yolks produces a shine to the hair, the egg white in raw form interferes with the absorption of biotin, a necessary vitamin. Therefore, eggs should be hard-boiled or cooked before serving to a dog.

Care must be taken when feeding bones to dogs. Large bones such as the knuckle and large marrow bones are alright for dogs to chew on if the dog does not rapidly gnaw them to a small size. Bones that splinter, such as chicken, turkey, or pork bones, should never be given to dogs. Splinters from these bones may get lodged in the digestive tract and cause disastrous results. Older, larger dogs have powerful jaws and can easily splinter most bones.

Exercise

Exercise is necessary for all dogs if they are to remain healthy. Dogs are like people in that some need more exercise than others. The amount of exercise varies according to the size of the animal, age, breed, and what the dog is being used for. Even small apartment dogs should be allowed some freedom to move around and exercise. Care must be taken when exercising young puppies; their bones may not be developed, and permanent injury could result. Older dogs need exercise also, but caution should be taken that they are not overworked. Short and more frequent walks are more suitable for older animals.

A few guidelines should be followed when exercising your dog:

1. Do not exercise any dog strenuously within 2 hours after it has eaten a large meal. Eating rapidly or exercising soon after a meal increases the risk for GDV (gastric dilatation-volvulus). This is an extremely serious and life-threatening condition. The condition is also known as bloat, stomach torsion, or twisted stomach. In bloat, the stomach fills with air and puts pressure on the internal organs. The animal has trouble breathing and the blood vessels are compressed. Once the stomach is filled with air it can then rotate, pinching off the blood flow. Great Danes and Saint Bernards are at greatest risk for GDV.

2. Dogs should receive a physical examination to rule out cardiac, circulatory, or skeletal and joint problems.

3. Before taking a dog out for strenuous exercise or strenuous hunting trips, the dog should be conditioned over time. Dogs are like athletes and need proper conditioning to strengthen muscles,

joints, and cardiac systems, as well as to toughen foot pads. This is especially important for obese dogs.

4. Dogs are eager to please their masters. They will not stop or give up a challenge, and one must be alert to signs of fatigue. Experience and working with the dog should allow the owner to note excessive panting, changes in color of the lips and face, vacant stares, and changes in behavior that might signal fatigue.

TRAINING

The amount and type of training a dog receives depends on what the dog is going to be used for. Almost all dogs should learn five basic commands: heel, sit, down, stay, and come. Most owners would like a dog that will respond and be under control. Dogs that are well trained can provide many hours of enjoyment and be great companions, whereas untrained dogs can be a nuisance.

Training should begin as soon as the new puppy arrives home. The puppy should learn its name, the basic corrective word *no*, and the praise words *good boy* or *good girl.*

The puppy must learn what types of behavior are acceptable and what types are not. The owner's voice is the best corrective tool. A strong, authoritative "no" should be all that is necessary. "Good dog," "good boy," or "good girl" should be used when the puppy does something correctly. The puppy will soon learn that it will be corrected if it does something wrong and praised when it does something right.

The puppy should never be hit or threatened with a hand or an object. A puppy that gets hit will soon shy away from its owner for fear of being beaten. When this occurs, the owner has a nuisance dog. When the puppy does something wrong, it must be corrected immediately so that it will associate the mistake with correction. If the correction cannot be made immediately, it should not be made at all. Puppies are babies and are quick to forget. If they are punished for an inappropriate behavior that occurred several minutes earlier, the puppy may not realize why it is being corrected.

House-Breaking or House-Training

To be an enjoyable companion and member of the family, a dog should be house-trained. House-training can be accomplished by taking the new puppy outside shortly after eating or several times during the day. Praise the puppy when it eliminates and bring it back inside. If an occasional accident occurs, use the words *bad dog* and immediately take the puppy outside. Again, it is important that the owner not hit or spank the puppy. Patience, correction, and praise will soon accomplish the objective. As the puppy gets older, it will make its way to the door and "ask" to be allowed outside.

Another method used is known as paper training. This involves laying several layers of newspaper down on the floor. Try to confine the puppy to the newspaper until it relieves itself. If a mistake occurs, use the words *bad dog* and place the puppy on the newspaper. The puppy soon will learn to associate the newspaper area with elimination. As the training progresses, the area covered by the newspaper can be made smaller and moved closer to the door. Eventually, the newspaper can be moved outside and the puppy allowed to go outside to relieve itself.

Obedience Training

Training periods should be held once or twice a day, be short in the beginning, and gradually increase to 15 to 30 minutes at a time. Longer training periods will tire the dog and may bore it. Do not forget to correct the dog immediately and be generous with praise when it performs well.

Equipment

Before beginning training, one may choose to purchase a leather or nylon collar and a leash that can be used to aid in the training of one's puppy. Other options are the use of halters or head collars. Most trainers using the traditional method of obedience training recommend the use of a training collar. Also know as choke collars, these collars can be dangerous if left on the dog after a training session is over. During a training session, the momentary tightening of the training collar, as well as the subsequent loosening, help the handler reinforce verbal commands.

It is common for novice trainers to misuse training collars and other training equipment. First-time trainers should seriously consider enrolling themselves and their dogs in a class taught by an experienced trainer.

Basic Commands

The following is a description of training basic obedience commands using the traditional method. *Clicker training* is another popular method that uses rewards, but no corrections or physical manipulation, to train basic commands. To heel is to walk close to the side of the handler. The handler will want the dog to walk on his or her left side. As the handler starts to walk, he or she should use the dog's name and give the command to *heel*. The handler may have to give a slight pull on the lead to persuade the dog to follow. The handler should practice with the dog two or three times a day until the dog follows on its own accord.

Teaching the dog to sit is usually part of the process in teaching the dog to heel. When the handler stops walking, he or she would want the dog to sit on the left side. At first, when the handler stops walking, he or she should use the command *sit*. As the handler gives this command, he or she will pull up on the lead and, with the free hand, push down on the dog's rump. Gradually, the dog will learn to sit without command when the handler stops walking.

The next step is to have the dog remain in the "sit" position until it is called. When the handler has the dog in the sit position, he or she will give the command *stay*. As the command is given, the handler will place his or her left hand in front of the dog's muzzle and step forward with the right foot first. The handler will repeat the command and hold the dog if necessary until it gets the idea. Gradually, the handler will increase the time for the dog to stay and the distance walked away. A well-trained dog should stay on command for at least 3 minutes.

The handler may also want to teach the dog to stay in a standing position. While the dog is heeling, the handler will slow down, stop, and give the command *stand*. The dog may want to sit, because that is what it has been taught. The handler will not let it sit, but will simply start walking again and give the command *heel*. After a few steps, the handler will give the command *stand;* the handler will also give the command *stay*. Gradually, the dog will associate the two terms, *stand* and *stay*. The dog may get confused, so patience is necessary and the handler should remember to be generous with praise.

To teach the dog to lie down on command, the dog should be put in the sit position. The handler will kneel down beside the dog and reach over it with his or her left arm and grasp the dog's left front leg. The handler will grasp the dog's right front leg with the right hand and gently lift the dog's front feet off the ground and ease the dog down into a lying position. The handler will practice the *down* command along with the *stay* command until the dog will lie down and stay until it is called.

The first command is *come* and is usually eagerly obeyed. By this stage, the dog should obey the commands *sit, stand,* and *stay,* so one simply calls the dog with the command *come*. In the beginning, if the dog hesitates, the handler should give a couple of tugs on the lead and the dog will come.

The handler should use all of the commands together during training so that the dog will obey them without the use of the collar and lead. If the dog hesitates, appears confused, or does not readily obey, the collar and lead will be put back on and the training routine continued.

The training steps listed here are brief. For questions about how to proceed, consult with a professional trainer or purchase training manuals that are readily available from pet shops and bookstores. Once the dog has mastered these basic commands, it is ready to go on to advanced training, depending on the plans for the dog. Specific training books on the various breeds are also readily available.

GROOMING AND CARE

Before purchasing a dog, it is important to consider the amount of time one has not only to train the dog but also to attend to its basic grooming and health care needs. Time and attention need to be given to the care of the dog's coat, nails, ears, eyes, and teeth.

Hair Coats

The amount of time spent on grooming depends on the type of hair coat; daily brushing is recommended to remove dead hair and distribute the skin's oils. Brushing down to the skin will help to remove flakes of dead skin and dandruff.

In addition to regular combing and brushing, long-haired dogs need to be checked for mats of hair, which commonly occur behind the ears and under the legs. These masses of hair can usually be teased out with a comb, but they occasionally need to be clipped out. To avoid cutting the skin, electric clippers with a number 10 blade are used. Long-haired dogs also have problems with burrs from plants. When cutting the hair, one should slide the comb down under the mat or burr and cut the hair on the outside of the comb. This method will be less likely to injure the dog.

The Terriers and wire-haired breeds need their coats plucked to remove dead hair. Plucking is accomplished by using a stripping knife, grasping a section of hair between knife and thumb, and pulling the knife away with a twisting motion. This will remove dead hair and trim live hair.

Many different kinds of equipment are available to groom a dog. The equipment needed depends on the breed, coat texture, and what the dog is being trained for. The local veterinarian, pet shop, or grooming parlor can give advice.

Bathing

Dogs should be bathed only when they become dirty; frequent bathing removes natural oils and causes the coat to become dry and harsh. When bathing becomes necessary, pH-balanced shampoo especially developed for dogs should be used. (pH is a measurement of the acidity level.) The recommended pH for shampoo is between 6.2 and 8.2, depending on the breed. Many shampoos used for human use have too high of a pH level and will dry the skin out.

Detergent soaps should never be used because some dogs may have skin reactions to them. Many of the shampoos for pets are medicated and help prevent external parasites. When washing, one must be careful not to get shampoo in the dog's eyes. After bathing, it is important that the dog not become chilled. One should be careful of the temperature if bathing outside, and when bathing inside, one must make sure the dog is completely dry before letting it outside.

Nails

The dog's nails will need to be trimmed occasionally. How often depends on the surface the dog is kept on. Outside dogs usually wear their nails down naturally on the soil surface; inside dogs need to be checked more often. Nails can be clipped at home using clippers obtained from a veterinarian or pet shop.

Clippers should be sharp so that they cut the nail and not crush it; crushing the nail may cause pain to the nail bed. One must be extremely careful not to cut into the nail bed because this will cause bleeding and pain. With clear or white nails, you can easily see the nail bed. Extreme caution needs to be taken when clipping dogs with black nails. Bleeding can be stopped using a **styptic** powder. One should check the dew claws on the inside of the leg; these will not wear down and many times are overlooked.

Sharp, pointed scissors should never be used to trim nails because they may injure the dog if it should happen to move suddenly.

Ears

During routine grooming, the ears should be checked; a dog's ears need to be cleaned about once a month. A cotton swab or soft cloth soaked in mineral oil, hydrogen peroxide, or alcohol can be used. *See Figure 7–65.* Only the part of the

Figure 7–65 One should follow the veterinarian's instructions when applying ear drops or ointments in the ears of the pet.

ear that can be seen should be cleaned. Never poke any sharp or hard object into the ear; a quick head movement could cause serious injury. *See Figure 7–66.*

Spaniels and other dogs with long hair need to be checked for mats and burrs that may block air movement and cause infections.

Eyes

Irritating substances can be removed from the eyes with contact lens solutions or other eyewash solutions. Dogs with large, protruding eyes and long hair need special care to keep hair from rubbing the eyeball. Hunting dogs and other field dogs should be checked carefully after each outing for irritating substances. Serious irritations and injuries need to be brought to the attention of the veterinarian. *See Figure 7–67.*

Teeth

A dog's teeth are usually not prone to decay. However, dogs do develop plaque and tartar that can lead to periodontal disease. Periodontal disease is painful and may lead to premature tooth loss. A dog's teeth should be cleaned once or twice a week. This can be accomplished using a small toothbrush or a gauze pad. Toothpaste made for dogs, salt water, or an equal mixture of salt water and baking soda can be used. The teeth should be scrubbed from gum

Figure 7–66 After applying ear drops or ointments, gently massage the ear.

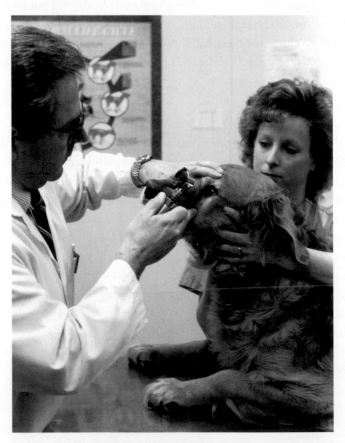

Figure 7–67 A veterinarian check-up will normally include checking the animal's eyes. *Courtesy of Brian Yacur and Guilderland Animal Hospital.*

Figure 7–68 During the regular visit, the veterinarian also checks the animal's teeth and gums. *Courtesy of Brian Yacur and Guilderland Animal Hospital.*

to crown; regular cleaning will prevent many problems. If the plaque is not removed, it will develop into a hard yellow-brown or gray-white deposit on the teeth called tartar. Periodically, the dog will need to have its teeth professionally cleaned at the veterinarian's to remove plaque and tartar both above and below the gum line. *See Figure 7–68.*

Feeding the dog dry dog food and some of the various treats on the market also helps prevent the buildup of tartar. Dogs love to chew, and chewing will keep teeth clean and sharp. Rawhide, synthetic bones, and hard chew toys are pleasurable and beneficial for the dog.

COMMON DISEASES

Unfortunately, even the most well-cared-for dog may come down with some type of disease, parasite, or ailment. With regular grooming and care, an owner can be attuned to early warning signs. Some of these early signs are constipation or diarrhea, shivering, fever, watery eyes, runny nose, coughing, loss of appetite, vomiting, loss of weight, ravenous eating without gaining weight, increased urination, restlessness, straining to urinate, labored breathing, increased water intake, lameness, paralysis, obvious pain, or nervous symptoms. If any of these signs are observed, it is important that the dog be taken to a veterinarian as soon as possible. *See Figures 7–69* and *7–70.* Prompt diagnosis and treatment may prevent more serious conditions from developing.

This unit of the text is broken down into six areas: infectious diseases, noninfectious diseases, fungal diseases, internal parasites, external parasites, and poisonings. More emphasis is placed on the more common and serious problems that occur. The cause, signs, treatment, preventions, and control

Figure 7–69   Pets should make regular visits to the veterinarian for examination and any vaccinations that may be needed. *Courtesy of BrianYacur and Guilderland Animal Hospital.*

Figure 7–70   The veterinarian can easily check to see whether an animal is overweight.

measures are listed when appropriate. This unit is intended as a general information unit and not as a diagnosis unit. It must be emphasized again that if any of the early warning signs appear, the dog should be taken to a veterinarian for diagnosis and treatment.

Infectious Diseases

This group of diseases is caused by pathogenic microorganisms such as viruses, bacteria, protozoa, and fungi that are capable of invading and growing in living tissues.

Canine Distemper

Canine distemper is caused by inhalation of an airborne virus that spreads throughout the body. The condition most often occurs in young dogs between 3 and 6 months of age, although older dogs may be susceptible. Early signs are vomiting and diarrhea. Some improvement may be seen after the first signs appear, but weeks or months later, signs that show nervous system damage appear. These can range from tremors to full seizure activity. The dog will eventually die or have permanent nervous system damage. Treatment is supportive: antibiotics to prevent secondary infection, cough suppressants, vomiting and diarrhea treatments, and anticonvulsants are used to try to keep the animal alive until it can recover from the virus. Canine distemper can easily be prevented with vaccination. Vaccine protocols are evolving, so you should follow your veterinarian's advice regarding the frequency and timing of distemper vaccinations. Periodic revaccination is recommended throughout the dog's life because of loss of antibody protection.

Infectious Canine Hepatitis

Infectious canine hepatitis is a highly contagious virus disease of dogs. The disease is caused by canine adenovirus type 1 (CAV-1). The virus is found worldwide and is spread from dog to dog through the urine, feces, or saliva of infected dogs. The virus targets the liver, kidneys, spleen, and lungs. Long-term kidney damage can occur, and the cornea of the eyes may become clouded. Dogs that recover usually shed the virus in their urine for up to 6 months. Signs vary from slight fever to death. The disease is seen most often in dogs less than 1 year of age; affected animals appear dull and apathetic. They usually refuse to eat and may have an intense thirst. The body temperature increases, and swelling may occur in the head, neck, and lower portion of the abdomen. The dog may moan in pain when eliminating. The gums may be yellow, and sometimes hemorrhages appear on them; the tonsils are enlarged and painful. Most dogs recover or are dead within 36 hours after the initial signs. Vaccinations should be given at the same time as canine distemper.

Leptospirosis

Leptospirosis is caused by bacteria and is a disease of major importance because dogs and other animals can act as reservoirs for human infections. The most common method of infection is through contact or ingestion of contaminated food and water. Outbreaks have occurred in humans and dogs after floods or after swimming in or drinking water contaminated by urine of rodents or other animals.

Early signs of the disease are high fever, loss of appetite, vomiting, and diarrhea; dehydration and depression may occur. Reddening of the membranes of the eyes and mouth are common. In some cases, yellowing of the gums and whites of the eyes may occur. The disease attacks the liver, kidneys, and gastrointestinal tract, causing pain and discomfort. Death usually occurs from severe damage to the liver and kidneys.

If any of the signs occur, your veterinarian should be consulted immediately. The bacteria respond to antibiotics, so if treatment is prompt, there is a chance for recovery. Vaccinations are available and should be given when the dog receives its distemper and hepatitis vaccinations. Annual revaccination is recommended for effective immunity.

Canine Parvovirus Infection

Canine parvovirus infection is a disease that appeared in the United States in 1977. The disease is caused by a small DNA-containing virus that requires rapidly dividing cells for growth to occur. The main source of infection is ingestion of materials contaminated with feces from infected dogs.

There are two forms of the disease; one involves the intestines and the other the heart. In the intestinal form, the virus grows in the epithelial cells of the small intestine and rapidly destroys them. Early signs are vomiting followed by diarrhea, refusal to eat, and severe dehydration. The feces appear yellow-gray and are often streaked with or darkened by blood. Dogs may vomit repeatedly and have projectile and bloody diarrhea until they die. Most deaths occur within 48 to 72 hours following onset of clinical signs.

The heart form may be preceded by the intestinal form or may occur without apparent previous disease. In this form, the virus rapidly attacks the muscle cells of the growing heart. Pups may act depressed and stop suckling shortly before they collapse, gasping for breath. There is no specific treatment. Pups that survive may have damaged hearts and die from heart failure weeks or months later.

Veterinarians recommend a series of vaccinations for puppies, with the last one occurring after 16 weeks of age. Annual revaccination is recommended for effective immunity. Dogs in high-risk areas or activities may need revaccinations at 6-month intervals. Pregnant females should receive a vaccination 2 weeks before whelping so that protection can be passed on to the puppies.

Proper cleaning and disinfection of the kennel and other areas where dogs are housed are essential to controlling the spread of the virus. The virus is capable of existing in the environment for many months. A solution of 1 cup chlorine bleach in 1 gallon of water is an effective disinfectant.

Infectious Tracheobronchitis

Infectious tracheobronchitis is also known as bordetellosis, vordetella, or most commonly, *kennel cough*. This disease is the most common upper respiratory ailment of dogs in the United States. The disease is marked by coughing and in some cases fever, loss of appetite, and pneumonia. Any one of several agents can cause signs of the disease, but normally several are involved. Among the causal agents are parainfluenza, *Bordetella bronchiseptica,* and mycoplasma. Other possible agents are canine adenovirus, reovirus, and canine herpes virus.

This disease occurs primarily when dogs of varying ages and susceptibility are congregated under conditions that are less than ideal, such as pet shops, dog shows, kennels, dog pounds, animal shelters, or boarding and training kennels. The kennel-cough agents are quickly spread when an infected dog coughs.

The most prominent sign of disease is the cough. The cough is usually dry and hacking, followed by gagging or expectoration of mucus. The most severe form of kennel cough is seen in dogs that are 6 weeks to 6 months of age. When there is lack of appetite, nasal or eye discharge, and depression, the disease is usually severe and a veterinarian should be consulted. Left untreated, the disease could cause damage to the respiratory system in the form of bronchitis or pneumonia and eventually death.

Vaccines are available to protect against the main organisms responsible for kennel cough. One should consult a veterinarian for the appropriate types and schedule of vaccinations.

Rabies

Rabies is a viral disease of humans and other mammals. The rabies virus attacks the central nervous system, and if the disease is not prevented, it causes death.

All warm-blooded animals can transmit rabies. The majority of all animal rabies cases in the United States are found in wildlife, particularly in skunks, raccoons, fox, and bats. Among domestic pets, dogs and cats are the most commonly infected species. (Refer to Chapter 2 for more information.)

Rabies is usually transmitted from animals to humans by a bite from a rabid animal. The wound is contaminated with the virus found in the saliva of the infected animal. Signs of rabies develop within 2 weeks to 3 months after the bite has occurred. If the bite is on the head or face and is severe, symptoms may appear in 10 days or less.

There are two types of clinical rabies in animals: *dumb* and *furious*. During early stages of furious rabies, the animal may act strangely. It may be unnaturally withdrawn or abnormally affectionate. This stage usually lasts about 2 days and then is followed by the "mad" stage, in which the animal will attack and bite anything in its path. A common symptom of this stage is frothing at the mouth, which is caused by difficulty in swallowing. Eventually the animal lapses into a paralytic state and then dies.

Dumb rabies differs in that there is no wandering or "mad" stage. Paralysis of the lower jaw is usually one of the first signs, followed by paralysis of the limbs and vital organs, resulting in death.

When an animal is bitten, the virus is drawn to the nerves and follows the nerve fibers to the brain and the salivary glands. After the virus infects the brain, it reproduces rapidly, causing severe brain damage. The brain lesions lead to altered behavior, aggressiveness, progressive paralysis, and death.

To prevent and control rabies, all pets should be routinely vaccinated. Regular vaccinations every 1 to 3 years should be part of a pet's health program.

Coronavirus

Coronavirus is a disease that causes severe vomiting and diarrhea. It is primarily a disease of puppies. Older dogs carry antibodies for this disease. It is a highly contagious viral disease that spreads rapidly by way of contaminated feces. The disease was first isolated in 1971 from the feces of military dogs. The feces are usually orange, foul smelling, and may be streaked with blood. Treatment is usually of a supportive nature, making sure the puppy gets plenty of water to prevent dehydration. A veterinarian should be consulted for proper treatment.

Canine Brucellosis

Canine brucellosis is caused by the bacteria *Brucellosis canis*. The disease occurs worldwide and has been reported in every state in the United States.

The most common sign of brucellosis infection in females is abortion 45 to 55 days into the gestation period; this occurs in 75 percent of the cases. Other signs are early embryonic death and resorption, or abortion at 10 to 20 days after mating. In male dogs, the scrotum and testicles swell, and one or both testes may also atrophy. Infected animals may show no sign of the disease; infection may occur readily across mucous membranes. Breeding of infected animals is a common method by which the disease is spread; all breeding animals should be blood-tested. Infected animals should be neutered or spayed to reduce the spread of the disease through breeding. Infected animals should be isolated and euthanasia of infected animals should be considered.

The potential of canine brucellosis being transmitted to humans should be considered, and all possible precautions should be taken when treating or handling infected dogs.

Canine Herpes Virus

Canine herpes virus causes severe illness and death in puppies less than 6 weeks old. The infectious organism is a virus, and only dogs are known to be susceptible.

Spread of the disease is by direct contact. Adult dogs show no signs of illness but can pass the virus in oral, nasal, or vaginal discharges. The organism can also be spread by way of saliva, feces, and urine. Unborn puppies may be infected in the mother's womb by primary infection of the mother or during passage through the birth canal.

Illness in young puppies usually occurs between the 5th and 18th day after birth. Symptoms include a change in the color of feces, difficulty in breathing, abdominal pain, ceasing to nurse, and constant crying. Affected puppies die shortly after the symptoms appear. No vaccines are available for this disease.

Salmonellosis

Salmonellosis is caused by a bacterial organism that is found in both wild and domestic animals. It is spread by ingestion of the organism in food contaminated by infected feces.

Infected animals experience fever, diarrhea, and vomiting. Dehydration and severe inactivity usually accompany these signs of infection. Salmonellosis can be confused easily with other gastrointestinal diseases, such as canine distemper and parvovirus infections.

People who handle animals may be susceptible to the infection from animals or may transmit the infection to animals because they are active carriers of the organisms.

Vaccination of dogs and cats is not done at present. Strict attention should be paid to disinfecting of facilities and to the personal hygiene of individuals working with animals. Infected animals can be treated effectively with antibiotics.

Haemobartonellosis

Haemobartonellsis is a tick and flea-transmitted disease. Haemobartonellis is caused by *Mycoplasma haemocanis bacteria*. This disease targets the red blood cells. Signs include depression, loss of appetite, weight loss, and fever. Most animals survive without permanent damage. Flea and tick control is the basis for preventing this disease from occurring.

Gastrointestinal Campylobacteriosis

Gastrointestinal campylobacteriosis is caused by the two bacteria of the *Campylobacter* genus. This disease can be transmitted to humans. Dogs can serve as sources for human infection.

Exposure to the feces of infected animals and contaminated food and water are the common means of transmission to dogs. Another source could be the consuming of uncooked poultry and meat. Signs of infection include loss of appetite, inactivity, and diarrhea. Simple hygienic measures are important to prevent the spread of the disease. A veterinarian can test the dog's feces and blood for evidence of infection.

Noninfectious Diseases

This group of diseases is caused by physical injuries or genetic defects; these diseases are not contagious.

Heart Disease

Heart disease in dogs and cats occurs with moderate regularity. There are two types: **congenital** defects, which are present at birth, and acquired heart disease. The most common signs include coughing at night during sleep, coughing during exercise, inability to exercise, fainting spells during exercise, development of blue gums during exercise, open-mouth breathing at rest (not panting, which is the dog's normal way of perspiring), reluctance or inability to breathe when lying down, and development of a large, pendulous abdomen because of accumulation of fluid.

There are several causes of acquired heart disease. These abnormalities usually begin slowly and increase in severity as the animal ages. Among the causes are degenerative disease of the heart valves, bacterial infections of the heart valves and muscles, cancerous tumors, degenerative diseases of the heart muscle, and heartworms. These can eventually lead to congestive heart failure.

Cataracts

Cataracts in dogs cause a cloudy, white opacity of the lens. Cataracts may be hereditary or nonhereditary and usually cause blindness when fully developed. Removal of the cataract surgically is the only effective treatment.

Glaucoma

Glaucoma is a serious disease involving an increase in pressure within the eyeball caused by a variety of problems with the production, transport, and absorption of aqueous humor, the fluid within the eye. This pressure causes irreversible damage to the retina and optic nerve. Drops may be administered or surgery may be necessary.

Progressive Retinal Atrophy (PRA)

PRA is a genetic disease in which the cells of the retina gradually degenerate, leading to the loss of sight. The age of onset of the disease varies with the breed and is usually breed specific. The first sign of the disease is loss of night vision; dogs undergo behavioral changes, especially when light is limited. There is no treatment for this disease.

Cherry Eye

Cherry eye is caused by a prolapse and enlargement of the tear gland on the inner surface of the third eyelid. Cherry eye appears as a red, cherry-like growth. Antibiotics and anti-inflammatory drugs help relieve mild cases, but surgery is necessary to remove more serious cases.

Hip Dysplasia

Congenital hip dysplasia (CHD) is an inherited condition. The condition affects the hip joints and occurs more commonly in large-breed dogs. This widespread condition causes hind-leg lameness as the result of pain originating in the malformed ball-and-socket joint. Particularly in damp, cold weather, dogs with dysplasia exhibit pain on attempting to rise and when moving. Environmental conditions such as too much exercise, especially at a young age; rough play; jumping; excess weight gain; and rapid growth can contribute to the age at which the condition occurs and to the severity of the condition.

Treatment is usually either nonsurgical or surgical. Nonsurgical treatment involves proper diet, exercise, supplements, anti-inflammatory drugs, and pain relief. By using nonsurgical treatment, one may be able to decrease the progression of degenerative joint disease. Surgical treatment includes joint fusion, joint replacement, cutting the joint, and amputation. Before surgery, one needs to consider the age and weight of the animal, how extensive the surgery is, and the function of the animal. Cost may also be a consideration.

Arthritis

Arthritis is a degenerative joint disease that causes pain, lameness, and stiffness in the joints. The disease is usually associated with old age. Large dogs are more often victims, and the symptoms are more severe in obese dogs. Drug therapy can help relieve pain, and moderate activity is recommended.

Tetanus

Tetanus in dogs is caused by a bacterial spores present in the soil; most cases result from contamination of small puncture wounds and lacerations. The organisms produce toxins that cause overstimulation of the dog's nervous system.

Infected dogs usually experience spasms of the facial muscles. Spastic paralysis, lockjaw, inability to stand, prolapse of the third eyelid, and over-extended head, neck, and legs may be seen as the disease progresses. Death usually occurs because the disease affects the respiratory muscles and other complications set in.

Botulism

Botulism in dogs results from ingestion of a toxin-producing bacterium. Animals usually acquire the toxin by ingesting contaminated material such as rotting carcasses or garbage. Botulism organisms have a paralyzing effect at the junctions of the muscles and nerves. The animal usually loses total muscle function and dies.

Anal Sacs

Anal sacs are located at the four o'clock and eight o'clock position on either side of the anus between the external and internal muscular rings of the rectum. These sacs contain a substance that allows the dog to mark its territory; occasionally, these sacs become blocked and need emptying. They may become swollen and painful, and the dog may be seen licking the area and dragging his or her rear on the ground. A veterinarian should empty them the first time. After observing the procedure the veterinarian uses, the owner may want to attempt the procedure the next time the problem occurs.

Fungal Diseases

Ringworm

Fungal diseases are found throughout the world; ringworm is one of the most common. Ringworm in dogs is caused by three fungal organisms: *Microsporum canis, Microsporum gypseum,* and *Trichophyton mentagrophytes.*

The natural reservoir for *Microsporum canis* is cats and dogs. Infection is usually a result of direct contact with an infected animal. *Microsporum canis* can infect humans and cause tinea capitis, a fungal infection of the scalp, eyebrows, and eye lashes, and can cause tinea corporis, or ringworm of the body.

Microsporum gypseum normally grows in the soil. Animals become infected by digging in or making contact with infected soil.

Trichophyton mentagrophytes are commonly found on wild rodents but can also be found on many species of animals. Dogs and cats become infected when they dig in the burrows of infected rodents.

Symptoms of ringworm begin as broken hairs around the face, ears, or feet. Reddened and scaly skin develops; in severe cases, scaling and crusting accumulate and the skin becomes thickened and itchy. A characteristic redness develops around a healing center as the disease progresses. Lesions in dogs are usually more severe than in cats.

In treating ringworm, the skin in and around the lesions should be clipped. Local treatment with baths, dips, creams, or lotions containing antifungal medications should be applied. Oral medications may also be used. Treatment should continue for 6 weeks or longer and under the supervision of a veterinarian.

Systemic Fungal Infections

Systemic fungal infections are so called because they may involve one or all of the body systems.

Blastomycosis

Blastomycosis is caused by the fungal organism *Blastomyces dermatitidis.* The organism prefers moist soil enriched with bird or bat droppings; infection is caused by inhaling infected spores. Common signs of infection are coughing, rapid breathing, lack of appetite, eye problems, lameness, and skin problems. Many dogs with blastomycosis will have lung lesions, and many will have skin lesions. Several treatment options are available, and one needs to consult with a veterinarian for proper treatment.

Histoplasmosis

Histoplasmosis is caused by the organism *Histoplasma capsulatum.* This organism is also found in soil enriched with bird or bat droppings. Signs are similar to those in blastomycosis. If the organism affects the lungs, there will be coughing, respiratory difficulties, fever, and depression. If the organism affects the intestines, the signs will be diarrhea, weight loss, and fever. Several treatment options are available; consultation with a veterinarian is necessary for proper treatment.

Coccidioidomycosis

Coccidioidomycosis is caused by the organism *Coccidioides immitis.* This organism prefers hot, dry, alkaline soils. Signs are similar to those in blastomycosis and histomycosis. Early signs are lack of appetite and loss of weight. Later on, as the disease progresses, signs include lameness, bone pain, spinal pain, and coughing. The early form of the disease affects the lungs and then moves to the bones. One needs to consult a veterinarian for proper treatment.

Treatment of animals for all three fungal systemic diseases requires months of therapy with expensive and potentially toxic drugs.

Internal Parasites

Roundworms

Ascarids *(Toxocara canis)* of dogs, commonly called *roundworms,* are large worms that reach 4 to 8 inches in length when mature. As egg-laying adults, they live in the small intestines of dogs.

The major means of infection is by ingestion of eggs or by the transmission of infective larvae from mother dog to her unborn or nursing pups. The small larvae, which hatch in the dog's stomach, grow to adult worms that begin to produce eggs in about 4 weeks. Older dogs develop a resistance to the adult worm, so if they ingest an infective egg, the hatched larvae stage may not grow into an adult worm but may migrate to the muscle and remain inactive.

In male dogs, these larvae have no escape, but in females that become pregnant, the larvae become active and migrate via the umbilical cord to the liver of the fetuses. From the liver, the larvae make their way to the lungs, are coughed up, are swallowed, and make their way to the small intestine, where they grow to adult worms. The larvae can also be passed from the mother to the pups in the milk, although only about 5 percent are passed this way.

Clinical signs depend on the number of worms present. If only a few worms are present, there may be no evidence of infection, but if several hundred worms are growing to adults, the abdomen will distend, giving the pup a pot-bellied appearance. These worms deprive the pup of nutrients.

It is important to treat pups early so that the worms are expelled before they become egg-laying adults. There are several drugs available that a veterinarian can prescribe. Treatment of the mother dog involves destroying the larvae ascarids in her small intestine before the pups are born. Treatment must begin on about the 40th day of pregnancy and continue daily until 2 weeks before whelping, or about 37 consecutive days. Consult with a veterinarian for proper treatment.

Roundworms are a danger to children. Infected dogs should be kept away from children and their play areas. (See Chapter 2 for further information.)

Hookworms

Hookworms are bloodsucking parasites. The most common species is *Ancylostoma caninum*. The adult worm is about 1 inch long and has a large mouth that attaches to the small intestine and digests a plug of tissue; this causes a small bleeding site. When 50 to several hundred worms are present, the loss of blood can be dangerous.

Dogs heavily infected with hookworms develop **anemia.** The loss of blood into the intestine results in dark-colored feces. Gums of the animal will be white, and the animal will appear weak and listless.

Both young and adult dogs can have infections of adult hookworms, which shed eggs in the dog's feces. Once passed in the feces, hookworm eggs hatch into larvae and can infect dogs either by penetrating the skin when dogs lie down in damp, contaminated places, or by being ingested. Older dogs can harbor both adult hookworms and immature larvae in their intestines and mammary glands. Immature larvae can be passed to nursing pups by the lactating mother.

Larvae grow to egg-laying adults in about 2 weeks. If the number of hookworms is great, pups begin showing clinical signs of anemia at about 2 weeks of age and may die before 3 weeks of age.

Treatment is the same as that listed for ascarids. Infected dogs should be kept away from children and their play areas. (See Chapter 2 for further information.)

Whipworms

Whipworms *(Trichuris vulpis)* live in the lower digestive tract of dogs, especially in the cecum. The adult worm is shaped like a whip, broad at one end and narrow at the other. It is about 2.5 inches long.

These worms thread their narrow end into the lining of the cecum to hold on. They produce football-shaped eggs that are passed in the feces of infected dogs. These eggs, protected by a thick shell, can live in the soil for several years.

Dogs with heavy infections will have irritation to the gut lining, causing loss of tissue fluids, which results in watery feces usually tinged with blood. Dogs will become dehydrated and, if not treated, will die. Consult a veterinarian for appropriate treatment.

Dogs should not be returned to the same yard pen if possible because they will become reinfected and require further treatments.

Tapeworms

Tapeworms are flat, segmented worms, usually 1 foot or more in length, that live in the small intestine. Four types infect dogs: *Diplylidium caninum, Taenia pisiformis, Echinococcus granulosus,* and *Echinococcus multilocularis. See Figure 7–71.*

Figure 7–71 This type of tapeworm infects dogs and cats. Note the segmented body. These segments break off and pass out of the animal's body. These segments look like pieces of rice around the anal region and in the feces. *Photo reprinted with permission of Miles, Inc., Animal Health Products.*

Tapeworms shed their egg-filled terminal segment in feces of the infected animal. These segments are about the size of a grain of rice and may be seen crawling on the surface of feces.

An initial host is required for development of the young tapeworm to a stage that it will be infective. In the case of *Diplylidium,* the initial host is the flea; and in *Taenia,* wild rabbits serve as the initial host. *See Figure 7–72.*

Tapeworms are not harmful to the dog, but the two species *E. granulosus* and *E. multilocularis* can cause serious illness to humans (refer to Chapter 2). Drugs are available on the market that are effective in treating tapeworms. For complete control we must also control the fleas and prevent dogs from eating wild rabbits that serve as the initial hosts. *See Figure 7–73.*

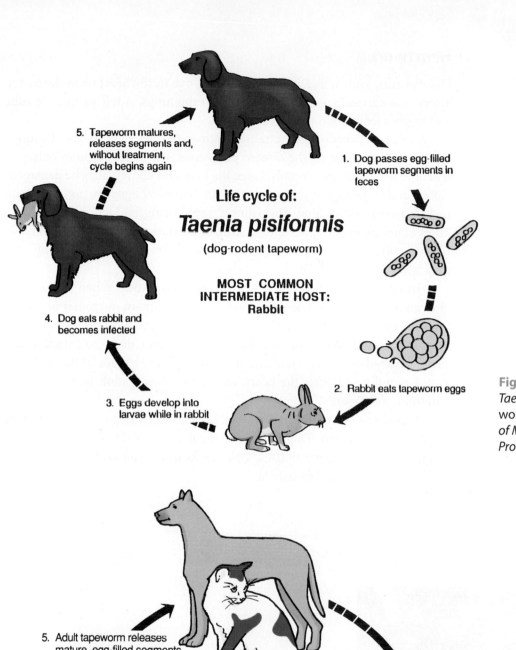

5. Tapeworm matures, releases segments and, without treatment, cycle begins again

Life cycle of:
Taenia pisiformis
(dog-rodent tapeworm)

MOST COMMON INTERMEDIATE HOST:
Rabbit

1. Dog passes egg-filled tapeworm segments in feces

4. Dog eats rabbit and becomes infected

3. Eggs develop into larvae while in rabbit

2. Rabbit eats tapeworm eggs

Figure 7–72 Life cycle of *Taenia pisiformis, dog/cat tapeworm. Reprinted with permission of Miles, Inc., Animal Health Products.*

5. Adult tapeworm releases mature, egg-filled segments and, without treatment, the cycle begins again

4. Tapeworm matures in pet

Life cycle of:
Dipylidium caninum
(dog/cat-flea tapeworm)
INTERMEDIATE HOST:
The Common Flea

An important point to remember about this tapeworm species is that any pet that has fleas also has a very real possibility of having tapeworms.

1. Pet passes egg-filled tapeworm segments in feces

3. During grooming or biting, pet ingests tapeworm-carrying fleas

2. The segments rupture, releasing individual eggs that are eaten by flea larvae

Figure 7–73 Life cycle of *Dipylidium caninum. Reprinted with permission of Miles, Inc., Animal Health Products.*

Heartworms

Heartworms, known as *Dirofilaria immitus,* live in the heart or in the major artery that carries blood from the heart to the lungs. Adult worms are thin and reach a length of 6 to 14 inches.

Adult heartworms cause extensive damage to the vital organs; the lungs are the first affected. As the disease progresses, the heart becomes enlarged and does not function normally. Later, the liver and kidneys may be damaged. Considerable damage may occur before any outward signs are noticed. Signs include frequent coughing, sluggishness, rapid tiring, and labored breathing. These signs are easily noticed in hunting or working dogs. When the disease reaches critical stages, the animal is usually weak, has difficulty breathing, and may faint. *See Figure 7–74.*

Heartworm infection is transmitted by mosquitoes. When a mosquito bites an infected dog, it takes up blood that may contain microfilariae. These incubate in the mosquito for about 2 weeks, during which time they become infectious larvae. When the mosquito bites another dog, the infective larvae are passed on. These larvae migrate through the tissues of the body for 2 to 3 months and enter the heart, where they reach adult size in another 3 months. *See Figure 7–75.*

Treatment for heartworm can be successful if the disease is detected early, but it is very expensive. Dogs should be regularly treated and be kept on prescription heartworm preventives. One needs to consult with a veterinarian concerning testing and treatment.

Figure 7–74 Heartworms can plug up the heart of an animal. *Courtesy of Stacey Riggert.*

Figure 7–75 Heartworms can be prevented by giving heartworm pills (DEC) daily during the mosquito season and for 2 months after the mosquito season. Another drug, Ivermectin, can be given monthly.

External Parasites

Fleas

Fleas are wingless, brown, bloodsucking insects that infest warm-blooded animals. Young animals may die from severe blood loss.

Both the cat flea *(Ctenocephalides felis)* and the dog flea *(Ctenocephalides canis)* infest dogs. Fleas move rapidly on the skin and are most commonly found on the rump and groin area.

Flea eggs laid on the host are smooth and quickly fall off into the animal's environment. The eggs are oval, white, and glistening.

Small larvae hatch from the eggs and feed on the feces from adult fleas. After several molts, the last larvae stage forms a pupal case. The adult flea emerges from the pupal case and searches for an animal. Time required for the flea to develop from egg to adult may be as short as 16 days.

Each bite a flea takes on the host animal allows a small amount of saliva to enter the skin and causes an irritating and allergic reaction that leads to severe itching. Biting and scratching around the rump and groin areas are the most common signs of flea allergy. Hair loss, a red rash, and thickening of the skin are also commonly seen.

Flea feces can confirm the presence of fleas and consist of dried blood from the host animal. Mixed with a small amount of water, it will form a blood-red solution.

To control fleas, both the animal and its environment must be treated. Many products to control fleas are available. Among the products are flea shampoos, powders, rinses or dips, sprays, foams, spot-on products, oral insecticides, flea collars, foggers, and house and yard sprays. Read the labels on the products carefully and use them according to the directions on the labels.

Ticks

Ticks are bloodsucking arthropod parasites of the skin. Severe blood loss may result from heavy infestations. Ticks have a four-stage life cycle and, depending on the type, may require one, two, or three hosts to complete their life cycle.

Hard ticks *(Ixodidine)* and soft ticks *(Argasidae)* are the two main families of parasitic ticks. Hard ticks have hard shields on their backs, distinguishing them from soft ticks. Most hard ticks require three different hosts to complete their life cycle; each stage feeds only once.

Thirteen species of hard ticks are of importance; two of these are major problems in dogs. The brown dog tick *(Rhipicephalus sanguineus)* can survive indoors and infest kennels and households, and the American dog tick *(Dermacentor variabilis)* lives only in grasses and shrubs.

The brown dog tick is a three-host tick that attacks cats, horses, rabbits, and humans. The larval and nymph stage of the American dog tick parasitizes field mice. Hard ticks are normally found securely attached by their head to the skin. Before removal, they should be sprayed with insecticide safe for use on animals or soaked with alcohol. Tweezers should be used to grasp the head to remove the tick. Cigarettes, lighters, gasoline, and kerosene can severely injure the skin and should not be used. In the case of the brown dog tick, the kennel or household premises where the dog stays may need to be treated.

Soft ticks have a leathery outer covering. The spinose ear tick *(Otobius megnini)* is the only medically important soft tick. The young stages (larvae and nymphs) live in the outer ear canal of dogs, cats, cattle, and horses. Adult soft ticks do not live on the animals. The larvae and nymphs can cause severe irritations to the ear canal and may occasionally cause paralysis and seizures in some animals. Care must be taken when removing these ticks because damage to the ear canal can occur. Control and treatment of ticks is similar to controlling fleas.

Lice

Lice are wingless insects that are uncommon parasites of dogs. Lice spend their entire life cycle on the host. Lice are usually host specific.

The dog has one common biting louse *(Trichodectes canis)* and one common sucking louse *(Linognathus setosus)*. Female lice attach their eggs to the hair of the host. The young lice undergo several molts before becoming adults; development from egg to adult takes about 19 to 23 days. Lice usually cause severe hair loss from scratching and rubbing. Dogs may be treated with dips, washes, sprays, or dusts. Two treatments about 12 days apart are usually necessary.

Mites

Mites are tiny, sometimes microscopic, arachnids that are parasitic on many animals. There are five types of mites important to dog owners.

Demodectic mites *(Demodex* species) live in hair follicles. They are present on the skin of all normal animals and usually exist in small numbers that do not cause a problem. The problem occurs when the population of the mites on an animal increases dramatically. Because all animals have demodectic mites, the mites are not considered contagious.

Demodectic mites cause demodicosis, which is a potentially serious disease in the dog; there are two forms: localized and generalized. Patchy hair loss on the head, forelegs, and trunk is called localized demodicosis. In generalized demodicosis, hair loss, reddening, and crusting may involve the entire body. Dogs with generalized demodicosis may also develop bacterial infections in the skin.

The increase above normal numbers of mites is believed to be due to a deficiency in the immune system, and some breeds are more susceptible to the disease.

Sarcoptic mites *(Sarcoptic* species and *Notoedres* species) are highly contagious mites that affect many species of animals. This family of mites burrows within the outer layers of the skin, and the entire life cycle is spent on the host animal. The common skin disease, sarcoptic mange or scabies, is caused by sarcoptic mites. Sarcoptic mange is an intensely itchy disease that causes the animal to scratch, chew, and rub constantly. The constant scratching, chewing, and rubbing of the affected areas produces reddening, scaling, and crusting of the skin. The head, ears, and abdomen are most usually affected, but the entire body may become infected. Many times, animals will cause self-inflicted injuries as they scratch, chew, and rub the skin.

If sarcoptic mange is suspected, a veterinarian should be consulted. All animals that have been in contact with the infected animal should be examined.

Ear mites *(Otodectes cyanotis)* are a common problem in dogs. These highly contagious mites are usually found in the outer ear but may be found on other areas of the animal. Ear mites feed on the wax and other materials found in the ear. Symptoms may vary from no signs to a severe irritation with thick, dry, black crusts in the ear canals. Dogs that are severely infected may shake their heads.

Ear drops containing an insecticide are one means of treatment. To be totally effective, the entire body should be treated with flea products, and all pets in the household should be treated.

Cheyletiella mites *(Cheyletiella yasguri)* are contagious mites that live on the surface of the skin. These mites cause cheyletiellosis, a condition referred to as walking dandruff, which produces severe scaling, usually on the back of the infected animal. Some itching may occur, but not nearly as bad as from the other types of mites that have been discussed. Treatment for cheyletiella mites is the same as for sarcoptic mange mites.

Chiggers

Chiggers are larvae stages of the *Trombiculid* mites; only the larval stages are parasitic. The North American chigger *(Trombicula alfreddregesi)* is the most common chigger that affects animals and people. On animals, they cause an itchy, red rash on the belly, face, feet, and legs. Chiggers are orange-red in color and are usually picked up from contact with heavy underbrush. The larval mites remain attached to the skin for only a few hours, so the only treatment needed is something to stop the itching.

Poisonings

Several groups of chemicals can cause poisoning in dogs and cats. These groups include insecticides, plants, household products, **rodenticides**, herbicides, human and veterinary medicines, metals, and miscellaneous compounds.

Insecticides

Insecticides such as the organophosphate and pyrethroid compounds produce most of the insecticidal poisonings in dogs and cats. Boric acid used in roach baits and arsenic used in ant traps also cause significant numbers of poisonings.

Plants

Dogs that eat plants containing insoluble calcium oxalate crystals, such as in philodendron, dieffenbachia, pothos, and caladium, experience severe irritations of the mouth and intestinal tract.

There are too many toxic plants to list here. Before purchasing house or landscaping plants, consult the ASPCA animal poison center Web site or other similar references.

Household Chemicals

Cleaners and household chemicals containing bleach, ammonia, borates, hydroxides, pine oil, and phenol cause poisonings in pets.

Strychnine, a rodenticide used to kill rats and mice, is often ingested by pets. Rodenticides such as warfarin, brodifacoum, and diphacinone contain anticoagulants and are also ingested by pets. Anticoagulants cause bleeding in animals, usually more than 2 days after ingestion.

Herbicides

Herbicides used to kill weeds in lawns are also responsible for poisonings. Glyphosate, paraquat, and arsenic-based herbicides have been known to cause poisonings in dogs and cats.

Medications

Medications that are normally given to people and pets can be poisonous if left around and then ingested in large amounts or taken improperly. Many times, pet owners will give medications to pets that are not prescribed by veterinarians. Medicines that are safe for dogs may poison cats or puppies.

Metal Poisonings

Metal poisonings are rare but can occur. The most common type is lead poisoning from animals ingesting old paint chips, fishing and drapery weights, roofing shingles, and used motor oil.

Antifreeze

Antifreeze used in radiators is a common source of poisoning in both dogs and cats; it has a sugary taste and is readily ingested. Even small amounts cause animals to appear drunk or depressed and can cause death. Treatment is most successful if begun before symptoms occur. Call a veterinarian immediately if you suspect antifreeze ingestion.

Treating animals that have been poisoned must begin as soon as a pet owner recognizes a poison is involved; prompt diagnosis and treatment is extremely important. A veterinarian should be consulted immediately. If the cause of the poisoning is known, this can save valuable time in getting the correct treatment started.

REPRODUCTION

A female dog usually has her first **estrus** (**heat period** or season) between 6 and 12 months of age. Smaller toy dog breeds and terriers usually begin at 6 and 9 months, but some of the larger breeds may not begin until 2 years of age.

The estrus period occurs at intervals of 6 months on average and is preceded by a **proestrus** that lasts about 9 days. During this period, the female is attractive to the male and has a bloody vaginal discharge and a swollen vulva. However, the female will not accept the male for mating during this time.

The estrus period follows and usually lasts 9 days. Ovulation usually occurs in the first 48 hours. Breeders differ in their opinion as to the best time to breed. From 10 to 12 days after the beginning of discharge (proestrus) is the general period recommended. A good compromise is a double mating on the 10th and 12th days, or on the 11th and 13th days.

Females are usually taken to the stud for mating. After mating, the dogs may be locked together. This is called a "tie" and is perfectly normal. It results from a swelling of the bulbus glandis. During the tie, the male may move around so that they are positioned rear to rear. Do not try to separate them during the tie because doing so may injure both animals; they will separate naturally in 10 to 30 minutes.

Pregnancy

Pregnancy, or **gestation,** in dogs usually lasts an average of 63 days, with a variation from 56 to 70 days. Signs of pregnancy include increase in appetite, weight, and size of the mammary glands. Some females may show very little weight or increase in the mammary glands; a change may not be noticeable until around the 28th day.

A whelping box provides a nice place for the female to have her puppies. She should be introduced to her box about 3 weeks before whelping so that she can get comfortable with it. The bottom of the box should be covered with newspapers.

As the whelping date approaches, the female's temperature will be slightly below normal (100.4°F). About 24 hours before whelping, her temperature will drop further to about 98°F. Most females will refuse meals in the last 24 hours.

A normal, healthy female usually gives birth easily, without need of additional help; however, one should stand by and give assistance if necessary. Each puppy will be contained in its own **placental membrane,** which must be removed before the puppy can breathe. The mother will do this by eating the membrane and severing the umbilical cord. She will be busy licking the puppy to stimulate respiration, cleaning it, warming it, and allowing it to suckle.

A placenta will follow each puppy after a few minutes. Count them to make sure the number of placentas matches the number of puppies; a retained placenta can cause problems later.

One must be prepared to help the mother remove a puppy from its placental membrane, to clean it, and encourage it to nurse. When cleaning a membrane off a puppy, the owner may also have to sever the umbilical cord. A piece of cotton thread should be tied around the cord about ¾ to 1 inch away from the puppy's body; the cord should then be cut ¼ to ½ inch beyond the thread, taking care not to pull on the cord and making sure hands are clean. An application of iodine to the naval cord is recommended. One must be careful of the mother also because she may be protective of the litter and may bite.

The first milk produced by the mother is called **colostrum.** It is important that each puppy receive colostrum because it contains immunoglobulins that will help protect it from infectious disease.

Eclampsia is a calcium deficiency that may be observed shortly after whelping and during the first month of lactation. It usually occurs among the toy breeds and females with large litters that require a large milk production.

Symptoms include restlessness, panting, whining, and getting up and down. As the problem worsens, the mother will become stiff, with twitching muscles and lack of coordination. If not treated, she will collapse into convulsions and die. A veterinarian must be consulted immediately to give her a calcium injection. Prevention is easy by ensuring that the pregnant female gets proper vitamin and mineral supplements, especially calcium.

SUMMARY

The dog was probably the first animal domesticated by humans. The domesticated dog originated about 12,000 to 14,000 years ago in Europe and Asia and was domesticated about 10,000 years ago.

The dog traces its ancestors back about 40 million years ago to a tree-climbing mammal called *Miacis*. *Miacis* was a small animal with an elongated body that probably spent most of its time in the lush forest areas.

Today, there are more than 400 breeds of dogs in the world. The American Kennel Club recognizes 155 different breeds in seven groups. The seven groups are sporting dogs, hounds, working dogs, terriers, toys, herding dogs, and nonsporting dogs.

Bringing a puppy or dog home into the family can be an enjoyable experience for everyone, but unfortunately, many dogs end up in animal shelters, abandoned to run through the streets, dumped along the roadside, or destroyed. Time should be given to studying, researching, and fully considering the decision before bringing a dog home.

To keep an animal healthy, an adequate diet and clean water are essential. Several types of commercial foods are available to choose from. Although one can mix his or her own rations, these commercial foods save time and are very convenient to use.

The amount and type of training a dog receives depends on what the dog is going to be used for. Almost all dogs should learn five basic commands: heel, sit, down, stay, and come. Most owners would like a dog that will respond and be under control. Dogs that are well trained can provide many hours of fun and enjoyment and can be great companions, whereas untrained dogs can be a basic nuisance.

Before purchasing a dog, it is important to consider not only the amount of time one has to train the dog, but also the time needed to attend to its basic grooming and health care needs. Time and attention need to be given to the care of the dog's coat, nails, ears, eyes, and teeth.

Unfortunately, even the most well-cared-for dog may at times come down with some type of disease, parasite, or ailment. By regular grooming and care, an owner can be attuned to the early signs that may alert one to a problem. Some of these early signs are constipation or diarrhea, shivering, fever, watery eyes, runny nose, coughing, loss of appetite, vomiting, loss of weight, ravenous eating without gaining weight, increased urination, restlessness, straining to urinate, labored breathing, increased water intake, lameness, paralysis, obvious pain, or nervous symptoms. If any of these signs are observed, it is important that the dog be taken to a veterinarian as soon as possible. Prompt diagnosis and treatment may prevent more serious conditions from developing.

A female dog usually has her first estrus between 6 and 12 months of age. Smaller toy dog breeds and terriers usually begin at 6 and 9 months, but some of the larger breeds may not begin until 2 years of age. This estrus period occurs at intervals of 7 months on average.

Pregnancy, or gestation, in dogs usually lasts an average of 63 days, with a variation from 56 to 70 days. Signs of pregnancy include increase in appetite, weight, and breast size. Some females may show very little weight and breast increase; a change may not be noticeable until around the 28th day.

The first milk produced by the mother is called *colostrum,* and it is important that each puppy receive this because it contains immunoglobulins that help protect the puppy from infectious disease.

DISCUSSION QUESTIONS

1. What is the history of the dog? How has the modern dog evolved?

2. What are the seven major groups of dogs? How do they differ?

3. List the different breeds of dogs. What are the characteristics of the different breeds?

4. List the various terms that describe the behavior of the different breeds. What do these terms mean?

5. What aspects should be considered before selecting a breed of dog?

6. What are the various feeds available for dogs?

7. How does the feeding of pregnant and lactating females, of puppies, and of older dogs differ?

8. What are some of the general grooming and health care practices necessary to keep a dog clean and healthy?

9. What are the six groups of diseases that affect dogs?

10. List the diseases belonging to each of the six groups. What are the symptoms of each disease?

11. What are some of the general care practices necessary for a pregnant female before, during, and after whelping? What are some practices for the general care of the new litter?

SUGGESTED ACTIVITIES

1. With the approval of the instructor, bring in dogs of various breeds and talk about the characteristics of the breeds.

2. Bring the labels from various dog foods and compare the ingredients that are used.

3. With the approval of the instructor, invite dog trainers into the class and ask them to explain the methods they use in training their animals.

4. With the approval of the instructor, invite local dog groomers to give demonstrations on the methods used to groom various breeds of dogs.

5. With the approval of the instructor, invite the local veterinarian into the class to talk about some of the common diseases of dogs.

6. Visit a dog show. Watch how the various groups of dogs are judged and watch what criteria are used to select the winners.

ADDITIONAL RESOURCES

1. American Kennel Club
   http://www.akc.org/

2. American Veterinary Medical Association
   http://www.avma.org

3. Animal Clinic
   http://www.animalclinic.com

4. Dog Owner's Guide
   http://www.canismajor.com/dog

5. Net Vet Veterinary Resources
   http://netvet.wustl.edu

6. United Kennel Club
   http://www.ukcdogs.com

7. The Pet Center
   http://www.thepetcenter.com

8. Pet Planet
   http://www.petplanet.co.uk

9. Your Purebred Puppy: An Honest Guide to Purebred Dogs
   http://www.yourpurebredpuppy.com

10. Pet Education.com
    http://www.peteducation.com

11. The Dog Breed Info Center
    http://www.dogbreedinfo.com/americancaninehybridclub.htm

12. The Dog Pages
    http://www.dogpage.us/designer_dog_breeds.html

END NOTES

1. From the *Encyclopedia International,* 1979 edition. Copyright 1979 by Lexicon Publications, Inc. Reprinted by permission.

2. American Kennel Club [Online March, 2007] AKC Dog Registration Statistics: 2006. Available at http://www.akc.org/reg/dogreg_stats.cfm

CHAPTER **8**

Cats

OBJECTIVES

After reading this chapter, one should be able to:

- 🐾 discuss the history of the cat.
- 🐾 list the major groups and breeds of cats.
- 🐾 discuss the proper methods of feeding cats.
- 🐾 discuss the general grooming and health care of cats.
- 🐾 discuss the various diseases of cats.

CLASSIFICATION

Kingdom	—	Animalia
Phylum	—	Chordata
Class	—	Mammalia
Order	—	Carnivora
Family	—	Felidae
Genus	—	*Felis*
Species	—	*F. catus*

TERMS TO KNOW

agouti

cochlea

colorpoint

coronavirus

ELISA (enzyme-linked immunosorbent assay)

feral

IFA (indirect immunofluorscent antibody assay)

Jacobson's organ

jaundice

nictitating membrane

occlusion

olfactory mucosa

papillae

points

primary viremia

secondary viremia

HISTORY

Cats are descendants of the same carnivorous, tree-climbing mammal *Miacis* from which dogs descended. Cats, like dogs, trace their lineage through *Cynodictis*. The skeleton of these early primitive carnivores was much like that of a civet or a weasel. The body was long and flexible, the limbs short. All five toes were present, which appear to have been armed with retractile claws. These genera are customarily considered primitive dogs, but their general characteristics would make them ideal ancestors of the whole series of Carnivora.

Proailurus, the first normal or feline cat, appeared about 35 million years ago. The exact evolution of the cat from that time forward is not known, but modern cat types, such as *Felis* and related genus groups with "normal" canine teeth, appeared about 7 million years ago. Zoologists do not know whether cats evolved from long-tusk or saber-tooth felines or whether they evolved as a separate and distinct genus.

In the last chapter, we learned that dogs were domesticated about 10,000 years ago. The cat, however, was not domesticated until much later; this was probably because of its independent, solitary nature. It is believed that domestication of the cat began around 4,000 years ago, although there is no clear documentation until around 2500 BC. Early Egyptian paintings and inscriptions suggest that the cat was kept in captivity, tamed, and eventually revered and protected. The cat was used in religious ceremonies, and many mummified cats have been found in Egyptian tombs.

As civilization spread from Egypt and the Middle East, domestication of the cat spread also. The Romans introduced the cat into Europe, and European explorers, travelers, and traders carried the cat to all parts of the world. Today, the domesticated cat can be found on every continent except Antarctica.

Cats have long been valued by farmers, ranchers, and home owners for their use in controlling rats and mice; they were also carried aboard ships for this purpose.

It is believed that the domesticated cat of today originated from the African Wild Cat *(Felis libyca).* This cat was common to all areas of Asia and North Africa. The African Wild Cat had tabby or striped markings similar to the tabby today. It was larger than today's domestic cat but was a very quick, agile animal.

Two other cat species were the Jungle Cat *(Felis chaus)* and the European Wild Cat *(Felis silvestis).* Both had varying tabby markings and were known to have been kept by the Egyptians and Romans.

Zoologists disagree on the exact number of cat species; there are about 36 or 38 species depending on the method of classification.

Cats have erect ears and large eyes with vertical-slit pupils. They have 28 to 30 teeth, five toes on each front foot, and four toes on each hind foot. All cats, except the cheetah, can retract their claws.

Cats are usually divided into three genus groups. One group is the large, roaring cats, *Panthera,* which includes the lion, leopard, tiger, and jaguar. Second are the cats that cannot roar, *Felis,* which include the small cats and the domesticated cat. The third group, *Acinonyx,* includes only one member, the cheetah. The cheetah belongs to a special group because it has claws that do not fully retract.

Domesticated cat breeds are commonly divided into two major groups: the short-haired breeds and the long-haired breeds. A miscellaneous or third group includes the breeds that are semi-long-haired and those breeds that may be either short haired or long haired. There are 12 long-haired breeds, 26 short-haired breeds, and 5 breeds in the third group listed in this text. Not all breeds are recognized by the popular cat clubs.

GROUPS AND BREEDS

Short-Haired Breeds

The Abyssinian

The Abyssinian is believed to be a direct descendant of the Sacred Cat of Egypt, which was worshiped and held sacred; however, its exact origin is not known. Modern-day Abyssinians come from several cats that arrived in England around 1868. It is believed that the cats were brought to England by soldiers returning to England from the Abyssinian War. (Abyssinia is now known as Ethiopia.) *See Figure 8–1.*

There are three recognized breed colors: ruddy, red, and blue. Ruddy is an orange-brown color, ticked with dark brown or black. The red is usually ticked with chocolate-brown, and the blue is a blue-gray, ticked with slate-blue. (*Ticked* refers to the darker colors that are found on the tips of each hair.)

Abyssinians are considered extremely affectionate, quiet, highly intelligent, and alert. The breed is very active, can learn to retrieve, and is fond of water.

The Abyssinian is medium sized with a firm, muscular, well-balanced body. Its head is slightly rounded and wedge-shaped with large ears and large, almond-shaped eyes.

Figure 8–1   Abyssinian. *Photo by Isabelle Francais.*

The American Shorthair

The American Shorthair is a descendant of short-haired cats brought to the United States from England. They arrived with the early settlers and have been a part of our lives ever since. There are 34 recognized colors and patterns, with the classic tabby pattern being the most common. The American Shorthair is extremely affectionate and makes a great lap cat; it is easily disciplined and trained. *See Figure 8–2.*

This cat is a medium- to large-sized cat. It is heavily built and well muscled, with a broad chest, heavy shoulders, and thick hind legs. It has an oblong head, wide face, and wide-set ears. Females of the breed can reach 10 pounds, with males reaching 14 pounds.

Figure 8–2   American Shorthair. *Photo by Isabelle Francais.*

The American Wirehair

The American Wirehair originated from a litter discovered in a barn in upstate New York in 1966; the breed is a mutation of the American Shorthair. The general characteristics of this breed are the same as the American Shorthair; however, the hair coat is very dense, resilient, and coarse, with a curly appearance. *See Figure 8–3.*

Figure 8–3   American Wirehair. Note the curly whiskers and rough, wiry hair coat that is a characteristic of the wirehair breed. *Photo by Isabelle Francais.*

All the hair on the animal's body is curly, including the hair on the toes, ears, and whiskers. Extra care must be taken not to damage the wiry-hair coat. The coat should never be brushed or combed because this will rip out the hair. These hairs easily pick up dirt, dust, and lint. Regular bathing at 3-week intervals may be necessary to keep the coat clean.

The Bombay Breed

The Bombay breed was produced in 1958 by crossing a sable Burmese with a black American Shorthair; the result is a jet-black cat with gold to copper eyes. The Bombay is considered a graceful, charming cat that gets along well with others, even with strangers. *See Figure 8–4.*

The Bombay is a medium-sized cat, with a rounded head and broad forehead. The ears are rounded and tilt slightly forward, and there is a deep nose break. The Bombay may carry genetic abnormalities that are common to the breed.

Figure 8–4 Bombay. *Photo by Isabelle Francais.*

The Burmese

The Burmese breed developed in the United States in the 1930s; Dr. Joseph Thompson crossed a brown female Burma with a Siamese. The cross produced kittens with Siamese coloring, dark-brown kittens, and intermediate kittens. The dark-brown kittens were retained for breeding purposes. *See Figure 8–5.*

The Burmese is considered to have a sweet disposition and enjoys being held. However, the breed can become bossy, stubborn, and angry. The Burmese is a medium-sized cat with a solid, short body. The breed has a rounded head with a broad, short muzzle and a distinct nose break. They are sable-brown in color with gold eyes.

Upper respiratory diseases are common in kittens, and inbreeding has increased the incidence of heart defects, mutative malformations of the face and head, and other abnormalities.

Figure 8–5 Burmese. *Photo by Isabelle Francais.*

The British Shorthair

The British Shorthair is the oldest natural English breed and probably traces its lineage back to the domestic cats of Roman times. The British Shorthair is considered quiet and easygoing. It sleeps a lot and is content to lie around the house. *See Figure 8–6.*

There are 18 recognized colors and patterns of the British Shorthair cat, with most colors having gold or copper eyes. Although it is related to the American Shorthair, the British Shorthair is larger and taller and has a large, rounded head with big, round eyes.

Figure 8–6 British Shorthair. *Photo by Isabelle Francais.*

The Chartreux

The Chartreux is among the oldest natural breeds. These cats are believed to have been brought to France from the Cape of Good Hope by Carthusian monks during the seventeenth century. It is not known whether the monks gave the cat its name or whether people associated the cat with the monastery and referred to them as Chartreux. The name of the monastery was known as "Le Grand Chartreux." *See Figure 8–7.*

The Chartreux is a large, robust cat with a broad chest and powerful hind legs. Males may reach a weight of 14 pounds. The head has a wide forehead, extremely full cheeks, and narrow but not pointed muzzle. The coat is gray-blue, and the eyes are gold to copper.

The Chartreux is considered warm, friendly, and highly tolerant of children and dogs. The breed is easy to train and is rarely heard. Males are considered clean, nonaggressive, and rarely spray or leave an odor.

Figure 8–7   Chartreux. *Photo by Isabelle Francais.*

The Colorpoint

The Colorpoint Shorthair was developed in England in 1947. **Colorpoint** gets its name from the color of the extremities: nose, ears, feet, and tail. Colorpoint cats have one general body color with darker extremities or **points.** Standard-point or regular-pattern Siamese were bred with red, domestic cats. This cross produced solid-red offspring, which were bred back to Siamese, producing tortoiseshell-pointed Siamese. When these cats were bred to one another and their offspring bred back to standard-point Siamese, the results were cats with red, cream, and tortoise points, as well as standard colorpointed offspring. Siamese breeders did not want to recognize any colors beyond the classic four point colors (seal, blue, chocolate, or lilac); thus, the new colors were given a separate breed identity. *See Figure 8–8.*

The breed is considered bold, very demanding, vocal, and full of nonsense. Colorpoints are considered very intelligent and loving. The Colorpoint is genetically Siamese and possesses the same characteristics, with an elongated, angular body and wedge-shaped head. The breed is medium in size, and the eyes, like those of the Siamese, are a vivid blue.

Figure 8–8   Colorpoint Shorthair. *Photo by Isabelle Francais.*

The Cornish Rex

The Cornish Rex originated in Cornwall, England, in 1950 as a result of mutation and inbreeding. The main characteristic of the Rex is its curved, rippling hair coat that is lacking in guard hairs. *See Figure 8–9.*

The Cornish Rex is very affectionate, extroverted, persistently inquisitive, and energetic. The breed is playful and excellent with children; it is considered a lap cat and should not be an outdoor cat.

Figure 8–9   Cornish Rex. Rex cats have extreme curl to their whiskers and curly hair coats. *Photo by Isabelle Francais.*

The Cornish Rex is a small- to medium-sized cat with a long, narrow head, rounded forehead, and Roman nose. The ears are large, flared, erect, and set high on the head. The eyes are oval and give the cat a look of surprise. The body of the Cornish Rex arches like that of a greyhound; there are 28 recognized colors and patterns.

Some care should be taken when brushing the cat, because the curly coat can be damaged easily. One should use a soft brush and avoid overbrushing.

The normal body temperature of the Cornish Rex is one degree higher than other breeds and may be confusing to veterinarians not familiar with the breed. Because Rex cats lack guard hairs, they need to be protected from extreme cold and heat.

The Devon Rex

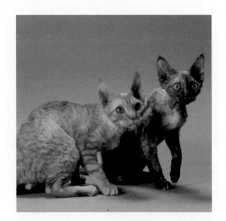

The Devon Rex is a result of mating a **feral** (wild) domestic cat to a cared-for stray cat; although the breed has a wavy hair coat similar to the Cornish Rex, the two breeds are different. *See Figure 8–10.*

The Devon Rex is a medium-sized cat, with a wavy coat that is longer and less curly than the Cornish Rex. The face is short and wedge-shaped with pronounced cheekbones, short nose, and prominent whisker pads. The large, wide-based ears set low on the head are an outstanding characteristic of this breed.

There are 34 recognized colors and patterns of the Devon Rex, with most having gold eyes. Like the Cornish Rex, grooming should be done carefully. This breed likes people, can be taught to retrieve, and is an excellent climber.

Figure 8–10 Devon Rex. *Photo by Isabelle Francais.*

The Egyptian Mau

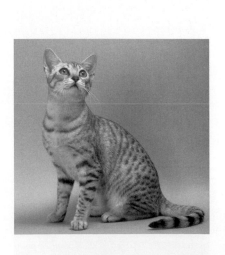

The Egyptian Mau traces its ancestry back to 1400 BC; *Mau* is the Egyptian word for cat. The breed was first imported into the United States in 1953; the most important characteristic is its spotted pattern. There are three colors: silver, bronze, and smoke; the eyes are gooseberry-green. *See Figure 8–11.*

The character of the Mau varies considerably. Some make great pets, are patient with children, and enjoy being handled; others are excitable, unpredictable, and react adversely. They tend to become more friendly with one or two people. Bloodlines should be considered when purchasing an Egyptian Mau because of character trait differences.

The Egyptian Mau is a medium-sized cat. Its head is a gently contoured wedge with almond-shaped eyes; the ears are broad at the base and erect.

Figure 8–11 Egyptian Mau. *Photo by Isabelle Francais.*

The Exotic Shorthair

The Exotic Shorthair was developed by crossing Persians with American Shorthairs. This cross produced a breed that has a compact, short-legged build, with the short, pushed-in face characteristic of the Persians and the short, thick hair coat of the American Shorthair. *See Figure 8–12.*

This breed makes an ideal pet. It is calm, quiet, and good with children. The Exotic Shorthair is a medium- to large-sized cat and resembles the Persian, except for the short hair. It has a massive, round head, with a tremendous skull and small, rounded ears. The eyes are large, round, and set apart; the nose is short and has a deep break.

Colors are the same as the American Shorthair with Persian color additions; the eye color is primarily copper.

Figure 8–12　Exotic Shorthair. *Photo by Isabelle Francais.*

The Havana Brown

The Havana Brown is a result of crossing a black domestic Shorthair with a seal-point Siamese carrying the brown gene. The resulting solid-colored brown cat was further developed into two separate breeds: the chestnut Oriental Shorthair and the Havana Brown. The name *Havana Brown* comes from the color, which is similar to rich Havana tobacco. *See Figure 8–13.*

The Havana Brown is an excellent companion cat that enjoys lots of attention. It is a medium-sized cat with a distinctive rectangular muzzle and is brown or mahogany in color. The ears are very large, stand erect, and are pointed slightly forward. There is very little hair on or in the ears; the eyes are a vivid green.

Figure 8–13　Havana. *Photo by Isabelle Francais.*

The Japanese Bobtail

The Japanese Bobtail cat is native to Japan. Paintings, wood carvings, and statues date it back to very early times. In Japan, this breed is known as Mi-Ke and is a symbol of good luck; *Mi-Ke* in Japanese means "three fur." *See Figure 8–14.*

The Japanese Bobtail is affectionate, is generally sweet-tempered, and enjoys human companionship. The Japanese Bobtail enjoys water and can be taught to retrieve.

The Japanese Bobtail is a medium- to large-sized cat; its distinctive, short tail is kinked. The hair on the tail is usually longer and forms a pompon; people sometimes refer to it as a bunny tail. The head forms an equilateral triangle with high cheekbones; the eyes are slanted and luminous.

The preferred color and pattern for the Bobtail is tricolored with a red, black, and white calico pattern. Bicolors of black and white or red and white are next in preference. There are 23 possible color combinations accepted for shows. The eyes may be gold, green, or blue. White cats may have one blue and one gold eye. Care must be taken in handling the Bobtail to prevent injury and pain to the sensitive bobtail.

Figure 8–14　Japanese Bobtail. Note the short "bobtail" of this breed.
Photo by Isabelle Francais.

The Korat

The Korat is a centuries-old breed native to Thailand. It is rare even in Thailand, where they are highly prized and rarely sold. The Korat is considered intelligent and active. It makes an excellent companion cat, is protective of family members, and is reserved with strangers. *See Figure 8–15.*

The distinctive characteristic of this medium-sized cat is its silver-blue fur that is tipped with silver, producing a sheen or halo effect. The head is heart-shaped, and the eyes are very large, slanted, and luminous and appear oversized in proportion to the face. The preferred eye color is luminous green. Korat cats may be prone to upper respiratory viruses, and routine vaccinations are important.

Figure 8–15   Korat. *Photo by Isabelle Francais.*

The Malayan

The Malayan is an "other-colored" Burmese. The Burmese, as we learned earlier, is sable-brown in color. In the breeding that produced the Burmese, the colors blue, champagne, and platinum were also produced. These colors were rejected by Burmese breeders, but fanciers worked hard to establish them as a separate breed now recognized as the Malayan. The characteristics of the Malayan are basically the same as the Burmese; although stubborn and bossy, they can be outstanding companions. *See Figure 8–16.*

Figure 8–16   Malayan. *Photo by Isabelle Francais.*

The Manx

The Manx breed is native to the Isle of Man, an island between England and Ireland. The breed is the result of a genetic mutation and is among the earliest breeds registered in Europe, Great Britain, and North America. *See Figure 8–17.*

Figure 8–17   Manx "rumpy" cats do not have a tail; they have a dimple where the tail should be. *Photo by Isabelle Francais.*

The Manx is affectionate but leans toward one-on-one relationships. The breed is alert, intelligent, healthy, and powerful; it can also be devilish and stubborn. The distinctive characteristic of the Manx is the complete lack of a tail. The perfect show animal is a "rumpy" and has a dimple where the tail should be. Two other types are the "stumpy" with a short tail and the "longy" with a complete tail. *See Figure 8–18.*

Figure 8–18   Manx "stumpy" cats have a short tail. Manx "longy" cats have a complete tail.

The Manx is a solid, compact, medium-sized cat. It has a round head and muzzle, prominent cheeks, and a full face. The ears are of medium size and rounded at the tip. The rump of the Manx is considerably higher than the shoulders. There are many colors and patterns; therefore, this is not an important show characteristic. The eyes are round and primarily copper, but other eye colors are possible.

The Manx breed carries a lethal gene; its lack of a tail is a genetic defect and is related to possible spinal deformities and cartilage weaknesses. These may affect the legs and neck, causing paralysis.

The Ocicat

The Ocicat is a result of second-generation crossings of Siamese and Abyssinians; it was first developed in 1964. The first kittens resembled baby ocelots *Felis paradalis,* which are large, wild cats native to North and South America. They have a yellow or gray fur coat with black spots. The original breeders gave the cats the name Ocicat because of the similarity of the markings. *See Figure 8–19.*

The Ocicat is considered gentle, calm, friendly, and intelligent with a sweet temperament. It has a spotted, tabby-patterned coat. There are eight recognized color patterns. The hairs show bands of color in **agouti** pattern; where the bands fall together, a dot or dark area is formed. Agouti is a color pattern where one color is streaked or intermingled with another color.

The main characteristic of the Ocicat is its wild, feral appearance. Females of the breed are medium size, and the males are very large; males may weigh 12 to 15 pounds. The head is oval-shaped, with almond-shaped eyes that angle downward toward a broad nose. The ears are moderately large and sometimes tufted.

Figure 8–19 Ocicat. *Photo by Isabelle Francais.*

The Oriental Shorthair

The Oriental Shorthair resulted from the breeding that produced the Colorpoint Shorthair. The original crossings produced not only the color-pointed patterns of the Colorpoint Shorthairs, but also solid and full-colored patterns. These cats became known as the Oriental Shorthair breed. *See Figure 8–20.*

The Oriental Shorthair is a talkative companion. The breed is considered intelligent, witty, graceful, and elegant. It is extremely affectionate and dependent on people; it is also playful and agile.

The Oriental Shorthair is a medium-sized breed and almost identical to the Siamese standard. The breed is long, slender, and fine-boned. The head is long, tapering, and slightly wedge-shaped. The ears are large and flared; the eyes are almond-shaped.

There are numerous colors of Oriental Shorthairs. They are found in solid, shaded, smoke, tabby, and multiple colors. The eyes of the Orientals are green. White cats may have blue or green but not odd eyes (one blue and one green). Oriental Shorthairs are usually of good health.

Figure 8–20 Oriental Shorthair. *Photo by Isabelle Francais.*

The Russian Blue

The Russian Blue was brought into Europe from the White Sea port of Archangel in the 1860s. They have been called Archangel, Spanish Blue, Maltese Blue, and English Blue. This is a quiet, gentle, and somewhat reserved breed. The Russian Blue is fine with children and dogs; it is very affectionate with its owner but conservative with strangers. *See Figure 8–21.*

The main physical characteristic of the Russian Blue is its double plushy coat. The hair is short, dense, and fine. The Russian Blue is a medium-sized cat with a smooth, medium wedge-shaped head. The ears are large, wide at the base, and more pointed than round; the eyes are vivid green.

Figure 8–21   Russian Blue. *Photo by Isabelle Francais.*

The Scottish Fold

The Scottish Fold is a natural mutation. The first cat with folded ears was discovered on a Scotland farm in 1961. The Scottish Fold is an excellent family cat; it is reserved and rather shy, but affectionate. It has a sweet nature and is friendly. It is a quiet breed and gets along well with other cats and dogs. *See Figure 8–22.*

The main characteristic of the Scottish Fold is the ears that are folded forward and downward. This is a medium-sized cat with a rounded head, prominent cheeks, and prominent whisker pad. The eyes are large and round. There are 23 colors and patterns of the Scottish Fold. They are found in solid, shaded-tipped colors, tabby, and multiple colors.

Although generally healthy, the dominant gene that produces folded ears can cause crippling arthritis when two Scottish Folds are bred together.

Figure 8–22   Scottish Fold. Note the folded ears that are characteristic of this breed. *Photo by Isabelle Francais.*

The Siamese

The actual origin of the Siamese cat is unknown. Pictures from Atutthaya, the old capital of Siam, date back to 1350 AD, which show Siamese cats kept by royalty. *See Figure 8–23.*

The Siamese is a one-on-one cat and usually expresses devotion to one particular family member. It is described as a prima donna: extraordinarily intelligent, precocious, talkative, loyal, fearless, and obstinate; it is also totally unpredictable.

The Siamese is a medium-sized cat with long, refined, elegant tapering lines. The head of the Siamese forms an equilateral triangle from its nose to the tip of the ears. The eyes are almond-shaped and slanted toward the nose. The bones are fine, the body is slender and tube-shaped, and the legs are long.

The general body color of the Siamese shows very little pigmentation. The pigmentation is darker at the points, and color is restricted to the points: mask, ears, tail, and feet. The point colors are seal, chocolate, blue, or lilac. The eye color is always deep, vivid blue. Siamese inherit heart problems.

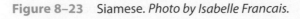

Figure 8–23   Siamese. *Photo by Isabelle Francais.*

The Singapura

The first Singapuras were imported into California in 1975 from Singapore. In Singapore, these were primarily street cats. Because of this background, the Singapura may show some shyness, but it warms readily to people who show it no harm. The Singapura is a quiet cat that adapts quickly, is very playful, and becomes very sociable around people. *See Figure 8–24.*

The Singapura is a small breed. Males weigh no more than 6 pounds, and females usually weigh less than 4 pounds. It has a rounded head with large eyes and ears. The nose is blunt on a broad muzzle with a slight stop below eye level.

The main characteristic of the Singapura is its ticked coat. The colors are usually yellow or ivory, with light brown hairs interspersed; there are darker bars on the inner front legs and back knee. The eyes are hazel-green or yellow and are described as being extremely beautiful and expressive.

Figure 8–24 Singapura. *Photo by Isabelle Francais.*

The Snowshoe Breed

The Snowshoe breed originated in the Kensing Cattery in Philadelphia, where two Siamese cats produced three females with white feet; the distinctive white feet led to its name. Snowshoe cats are considered people-oriented and will follow their owners everywhere. They are rather docile and make excellent house pets. *See Figure 8–25.*

The Snowshoe is medium to large in size. The body type is similar to the American Shorthair. It is medium-boned, well-muscled, and has a solid body with a medium neck in proportion to the body; the head forms a modified wedge with high cheekbones. The nose rises slightly at the bridge, and the ears are large, broad at the base, and pointed. The eyes are large, oval, and slant toward the nose.

The Snowshoe has white-tipped front paws and a white chest, chin, and muzzle; the hind legs are white to the hock. The mask, tail, ears, and legs are usually of defined seal or blue, with masking covering the entire face, except for an inverted white V-shaped pattern. The eyes are a bright, sparkling blue.

Figure 8–25 Snowshoe cats have distinctive white feet.
Photo by Isabelle Francais.

The Sphynx

The Sphynx is a hairless cat that was first introduced by breeders in Canada in the late 1960s. Some evidence shows that hairless cats existed at the time of the Aztec Indians, and the Mexican Hairless was somewhat popular in the 1800s. *See Figure 8–26.*

The Sphynx breed, because of its lack of fur, is very dependent on people for survival. It cannot survive outdoors and may even become ill if exposed to drafty conditions. It is very affectionate, has a sweet nature, and adores body contact. It purrs constantly.

Figure 8–26 Sphynx. Note the lack of normal fur and whiskers of this breed.
Photo by Isabelle Francais.

The main characteristic of the Sphynx is the tough, wrinkled, hairless skin. The breed has a sweet facial expression, expressive eyes, and large ears. The body type is similar to that of the Devon and Cornish Rex.

The Sphynx cat must be protected from the cold at all times; sweaters are recommended to keep the animal warm. The normal body temperature is four degrees higher than that of most other breeds.

The Tonkinese

The Tonkinese was developed during the 1950s through the 1970s and is a hybrid cross between the Siamese and Burmese. The breed was recognized by the Canadian Cat Association in 1974 and by the Cat Fanciers Association in 1978. Tonkinese bred together usually produces a litter that is one-half Tonkinese, one-fourth Siamese, and one-fourth Burmese. *See Figure 8–27.*

The breed is said to be a combination of the best of Siamese and Burmese. It is considered clever, active, intelligent, fearless, and obstinate. It is also mischievous and a prankster. The Tonkinese is of medium size; the head is of modified wedge shape with a slight dip between the eyes. The body is of medium length, as is the tail and legs.

The Tonkinese are Siamese-patterned cats in five unique breed colors. The densely colored mask, ears, feet, and tail gradually lighten as you move from the points. The coat color, unlike the Siamese, is a dilution of the point color. The five breed colors are natural mink, blue mink, champagne mink, platinum mink, and honey mink; all colors have characteristic aqua eyes.

Figure 8–27   Tonkinese. *Photo by Isabelle Francais.*

Long-Haired Breeds

The Balinese and Javanese

The Balinese and Javanese have been developed from litters of registered Siamese. Except for the coat length, the breeds maintain the same body style as the Siamese. The name *Balinese* is derived from the South Sea island of Bali, and the name *Javanese* is derived from Java. *See Figures 8–28* and *8–29.*

Figure 8–28   Balinese. *Photo by Isabelle Francais.*

The Balinese and Javanese are considered warm, eager, inquisitive, and enthusiastic. They are also vocal and persistent. The Balinese and Javanese are long-haired cats with Siamese coat patterns. Balinese colors are seal point, chocolate point, blue point, and lilac point. Javanese colors include red, tortoise, and lynx point patterns. Color on the diamond-shaped mask, ears, legs, feet, and tail strongly contrasts with the pale, even body color. The eyes are deep, vivid blue.

Although these two breeds are generally healthy, certain lines may have weak hind legs. Nasal obstruction or poor **occlusion** (upper and lower teeth not fitting together properly) can cause breathing through the mouth.

Figure 8–29   Javanese. *Photo by Isabelle Francais.*

The Birman

The Birman is known as the sacred cat of Burma; the exact origin of the breed is not known. The breed first became known outside of Burma in 1898. The Birman is a very sociable cat that needs the company of others. It is considered a sweet-natured, charming, and intelligent cat. *See Figure 8–30.*

The breed is medium to large in size. The Birman is a sturdy, stocky cat with a long body. The head is strong, broad, and rounded, with a strong, rounded muzzle and low-set nostrils. The eyes are round, and the ears are medium in size with rounded tips.

The coat of the Birman is long and silky. The colors are seal point, chocolate point, blue point, and lilac point, all with white gloves (white paws). The front paws are white to the third joint, and the back feet are white up to the hocks. The long, silky coat of the Birman does not mat and requires little care.

Certain lines of the Birman may show genetic defects in the form of weak hind legs, nasal obstruction, and poor occlusion, which may cause breathing through the mouth.

Figure 8–30   Birman. *Photo by Isabelle Francais.*

The Cymric

The Cymric (pronounced Kim-rik) is a long-haired Manx. During the original breeding of the Manx cats, some long-hair cats occasionally resulted due to recessive genes. Both Shorthair Manx parents must have the long-hair gene to produce Cymric offspring. Cymrics bred to each other produce only long-haired kittens. *See Figure 8–31.*

The Cymric is alert and sweet-natured. It is considered very quiet, extremely intelligent, and observant. It is excellent with other pets and with children.

The Cymric is a large, tailless breed with moderately long hair. The head is round with a broad muzzle and prominent cheeks. The front legs are short, balanced, substantial, and contrast with the higher, heavily muscled hindquarters. Color is not very important for show points. Taillessness in the Cymric can be associated with defects not always apparent in kittens. Be alert for signs of weakness in hindquarters such as unusual gait or limp.

Figure 8–31   Cymric. The long-haired Manx. Note the lack of a tail.
Photo by Isabelle Francais.

The Himalayan and Kashmir

The Himalayan and Kashmir were developed in Britain and North America in the 1930s. The crossing of Siamese, Persian, and Birman produced a Persian-type cat with a long, flowing coat and a Siamese pattern. During the development of the chocolate-point and lilac-point Himalayans, solid-chocolate and solid-lilac kittens often showed up in litters. These solid-colored Himalayans are now called Kashmirs. *See Figures 8–32 and 8–33.*

The Himalayan and Kashmirs are easygoing breeds that fit well into the family lifestyle. They like attention, purr loudly, and are considered intelligent and outgoing. Their voices are louder than that of the Persian but much quieter than that of the Siamese.

Figure 8–32 Himalayan. *Photo by Isabelle Francais.*

The Himalayan and Kashmirs are medium- to large-sized cats; the body is short and compact. The head is large on a short neck, and the nose is snub-like with good break. The ears are set low on the head; the eyes are large, round, and expressive. The eyes of the Himalayan are a vivid blue, whereas the eyes of the Kashmir are copper-colored. The very short nose of the Himalayan and Kashmir may be associated with tearing because of malformed tear ducts; breathing problems may also be possible in some lines.

Figure 8–33 Kashmir. *Photo by Isabelle Francais.*

The Maine Coon

The Maine Coon is considered to be the native American long-hair cat that first appeared on the rocky coast of New England. It is the oldest natural breed in North America. The Maine Coon is considered lovable and friendly and enjoys the companionship of another pet or person. The breed has a tiny voice with squeaks and chirps. *See Figure 8–34.*

The Maine Coon is a large-sized cat, often reaching 12 to 18 pounds. It has a medium-wide head, medium-long nose, and large tufted, tall ears. The fur on the tail is especially long compared with the fur covering the body. The brown tabby is the best-known color and pattern for the Maine Coon, but a wide range of colors and patterns exists. Eye color varies from green to gold to copper; white cats have blue eyes.

Figure 8–34 Maine Coon. *Photo by Isabelle Francais.*

The Norwegian Forest Cat

The Norwegian Forest Cat is a breed native to Norway and is believed to be descended from numerous cats that lived in the Norwegian woodlands. This cat is considered exceedingly alert and intelligent. The breed is responsive and loving to its owner and must have companionship. *See Figure 8–35.*

The Norwegian Forest Cat is medium to large in size, with a moderately long body and powerful appearance due to a full chest and considerable girth. The tail is long and flowing. Almost all color combinations are acceptable.

Figure 8–35   Norwegian Forest Cat. *Photo by Isabelle Francais.*

The Persian Breed

The Persian breed is one of the oldest and most popular breeds. The exact origin of the breed is not known, but the present Persian cat is believed to be descended from long-haired cats introduced into Italy from Asia and cats introduced into France from Ankara, Turkey. The Persian cat is well mannered, easygoing, and quiet. It is an excellent companion and adapts well to apartment living. *See Figure 8–36.*

The breed is medium to large in size and has a short, compact body. The head is large with ears that are small and round-tipped. The nose is short, broad, and stubbed with a deep break; the eyes are large and round, giving the cat a sweet expression.

Figure 8–36   Persian. *Photo by Isabelle Francais.*

The Persian breed is divided into six groups for show purposes: solid colors that usually have copper eyes, except for the white, which may have blue, copper, or odd eyes; shaded colors that have green, blue-green, or brilliant copper eyes; smokes that have copper eyes; tabby colors that have brilliant copper eyes, except for the silver tabby, which has green or hazel eyes; multicolors that have copper eyes; and the point-restricted colors that have vivid blue eyes.

Some white Persian cats are born deaf; this usually occurs in cats with blue or odd eyes. Odd-eyed cats, having one blue eye and one copper eye, may be deaf only on the blue-eyed side. Some lines with very flat faces may experience breathing problems or develop clogged tear ducts.

Persians usually require daily grooming. Cats that do not receive regular grooming develop tangles and knots in their fur. Regular bathing is necessary to remove excess oil from the coat.

The Ragdoll

The Ragdoll resulted from the mating of a white Persian and a seal-point Birman. The offspring of the cross were then mated to a sable Burmese; the breed was developed in the 1960s. It takes its name from the limp and floppy posture taken when being handled, and is said to feel no pain or fear. The Ragdoll is considered docile, quiet, and composed. The breed is very intelligent and enjoys retrieving. *See Figure 8–37.*

The Ragdoll is exceptionally large and heavy. The head is moderately wedge-shaped, and the ears are of medium size. The eyes are large, oval, and blue in color. Four colors are common: seal point, chocolate point, blue point, and lilac point; these colors are found in three distinct patterns. In the first pattern, there can be no white. In the second, or mitten, pattern, there are white gloves or paws in addition to the color points and a white abdomen. The third, or bicolored, pattern has colored points with white on the head, body, and legs.

Figure 8–37 Ragdoll. *Photo by Isabelle Francais.*

The Somali

The Somali was produced in the 1960s when Abyssinian parents carrying the recessive gene for long hair were mated. The Somali is considered friendly, affectionate, companionable, playful, intelligent, and very active. *See Figure 8–38.*

The breed is medium in size. The body characteristics are basically the same as those for the Abyssinian, but with a medium-long coat and a fox-like, bushy tail. There are two special colors: ruddy, an orange-brown ticked with black; and red, ticked with chocolate-brown. Both breed types have gold or green eyes.

Figure 8–38 Somali. *Photo by Isabelle Francais.*

The Turkish Angora

The Turkish Angoras were originally known as Ankara cats because the breed originated in the province of Ankara, Turkey. They probably share the same distant ancestors as the Persian but developed from a different breeding population. *See Figure 8–39.*

This breed is considered polite, intelligent, and responsive. It is easily trained to retrieve and perform tricks. The Turkish Angora is fastidious and prefers a clean, spotless environment.

The Turkish Angora is a small- to medium-sized cat with a long, slender, graceful torso on a light frame. The ears are large and erect on a small- to medium-sized, wedge-shaped head.

The breed was originally bred only in white, but many colors and patterns exist today. The fur is medium in length and silky. The breed lacks the thick underfur that its Persian cousin has. The eyes are large, almond-shaped, and show radiant, deep color.

Figure 8–39 Turkish Angora. *Photo by Isabelle Francais.*

Miscellaneous Breeds

American Curl

This is a relatively new breed. The first of the breed appeared on the doorstep of a family in Lakewood, California, in June, 1981. The popularity of the breed has increased tremendously since its beginnings.

The cats appear normal in every way except for their ears, which curl back from their heads. Studies have determined that the curled ear was a spontaneous mutation, and the gene that causes the ear to curl is a dominant gene. When the American Curls are born their ears are normal, but within 2 to 7 days the ears begin to curl back.

American Curls are people oriented. They are very affectionate and social. They tend to keep their youthful playfulness into their more mature years. The breed is considered very intelligent and curious.

Variations in ear type, head shape, and body length occur in this new breed. These differences are probably due to the non-pedigree domesticated cats that have been used in their breeding program. As more guidelines are used in their breeding programs, more definite standards will be developed. *See Figure 8–40.*

Figure 8–40   American Curl Longhair. *Photo by Isabelle Francais.*

Turkish Van

This is a rare and ancient breed of cat developed in Central and Southwest Asia. The term "Van" in the local language is given to several towns and villages in the region. The breed arrived in England in 1955 and in the United States in 1982.

The Turkish Van is white in color with color markings on the head and tail. They are semi-long-haired cats. The coat of the Turkish Van lacks an undercoat and has a unique, cashmere-like texture. This characteristic makes the coat water resistant. They love water and were referred to as "Swimming Cats" in their native land. The breed is considered intelligent, curious, and makes good companions. *See Figure 8–41.*

Figure 8–41   Turkish Van. *Photo by Isabelle Francais.*

LaPerm

This breed originated from a 1982 mutation in a "barn cat" on a farm in Oregon. The original kitten was born bald, but began to develop soft, curly hair at about 8 weeks of age. The name LaPerm means wavy or rippled.

The LaPerm can have either long hair or short hair. The coat can be wavy to ringlet-type curls. The curls can lie in tight ringlets to long corkscrews. The tightest curls will be on the underside, the throat, and at the base of the ears. The coat can be in every recognized color and pattern. Kittens can be born either bald or with short wavy hair. Some kittens born with hair may go bald shortly after birth, but the hair will come back in about 2 weeks.

LaPerm cats are considered gentle and affectionate. They can be very active, but are content being lap cats. The breed has a unique combination of appearance and personality. *See Figure 8–42.*

Figure 8–42   LaPerm. *Photo by Isabelle Francais.*

Selkirk Rex

This breed originated from a mutation in a house cat found in a Montana shelter. The Selkirk Rex is being developed as a large, heavy-boned cat. The coat has a curl that is plush and loosely curled. The curls are most pronounced in the long-hair variety. The most curls appear on the underside, around the neck, and on the flanks. The least amount of curl is on the back. Kittens with the curly coats at birth will also have curly whiskers. The whiskers are brittle and may break as they grow longer.

The breed is considered patient, loving, and tolerant. All coats can be in any of the recognized colors and patterns. *See Figure 8–43.*

Figure 8–43 Selkirk Red. *Photo by Isabelle Francais.*

Siberian

This breed is native to Russia and dates back more than one thousand years. The Siberian was first imported into the United States in 1990 and is fairly rare in this country.

The Siberian has a semi-long-haired coat and can be found in all recognized colors and patterns. Their coat is soft and plush. They become loyal to their adopted families and are considered intelligent, playful, and affectionate. The Siberian is very agile and can leap great distances and heights. They are strong and powerfully built. Males typically weigh between 15 and 20 pounds and females 10 to 15 pounds. *See Figure 8–44.*

Figure 8–44 Siberian. *Photo by Isabelle Francais.*

ANATOMY

The anatomy and skeletal structure of the cat are identified in *Figures 8–45* and *8–46.*

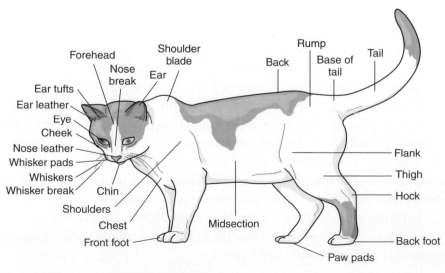

Figure 8–45 The anatomy of the cat.

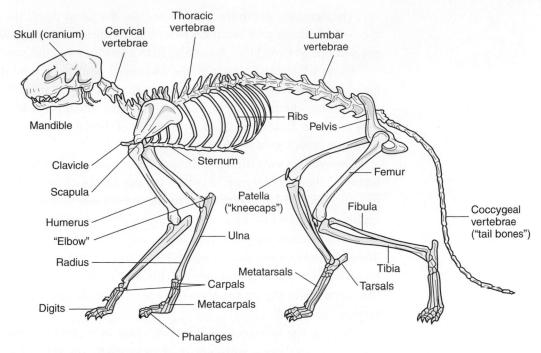

Figure 8–46 The skeletal structure of the cat.

CHARACTERISTICS

The body of the cat has developed for speed and flexibility. The bones, joints, and muscles form together to give the animal its size, shape, speed, and strength.

The skeleton of the cat contains 244 bones, which is about 40 more than a human; most of these extra bones are found in the spine and tail. Spread over this skeletal structure is approximately 500 separate muscles, compared with more than 650 in the human body. The largest muscles on the cat are found on the rear legs, which gives the animal its tremendous power for running and jumping.

One of the most obvious characteristics of the cat is its beautiful coat, which provides the cat with protection from excessive water loss, heat, cold, excessive sunlight, and physical injuries, and guards against the invasion of germs and parasites. Cats are carnivores that use their forelegs for grasping prey. They also have whiskers and eyelashes that are sensitive to touch.

The sense of touch, although important in the cat, appears to be one of its less important senses. Specialized nerve endings that respond to touch are found throughout the skin. Petting the animal usually brings about a relaxing purr or, if not appreciated, results in the cat extending its claws and teeth.

The cat's nose and the paw pads are especially sensitive. Cats often extend a paw to investigate food, water, or unfamiliar objects. The nose and the tongue are then used to further determine smell and taste.

The ears of the cat are supersensitive. Being a hunter, the cat can detect the slight sounds of mice and other prey. The ears can detect sounds in the ultrasonic range, far higher in pitch than the human ear.

The structure within the ear that enables the cat to detect these sounds is the **cochlea;** its range of hearing is from 30 cycles per second up to 65,000 cycles per second (30 to 65 kHz). It can also detect and locate sound very efficiently.

Some white cats, especially the blue-eyed whites, may be deaf. This is caused by degenerative changes in the cochlea starting at about 5 days of age.

Another characteristic of the cat is its eyes, which point forward to provide for three-dimensional vision. The eyes of the cat function in almost total darkness as well as in bright daylight.

Cat eyes are very similar to those of other animals but have some special features. They are designed to collect the maximum amount of light; the cornea and lens are large in relation to the overall size of the eyeball. They are highly curved, and the lens is farther back in the eyeball than in humans. The tapedum lucidum is a membrane inside the back of the eye that reflects light. These differences give the cat the ability to project an image on the retina five times brighter than the human eye. This, along with the greater sensitivity of the retina, gives the cat the ability to see in light six times dimmer than the human eye can see.

The cat eye has three eyelids; two close to protect the eye if anything touches the eyelashes, whiskers, or the eye itself. The third is known as the **nictitating membrane,** or third eyelid, and moves diagonally across the eye under the eyelid to help lubricate the cornea.

The eyes of kittens remain closed for about 10 to 12 days after birth. It takes about 3 months before the eye is completely developed.

Smell receptors located in the lining of the nose enable the cat to pick up the smell of airborne substances. Taste receptors located on the tongue allow the cat to detect and recognize different substances it ingests.

Cats differ from humans in that they have a third set of receptors. These receptors, called **Jacobson's organ,** are located in the roof of the mouth. The Jacobson's organ is stimulated by odors that are picked up by the tongue from the air and transferred to the organ when the cat presses its tongue against the roof of the mouth.

The **olfactory mucosa** is the nasal lining that picks up smells from the air. This area is almost twice the size of the olfactory mucosa of humans and indicates the importance of smell to cats.

The cat's tongue has several functions and is lined with knobs called **papillae.** The papillae in the center of the tongue form backward-facing hooks, which help hold food and are used to lick or scrape meat off bones. Papillae along the front and sides of the tongue carry the taste buds; large papillae located at the back of the tongue also carry taste buds.

The tongue is spoon-shaped and enables the cat to lap up liquids; the cat swallows after every four or five laps. The tongue is also used for grooming.

CHOOSING A CAT

Many people select cats for pets for their beauty and personality. Cats are independent and self-sufficient. They make excellent companions, especially for the elderly, and require less care than dogs. Cats fit well into apartment living and the less active lifestyle of the elderly person. The initial purchase price and the cost of feeding and keeping a cat is usually less than that for a dog. Cats are clean, generally requiring very little or no house-training. Depending on the circumstances, cats can be used to control mice and other rodents.

For the busy family or elderly person, it may be best to purchase an adult cat. It will not require as much time to care for as a young kitten. Cats should be taken to a veterinarian soon after obtaining them, because they may have parasites. This is especially true if the cat is a stray. Older cats may not live as long, which could be emotionally upsetting. However, many cats live into their late teens and early twenties—considerably longer than the life span of the average dog.

Young kittens adapt quickly to a new family; they can be fun to watch as they grow. The more attention a kitten receives, the closer it becomes attached to the family.

If one does not intend to breed the cat, it is best to have the animal spayed or neutered. It makes very little difference if the pet cat is male or female; both are equally intelligent, affectionate, and playful.

A male kitten reaches sexual maturity at about 6 or 8 months of age; even if the owner lets it outside, it will start spraying areas to mark its territory. The smell is very strong and unpleasant; if one decides to keep a male cat, special housing may need to be provided.

A female kitten becomes sexually mature as early as 5 months of age. She usually comes in heat about every 3 weeks; unwanted litters may be expected by keeping an unspayed female cat.

If one is just looking for a pet, the common domestic, mixed-breed house cat will probably make a suitable selection. These animals have traits from several breeds and make affectionate, friendly pets.

Pedigree animals usually have both good and bad characteristics; one needs to be aware of these traits before purchasing a particular breed. Some breeds are basically inactive, solitary animals, whereas others are more active, playful, and demanding. Long-haired cats require more time for grooming. Also, the climate of the area is important; unless they have air conditioning, long-haired cats may become uncomfortable in high temperatures.

Mixed-breed animals can usually be obtained free from neighbors or friends or by checking advertisements in newspapers. To obtain some of the specialty breeds, one may have to contact local or state cat fancier clubs. Pet stores also offer cats, but it is important to make sure the animal has received vaccinations, because diseases may be a problem in some pet stores. Cats may also be obtained from animal shelters.

When selecting individual animals, one should be alert to signs of illness. Gums should be pale pink and the teeth white; older cats may show some yellowing of the teeth because of tartar buildup. The eyes should be bright and clear. Animals that have discharge emanating from the eyes or nose should be avoided. An extended, red third eyelid is usually a sign of illness. One should check the ears for a buildup of dark brown discharge, which may indicate ear mites. The coat should be clean, glossy, and free of mats; one should check for traces of fleas and other parasites. A potbellied condition may indicate an animal with worms. One should feel the animal to check for lumps, abscesses, or other abnormalities. The rear area of the animal should be checked for signs of diarrhea or worms. The fur should be clean and free of mats.

Kittens are best purchased at 8 to 12 weeks of age. One should select a kitten that is playful, alert, and lively and that does not shy away.

When selecting animals for breeding and show, one also needs to check for correct marking, eye color, coat color, and conformation. Be aware that adult characteristics may not develop for several months.

FEEDING AND TRAINING

Cats are true carnivores and require almost twice as much protein in their diet as dogs. The best source of this is from animal products; 30 to 40 percent of the cat's diet should be animal-type proteins (meat, meat by-products, fish, eggs, and milk). *See Figure 8–47.*

About 10 percent of the diet should consist of fat; fat provides calories and the essential fatty acids. A cat should be fed using one of the many commercial cat foods. If feeding fresh foods, it is important to provide a variety. Strictly feeding meat, chicken, fish, and other muscle meats may cause bone disease, stunted growth in kittens, poor eyesight, and other problems due to the lack of calcium and vitamin A. Too much liver can cause vitamin A poisoning. Milk is a good source of calcium, but it may cause gas and diarrhea in adult cats. Calcium can be supplemented by adding sterilized bone flour, calcium phosphate, calcium lactate, or calcium carbonate, but too much calcium can also induce health problems. Feeding too much of some oily fish such as tuna may, through oxidation, destroy vitamin E and lead to a deficiency called steatitis or yellow fat disease.

Raw egg white contains a substance that destroys the B vitamin biotin, but egg yolk and cooked egg white provide valuable protein, fat, and vitamins.

When feeding meat, it is important that all of the bone has been removed or chopped to prevent pieces of bone from becoming lodged in the throat or digestive system. Cats should never be given chicken bones. Preparing a homemade diet that is complete and balanced is difficult and time-consuming. Commercial foods are a better choice for most cat owners.

Canned foods contain more animal protein than the other commercial rations, have a higher fat content that makes them more palatable, and contain about 75 percent water. Because of the high water content, cats may not drink as much water, but water should be available at all times. Labels should be checked because some of the commercial rations may be nutritionally incomplete. Fresh

GUARANTEED ANALYSIS

Crude Protein . . . . .not less than 30.00%	Ash . . . . .Not more than 6.50%
Crude Fat . . . . .not less than 14.00%	Magnesium . . . . .not more than 0.10%
Crude Fiber . . . . .not more than 2.00%	Taurine . . . . .not less than 1500 ppm
Moisture . . . . .not more than 10.00%	Linoleic Acid .not less than 4.00%

INGREDIENTS

Chicken Meal, Ground Rice, Rice Gluten, Whole Dried Egg, Rice Flour, Poultry Fat (preserved with Tocopherols, a source of Natural Vitamin E), Rice Bran, Lamb Meal, Dried Brewers Yeast, Sunflower Oil (preserved with mixed Tocopherols, a source of Natural Vitamin E), Salt, Potassium Chloride, Natural Flavoring, DL-Methionine, Monosodium Phosphate, Taurine, Garlic, Ferrous Sulfate, Zinc Sulfate, Copper Sulfate, Cobalt Carbonate, Choline Chloride, Vitamin A Supplement, Vitamin D3 Supplement, Vitamin E Supplement, Niacin, Calcium Pantothenate, Riboflavin Supplement (source of Vitamin B2), Thiamine Mononitrate (source of Vitamin B1), Pyridoxine Hydrochloride (source of Vitamin B6), Folic Acid, Biotin, Inositol, Potassium Iodide, Vitamin B12 Supplement.

Figure 8–47 Cats require a high-protein diet. One should always check the label when purchasing pet food.

and canned foods should not be fed straight from the refrigerator; these foods should be allowed to warm to room temperature before being served.

Dry foods contain about 10 percent water and less fat and protein than semi-moist foods. Cats on dry diets should have plenty of water available. Milk, water, or gravy can be mixed with the food to improve palatability and to ensure that the cat gets adequate water intake. One may wish to feed canned foods occasionally to get the cat used to different types and textures of foods. Dry foods do have the advantage of helping to clean the teeth and prevent the buildup of tartar.

The amount of food one gives depends on the cat's age, weight, breed, condition, and amount of activity it gets. Cats and young kittens will not consume enough food in one meal to last 24 hours. Two meals or leaving out dry food all day are recommended, and young kittens and females that are pregnant or nursing require frequent feedings.

Cats should never be given a diet of dog food because it contains large amounts of cereals and vegetables. Because of this, the cat may not get enough animal protein. Dog food also lacks necessary amounts of vitamins A and B and some essential fatty acids. Many times cats will be seen eating grass. The exact reason for this is not known, but it may be an attempt to increase roughage in the diet or to eliminate a hairball.

The amount of exercise a cat needs varies considerably depending on the breed and where its home is. A cat living inside may get little exercise, whereas one allowed to run free may be more active. Indoor cats should be furnished with toys, cardboard tubes, or other play equipment to provide them with a means to exercise. Owners of valuable purebred or show cats may not want their animals to run free where they risk injury, loss, or unwanted litters. Outdoor cats get plenty of exercise; however, they run a greater risk of injury from fights, of death or injury on the roadways, and of contracting diseases and parasites. Cats living indoors should have a clean litter box and plenty of water. Owners should be aware of plants that may be poisonous to their pets. To protect the furniture, cats should be trained to use a scratching post.

Cats scratch to sharpen their claws, to remove loose scales and fragments of dry skin, and to leave a mark for other cats. Kittens should be trained to use a scratching post as soon as they are weaned. The kitten should be held near the post and its claws placed on the post. The kitten will soon learn what to do and will usually continue to use the same post. A cloth-covered post may offer the cat an alternative to clawing soft furniture and draperies.

Correct toilet training is easier with cats than with dogs because cats naturally cover their urine and feces. If a mother cat does a good job of raising her litter, she will probably train them to use a litter box. If a kitten does not know what the litter box is for, one can train it easily. Holding its front paws, the owner should show it how to scratch in the litter material. Every time the kitten appears to be looking for a place to urinate or defecate, it should be placed in the litter box.

Litter-trained cats may spray or urinate in places the owner deems inappropriate. Both sexes may spray urine to mark territory. This behavior is most common in intact (un-neutered) males, but any cat may spray if it is upset by changes in the household, or other cats. Spraying must be distinguished from urinating. When a cat sprays it typically backs up to a vertical surface.

Cats squat to urinate on horizontal surfaces. Inappropriate urination is often a symptom of urinary disease and warrants a quick trip to the vet. Successful treatment of inappropriate elimination requires accurate diagnosis of the underlying cause. Confine the cat to a large carrier or bathroom (containing a litter box, food, and water); clean soiled areas with an odor eliminator (available from pet stores); and consult a veterinarian for advice.

Some cats can be trained to sit, beg, jump through hoops, and do other tricks. Training should be similar to the clicker training methods used with a dog, with short training periods, firm verbal commands, and instant rewards. Whether a cat learns any tricks depends on its intelligence and whether it wants to oblige. In any case, training should not be forced on a cat that does not appear interested.

GROOMING AND CARE

A sleek, glossy hair coat is an indication of a cat's general health and care. All cats benefit from grooming, although the long-haired breeds require more attention. A cat's fur usually sheds in spring and fall, although some shedding may constantly occur. Grooming removes old, dead hair and lessens the risk of hairballs. Regular grooming gives the opportunity to check for parasites, skin disorders, and eye and ear problems. *See Figure 8–48.*

Long-haired cats should receive daily care. If neglected, the hair will tangle and mat. Removal of these tangles and mats may be difficult and an unpleasant experience for the cat. *See Figure 8–49.*

Equipment for long-haired cats should include a comb with two sizes of teeth, a fine-tooth or flea comb, nail clippers, a grooming brush made with natural bristles (nylon may cause excessive static), and grooming powder (baby powder, talcum powder, or cornstarch).

Figure 8–48 A large selection of combs, brushes, and pet care products is available for grooming pets. *Courtesy of Brian Yacur and The Feedbag Plus.*

Figure 8–49 Regular grooming of one's pet will keep the pet looking nice and gives one the chance to examine the animal. *Courtesy of Sandy Clark.*

Using a wide-toothed comb, comb all areas of the animal. One must be careful of the sensitive areas of the stomach, the insides of the legs, and under the tail. If the coat is free of tangles and mats, the fine-toothed part of the comb should be used. The skin and not just the outer fur should be combed; next, the fur should be brushed out. One should brush in the opposite direction to which the hair naturally lies and occasionally sprinkle grooming powder into the fur.

If the fur has become badly tangled and matted, scissors should be used to cut the mats out, being careful not to injure the animal. Blunt-ended scissors—never sharp-pointed scissors, knives, or razor blades—should be used. Do not cut parallel to the skin; split mats into strips by cutting perpendicularly away from the skin. A sudden movement by the cat could cause serious injury. Badly matted cats may require sedation and clipping by a vet.

Grooming short-haired cats can usually be accomplished with a fine-toothed or flea comb. In many cases, hand grooming is sufficient to remove dead hair. A rubber grooming brush is also very effective, but it must be used carefully because good hair may also be removed. The use of a soft chamois, silk, or nylon pad causes some static in the coat and helps it cling tightly to the body. This is an especially effective way to complete the grooming for Siamese, Oriental Shorthairs, Colorpoint Shorthairs, and Burmese. Other breeds enjoy a light brushing with a soft brush.

The coat of the Cornish and the Devon Rex is delicate and easily damaged; a soft brush should be used occasionally, avoiding friction or over-brushing. Friction and overbrushing may cause the hair to break and may even cause bald spots.

Occasionally, it may be necessary to give a cat a bath. In the case of show animals, more frequent bathing is necessary. Cats should be accustomed to bathing at about 4 months of age. Cats get nervous from the sounds of the water, and unnecessary splashing should be avoided. The tub should be filled with about 4 inches of warm water. The cat should be gently lowered into the water and wet all over, being careful not to get water in the eyes or ears. If the cat will allow it, a piece of cotton should be placed in its ears. After the coat is thoroughly wet, one should apply shampoo and lather the coat, followed by a thorough rinse to remove all traces of the shampoo. The cat should be wrapped in a towel and patted dry or blown dry using a hair dryer on the warm setting. Excessive rubbing should be avoided because this causes the hair to tangle. The cat should be dry before letting it outside, including the insides of the ears. Many pet shampoos on the market are suitable; some of these are medicated to help control fleas and other parasites, but make sure they are labeled for use in cats. Products for dogs may poison cats. Baby shampoo is also suitable. *See Figure 8–50.*

During regular grooming, the ears should be checked for mites; signs are a dark, crumbly residue inside the ear. Ear mites can be controlled with ear drops available at any pet store and frequent ear cleaning; veterinarians can prescribe more effective treatments. *See Figure 8–51.*

The eyes of the cat should be bright and clear; any discharge may indicate illness or infections. White cats, especially Persians, may show staining around the eyes; this staining can be removed by careful bathing. *See Figure 8–52.*

Teeth and gums should be observed during regular grooming and should be free of any soreness. Use of dry food helps clean the teeth. Tartar needs to be removed periodically by a veterinarian to prevent tooth loss and gum disease. *See Figure 8–53.*

Figure 8–50 A couple of a wide range of pet care products that are available. *Photo by Dean Warren.*

Figure 8–51 Ear drop products for the control of mites are readily available at most pet stores and discount stores. *Photo by Dean Warren.*

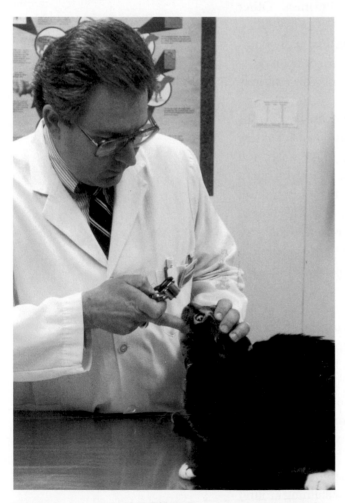

Figure 8–52 Cats should be taken to a veterinarian for regular checkups and vaccinations. The veterinarian will examine the animal's eyes as part of a regular checkup. *Courtesy of Brian Yacur and Guilderland Animal Hospital.*

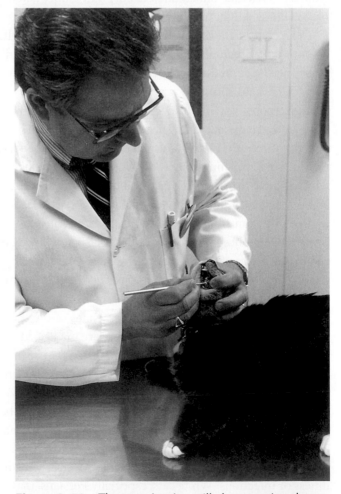

Figure 8–53 The veterinarian will also examine the animal's teeth and gums. *Courtesy of Brian Yacur and Guilderland Animal Hospital.*

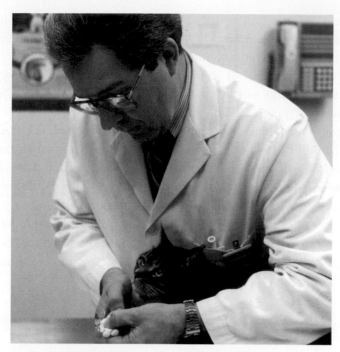

Figure 8–54   Although the veterinarian can trim the animal's claws during the regular checkup, cats that are using a scratching post will probably never need their claws trimmed. *Courtesy of Charles Miller/Alamy.*

If claws require trimming, one should be careful not to cut into the pink area of the claw because this causes pain and bleeding. Specially designed clippers should be used—never use scissors. Cats that are using a scratching post will probably never need their claws trimmed. *See Figure 8–54.*

COMMON DISEASES

This unit is broken down into six parts: infectious diseases, noninfectious diseases, internal parasites, external parasites, fungal diseases, and poisonings.

General symptoms of illness that should alert the owner are loss of appetite, sudden loss of weight, changes in behavior, lethargy, neglect of grooming, dull fur, loss of hair, itching or scratching, swelling of the body, frequent vomiting, increased thirst, and diarrhea. If any of these signs are observed, the animal should be taken to a veterinarian for diagnosis and treatment.

Infectious Diseases

Feline Panleukopenia

Feline panleukopenia, also referred to as feline infectious enteritis or cat distemper, is caused by a parvovirus or DNA virus. This is primarily a disease of young cats. Kittens less than 16 weeks of age may die at a rate of 75 percent. Older cats may have acquired some immunity and are less likely to die.

Feline panleukopenia is spread by direct contact among infected cats. Infection may also occur from infected food and water dishes, bedding, and litter boxes. Infected cats usually show signs of depression, loss of appetite, high fever, lethargy, vomiting, dehydration, and hanging over the water dish. Diarrhea may show large amounts of liquid, mucus, and blood.

The virus is capable of passing the placental barrier in pregnant cats. Infection of the fetus usually results in abortion, stillbirth, early fetal death, or permanent structural brain damage.

Recovery for kittens less than 8 weeks of age is poor; there are no antibiotics that kill the virus. The veterinarian will try to combat the dehydration, provide nutrients, and prevent secondary infection with antibiotics. Immunity in kittens is possible from their mothers; this passive immunity is temporary and is usually ineffective after 12 weeks of age.

Vaccines offer the safest and most effective means of protection. Kittens should be vaccinated and get booster shots according to the schedule recommended by the veterinarian. Annual revaccinations are usually recommended throughout the cat's life.

Feline Herpes Virus and Feline Calicivirus

Two respiratory virus infections are of primary importance in cats: feline herpes virus (FHV) and feline calicivirus (FCV). The herpes virus is a DNA virus, and the calicivirus is an RNA virus. The viruses of both can be shed in discharge from the nose, eyes, and throat. Transmission results mainly from direct contact between animals.

The virus may persist in cats for a long period of time after infection, and the cat can become a carrier of the virus. Early signs of infection are depression, sneezing, and coughing. As the infection progresses, severe eye and nasal discharges occur, with an increase in temperature. Both viruses may cause ulcers in the mouth, but corneal ulceration appears only in feline herpes virus infections. Vaccinations are available for both feline herpes virus and feline calicivirus.

Feline Rhinotracheitis

When herpes virus infection is confined to the respiratory tract, it is generally referred to as feline rhinotracheitis (PVR). Diagnosis of these diseases is based on symptoms, especially sneezing and discharge from eyes and nasal passages. Definite diagnosis depends on isolation and identification.

Vaccinations for respiratory viruses are usually included in combination with the panleukopenia virus vaccination. At least two vaccinations are required for all ages. Annual revaccination is recommended.

Feline Infectious Peritonitis

Feline infectious peritonitis (FIP) was first observed in the early 1950s. The disease is caused by a **coronavirus.** Coronavirus infections are relatively common in domestic cats, but the majority of these do not produce signs of disease. Although a cat may show no signs of disease, it may serve as a carrier. The normal means of transmission is by direct contact.

The disease occurs in two forms. In one form, the abdomen and chest accumulate fluid. The cat experiences a fever, refuses to eat, appears depressed, and loses weight. As the disease progresses, the body cavities continue to fill with fluid, organ systems are affected, and signs of specific organ failure occur. Death may eventually occur as a result of organ failure.

The second form is usually associated with specific organ failure. Lesions affecting the cat's eyes are very common and may be the first indication of the disease. In this form, the disease may progress to the point that the cat dies from organ failure. Diagnosis is made from symptoms and laboratory results. Routine immunization against feline infectious peritonitis is not possible at this time. A vaccination for FIP was developed in 1991. One should consult a veterinarian about FIP vaccinations.

Feline Leukemia Virus

"Feline leukemia virus (FeLV), a retrovirus, is so named because of the way it behaves within infected cells. All retroviruses, including feline immunodeficiency virus (FIV) and human immunodeficiency virus (HIV), produce an enzyme, reverse transcriptase, which permits them to insert copies of their own genetic material into that of the cells they have infected."[1]

FeLV is one of the most important causes of illness and death in cats. The disease suppresses the immune system and decreases the body's resistance to other diseases. Some FeLV related ailments include suppression of the immune system, cancers, anemia, reproductive problems, and inflammation of the digestive system.

FeLV is spread through saliva, feces, and blood. FeLV can also be transmitted to kittens during gestation and during nursing. Not all cats exposed to FeLV become infected. Some cats have strong immune systems that can destroy the virus. Cats that live outdoors, cats living with infected cats, and kittens are at highest risk. FeLV infected cats are found worldlife. It is estimated that 2 to 3 percent of cats in the United States are infected with FeLV.

After a cat is infected, the virus replicates in the tonsils and lymph nodes. It then spreads by the blood to other parts of the body. FeLV is present in the blood during two stages of infection. Primary viremia is the early stage. It is during this early stage that the immune system of some cats can destroy the virus. If the virus is not destroyed during the primary stage, it will progress to the secondary stage. Cats with secondary viremia will probably be infected the rest of their lives.

Some early symptoms of the disease may include fever, swollen lymph nodes, weakness, jaundice, weight loss, and cats may appear listless and lethargic. There are no signs specific to FeLV infection. Cats can have the virus for years before showing signs of the disease.

Two FeLV blood tests are commonly used to detect the virus in the blood. The ELISA (enzyme-linked immunosorbent assay) test can be performed in a veterinarian's office. IFA (indirect immunofluorscent antibody assay) tests are sent to a diagnostic laboratory and are used to detect secondary viremia only. A positive test for FeLV does not mean immediate death for the cat. An infected cat can live several months or years if precautions are taken. Infected cats should be: (1) kept indoors, (2) treated at the first sign of illness, (3) fed a high quality diet, and (4) provided vitamins and medications as recommended by a veterinarian. New cats should not be introduced to the household and frequent visits should be made to the veterinarian for examination.

There is no cure for FeLV once a cat is infected. There are a number of FeLV vaccinations available, none provide 100 percent protection. One should consult with a veterinarian on the prevention, diagnosis, and treatment of FeLV and related diseases.

Feline Pneumonitis

Feline pneumonitis is caused by infection with *Chlamydia psittaci*. Runny eyes and noses are the most characteristic signs of infection. Early immunity is present in the colostrum, and vaccinations are available for further immunity. Annual revaccination is recommended.

Rabies

Rabies is a viral infection of warm-blooded animals. Three hundred and eighteen cases of rabies in cats were reported to the Centers for Disease Control and Prevention (CDC) in 2006. During 2006, only 79 cases of rabies in dogs were reported to the CDC. (Refer to Chapters 2 and 7 for more information on rabies.) Cats should be vaccinated against rabies; however, not all states require rabies vaccination for cats. A first vaccination should be given at 12 weeks of age and a second vaccination at 18 weeks of age. A booster should be given at 1 to 3 years, depending on state law. One needs to check with a veterinarian to determine the need and frequency of rabies vaccinations.

Noninfectious Diseases

Feline Urologic Syndrome

Feline urologic syndrome (FUS), also referred to as feline lower urinary tract disease (FLUTD), is a urinary problem in cats. The condition may range from mild inflammation to a blockage of the urethra, uremic poisoning, and death. The condition is believed to be brought on by an improper diet and reduced water intake. If an animal does not have sufficient water intake, the urine can become concentrated and various salts will settle out and form crystal stones. A diet high in magnesium and phosphorus is implicated. If resulting stones block the urethra, prompt treatment is necessary to prevent uremic poisoning and ruptured bladder. Signs of feline urologic syndrome include straining to urinate, small amounts or no urine at all, failure to use litter box, and blood in the urine.

Wet Eyes

Wet eyes is a condition of excessive tear production or blockage of the canals that normally drain tears into the nasal cavity, causing tears to overflow at the inner corners of the eyes. This condition appears more frequently in the Persian and related breeds.

Internal Parasites

Toxoplasmosis

Toxoplasmosis is caused by infection with *Toxoplasma gondii,* a single-celled protozoan parasite. This disease occurs throughout the world and has been observed in a wide range of birds and mammals. Cats are frequently infected. The infectious agent may be picked up by eating raw meat or contaminated feces.

Infected cats may develop fever, **jaundice,** enlarged lymph nodes, difficulty breathing, anemia, eye inflammation, encephalitis, and intestinal disease. Abortion is also seen in pregnant cats that are infected. Cats may also develop stiff, painful muscles and have difficulty moving or be unable to move.

Accurate diagnosis is possible through laboratory tests; because symptoms are so widely varied, they are not reliable aids in diagnosis of the disease. Vaccination of animals against toxoplasmosis is not possible at this time.

Humans can become infected with *Toxoplasma gondii parasites.* Transmission is usually through the ingestion of infected raw or undercooked meat, the accidental ingestion of aged cat feces, or drinking water contaminated with *Toxoplasma.* Accidental ingestion of cat feces can occur if one accidentally touches his hands to the mouth after working in the garden or cleaning the cat's litter box. Cats, which can pass the parasite in their feces, often use gardens and sand boxes as litter boxes. The organism must be exposed to the environment for 24 to 48 hours before it can cause disease. Therefore, daily cleaning of litter boxes will stop transmission of toxoplasmosis from pets to people.

Pregnant women should be especially concerned because unborn fetuses appear to be at risk for toxoplasmosis during early pregnancy. Pregnant women should avoid handling litter boxes.

To avoid human infection, meat should be cooked thoroughly; children should be especially careful and wash their hands before eating. Gloves should be worn when working in areas where cats defecate. Children's sand boxes should be covered when not in use so that cats cannot use them as litter boxes. Litter boxes must be changed frequently and feces disposed of so that animals and humans cannot come in contact with the feces.

Ascarids and Hookworm

Cats can be infected with ascarids and hookworms. The common ascarid *(Toxocara cati)* and hookworm *(Ancylostoma tubaeforme)* of cats are different species from those that attack dogs.

Cats acquire ascarid infections from ingesting eggs that have been passed in the feces of an infected animal. Nursing kittens may become infected with ascarid larvae through the milk of the mother cat.

Hookworm infection occurs when the larvae penetrate the skin or when the larvae are ingested in contaminated food or water. The larvae are not passed in the mother's milk as in the case of dogs.

Symptoms of infection by ascarids and hookworms are basically the same as those described in dogs. Symptoms may not occur in cats because cats do not become as heavily infected as dogs. Several medications are available to treat cats infected with ascarids and hookworms. One should consult a veterinarian for the proper medication and treatment. *See Figure 8–55.*

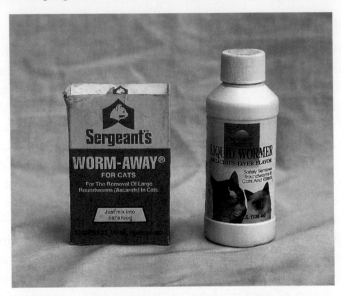

Figure 8–55 Roundworms, hookworms, and tapeworms can be prevented with products available at pet stores and discount stores.

Tapeworms

Tapeworms are also found in cats. One species, *Dipylidium caninum,* infects both dogs and cats. Another species, *Taenia taeniaeformis,* is sometimes found in cats.

An initial host is required for development of the young tapeworm. The flea serves as the initial host of *Dipylidium,* and rats or mice serve as the initial host for *Taenia* in cats. *See Figure 8–56.*

Tapeworms do not cause major harm to cats. The tapeworms consume nutrients from the cat and deprive the cat of needed nutrients. Infested cats will usually lose weight and have a rough hair coat. One should consult a veterinarian for proper medications and other treatment. To prevent continued reinfection, one must also eliminate fleas and prevent the cat from eating rats and mice.

External Parasites

Lice

Cats are infected by one common species of biting louse *(Felicola subrostratus).* Healthy and well-cared-for cats are unlikely to be troubled by lice. When a cat is infected, the lice are usually found around the head. Many pesticides are available to control adult lice; pesticides can be toxic to cats. Always make sure the product is labeled for cats and carefully follow the manufacturer's directions.

5. Adult tapeworm releases egg-filled segments and, untreated, the cycle begins again

1. Cat passes egg-filled tapeworm segments in feces

Life cycle of:

Taenia taeniaeformis
(cat-rodent tapeworm)

MOST COMMON INTERMEDIATE HOST: Rodents

4. Cat eats infected rodent and tapeworm larvae mature in cat

2. Individual tapeworm eggs are released as segments rupture

3. Rodent eats tapeworm eggs

Figure 8–56 The life cycle of *Taenia taeniaeformis*, cat-rodent tapeworm. *Reprinted with permission of Miles, Inc., Animal Health Products.*

Mites

Mites are rare in cats; however, there are two types of demodectic mites: *Demodex cati* and an unnamed *Demodex* species. In demodicosis, there is patchy hair loss, reddening, and occasionally crusting on the neck, ears, and head. Severe cases may involve the entire body.

Feline Scabies

Feline scabies, or notoedric mange, caused by the mite Notoedres *cati,* is a skin disease that occurs worldwide. In the United States it appears to be more prevalent in Southern California and along the Gulf Coast. These mites are highly contagious to other cats. They cause hair loss and thickening and crusting of the skin, usually affecting the head and forelegs. There are several options for treatment. One should consult a veterinarian for advice on the treatment and control of these mites. *Notoedres cati* mites can infect humans, causing temporary itching.

Other External Parasites

Ear mites, walking dandruff mites, chiggers, fleas, and ticks that infect dogs also infect cats. These parasites were covered in Chapter 7; refer to that unit for more detailed information on external parasites. There are several options available for treating these parasites. Consult with your veterinarian on the proper medications for treatment and prevention of parasites.

Fungal Diseases and Poisonings

These diseases and problems were also covered in Chapter 7; they are problems common to dogs also. Please refer to Chapter 7 for more detailed information.

REPRODUCTION

The age at which female cats (queens) first come into heat varies with breeds. The Siamese may come into heat as early as 5 months of age, whereas most long-haired breeds of the Persian type may not come into heat until 10 months of age or more.

Estrus in cats is induced by lengthening daylight and rising temperatures of spring. The heat period lasts for 9 or 10 days. If the cat is not bred during the heat period, she will come back in heat in 15 to 21 days.

A female cat in heat is restless and unusually friendly, rubbing up against objects, furniture, and the owner's legs. She usually has a decreased appetite and urinates more frequently. A repeated monotone howl or call is made, and she rolls around on the floor. Any touch to the back of the tail is likely to stimulate a crouch position, and further stroking at the back and base of the tail exaggerates this receptive position.

In response to the call of the female, intact males (toms) also emit courtship cries and spray urine to leave scent marks to attract the female's attention. Immediately after mating occurs, the female releases a loud, piercing cry. The male's penis is barbed, and upon withdrawal the female cat will often scream, turn, and attack the male. This aggressive behavior may be caused by the withdrawal of the penis, which causes an intense and possibly painful stimulation of the female's vagina. This stimulation is believed necessary to induce ovulation (release of the egg cells) approximately 24 hours after mating occurs.

Pregnancy in cats averages about 65 days but may vary from 51 to 68 days. At 3 weeks into the gestation period, the nipples of the female become pink and the hair recedes slightly. At 6 weeks, pregnancy is obvious, as the female's abdomen shows an increase in size.

About 2 weeks before the kittens are due, the female becomes restless and searches for a quiet place to give birth. If one offers her a well-prepared box, she will usually take to it.

During the first stage of labor, the female's rate of breathing increases, and she may start to breathe through her mouth. As labor proceeds, there will be some discharge, usually colorless at first, but later becoming yellow or straw-colored.

The second stage of labor begins with the contractions of the uterus. At first, these contractions may occur once an hour; just before delivery they occur every 30 seconds.

The female repeatedly licks her genital area and may show signs of agitation. The vaginal opening dilates and the first amniotic sac appears. The sac and the enclosed kitten are usually expelled within 15 to 30 minutes. As soon as the kitten is expelled, the female ruptures the amniotic sac. She cleans the kitten; this persistent licking stimulates the kitten to breathe.

Next, she severs the umbilical cord with her teeth. The placenta is soon expelled, and the female promptly eats it. This natural instinct provides the female with a source of nutrients. The rest of the litter will follow at about 15-minute intervals.

The female may go out of labor for a period of time and then restart contractions, giving birth at intervals of perhaps 1 or 2 hours. In some cases, she may rest for a period of 12 to 24 hours, nursing her first kittens before going back into labor and delivering the rest of the litter.

Occasionally, a female does not attempt to remove the sac from the newborn kitten. She may not know what to do or may be too busy with the next kitten. In this case, one must remove the membrane. If the female still does not show interest, one must carefully sever the umbilical cord, rub the kitten dry, and stimulate it to breathe. Then, the kitten should be placed close to the mother's nipples. Hopefully, the kitten will begin to nurse and arouse maternal instincts in the female.

The first milk produced by the mother contains colostrum. Kittens receive temporary immunity from many diseases from the antibody-rich colostrum in their mother's first milk.

SUMMARY

Cats are descendants of the same carnivorous, tree-climbing mammal *Miacis* from which dogs descended. Cats, like dogs, trace their lineage through *Cynodictis*. The skeleton of these early primitive carnivores was much like that of a civet or a weasel.

Modernized cat-like types, *Felis* and related genera, with "normal" canine teeth appeared early in the Pliocene epoch. Their origin has been much debated, for all the better-known older felids have much larger canines and hence were saber-tooths of one sort or another.

In Chapter 7, we learned that dogs were domesticated about 10,000 years ago. The cat, however, was not domesticated until much later. This was probably because of its independent, solitary nature. It is believed that domestication of the cat began around 4,000 years ago.

Cats are usually divided into three genus groups. One group is the large, roaring cats, *Panthera,* which includes the lion, leopard, tiger, and jaguar. Second are the cats that cannot roar, *Felis,* which includes the small cats and the domesticated cat. The third group, *Acinonyx,* includes only one member, the cheetah. The cheetah belongs to a special group because it has claws that do not fully retract.

The domestic cat breeds are commonly divided into two major groups: the short-haired breeds and the long-haired breeds. A miscellaneous or third group includes breeds that are semi-long haired and those breeds that may be either short haired of long haired. There are 13 long-haired breeds and 25 short-haired breeds, and 5 breeds in the third group listed in this text.

The body of the cat has been developed for speed and flexibility. The bones, joints, and muscles give the animal its size, shape, speed, and strength.

The skeleton of the cat contains 244 bones, which is about 40 more than a human; most of these extra bones are found in the spine and tail. Spread over this skeletal structure is approximately 500 separate muscles, compared with more than 650 in the human body. The largest muscles on the cat are found on the rear legs, which gives the animal its tremendous power for running and jumping.

Cats require almost twice as much protein in their diet as dogs. The best source is from animal products; 30 to 40 percent of the diet should be animal-type proteins.

A sleek, glossy hair coat is an indication of a cat's general health and its care. All cats benefit from grooming, although the long-haired breeds require more attention. A cat's fur usually sheds in spring and fall, although some shedding constantly occurs. Grooming removes old, dead hair and lessens the risk of hairballs. Regular grooming gives the opportunity to check for parasites, skin disorders, and eye and ear problems.

Common diseases in cats can be divided into six groups: infectious, noninfectious, internal parasites, external parasites, fungal, and poisonings. Common symptoms of illness that may occur are loss of appetite, sudden loss of weight, changes in behavior, apathy, neglect of grooming, dull fur, loss of hair, itching or scratching, swelling of the body, constant vomiting, increased thirst, and diarrhea. If any of these signs are observed, the animal should be taken to a veterinarian for diagnosis and treatment.

The age at which female cats first come into heat varies with breeds. The Siamese may come into heat as early as 5 months of age, whereas most long-haired breeds of the Persian type may not come into heat until 10 months or more. Estrus in cats is induced by lengthening daylight and rising temperatures of spring. The heat period lasts for 9 or 10 days. If the cat is not bred during the heat period, she comes back in heat in 15 to 21 days. Ovulation is induced when the female is bred.

Pregnancy in cats averages about 65 days but may vary from 51 to 68 days.

DISCUSSION QUESTIONS

1. What is the history of the cat?

2. What are the two major groups of domesticated cats?

3. List the different breeds of cats. How do they differ?

4. What is meant by the term *colorpoint*?

5. How should long-haired cats be groomed? Do short-haired cats need grooming?

6. What is the most important thing to remember in feeding a cat?

7. What are the six groups of diseases that affect cats?

8. List the diseases in the six groups that affect cats. What are the symptoms of these diseases?

9. List the general health care practices for a pregnant cat before, during, and at birth.

10. What health care practices should be observed with the new litter?

SUGGESTED ACTIVITIES

1. Zoologists disagree on the evolution of the cat. Visit the library and research the evolution of the cat. Could you find a more detailed path of evolution than the one in this text? Did the modern cat evolve from the saber-tooth? Did the modern cat and saber-tooth evolve separately?

2. Bring the labels from various cat foods and compare the ingredients that are used. Compare these with labels obtained from dog foods.

3. With the approval of the instructor, invite the local veterinarian to talk about some of the common diseases of cats.

4. Visit a cat show. Watch how the show is conducted and how the cats are presented.

5. With the approval of the instructor, have local cat breeders bring their cats to the class and discuss characteristics of the breed.

6. Investigate other methods for training cats to use a scratching post. What types of materials are attractive to cats for scratching?

7. Investigate "clicker training." See if you can train a cat to sit up on its hind legs.

ADDITIONAL RESOURCES

1. Cat Fanciers
   http://www.fanciers.com

2. The Cat Fanciers Association (CFA)
   http://www.cfainc.org

3. Purina
   http://www.purina.com

4. The International Cat Association
   http://www.tica.org

5. Feline Veterinary Medicine
   http://www.fanciers.com/vetmed.html

6. The American Veterinary Medical Association, Feline Health
   http://www.avma.org/care4pets/animfelh.htm

7. Wellsboro Veterinary Hospital and Reptile and Bird Clinic
   http://www.drkreger.com

8. MarVista Animal Medical Center
   http://www.marvistavet.com/index.html

9. Pet Education.com
   http://www.peteducation.com/default.cfm

END NOTE

1. Cornell University College of Veterinary Medicine, Feline Health Center. "Feline leukemia virus." 6 January 2008. http://www.vet.cornell.edu/fhc/brochures/felv.html.

CHAPTER 9

Rabbits

OBJECTIVES

After reading this chapter, one should be able to:

- 🐾 discuss the history of the rabbit.
- 🐾 discuss uses for rabbits.
- 🐾 list the five weight categories of rabbits.
- 🐾 list the breeds of rabbits.
- 🐾 discuss the housing and equipment needs of rabbits.
- 🐾 discuss the nutrient requirements of rabbits.
- 🐾 discuss various diseases that affect rabbits.

CLASSIFICATION

Kingdom	—	Animalia
Phylum	—	Chordata
Class	—	Mammalia
Order	—	Lagomorpha
Family	—	Leporidae
Genus	—	*Oryctolagus*
Species	—	*O. cunniculus*

TERMS TO KNOW

conjunctiva	enteritis	oocysts
coprophagy	fly-back fur	roll-back fur
dew-drop valve	kindling	sore hocks
dewlap	malocclusion	

HISTORY

Rabbits were once classified in the order Rodentia, but they are now placed in a separate order, Lagomorpha. The chief difference between rabbits and rodents is that rabbits have four upper incisor teeth arranged one pair behind the other, whereas rodents have two chisel-like incisor teeth in each jaw.[1]

Fossil remains of Lagomorpha have been found in Oligocene deposits, which date back 30 to 37 million years. These fossil remains indicate that the animals have changed very little.

The order Lagomorpha includes the family Leporidae, the rabbits and hares; and the family Ochotonidae, which are the pika. Pika are small, furry animals that more closely resemble guinea pigs than rabbits or hares and are found in Asia, Europe, and Western North America.

Among the family Leporidae are the American cottontail, genus *Sylviagus;* the true hares, genus *Lepus;* and the European wild rabbit, genus *Oryctolagus,* species *cunniculus.*

The European wild rabbit is the species from which all domestic rabbit breeds have been developed. Although many of our current domestic breeds have been developed in the United States, the American cottontail was not used in the development of our domestic rabbit.

Hares, genus *Lepus,* are usually larger than rabbits and have longer, black-tipped ears and are slightly different in appearance. Hares do not build nests, and their young are born fully furred with their eyes open; hares live above ground and do not dig tunnels.

Rabbits were fairly widespread and abundant in Europe during the late Tertiary period and early Pleistocene epoch. Abnormal climates during the Ice Age drove them to the southern parts of Europe.

The first reports of rabbits were from Phoenician traders who visited the coast of Spain and islands along the coast during 1100 BC. The Phoenicians are probably responsible for transporting rabbits to other parts of the world.

Rabbits were of great economic importance. They were hunted for food, and their pelts were used to make clothing. Credit for domestication of the rabbit is given to French monks of the Middle Ages, who raised rabbits in walled cages kept in their monasteries. They served as an easily raised source of food, and the fur was used for clothing.

In the second half of the nineteenth century, the European wild rabbit was introduced into Australia and New Zealand, where it quickly spread and became a serious pest. Brought to Chile in the early twentieth century, it eventually spread over much of the South American continent. Some European wild rabbits were released on the San Juan Islands off Washington State in the United States at the beginning of the twentieth century and have flourished since, but the major rabbit of North America remains the cottontail.[2]

In wild populations, the European wild rabbit measures 13 to 18 inches long with a tail 1½ to 3 inches long, and weighs 3 to 5 pounds. The upper body is a mixture of black and light brown hairs; the ears are the same color except for the black-edged tip. The nape is buff colored, the collar is dark, the tail is white below and brownish-black above, and the underparts and inner surface of the legs are buffy-white. The hind legs are relatively long; the feet are well furred beneath and have large, straight claws. Many of the domestic forms bear little resemblance to the original wild stock.[3]

USES

Meat

Rabbits are unique among the animals covered in this book because they are not only used for pets but are also raised for meat, fur or wool, and laboratory use.

Those who encourage the use of rabbit meat stress the following advantages (see Tables 9–1 and 9–2):

- high in protein
- low in cholesterol
- low in fat

TABLE 9-1 Composition of Some Common Foods

Edible Portion Uncooked	Percent Protein	Percent Fat	Percent Moisture	Calories per 100 grams
Beef, good grade	16.3	28.0	55.0	317
Beef, choice and prime grade	13.7	39.0	47.0	406
Chicken fryers, total edible, including flesh, fat, skin, and giblets	20.0	11.0	67.6	179
Chicken fryers, flesh only	20.6	4.8	73.4	126
Lamb, intermediate	15.7	27.7	55.8	312
Lamb, fatty	13.0	39.8	46.2	410
Pork, medium	11.9	45.0	42.0	453
Pork, fatty	9.8	55.0	35.0	534
Rabbit, domesticated, total edible, including flesh, fat, and giblets	20.8	10.2	67.9	175
Rabbits, domesticated, flesh only	20.8	10.2	67.9	175
Salmon, Atlantic	22.5	13.4	63.6	211
Salmon, Pacific (Chinook and King)	17.4	16.5	63.4	218
Tuna, blue fin	24.8	5.2	69.1	146
Tuna, yellow fin	24.7	3.0	71.5	126
Turkey, medium birds, including flesh, skin, fat, and giblets	20.1	20.2	58.3	262
Turkey, flesh only	24.0	24.0	68.6	156
Turkey, light meat	24.5	4.6	69.2	139
Turkey, dark meat	23.2	9.4	68.0	177
Veal, medium	19.1	12.0	68.0	184
Veal, fatty	18.5	16.0	65.0	218
Whitefish	22.9	6.5	69.8	150

Information taken from United States Department of Agriculture Circular No. 549, *Proximate Composition of American Food Materials.*

TABLE 9-2 Cholesterol Content of Foods

Food	Cholesterol Content (milligrams per 100 grams)
Beef round (medium fat)	125
Beef round (lean)	95
Veal shank	140
Veal breast	100
Pork spare ribs	105
Rabbit	50
Chicken (light)	90
Chicken (dark)	60
Tuna	149–172
Salmon	60

Information taken from McLester, J. S., M.D., and W. J. Darby, M.D., Ph.D. 1953. *Nutrition and Diet in Health and Disease,* 6th Edition, Philadelphia: W.B. Saunders Company.

- low in sodium
- very palatable

Rabbit meat is primarily a white meat that is very fine in texture and has a very low fiber content. Because of the low fiber content, it is easily digested, which is desirable for individuals who may have difficulty chewing their food. Rabbit is a more popular meat in Europe than in the United States. Italy, Germany, France, and Spain are the world's largest producers and consumers of rabbits. The major reason for the low consumption of rabbit in the United States is probably because of the image we have of rabbits. We grow up with rabbits as friendly characters in cartoons and games. Only a small number of rabbits are produced for the meat, and the supply to the supermarkets is small. The two major breeds used for meat production are the New Zealand and the Californian.

Research

The number of rabbits used in research projects in the United States reached a high of approximately 554,000 in 1987. That number declined to about 240,000 in 2006. The decline in the number of rabbits used in research is due to the public protest in using live animals in research and the development of alternative methods for doing research. Rabbits have been used to produce disease-fighting antibodies, to study reproduction, and to research several human diseases. Rabbit blood is one of the best mediums for growing the AIDS virus.

Rabbits have been used in controversial tests. The Draize Eye Test has been used by many companies to test cosmetics and other products. Rabbits do not have tear ducts and cannot shed tears to dilute chemicals or products put into their eyes. Many people have called for a ban on this type of test, and many cosmetic companies no longer use it.

Skin irritation tests have also been used with rabbits. A small patch of fur is removed along the back of the animal, and the product being tested is applied to the bare area. Irritations or reactions to the products are then noted.

Fur/Wool

Rabbit fur is divided into four types: normal, rex, angora, and satin. Each type of fur can be used in the manufacture of clothing, toys, coats, hats, and gloves. White is the preferred color because it can be used to match almost any type of animal fur.

Rex fur is short, with guard hairs being of the same length; the hair stands up straight at a right angle to the skin. This combination of properties gives the fur a very soft, plush characteristic. Pelts from Rex rabbits sell for up to $10 above ordinary fur pelts.

Rabbit wool comes from Angora breeds. Garments made from Angora are softer, finer, warmer, and much lighter in weight than those made from sheep's wool. Rabbit wool does not cause the irritation and is not as scratchy as sheep's wool.

The Angora rabbit wool market in the United States is primarily with hand spinners. Angora rabbits are not bred in as great numbers as meat rabbits, and more time is necessary for their care.

The value of Angora wool depends on its thickness, quality, and length. The English Angora rabbits produce the finest-quality wool; these rabbits produce 8 to 17 ounces of wool per year and command the highest prices. French Angoras produce about the same amount of wool per year, but their wool is coarser and not as highly valued. The Giant Angora and German Angoras produce 18 to 36 ounces of wool per year, but their fur, too, is coarser and does not command high prices.

Satin fur is a mutation that first appeared in the 1930s. Satin fur has a smaller diameter and a transparent outer shell, which gives a more intense appearance to the color. The transparent shell also gives the fur its sheen, luster, and slick appearance.

Pets

Rabbits make very clean, gentle, and lovable pets that require little special care; some can become very territorial and aggressive and require careful handling. They can easily be trained to use a litter box. Because of the diversity in sizes, they can fit into many different home situations. The small and dwarf sizes have been the most popular.

BREEDS

There are about 60 different breeds of domestic rabbits; however, only about 45 breeds have been developed to the point that they are recognized by various rabbit organizations. These 45 breeds are divided into five weight categories. These categories are listed below, along with the breeds.

Dwarf or Miniature

- Britannia Petite
- Netherland Dwarf
- Himalayan
- Dwarf Hotot
- Polish
- Jersey Wooly
- Holland Lop
- American Fuzzy Lop
- Mini Rex

Small

- Dutch
- Tan
- Florida White
- Silver
- Havana
- Mini Lop

Medium

- English Spots
- Standard Chinchilla
- English Angora
- Lilac
- Silver Martin
- Belgian Hare
- Rhinelander
- Harlequin
- Sable
- Satin Angora
- French Angora
- Rex

Large

- Beveren
- Californian
- Hotot
- Palomino
- Satin
- Cinnamon
- Creme d'Argent
- Champagne d'Argent
- American
- American Chinchilla
- English Lops
- New Zealand
- Silver Fox

Giant

- Giant Angora
- French Lops
- Checkered Giant
- Giant Chinchilla
- Flemish Giant

The Britannia Petite

The Britannia Petite is an old English breed known in Britain as "Polish"; it is generally thought to have been developed from small common rabbits. The breed is currently recognized as the smallest of the standard breeds. This breed weighs less than 2½ pounds. The Britannia Petite is commonly recognized in the United States as being white with red eyes, although they are bred in many color varieties in Britain. *See Figure 9–1.*

Figure 9–1   Britannia Petite. *Photo by Isabelle Francais.*

The Netherland Dwarf

The Netherland Dwarf was originally developed in Holland during the early part of the twentieth century from the Dutch rabbit and a wild rabbit. After many years of selective breeding, it was exported to England during the 1950s. British breeders improved the rabbit's stamina and standardized the colors. The modern Netherland Dwarf, as it was developed in England, first appeared in the United States in the mid-1960s. It has the distinction of being recognized in more different colors than any other breed. The breed is small, stocky, and compact. The head is round with a broad skull and a nose that is distinctly rounded. The eyes are round, bold, and bright; the ears are short and erect. Netherland Dwarfs usually weigh about 2½ pounds. *See Figure 9–2.*

Figure 9–2   Netherland Dwarf. *Photo by Isabelle Francais.*

The Himalayan

The Himalayan is one of the oldest established breeds, with a wider distribution than any other breed. It is believed to have originated in the Himalayan Mountain area. Some breeders believe it is an albino from the old Russian Silver rabbit. The young rabbits are white and slightly tinged with silver-gray. As the rabbit matures, the silver-gray disappears and its coat becomes snow white with the nose, ears, feet, and tail becoming a deep black. The Himalayan has ears that stand erect, a nose that is short and rounded, and red eyes. At maturity, Himalayans weigh 2½ to 4½ pounds. *See Figure 9–3.*

Figure 9–3   Himalayan. *Photo by Isabelle Francais.*

The Dwarf Hotot

The Dwarf Hotot was developed in the 1970s by two different breeders in two different parts of Germany using two different approaches. A West German breeder tried to develop a Hotot marked Netherland Dwarf by crossing a Red-Eyed White and a Black Netherland Dwarf. An East German breeder crossed a Red-Eyed White Netherland Dwarf with a Blanc de Hotot. The two approaches merged as breeders exchanged stock, improving their size and vigor. Elizabeth Forstinger of Vista, California, imported the first Dwarf Hotots into the United States in 1980. *See Figure 9–4.*

The body of the Dwarf Hotot is short, compact, and well rounded. The entire body is white, the only markings being a very thin black eye band around each eye. This breed weighs a maximum of 3¼ pounds.

Figure 9–4   Dwarf Hotot. *Photo by Isabelle Francais.*

The Polish

The exact history of the Polish breed is not known; it is believed to have been developed as an albino Dutch rabbit. The breed has a short, rounded body with short limbs and fine bones; the eyes are large. The Polish breed is commonly recognized in five colors: Black with brown eyes, Blue with blue-gray eyes, Chocolate with brown eyes, White with ruby-red eyes, and White with blue eyes. *See Figure 9–5.*

Figure 9–5   Polish. *Photo by Isabelle Francais.*

The Jersey Wooly

The Jersey Wooly was developed in New Jersey in the late 1970s by Bonnie Seely in an effort to produce small animals to sell as pets. It is a cross between the Netherland Dwarf and Angoras. Jersey Woolies weigh less than 3½ pounds, and the ears are 2 to 3 inches long. The breed is shown in five color groups: Selfs, Shaded, Agouti, Tan, and Pointed. *See Figure 9–6.*

Figure 9–6   Jersey Wooly. *Photo by Isabelle Francais.*

The Holland Lops

Holland Lops were developed by Adrian DeCock of the Netherlands. They were developed by breeding Netherland Dwarfs to French Lops, then breeding their offspring to English Lops in order to enhance the ear carriage. They first appeared in the United States in 1976. The apple-shaped head of the Holland Lop is short and broad. The eyes are set wide, the muzzle is broad, and it has a short, rounded nose. The body of the Holland Lop resembles the French Lop but is smaller in scale. The breed presents a massive, thick-set appearance with broad, deep hips. The Holland Lop weighs less than 4 pounds. It is recognized in solid or broken varieties. *See Figure 9–7.*

Figure 9–7 Holland Lop. *Photo by Isabelle Francais.*

The American Fuzzy Lops

The American Fuzzy Lops were developed during the 1980s. In an effort to get the broken pattern into the Holland Lop gene pool yet maintain the **roll-back** or **fly-back fur** (fur that gently returns to the original position) of the Holland Lop, breeders crossed the French Angora with Holland Lops. Although they produced the desired broken pattern with gentle, fly-back fur, they also produced occasional wooled rabbits. These wooled rabbits became the American Fuzzy Lops. They are a small lop-eared breed weighing less than 4 pounds. *See Figure 9–8.*

Figure 9–8 American Fuzzy Lop. *Photo by Isabelle Francais.*

The Mini Rex

The Mini Rex was developed by crossing standard Rex rabbits with the Netherland Dwarf in an effort to get the smaller animals. The Mini Rex weighs less than 4½ pounds. It has the same dense, plush fur as the Standard Rex and is shown in 11 varieties: Blue, Californian, Castor, Chinchilla, Lynx, Opal, Red, Seal, Tortoise, White, and Broken. *See Figure 9–9.*

Figure 9–9 Mini Rex. *Photo by Isabelle Francais.*

The Dutch

The modern Dutch rabbit was developed in England, although it originated in Holland; several rabbits were imported into England in 1864. The English refined their size, shape, and pattern. They were called "Dutch" rabbits and became very popular in England as well as in the United States. *See Figure 9–10.*

Figure 9–10 Dutch. *Photo by Isabelle Francais.*

The breed has unique color markings. There are colored patches on each side of the head that encircle the eyes and ears. This leaves a white blaze on the forehead and down over the front part of the rabbit, including the front legs. The rear half of the animal is colored the same as the head patches. White also appears halfway up on the rear paws. The Dutch rabbit weighs 3½ to 5½ pounds. The animals are recognized in six color varieties: Black, Blue, Chocolate, Tortoise, Steel Gray, and Gray.

The Tan

The Tan breed of rabbit was originally known as the Black and Tan breed. They were originally black with the underside tan. Today, they are also recognized in Blue, Chocolate, and Lilac with tan undersides. The tan color also appears in a narrow ring around the eyes and nostrils; the toes are also tan. These Tan rabbits originated in England and are a cross between a Dutch rabbit and a wild rabbit. The bucks of this breed weigh 4 to 5½ pounds, and the does weigh 4 to 6 pounds. *See Figure 9–11.*

Figure 9–11 Tan. *Photo by Isabelle Francais.*

The Florida White

The Florida White was developed in Florida. Dutch, Polish, and New Zealand White rabbits were crossed in an effort to develop a small meat rabbit and a small laboratory rabbit. The breed is short and compact with well-rounded hips and hindquarters; the head is small and the ears are short. *See Figure 9–12.*

Figure 9–12 Florida White. *Photo by Isabelle Francais.*

The Modern Silver

The modern Silver breeds were developed in England, France, and Germany. They are believed to be descendants of silver-gray rabbits that originated in India. Portuguese sailors are credited with bringing the rabbits to Europe in the 1600s. Silvers are recognized in three varieties: Brown, Fawn, and Gray. The Brown variety was perfected in England with the introduction of Belgian Hare blood. The original Fawns came from France. The Gray-Silvers were also developed in England and were probably the first to be developed. The fur of the Silver has large numbers of pigment-free, silvery-white tips, giving the animal the silver color. This small breed of rabbit weighs from 4 to 7 pounds. *See Figure 9–13.*

Figure 9–13 Modern Silver. *Photo by Isabelle Francais.*

The Havana

The Havana rabbit is the result of a mutation that occurred in 1898. A chocolate-colored rabbit appeared in a litter from a common doe with markings resembling a black and white Dutch; the sire of the litter was unknown. The fur of the Havana is a rich, dark chocolate with intense luster and sheen. The eyes are dark brown but, under certain light conditions, show a ruby-red glow. Three color varieties are recognized: Black, Blue, and Chocolate. The Havana is a small, compact, meaty rabbit weighing between 4½ and 6½ pounds. *See Figure 9–14.*

Figure 9–14   Havana. *Photo by Isabelle Francais.*

The Mini Lop

The Mini Lop originated in Germany, where it is known as the Klein Widder. It was bred in two colors: Agouti and White. The breed is a cross between the German Big Lop and the Little Chinchilla rabbit. *See Figure 9–15.*

In 1972, Robert Herschbach sent back to the United States one pair of Agouti Lops and one White Lop doe, and he is given credit for developing the breed in the United States. The broken color variety was bred in the United States by mating a standard Chinchilla to a Broken French Lop. The Mini Lop is a scaled-down version of the French Lop. Mini Lops have a massive, thick-set body. They are recognized in solid or broken varieties and weigh 4½ to 6½ pounds.

Figure 9–15   Mini Lop. *Photo by Isabelle Francais.*

The English Spot

The English Spot has been bred extensively since 1880. It is a cross of the Flemish Giant, English Lop, Angora, Dutch, Silver, and Himalayan. They were introduced into the United States from England. The background color is white and is accepted in seven color varieties: Black, Blue, Chocolate, Gold, Gray, Lilac, and Tortoise. A unique chain of colored markings, about the size of peas, runs from the base of the ear to the rear flank, with an ideal chain consisting of two symmetrically arranged rows. The breed has a butterfly nose pattern, eye circles, cheek spots, and colored ears; other spots appear on the body. They weigh 5 to 8 pounds. *See Figure 9–16.*

Figure 9–16   English Spot. *Photo by Isabelle Francais.*

The Standard Chinchilla

The Standard Chinchilla was developed in 1913 by a French engineer, M. J. Dybowski. To get the original Chinchilla, he crossed a wild gray rabbit (the Garenne) with a Himalayan. He also crossed a wild gray rabbit with a blue. The two litters produced by these matings were interbred to produce the modern Chinchilla rabbit. The breed was introduced to England in 1917 and to the United States in 1919. Their popularity is probably due to the similarity of their fur to that of the Chinchilla of South America, genus *Chinchilla*. The undercolor is dark slate-blue at the base, the intermediate portion is pearl, and the top portion has a very narrow black band with a very narrow light band above that. This underfur is brightly ticked with black hairs in either a wavy or even pattern. The breed is of medium length, with a rather compact and chubby body. The bucks weigh 5½ to 7½ pounds, and does weigh 6 to 8 pounds. *See Figure 9–17.*

Figure 9–17   Standard Chinchilla. *Photo by Isabelle Francais.*

The Angora

The Angora rabbit originated in Ankara, Turkey, hundreds of years ago. There are four different breeds today: English, Satin, French, and Giant. The English Angora was developed from the French Angora and was primarily an exhibition animal; it is somewhat smaller than the French Angora.

The English Angora

The English Angora has a short, compact body and the head, ears, feet, and body are covered with wool. English Angoras weigh 5 to 7½ pounds. *See Figure 9–18.*

Figure 9–18   English Angora. *Photo by Isabelle Francais.*

The Satin Angora

The Satin Angora was created by L. P. Meyer of Canada. It is a cross between a copper Satin and a fawn French Angora. The wool of this breed has good length and density and a brilliant satin sheen. *See Figure 9–19.*

The body of the Satin Angora is of medium length, is firm, and resembles the French Angora. The wool is finer than normal Angora wool and has a soft, silky texture. Satin Angoras weigh about 8 pounds. They are recognized in six color groups: Agouti, Pointed White, Self, Shaded, Solid, and Ticked.

Figure 9–19   Satin Angora. *Photo by Isabelle Francais.*

The French Angora

The French Angora originated in the French areas of Vosges, Savoy, Magence, Normandy, and Jura. The French Angora is longer and narrower than the English and not covered as heavily by wool. The tail is covered by wool, but the feet and legs are furred only to the first joint, and the ears are not as heavily covered. The wool of the French Angora is coarse, and the ideal length is about 2¾ inches. Animals of this breed weigh more than 7½ pounds. *See Figure 9–20.*

Figure 9–20 French Angora. *Photo by Isabelle Francais.*

The Lilac

The Lilac is a diluted Havana Brown that sometimes appears in litters of Havana rabbits. Lilacs of various shades can also be obtained by crossing the Havana with blue rabbits and then interbreeding the young. The ideal color is a dove-gray with a slight pink sheen, which is generally referred to as pinky-dove. The body of the Lilac is the same as the Havana and weighs 6 to 7 pounds. *See Figure 9–21.*

Figure 9–21 Lilac. *Photo by Isabelle Francais.*

The Silver Martin

The first Silver Martin appeared about 50 years ago in the nest box of a Standard Chinchilla. As the rabbit matured, black became the predominant color. Others began to appear in litters of Chinchilla rabbits. When these were bred together, they bred true to color. As breeding continued, the eye circles of silver gradually became more distinct, the silver lining of the ears became more noticeable, and the black coat took on a new sheen. Today, the breed is recognized in the Black, Blue, Chocolate, and Sable varieties. The fur is about 1 inch in length and is interspersed with silver-tipped guard hairs along the sides, rumps, and hips. Bucks weigh 6 to 8½ pounds, and does weigh 7 to 9½ pounds. *See Figure 9–22.*

Figure 9–22 Silver Martin. *Photo by Isabelle Francais.*

The Belgian Hare

The Belgian Hare originated in Flanders in Belgium. It was introduced into England in 1874 and arrived in the United States in 1888. The English are credited with the selective breeding that produced the modern Belgian Hare. Contrary to its name, the Belgian Hare is a true rabbit; it has the normal rabbit gestation period of 30 to 32 days, and the young are born without fur and with eyes closed. The name is derived from the fact that the breed originated in Belgium and has always been bred by devoted fanciers to resemble the wild hare. *See Figure 9–23.*

Figure 9–23 Belgian Hare. *Photo by Isabelle Francais.*

The breed is known as the racehorse or greyhound of the rabbit family. It has a long, fine body and a high arch to its back. The legs are long, straight, and slender. The head is rather long and set on a slender neck. The color varies from a fox to a mahogany red, with chestnut shading on the sides. Brilliant black ticking appears on the body and along the edges of the ears. The fur is close lying and of rather harsh texture. The Belgian Hare weighs 6 to 9 pounds.

The Rhinelander

The Rhinelander was developed in Germany. The ancestry is not known but is thought to be a cross of the English Butterfly, Japanese, and Flemish Giant. The breed has a butterfly marking on the nose, cheek spots, and eye rings. The rest of the body is white with a herringbone pattern or unbroken stripe running from the neck down the back to the tip of the tail; there are also color markings on the hip and flank. The Rhinelander is a tricolored rabbit breed. The primary color is a white body with black and bright orange spots. There is also a white variety with black and blue spots.

The body type is rather unique in that it is rounded and almost cylindrical, avoiding any heaviness in either the shoulders or hindquarters. Bucks weigh 6½ to 9½ pounds, and does weigh 7 to 10 pounds. *See Figure 9–24.*

Figure 9–24   Rhinelander. *Photo by Isabelle Francais.*

The Harlequin

The Harlequin rabbit originated in Northern France and Belgium in the 1880s. It is believed it was a cross mutation between the Dutch and the common domestic rabbit. *See Figure 9–25.*

The markings of the Harlequin are its distinctive feature; each side of the "perfect" Harlequin has a different pattern. Each ear is a different color, and each side of the head has the opposite color. The colors should be distinctly separated with no mixing of the colors; no two Harlequins are marked the same. Harlequins are in two basic body colors: orange or white with black, chocolate, lilac, or blue markings. The orange color patterns are referred to as "Japanese," and the white color patterns are referred to as "Magpie." The eyes of both color patterns are dark brown. Harlequin bucks weigh 6½ to 9 pounds, and does weigh 7 to 9½ pounds.

Figure 9–25   Harlequin. *Photo by Isabelle Francais.*

The Sable

The Sable rabbit originated in 1919, when rabbits with a deep sepia brown color and delicate shading began to appear in litters of English Chinchilla rabbits. Breeders decided to save them and developed the Sable breed. The Sable rabbit has a medium-length body with a slightly arched back. The head, ears, and feet have a purple or a shaded brown sheen. Bucks weigh 7 pounds or better, and does weigh 8 pounds or better. *See Figure 9–26.*

Figure 9–26 Sable.

The Shorthaired Rex

The Shorthaired Rex rabbit began to receive attention in 1919. The mutation had appeared frequently in litters of common gray rabbits over the years, but no importance was placed on these mutations. In 1919, a farmer named Desire Caillon of Pringe Louche in the Sarthe area of France bred a couple of them and found they produced short-haired young. A. M. Amedee Gillet, of Coulange, France, foresaw the possibility of introducing a new class of rabbit. *See Figure 9–27.*

The main characteristic of the Rex rabbit is the short hair coat, with guard hairs being of the same length as the underfur. The fur also stands at right angles to the skin, is about $3/5$ inch long, and has a very soft, plush feel. Rex rabbits are recognized in 14 different colors: Black, Blue, Californian, Castor, Chinchilla, Chocolate, Lilac, Lynx, Opal, Red, Sable, Seal, White, and Broken. The bucks weigh 7½ pounds, and does weigh 8 to 10½ pounds.

Figure 9–27 Shorthaired Rex. *Photo by Isabelle Francais.*

The Beveren

The Beveren rabbit originated late in the nineteenth century in the city of Beveren, Belgium. The Blue Beveren was imported into England in 1915 and became very popular. Although blue rabbits have been bred for centuries, none became as popular as the Blue Beveren. Blue-eyed White and Black Beveren were developed later. The White is the most popular in the United States, with its blue eyes distinguishing it from all other white breeds. The body is of medium length with a broad, meaty, slightly arched back. The head is full with an outward curvature between the eyes and nose that is characteristic of the breed. The bucks weigh 8 to 10 pounds, and does weigh 9 to 11 pounds. The Beveren can be temperamental at times and may not make the best family pet. *See Figure 9–28.*

Figure 9–28 Beveren. *Photo by Isabelle Francais.*

The Californian

The Californian was produced in 1928 by George West in an effort to produce a better commercial rabbit. He crossed the Himalayan and Chinchilla and produced a chin-color buck that was crossed with several White New Zealand does. Since its development, it has become one of the most popular breeds. The rabbit is white with black-colored nose, ears, feet, and tail; the eyes are red. The bucks weigh 8 to 10 pounds, and does weigh 8½ to 10½ pounds. *See Figure 9–29.*

Figure 9–29 Californian. *Photo by Isabelle Francais.*

The Hotot

The Hotot originated in France, where it is known as the Blanc de Hotot (White of Hotot); the Baroness E. Bernhard originated the breed around 1912. It is believed that native French-spotted rabbits and the Giant Papillon Français (Checkered Giant) were used in developing the breed. It was imported into the United States in 1978. The color is a shiny white except for a thin black band surrounding the eyes. The bucks weigh 8 to 10 pounds, and does weigh 9 to 11 pounds. *See Figure 9–30.*

Figure 9–30 Hotot. *Photo by Isabelle Francais.*

The Palomino

The Palomino rabbit was developed by Mark Youngs of Washington. Several breeds were crossed to produce a breed with distinctive colors. Two colors are recognized: the Golden, which has light gold guard hairs and a white to creamy undercoat, and the Lynx, which has orange-colored guard hairs and a white undercoat. Bucks of this breed weigh 8 to 10 pounds, and does weigh 9 to 11 pounds. *See Figure 9–31.*

Figure 9–31 Palomino. *Photo by Isabelle Francais.*

The Satin

The Satin rabbit appeared as a mutation in a litter of Havana in 1930. Walter Huey, a Havana raiser in Indiana, noticed baby rabbits with especially shiny coats. Mr. Huey was interbreeding Chocolate Havanas in an effort to perfect the deep brown color and outstanding fur. The Satin mutation occurred sometime during the selection and eventually expressed itself. The Satin factor was later recognized as an entirely new coat mutation and was bred into a number of breeds, with the result that there now is a "Satin" counterpart to many of the solid and normal-furred breeds. All of the Satins have been combined into a new breed having a uniform description of type and fur. The Satin breed is recognized in 10 varieties: Black, Blue, Californian, Chinchilla, Chocolate, Copper, Red, Siamese, White, and Broken. Satin bucks weigh 8½ to 10½ pounds, and does weigh 9 to 11 pounds. *See Figure 9–32.*

Figure 9–32 Satin. *Photo by Isabelle Francais.*

The Cinnamon

The Cinnamon rabbit was developed by Ellis Houseman of Montana. New Zealand White, Chinchilla, Checkered Giant, and Californian breeds were used to develop this breed. The Cinnamon is a medium-length rabbit with a well-filled, meaty body. The color is a rust or cinnamon ground color; the fur is ticked with smoke-gray across the back. The sides of the rabbit show a smoke-gray shading, and the underfur is orange; the fur around the head and ears shows a darker shading. Bucks of this breed weigh 8½ to 10½ pounds, and does weigh 9 to 11 pounds. *See Figure 9–33.*

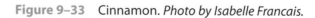

Figure 9–33 Cinnamon. *Photo by Isabelle Francais.*

The Creme d'Argent and the Champagne d'Argent

The Creme d'Argent and the Champagne d'Argent originated in France. The fur of the Creme d'Argent is a cream color with a bright orange underfur; the guard hairs are also orange. The Champagne d'Argent has a body color of bluish-white. The underfur is a dark, slate blue, and the guard hairs are jet black. Both breeds made their appearance in the United States around 1934 and are becoming very popular because of their luxurious and richly colored fur. Creme d'Argent bucks weigh 8 to 10 pounds, and does weigh 8½ to 11 pounds. Champagne d'Argent bucks weigh 9 to 11 pounds, and does weigh 9½ to 12 pounds. *See Figures 9–34 and 9–35.*

Figure 9–34 Creme d'Argent. *Photo by Isabelle Francais.*

Figure 9–35   Champagne d'Argent. *Photo by Isabelle Francais.*

The American

The American rabbit breed was developed in the United States. It is recognized in Blue and White color varieties. The background of the American White variety is a mystery but is believed to have developed from throwbacks, called "sports" of blue rabbits. The ideal body type is described as "mandolin" because the arch of the back from the hindquarters to the shoulders has a mandolin-shaped appearance to it. American rabbits were first shown in 1917 and became very popular during the 1920s and 1930s. Bucks weigh 9 to 11 pounds, and does weigh 10 to 12 pounds. *See Figure 9–36.*

Figure 9–36   American. *Photo by Isabelle Francais.*

The American Chinchilla

The American Chinchilla is a result of selective breeding for size from the Standard Chinchilla. It is larger than the Standard and has been referred to as the "Heavyweight" Chinchilla. The breed has the distinctive Chinchilla color pattern. Bucks weigh 9 to 11 pounds, and does weigh 10 to 12 pounds. *See Figure 9–37.*

Figure 9–37   American Chinchilla. *Photo by Isabelle Francais.*

The English Lop

The English Lop is perhaps the oldest breed of domestic rabbit known to humans. Its exact origin is not known, but it first appeared in Algiers, North Africa, and from there it spread to England. The breed became very popular in England, and breeders competed to increase the ear length. Ears measuring 23 to 24 inches were common, and the first Lop to have an ear length of over 24 inches appeared in 1885; the record is 28½ inches. They are recognized in self and broken varieties. The body is of medium length with a well-arched back and the head carried very low. Bucks weigh 9 pounds and over, and does weigh 10 pounds and more. *See Figure 9–38.*

Figure 9–38   English Lop. *Photo by Isabelle Francais.*

The New Zealand

The New Zealand breed originated in California and Indiana almost simultaneously. The Red variety was the first to appear and is a cross between a Belgian Hare and a White rabbit. The White variety appeared later and was the result of crossing a number of breeds, including the Flemish, American White, Angora, and perhaps the Red variety. The Black variety came much later, and it too resulted from various crosses. The New Zealand, especially the White, has been the leading choice for the commercial rabbit producer in the United States. The White is preferred because when it is dressed, the small hairs that remain cannot be seen; thus, it is easier to clean and package. This breed is primarily raised for meat production. Mature bucks weigh 9 to 11 pounds, and does weigh 10 to 12 pounds. *See Figure 9–39.*

Figure 9–39 New Zealand. *Photo by Isabelle Francais.*

The Silver Fox

The Silver Fox rabbit was originated and developed during the 1920s by W. B. Garland of North Canton, Ohio. The various crosses used in the development of this breed are not known, but it is believed to have resulted from crosses of Checkered Giants and English Silvers. The breed was originally known as the American Silver Giant. Because the fur closely resembles the real Silver Fox, the name "American Silver Fox" was given to the breed in December 1929. The unique features of this breed are its evenly silvered fur resembling fox fur, fur length (1 to 1½ inches long), and the lack of fly-back fur (the fur returning to its original position) when stroked. Bucks weigh 9 to 11 pounds, and does weigh 10 to 12 pounds. *See Figure 9–40.*

Figure 9–40 Silver Fox, Black variety. *Photo by Isabelle Francais.*

The Giant Angora

The Giant Angora was originated by Louise Walsh of Taunton, Maine. The main purpose of developing the breed was to produce an animal that would have a maximum amount of wool production. *See Figure 9–41.*

The Giant Angora has the body structure of a commercial animal and possesses a unique coat structure. The wool comprises three distinct types: the underwool, which is medium-fine, soft, and delicately waved; the awn fluff, a soft, wavy wool with a guard hair tip found between the underwool; and the awn hair, which is strong, straight hair (guard hair) protruding above the awn fluff.

Giant Angora bucks usually weigh more than 8½ pounds, and the does weigh more than 9 pounds. There are two varieties: the Ruby-Eyed White and the Blue-Eyed White.

Figure 9–41 Giant Angora. *Photo by Isabelle Francais.*

The French Lop

The French Lop rabbit was first bred in France around 1850 by a Frenchman named Conderier. It is believed to be a cross of the English Lop and the Butterfly rabbit of France. The Butterfly rabbit is still bred in France and closely resembles the Flemish Giant. The ears of the French Lop are much shorter than those of the English Lop. The ears hang down in a horseshoe shape from the crown approximately 1½ inches or more below the muzzle. The body has a very massive, thick-set appearance. The breed is broad, deep, and heavily muscled. It is shown in any solid color and in any color in combination with white. The bucks weigh more than 9 pounds, and does weigh more than 10 pounds. *See Figure 9–42.*

Figure 9–42   French Lop. *Photo by Isabelle Francais.*

The Checkered Giant

The Checkered Giant originated in Germany in 1904. Otto Reinhardt of Rheinpfalz, Germany, took a Rhenish Checker (something like an English Spot) and crossed it with a Black or Gray Flemish Giant. This mating produced the foundation stock for modern-day Checkered Giants. *See Figure 9–43.*

In 1910, the first Checkered Giants came to America. Other shipments followed, and American breeders have developed a breed that is more racy than the larger-boned, heavier-shouldered European type. The American Checkered Giant has a long, well-arched body that stands well off the ground. The body color is white with either black or blue color markings. The colored butterfly nose, eye circles, cheek spots, spine markings, tail, and side markings are well formed and distinct. The bucks weigh 11 pounds and over, and does weigh 12 pounds and over.

Figure 9–43   Checkered Giant. *Photo by Isabelle Francais.*

The Giant Chinchilla

The Giant Chinchilla is a result of selective cross-breeding and the addition of Flemish Giant blood; the breed was developed by Ed Stahl of Missouri. This larger variety of Chinchilla rabbit still maintains the distinctive Chinchilla color pattern. Bucks weigh 12 to 15 pounds, and does weigh 13 to 16 pounds. *See Figure 9–44.*

Figure 9–44   Giant Chinchilla. *Photo by Isabelle Francais.*

The Flemish Giant

The early development of the Flemish Giant is not known, but it is known that the breed originated in Belgium. The Dutch probably deserve credit for domesticating and improving the breed, but the Germans deserve much of the credit for perfecting the breed and increasing its length. *See Figure 9–45.*

Figure 9–45   Flemish Giant. *Photo by Isabelle Francais.*

The breed was imported into the United States about 60 years ago, and breeders in the United States are given credit for producing the different colors. Flemish Giants are now recognized in seven color varieties: Steel Gray, Light Gray, Sandy, Black, Blue, White, and Fawn. This is the largest of all domestic rabbit breeds. The bucks weigh more than 12 pounds, and the does weigh over 13 pounds; 15- to 18-pound Flemish Giants are common.

ANATOMY

The anatomy and skeletal structure of the rabbit are identified in *Figures 9–46 and 9–47.*

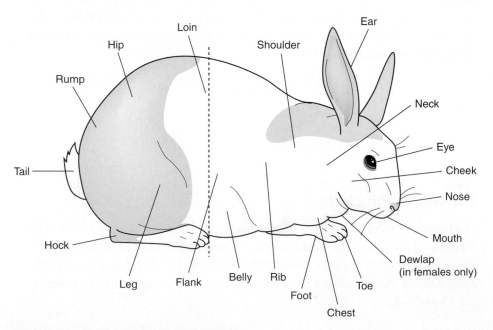

Figure 9–46   The anatomy of the rabbit.

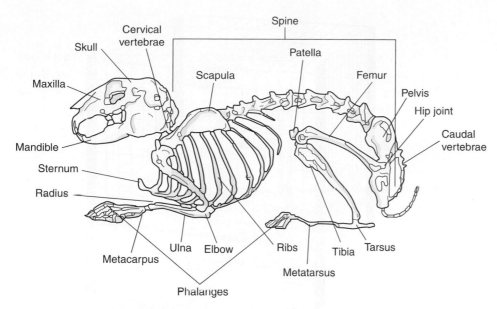

Figure 9–47 The skeletal structure of the rabbit.

HOUSING AND EQUIPMENT

Before purchasing a rabbit, some thought should go into the housing for the animal. Houses used to keep rabbits are referred to as *hutches*. The size of the cage or hutch depends on the breed that will be housed in it. A large cage is better as it gives the rabbit more room to move about. The miniature and small rabbits can be kept in wire cages that are 24 inches wide, 36 inches long, and 14 inches high; this size is common and can be purchased at most pet supply stores.

Cages for rabbits that weigh from 6 to 11 pounds should be 30 inches wide, 36 inches long, and 18 inches high. This size allows them plenty of space to move about and get exercise.

Rabbits that are 12 pounds and over need cages 24 inches wide, 48 inches long, and 18 inches high. Cages 30 inches wide, 48 inches long, and 24 inches high, minimum, should be provided for the Belgian Hare; cages 36 inches wide, 72 inches long, and 36 inches high are preferred for this active breed of rabbit. Another way to gage minimum cage size is the "rule of three." Cages should be at least three times the rabbit's length or allow for three hops in a straight line.

Wire cages are recommended for most rabbits; however, the larger breeds over 12 pounds should be kept on solid floors to prevent sore hocks. **Sore hocks** are ulcerations to the foot pads caused by the animal's body weight pressing down on the foot pads, which become irritated on wire cages. *See Figure 9–48.* All cages should include an area of solid flooring so the rabbit is not constantly on the wire.

Cages with wire floors have the distinct advantage of being easy to clean because the wire allows feces and urine to fall into trays below; these trays are easily emptied and cleaned. Wire cages are also easy to steam clean and disinfect.

Figure 9–48 Cages made from wire are very nice for rabbits. Feces and urine fall through the wire and are easily cleaned from the slanted surfaces below the cages.

The trays below the cages should be filled with wood shavings, kitty litter, or another type of absorbent material. Baking soda sprinkled in the corners helps cut down on the strong urine smell.

Wire used for the flooring of cages should have openings about ½ inch by 1 inch. The ½-inch side of the wire should be toward the animals' feet; this will cause less irritation.

Hutches with wood floors should be bedded with wood shavings, straw, or another type of absorbent material. It is important that this gets changed at least once a week. Rabbits bedded on wet bedding are subject to sore hocks, hutch burn, and other ailments. Sanitation is of utmost importance, especially if large numbers of animals are involved.

The miniature and small rabbits are very clean and almost odorless. They can be kept in the house and need their cages cleaned only once or twice a week. Larger rabbits kept indoors can create an odor problem, and their cages may need to be cleaned daily.

Several conditions are important in housing rabbits: temperature, humidity, ventilation, proper lighting, and the absence of drafts. Rabbits can be raised outside year-round. They can stand moderately cold temperatures as long as they are out of the wind; however, they have trouble surviving extremely warm temperatures. When the humidity reaches 60 percent, ventilation must be provided. If temperatures reach 80° F, rabbits will be very uncomfortable. High temperature and high humidity can lead to heat stroke. Temperatures above 80° F over an extended time can affect the semen quality of a male rabbit. High temperatures with low humidity can be tolerated if the air is moving.

Rabbits keep cool by the breathing process and by the air movement around them, especially past their ears. Fans can be installed, and a bottle of frozen water can be put in the cage for the rabbit to lie against.

Ventilation is also important. Fresh air is a must if rabbits are to be kept healthy and productive. Stagnant and stale air must not be allowed to build up because this will lead to respiratory problems and diseases.

Lighting is important in priming a rabbit's fur and is important in the receptiveness of a breeding doe. A doe must have a minimum of 16 hours of daylight for proper breeding.

Cool, damp drafts can cause animals to become chilled. Again, this condition can lead to colds, pneumonia, and other ailments.

Single rabbits can be watered with ceramic water bowls or with vacuum-type water bottles that hang on the outside of the cage. The disadvantage of the ceramic bowl is that it can be tipped over or become contaminated. Water bottles solve a lot of the problems associated with open bowl watering. Waterers can be constructed from plastic milk jugs and plastic soft drink bottles by using one of several water valves, such as the **dew-drop valve,** that are available on the market. All of these types of water systems are subject to freezing during cold weather. Several types of heating wires and heating tapes are available for outside water systems to keep them from freezing.

Nest boxes should be provided for the does to use when they are giving birth (kindling). They can be purchased from pet supply stores, or they can be constructed from metal, wire, or wood. The size of the nest box will vary depending on the size of the doe, but it should be about 12 inches long, 7 inches wide, and 6 inches high for the small breeds; 18 inches long, 10 inches wide, and 9 inches high for the medium breeds; and 22 to 24 inches long, 14 inches wide, and 12 inches high for the giant breeds. Nest boxes need to be clean and sanitized before use, or the doe may refuse to use the box.

FEEDING

Galvanized-metal self-feeders are the easiest to use and prevent a lot of waste. Metal feeders usually clip onto the outside of the cage and fit through an opening into the cage. They can be filled quickly and easily from the outside. *See Figure 9–49.*

Other types of feeders can be used, such as ceramic bowls and plastic bowls; however, most of these take up floor space and can be tipped over.

The easiest way to feed rabbits is to use one of the commercially prepared pellet-type feeds. Pellets made from timothy hay rather than alfalfa hay are better as a maintenance diet. Timothy hay has a better balance of fiber and protein. *See Figures 9–50 and 9–51.* These feeds are especially formulated for rabbits and will supply all of the necessary nutrients needed. Several companies have made studies in preparing pelleted feed for various stages of rabbit growth. As Table 9–3 shows, protein and energy requirements for maintenance are lower than the requirements needed for growth, gestation, and lactation.

Many owners like to supplement the diet with other types of foods. Several foodstuffs can be used as treats, but owners must be cautious because feeding too many fruits, vegetables, and green foods can cause severe gastrointestinal problems that may lead to fatal diarrhea. Any supplemented feeds should be kept to a minimum, especially if the animals are not used to receiving them.

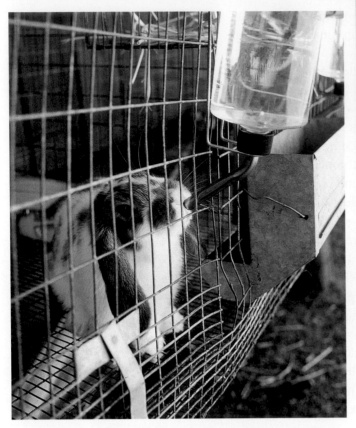

BUNNETS*–17%
with Aqua Vite*
* TRADE MARKS
A Complete Cereal, Vitamin and Mineral
Ration in Pellet Form for Rabbits

INGREDIENTS:

Dehydrated and Sun Cured Alfalfa Meal, Pulverized Oats, Soybean Oil Meal, Feeding Cane Molasses, Wheat Middlings, Wheat Bran, Vitamin A Acetate, D-Activated Animal Sterol, delphine Trophopheryl Acetate, Choline Chloride, Folic Acid, Riboflavin, Niacin, Vitamin B-12, Pyroxidine, Thiamin Hydrochloride, Ascorbic Acid, Carrot Oil Carotene, Soybean Oil, Sodium Chloride, DiCalcium Phosphate, Magnesium Oxide, Magnesium Sulfate, Iron Oxide from Sulfate, Copper Sulfate, Calcium Iodate, Cobalt Carbonate, Zinc Oxide, BHT (preservative), Ligmin Sulfonate, Sodium Selente, Calcium Carbonate, Verrets, Yucca, Extract, Sodium Benzonate, 6 Fermentation Product and Flavor.

GUARANTEED ANALYSIS:

Crude Protein, not less than	17.0%
Crude Fat, not less than	2.5%
Crude Fiber, not more than	25.0%

FEEDING DIRECTIONS:

The 17% protein ration is designed for rabbits in production, such as for rapidly growing rabbits, gestating or lactating does, and working bucks. It can be fed in either of two ways: free choice or limit feeding. Free choice feeding, where the feeders are kept full of pellets all the time, will result in faster gains of weaning rabbits, and is often used for does with litters. We recommend limit feeding, where only as much feed as will be cleaned up in a day is provided. This requires closer observation by the feeder, but problems are often discovered sooner, feed is kept fresher, and rabbits usually are healthier. Do not restrict feed to the degree that gains are severely reduced.

Do not overfeed does at kindling. Bring does into full milk production gradually by increasing feed as the litter grows and needs more milk.

Be sure to provide plenty of fresh clean cool water. Good quality hay can also be offered.

Figure 9–49 Galvanized-metal self-feeders are easily filled from outside the cage. *Courtesy of FFA.*

Figure 9–50 Protein content is important in feeding rabbits. Always check the labels. *Photo by Dean Warren.*

Some types of foodstuffs that can be added to the rabbit's diet are high-quality Timothy grass hay, corn, oats, oatmeal, and wheat germ. Other treats may include pieces of carrot, carrot tops, slices of apple, green beans, and pieces of banana, pineapple, or papaya. Any green foods or vegetables should be washed to eliminate any chemical residue.

Green lettuce or other types of leafy green vegetables should not be fed to rabbits, especially young rabbits. Green, leafy vegetables have a high water content and cause diarrhea and dehydration because the rabbit ceases to drink water. Clean, fresh water should always be available for rabbits; this is especially true during warm, summer months.

Coprophagy is the term associated with the eating of fecal material; rabbits consume their own feces. The fecal material of rabbits is of two types: a hard, dry type that is normally seen in the cage or under the wire flooring; and the soft form that the rabbit consumes as it is being expelled. Because this type of feces is normally consumed at night, it is often referred to as "night feces," although the rabbit may consume this during the day as well. Coprophagy allows rabbits to recover nutrients that were unabsorbed during the first pass through the digestive tract.

Figure 9–51 Pelleted feed is the easiest way to ensure that rabbits are receiving the proper diet. The main ingredient of these pellets is alfalfa meal. *Photo by Dean Warren.*

HANDLING

The correct way to handle a rabbit depends on the size of the rabbit; however, under no circumstances should the animal be picked up by the ears because this may cause injury to small blood vessels or cartilage in the ear. Also, picking the animal up by the nape of the neck is not recommended, because this too can cause damage to tiny blood vessels in the skin and could cause damage to the pelt.

When approaching a rabbit cage, one should walk slowly and speak to the rabbit. If the rabbit knows someone is close, it won't get as frightened and bolt to the corner of the cage.

Next, one should reach in and gently stroke the animal from front to rear and rub and stroke the animal's head. Then, the rabbit can be gently moved into a position where it can be picked up by slipping one hand in under the chest and belly and placing the other hand behind the rabbit. The rabbit can now be lifted up and removed from the cage. It is important to remove it from the cage tail first to prevent the rabbit's legs from getting caught in the wire floor or wire door of the cage.

Small rabbits can be supported with one hand under the body, and large rabbits can be supported on one arm with the other arm cradling the animal securely. If the animal feels secure and comfortable, it won't struggle or try to

TABLE 9–3	Nutrient Requirements of Rabbits Fed Ad Libitum (Percentage or Amount per Kilogram of Diet)			
Nutrients[a]	**Growth**	**Maintenance**	**Gestation**	**Lactation**
Energy and protein				
Digestible energy (kcal)	2500	2100	2500	2500
TDN (%)	65	55	58	70
Crude fiber (%)	10–12[b]	14[b]	10–12[b]	10–12[b]
Fat (%)	2[b]	2[b]	2[b]	2[b]
Crude protein (%)	16	12	15	17
Inorganic nutrients				
Calcium (%)	0.4	—[c]	0.45[b]	0.75b
Phosphorus	0.22	—[c]	0.37[b]	0.5
Magnesium (mg)	300–400	300–400	300–400	300–400
Potassium (%)	0.6	0.6	0.6	0.6
Sodium (%)	0.2[b, d]	0.2[b, d]	0.2[b, d]	0.2[b, d]
Chlorine (%)	0.3[b, d]	0.3[b, d]	0.3[b, d]	0.3[b, d]
Copper (mg)	3	3	3	3
Iodine (mg)	0.2[b]	0.2[b]	0.2[b]	0.2[b]
Iron	—[c]	—[c]	—[c]	—[c]
Manganese (mg)	8.5[c]	2.5[c]	2.5[c]	2.5[c]
Zinc	—[c]	—[c]	—[c]	—[c]
Vitamins				
Vitamin A (iu)	580	—[e]	>1160	—[e]
Vitamin A as carotene (mg)	0.83[b, c]	—[f]	0.83[b, c]	—[f]
Vitamin D	—[g]	—[g]	—[g]	—[i]
Vitamin E (mg)	40[h]	—[e]	40[h]	40[j]
Vitamin K (mg)	—[i]	—[i]	0.02[b]	—[j]
Niacin (mg)	180	—[j]	—[j]	—[j]
Pyridoxine (mg)	39	—[j]	—[j]	—[j]
Choline (g)	1.2[b]	—[j]	—[j]	—[j]
Amino acids (%)				
Lysine	0.65	—[g]	—[g]	—[g]
Methionine + cystine	0.6	—[g]	—[g]	—[g]
Arginine	0.6	—[g]	—[g]	—[g]
Histidine	0.3[b]	—[g]	—[g]	—[g]
Leucine	1.1[b]	—[g]	—[g]	—[g]
Isoleucine	0.6[b]	—[g]	—[g]	—[g]
Phenylalanine + tyrosine	1.1[b]	—[g]	—[g]	—[g]
Threonine	0.6[b]	—[g]	—[g]	—[g]
Tryptophan	0.2[b]	—[g]	—[g]	—[g]
Valine	0.7[b]	—[g]	—[g]	—[g]
Glycine	—[e]	—[g]	—[g]	—[g]

[a]Nutrients not listed indicate dietary need unknown or not demonstrated.
[b]May not be minimum but known to be adequate.
[c]Quantitative requirement not determined, but dietary need demonstrated.
[d]May be met with 0.5 percent Nacl.
[e]Converted from amount per rabbit per day using an air-dry feed intake of 60 g per day for a 1-kg rabbit.
[f]Quantitative requirement not determined.
[g]Probably required, amount unknown.
[h]Estimated.
[i]Intestinal synthesis probably adequate.
[j]Dietary need unknown.

Source: Reprinted with permission from *Nutrient Requirements of Rabbits*, 2d Edition. Copyright 1977 by the National Academy of Sciences. Courtesy of the National Academy Press, Washington, D.C.

Figure 9–52 Proper handling is important. Rabbits need to feel secure. Their powerful rear legs and sharp nails can inflict injury to the handler's arms.

escape. The head of the rabbit can be tucked in under the upper arm toward the elbow of the arm that is being used to support the rabbit. *See Figure 9–52.*

Correct restraint is important because a frightened rabbit can kick hard enough to fracture its own spine.

Small animals are easy to lift with the method just described; however, larger animals may pose a problem. This is especially true if they are not used to being handled. The handler should always wear long-sleeve clothing to prevent scratches.

When setting the rabbit down, one must do so gently and slowly, letting the animal see where it is going so that it will not be frightened. The handler should set the rabbit down, hind end first. A rabbit's foot is covered with fur, and when placed on slick surfaces, they have no traction. Rabbits can easily dislocate a hip or their spine when they try to hop or push off on their rear legs.

The handler should be careful when returning the rabbit to its cage. The rabbit, recognizing a familiar area, may try to jump from the handler's arms and may inflict an injury to the handler with its strong, powerful rear legs and sharp toenails.

DISEASES AND AILMENTS

Control of temperature, humidity, ventilation, lighting, and drafts is extremely important in preventing diseases in rabbits; proper cleaning and disinfecting of equipment are also important.

Enteritis

Enteritis is an inflammation of the intestinal tract and probably the most common cause of death in rabbits. There are four types of enteritis: mucoid enteritis, enterotoxemia, and Tyzzer's disease, which are caused by bacteria; and coccidiosis, which is caused by a protozal parasite.

The exact cause of mucoid enteritis is unknown. The disease expresses itself as mucous discharges from the rectum and causes the animals to appear potbellied or to be filled with water. Diseased animals fail to eat but drink a lot of water. If water is supplied in a bowl, a diseased animal often puts its feet in the water and grinds its teeth. Death usually occurs 4 days to 1 week following the onset of symptoms; very little can be done to save affected animals.

Mucoid enteritis may also be associated with stress, unsanitary conditions, and feeds with too high energy levels. Good management is important to reduce stress, and the animals should receive a ration containing less energy and higher levels of fiber.

Enterotoxemia is widely recognized as the leading cause of enteritis-type symptoms in young rabbits. It is caused when the balance of resident gastro-intestinal bacteria is disrupted and is associated with the ingestion of high-energy feeds that are not completely digested in the stomach and small intestine. As this undigested material reaches the cecum, high amounts of carbohydrates allow for an increase in the growth of some types of bacteria. As the growth of bacteria increases, there is an increase in toxins produced, which causes dehydration and shock. Severe diarrhea, which is a common symptom of this disease, is caused by the increased amount of fluid taken into the cecum and intestines in an effort to dilute the toxins. Other symptoms include rapid weight loss, rough appearance, and sunken eyes; there is almost 100 percent mortality in affected animals.

To prevent the disease, sanitation is extremely important, as well as reduced stress and the feeding of rations containing at least 18 percent fiber.

Tyzzer's disease is caused by the bacterium *Clostridium piliforme*. This disease causes severe diarrhea and death among young rabbits 6 to 7 weeks old. Other symptoms include rapid weight loss, dehydration, and death occurring within 1 to 3 days. Again, sanitation is especially important, and the control of rats and mice is important in preventing the organism from being carried into a rabbitry. Antibiotics are used to treat this condition.

Coccidiosis is caused by several species of the protozoan parasite *Eimeria*. Hepatic coccidiosis is caused by the protozoan *Eimeria stieda*, and intestinal coccidiosis is caused by several species of *Eimeria: E. magna, E. irresiduca, E. media,* or *E. perforans*.

Adult coccidia live intracellularly in the linings of the bile duct (hepatic coccidia) or the intestinal tract (intestinal coccidia). These damage the linings of the bile duct and intestinal tract and produce eggs, **oocysts,** which pass in the feces.

Oocysts mature for varying periods of time, depending on physical conditions, until they are "infective" or ready to grow to adulthood if ingested. The mature oocyst is ingested by the rabbit while consuming contaminated feed or when licking its feet or contaminated cages. Once inside the digestive system, the oocyst matures to the infective stage and enters the cells of the intestinal lining, where it matures to adulthood; thus, the cycle is completed. The infective stage of the hepatic coccidia makes its way from the cells of the intestinal lining to the cells of the bile duct.

Symptoms are more pronounced in young animals; older animals may become immune to the organism. Common symptoms include diarrhea stained with blood and mucus, potbellies, and rough hair coat. The infected animals stand around with arched backs.

The development of coccidiosis is greatly influenced by unsanitary conditions and stress. Birds are carriers of coccidia and should not be allowed access into the rabbitry.

Snuffles

Snuffles is a disease caused by *Pasteurella bacteria.* Symptoms of the disease include persistent sneezing with a white-colored nasal discharge; the fur on the front feet just above the paws will be matted from the rabbit constantly pawing at its nose. Abscesses caused by the *Pasteurella* organism may appear on the legs, neck, shoulders, and head of the animal. Affected animals are usually unthrifty and gain poorly. *See Figure 9–53.*

Many animals can carry the *Pasteurella multocida* organism in their upper respiratory tracts and do not show any signs of the disease. The disease usually appears when infected animals are placed under stress conditions such as poor sanitation, poor ventilation, changing temperatures, shipping, or the stress of show conditions.

Figure 9–53 This rabbit has snuffles. Note the nasal discharge and wet nose. *Reprinted from* Raising Rabbits, *copyright 1977 by Ann Kanable. Permission granted by Rodale Press, Inc., Emmaus, PA 18098.*

Mastitis

Mastitis is an inflammation of the mammary tissue. The mammary system becomes inflamed, swollen, hot, and dark red or blue in color; the condition usually occurs just after kindling or weaning. Any type of bacteria that enters the teat opening is capable of causing mastitis; however, the most common type of bacteria is the *Staphylococcus* species. Bacteria of the *Pasteurella* species may be present in the blood of an animal and circulate to the mammary tissue and cause mastitis as a result of stress caused by producing milk.

Treatment for mastitis must begin as soon as signs of the disease are noticed. A veterinarian should be consulted. In treating caked breast, all concentrates should be withheld for 72 hours, and only roughages and water should be fed. If a large number of does in the rabbitry become infected with mastitis, cages and nesting boxes should be cleaned and disinfected.

Weepy Eye

Weepy eye is an inflammation of the **conjunctiva,** the mucous membrane lining the inner surface of the eyelids and covering the front part of the eyeball. This inflammation is usually caused by a *Pasteurella multocida* infection. Affected rabbits rub their eyes with their front feet; there is a discharge from the eye, and the fur at the corner and lower lid of the eye becomes matted. One should consult a veterinarian and follow the treatment program recommended by the veterinarian. *See Figure 9–54.*

Figure 9–54 This rabbit is suffering from weepy eye, an inflammation of the conjunctiva of the eye. Note the discharge from the eye and the matting of the fur around the eyes. *Reprinted from* Raising Rabbits, *copyright 1977 by Ann Kanable. Permission granted by Rodale Press, Inc., Emmaus, PA 18098.*

Treponematosis (Vent Disease, Rabbit Syphilis)

Treponematosis is a venereal disease of domestic rabbits. The disease is also referred to as vent disease and rabbit syphilis. Signs first appear as bare areas around the genitals. Small blisters or ulcers usually form and eventually the area becomes covered with scabs; these scabs are usually confined to the genital areas but may appear on the lips and eyelids. Both sexes can become infected, and the disease is usually transmitted by mating infected animals. The specific organism responsible is the spirochete *Treponema cunniculi.* Infected animals should not be mated. The bacteria are not transmissible to other domestic animals or to humans.

Ear Mites

"Ear mites are a common parasite in rabbits. Mites irritate the lining of the ear and cause a watery fluid and thick brown crusts to build up, creating an 'ear canker.' Infested rabbits scratch at and shake their head and ears. They lose weight and may develop infections, which can damage the inner ear, reach the central nervous system, and result in torticollis or "wry neck" (a twisting of the neck to one side, resulting in the head being tilted). Your veterinarian will remove the brown crumbly discharge and then treat the affected ear with one of the drugs that are approved for use in dogs and cats. Products containing a substance that breaks down the waxy secretions in the ear are particularly useful in removing the heavy, crusted material. The medication should be applied within the ear and down the side of the head and neck. Ear mite infestations are less likely to occur when rabbits are housed in wire cages than in solid cages. The mite is readily transmitted by direct contact."[4] *See Figure 9–55.*

Infectious Myxomatosis

Infectious myxomatosis is a fatal disease of domestic rabbits that is primarily seen in the coastal areas of California and Oregon. Wild rabbits are resistant and do not get infectious myxomatosis. The California brush rabbit is the reservoir for the infection. The virus causing the disease is transmitted by mosquitoes, biting flies, and by direct contact. Incidence of the disease is greatest during the months of May to August, when mosquito populations are the greatest.

Papillomatosis

Papillomas are small warts that appear in or around the mouth and around the head, neck, and ears. Infected animals should be isolated and culled from the herd. *See Figure 9–56.*

Ringworm

Ringworm is a common disease of domestic rabbits usually associated with poor management practices. It is caused by *Trichophyton mentagrophytes,* which also affects guinea pigs, mice, and rats. Affected areas are circular, raised,

Figure 9–55 Wry neck is a condition in which the head is tilted to one side. *Courtesy of Eric Klaphake, DVM.*

Figure 9–56 Papillomatosis is characterized by warts or tumors. *Reprinted from* Raising Rabbits, *copyright 1977 by Ann Kanable. Permission granted by Rodale Press, Inc., Emmaus, PA 18098.*

reddened, and capped with white, flaky material. Transmission is by direct contact. The local veterinarian should be consulted. Infected animals should be isolated and treated. Disposable gloves should be worn when working with infected animals and equipment. Equipment and cages should be disinfected.

Fur Mites

"Fur mites are also common in rabbits. Because these mites live on the surface of the skin and do not burrow into the skin, they do not cause the intense itching seen with sarcoptic mange. Fur mite infestations usually do not cause any signs unless the rabbit becomes weakened due to age, illness, or other stress. The mites may be noticed as 'dandruff.' Scraping the dandruff from the skin onto a dark paper or background will demonstrate this and has led to the nickname 'walking dandruff' for this condition. Transmission is by direct contact. A diagnosis can be made by looking at skin scrapings under a microscope. Fur mites may cause a mild irritation or inflammation in humans. Weekly dusting of animals and bedding with permethrin powder can control mites."[5]

Mange Mites

"Rabbits are rarely infested with the mange mites that cause sarcoptic mange (canine scabies) or notoedric mange (feline scabies). These mites burrow into the skin and lay eggs. When infestation does occur, the rabbits are extremely itchy. It is difficult to get rid of these parasites on domestic rabbits. The condition is extremely contagious and can be transmitted to humans."[6]

Pinworms

Pinworms of the species *Passalurus ambiguus* may infect rabbits. If an animal is heavily infested, slow growth and poor coat condition will be a common sign. Sanitation and good management practices usually prevent infection from pinworms.

Rabbits are an intermediate host for the tapeworm species *Taenia pisiformis* and *Taenia serialis* in dogs, and *Taenia taeniaeformis* in cats. There are no visible signs of infection in rabbits, and evidence of infection is visible only upon necropsy or during processing. It is important to keep dogs and cats out of areas where food and nesting materials are stored.

Wet Dewlaps

Wet dewlaps is a condition caused by the rabbit dragging the **dewlap** into the water bowl; this causes the fur on the dewlap to become wet and matted. Infection of this area causes the skin to turn green and produce an odor.

The fur should be clipped from the infected area and an antibiotic ointment or cream applied. The use of water bottles or automatic waterers usually eliminates the problem. A rubber ball can be placed in the water bowl; this forces the rabbit to drink from the sides of the bowl and not drag the dewlap into the bowl.

Fur Chewing

Rabbits may pull fur from themselves or others in a pen and chew it. Fur chewing is thought to be caused by simple boredom, a low-fiber diet, or a nutrient deficiency. Increasing the fiber in the diet by feeding more Timothy hay may be effective in eliminating fur chewing.

Hutch Burn

Hutch burn or urine burn is a chapped or burning condition around the external genital area; the condition is caused by the animal sitting for long periods of time in a wet, dirty area. The chemicals in urine-soaked bedding cause a burn to the tender skin of the genital area; these burned areas may develop secondary infections if conditions are not corrected.

Dirty, wet hutches should be cleaned and fresh bedding used. Antibiotic creams or ointments should be applied to the infected areas.

Sore Hocks

Sore hocks are ulcerated areas on the bottom of the foot. Although the condition can affect all breeds, the larger breeds with extremely large bodies are more commonly affected. Rough, wire cage floors, nervousness of the animal, and matting of the fur on the foot are conditions that increase the likelihood of sore hocks. An accumulation of wet bedding, hair, and feces in the corners of the cage may also be a contributing factor.

Treatment should consist of putting the affected animal on a solid surface and applying medication to the sore foot.

Malocclusion

Malocclusion is a condition in which the upper and lower teeth grow to the extent that the animal cannot eat. The condition results because the teeth do not wear down properly, or it may be an inherited condition in which the teeth do not come together properly. A veterinarian can provide temporary treatment by anesthetizing the animal and trimming the teeth. Malocclusion is usually an inherited trait; affected animals should not be used for breeding.

REPRODUCTION

Rabbits become sexually mature at 4 to 8 months of age, depending on the breed. The miniature and small rabbits become mature at about 5 months. Males usually mature slower by 30 to 60 days.

Female rabbits do not have a regular cycle like most other animals. In nature, the female rabbit becomes sexually active based on the length of the day and the temperature, but rabbits kept inside under artificial light and warmth can be bred all year.

Even though the doe can be bred any time during the year, there may be times when the doe is more receptive. A breeder can check the external genital area or vulva of the doe. If the vulva is white, small, and dull, the doe will probably not be receptive; if it is mated, the chances of conception will be slight. If the vulva is pink, slightly swollen, and glistening in appearance, she will be more receptive and the chances of conception are better; if the vulva is reddish-purple and swollen, she will probably be most receptive and chances for conception are highest.

To mate a doe, she should be placed in the male's cage; the male should *not* be placed in the female's cage. If the male is put in the female's cage, he will probably be too busy examining the new cage, and the female is very territorial and will probably fight. After the female is placed in the male's cage, the rabbits should be observed to see that no fighting takes place and that actual mating occurs.

The act of mating will be over very quickly, and the male will fall backward or to the side as the act is completed. The actual mating causes the ovaries of the female to release eggs 8 to 10 hours after mating. The sperm will remain viable for 30 to 36 hours in the female's reproductive tract but can remain viable for up to 120 hours or longer.

Another mating 8 to 10 hours after the first stimulates the ovaries to release more eggs and thus increases litter size. The fertilized eggs make their way to the uterine horns and attach themselves within 4 days.

An experienced handler can palpate a doe at 12 to 14 days and determine whether the female is pregnant. Palpating is done by putting the doe on a flat surface and holding her gently with one hand and sliding the other hand under her belly. At 10 to 14 days, the fetuses feel like large marbles. The breeder should press slightly on both sides of the belly and back toward the groin area. If she is pregnant, birth will occur 30 to 32 days after mating.

Some breeders will put the doe back in with the buck 14 days after the first mating. If the doe accepts the buck and mating takes place, the handler can assume she was not bred the first time; if she grunts and runs or wants to fight him, the handler can assume she is bred.

Kindling is the term used for the birth process in rabbits. About 2 days before the doe is expected to give birth, she should be provided with a nest box; some straw or shavings should be placed in the nest box. About a day before she gives birth, the doe will pull fur from her belly and use it for the nest. This does not hurt the doe and is part of the nesting instinct.

The expectant mother should be left alone at this point. Does usually have no difficulty kindling and no assistance is needed. Too much activity around the nest box may cause the doe to become excited and nervous. She may not use the nest box and scatter her newborn all over the cage, and she may become so nervous that she kills the young as they are born.

About 24 hours after the doe gives birth, the breeder should check the nest to see whether the young are doing well. It is important to approach the nest quietly and slowly, speaking to the doe so she does not get excited. Any dead fetuses should be removed; the female will not remove any dead babies from the nest, and these can become a source of contamination from bacteria and flies.

The litter should be checked daily. A doe nurses her young once or twice a day. There is no reason to get excited if the doe is not in the nest; nursing usually takes place at night. If the litter is large, young kits can be successfully fostered off onto another doe during the first 5 days of life. If the doe does not accept and nurse them, they can be fed successfully by hand. When feeding by hand, feed twice a day. One may want to consult with a veterinarian on the type of formula to use. Replacement milk formulas are available.

Care should be taken not to force the formula into the mouth because it can enter the lungs and is almost sure to cause death. It is also imperative that the young be stimulated to urinate and defecate every 2 or 3 hours by wiping their genital area with a soft, cotton swab moistened with warm water.

Young rabbits' eyes should open at 12 to 14 days of age, and at 3 weeks, the young will start to venture from the nest, where they will start to nibble on hay and on feed pellets left for the doe.

The age at which the litter is to be weaned depends on the breed, the size of the litter, and the size of the individual kits. Usually, a litter can be weaned at 4 to 8 weeks of age. After weaning and separating the young kits from their mother, they should be left together for a period of 7 to 14 days to help reduce stress. Young rabbits raised for fryers should reach a marketable weight of 4 to 5 pounds by 8 to 12 weeks of age.

The doe can be rebred 3 days after kindling; however, this is not recommended. A better time is 2 weeks before weaning the litter; higher conception rates are usually achieved at this time.

SUMMARY

Rabbits were once classified in the order Rodentia but are now placed in a separate order, Lagomorpha. The chief difference between rabbits and rodents is that rabbits have four upper incisor teeth arranged one pair behind the other, whereas rodents have two chisel-like incisor teeth in each jaw.

Fossil remains of Lagomorpha have been found in Oligocene deposits, which date back 30 to 37 million years. These fossil remains indicate the animals have changed very little.

The European wild rabbit is the species from which all domestic rabbit breeds have been developed. Although many of our current domestic breeds have been developed in the United States, the American cottontail was not used in the development of our domestic rabbit.

Several conditions are important in housing rabbits: temperature, humidity, ventilation, proper lighting, and the absence of drafts.

The easiest way to feed rabbits is to use one of the commercially prepared pellet-type feeds supplemented with Timothy hay. These feeds are especially formulated for rabbits and supply all the nutrients needed.

Rabbits become sexually mature at 5 to 8 months of age depending on the breed; the miniature and small size become mature at about 5 months. Males usually mature slower by about 30 to 60 days.

Female rabbits do not have a regular cycle like most other animals. Rabbits kept inside under artificial light and warmth can be bred all year.

DISCUSSION QUESTIONS

1. What is the history of the rabbit?
2. What are the uses for rabbits?
3. What are the advantages of rabbit meat over other types of meat?
4. What are the five weight categories of rabbits?
5. List the breeds in each of the weight categories.
6. What are the houses for rabbits commonly called?
7. What are several important things to remember in housing rabbits?
8. Describe how a rabbit should be handled.
9. List and describe the common diseases and ailments of rabbits.

SUGGESTED ACTIVITIES

1. With the approval of the instructor, bring in breeds of rabbits or invite local breeders to bring in rabbits to the class and describe the characteristics of the breeds. The class can observe the animals. Correct methods of handling can be demonstrated.
2. Obtain tags from various rabbit feeds; bring them to class and explain the differences in the feed formulations.
3. With the approval of the instructor, invite a local breeder or veterinarian to demonstrate the correct methods of palpating a pregnant doe.
4. With the approval of the instructor, have a local veterinarian come to the classroom to discuss the common diseases and ailments of rabbits. The veterinarian can also discuss the prevention and treatment of rabbit diseases and ailments.

ADDITIONAL RESOURCES

1. House Rabbit Society
   http://www.rabbit.org/faq/index.html
2. Net Vet Veterinary Resources
   http://netvet.wustl.edu/rabbits.htm
3. The Rabbit and Cavy Directory
   http://www.rabbitandcavydirectory.com/index.htm
4. All About Domesticated Rabbits
   http://www.petcaretips.net/all-about-domesticated-rabbit.html
5. Hopper Home
   http://www.hopperhome.com
6. Pet Web Site
   http://www.petwebsite.com/default.asp

END NOTES

1. "Rabbit." 1979. *Encyclopedia International* 15:237. Copyright 1979 by Lexicon Publications, Inc. Reprinted by permission.

2. "Rabbit." 1984. *Encyclopedia Americana* 23:110. Copyright 1984 by Grolier Incorporated. Reprinted by permission.

3. Nowak, R. M., and J. L. Paradise. 1991. *Walker's Mammals of the World,* 5th Edition. Baltimore/London: Johns Hopkins University Press, p. 552. Reprinted by permission.

4. *The Merck/Merial Manual for Pet Health.* Copyright © 2007 by MERCK AND CO., INC. Whitehouse Station, NJ. USA. All rights reserved. Used with permission.

5. Ibid.

6. Ibid.

CHAPTER 10

Hamsters

OBJECTIVES

After reading this chapter, one should be able to:

- recognize and describe the common types of hamsters.
- list common management practices for maintaining health.
- list common diseases and ailments of hamsters.

TERMS TO KNOW

demodectic mites	meningitis	solitary
demodicosis	nocturnal	wet tail
estivation	rectal prolapse	

CLASSIFICATION

Kingdom	—	Animalia
Phylum	—	Chordata
Class	—	Mammalia
Order	—	Rodentia
Suborder	—	Myomorpha
Family	—	Muridae
Subfamily	—	Cricetinae
Genus	—	*Mesocricetus*
Species	—	*M. auratus*

HISTORY

The next six chapters cover common rodents that are kept as pets. These common rodents are hamsters, gerbils, rats, mice, guinea pigs, and chinchillas. Rodents evolved during the Paleocene and Ecocene epochs 55 to 65 million years ago. *Paramys* is considered the ancestor of the rodents and was a large squirrel-like animal with clawed feet for grasping and climbing and a long tail for balance.

Today, the order Rodentia contains 29 families, 380 genera, and 1,687 species and is further divided into three suborders based on the musculature of the jaw and other structures of the skull. These suborders are the Sciuromorpha (squirrel-like rodents), the Myomorpha (rat-like rodents), and the Hystricomorpha (porcupine-like rodents).

In studying small animals used for pets, two suborders are of primary importance to us: the suborder Hystricomorpha, because it contains the families Chinchillidae (chinchillas) and Caviidae (cavies or guinea pigs), and the suborder Myomorpha, because it contains the families Murinae (old-world rats and mice), Cricetinae (hamsters), and Gerbillinae (gerbils).

MAJOR GROUPS

The Golden hamster, *Mesocricetus auratus,* is the most abundant of the hamsters. This species has been most often used in research and is the species most commonly found in pet stores.

The Golden hamster is native to the desert areas of Syria and is sometimes referred to as the Syrian Golden hamster. It was discovered in 1930 by Professor I. Aharoni of the Department of Zoology, Hebrew University, in Jerusalem. Professor Aharoni was exploring an animal tunnel near Aleppo, Syria, when he came across a mother and her litter. Of the litter of 12, only a male and two females survived the trip back to the university. These three were later mated and are believed to be the foundation for all Golden hamsters in captivity today. *See Figure 10–1.*

Figure 10–1 (Left) A long-haired Golden hamster. (Right) A short-haired Golden hamster. *Courtesy of Stacey Riggert.*

In 1931, young hamsters were shipped to the United States Public Health Service Research Center at Carville, Louisiana, where they were used in medical research. While being used in research, it was discovered that the Golden hamster could be tamed and made into a pet.

An adult hamster is about 5 to 6 inches long with a short stump of a tail and weighs about 4 ounces.

A true Golden hamster has a rich mahogany or orange color on the back with a white or creamy-colored underside and legs; a black patch is usually present along the side of the cheeks. Hamsters have large black eyes and large erect ears. They have cheek pouches in which to store food.

Today, the Golden hamster can be found in several different varieties and many different colors and color combinations. Long-haired or "teddy bear" hamsters with long, silky fur are popular pets. *See Figure 10–2.* Golden hamsters can be found in single or solid colors, two-colored or banded patterns, and the speckled or "piebald" pattern. Golden hamsters can be found with regular fur or satin fur. In addition to the regular black eye color, red or ruby eyes are also common. Hamsters can occasionally be found with one ruby eye and one red eye.

Figure 10–2 A Teddy Bear hamster. *Courtesy of Stacey Riggert.*

Male hamsters have prominent scent glands on their flanks. These appear as dark pigmented spots on the skin and become more obvious as the animal ages. Secretions from these glands are used to mark their territory. The female hamsters are very aggressive and usually dominate the males.

Another species of hamster that can be found in pet shops is the small desert or dwarf hamster, *Phodopus sungorus*. The dwarf hamster is light gray with a dark stripe down the back. It is smaller than the Golden hamster, being about 4 to 4½ inches long. These small hamsters are very active, make friendly pets, and are fun to watch; however, if not handled frequently they may become irritable and can be aggressive. These small hamsters are not recommended for children because of their small size, which makes them harder to handle. *See Figure 10–3.*

Figure 10–3 A dwarf hamster. *Courtesy of Stacey Riggert.*

ANATOMY

Figure 10–4 shows the anatomy of a rodent.

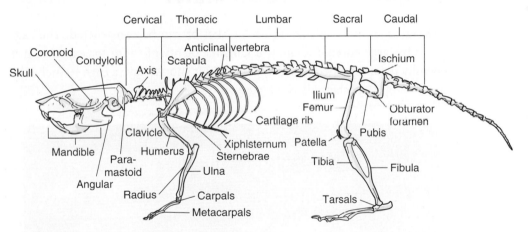

Figure 10–4 Anatomy of the rodent.

CHARACTERISTICS

Hamsters are **nocturnal** animals; they sleep during the day and are active at night. In the wild, hamsters travel long distances at night foraging for food, which they carry back to their burrows in cheek pouches. These cheek pouches are large and extend back to the front shoulders. The pouches can even be used by the female to carry her young. When food is plentiful, hamsters can carry and hoard large amounts in their burrows.

Hamsters are protective of their territory and hoarder of food. Even a tame hamster may bite when a finger is stuck into its cage and intrudes into its space. Once out of a cage, hamsters are usually more friendly and easy to handle.

Hamsters are **solitary** animals. Young can usually be kept together until they are about 5 weeks old, then they will begin to fight and may need to be separated. Cannibalism may also result if litters are left together for longer periods of time. Females usually tolerate the male only for breeding. In housing mature hamsters, it is best to keep a single hamster in a single cage.

The life span of hamsters is usually 1 to 3 years.

Hamsters prefer a temperature around 70°F. If the temperature gets much above 80°F, they go into a deep sleep-like state known as **estivation,** the method by which the furry hamsters originally survived the hot, dry desert environment. In captivity, the hamsters may go into this state of deep sleep, and it may take considerable effort to wake them. Sometimes, it appears that the hamster is dead, and it can be placed in the hand and gently bounced up and down before it awakens. To prevent estivation, room temperature should be maintained below 80°F.

If the temperature drops below 50°F, hamsters go into hibernation. Hamsters can develop respiratory problems or pneumonia if exposed to changing temperature or drafts; therefore, cages should always be kept out of drafts, and room temperature should remain constant.

HOUSING AND EQUIPMENT

Hamster cages are available on the market or can be constructed. The cage should be large enough to allow a hamster plenty of room for exercise. It is generally recommended that a single hamster should have a cage 10 inches by 16 inches by 10 inches tall. *See Figure 10–5.*

Figure 10–5 Wire cages with plastic bottoms are available in most pet and discount stores. *Photo by Dean Warren.*

A cage should be constructed of gnaw-proof materials or constructed such that the hamster cannot gnaw its way out. Most of the commercial types of cages are constructed with stainless steel side and top bars and have plastic floors. The plastic floors are usually constructed in such a way that the hamster cannot get its teeth into the plastic and start gnawing. These cages are usually easy to clean, allow the hamster to see out, and provide exercise because the hamster can climb up the sides and across the top bars.

Water bottles can be hung on the outside of the cage with the drinking tube extending into the cage through the side bars. By hanging the bottle on the outside, the hamster is prevented from gnawing the bottom of the water bottle.

Aquariums make nice cages; they are easy to clean and allow the hamster to see out. A water bottle can be hung from the top of one of the sides but may provide a means of escape, as the hamster can climb up the water bottle and over the sides; wire mesh tops solve this problem (*see Figure 10–6*). Water bottles hung on the inside of an aquarium must have a protective metal hanger to prevent the hamster from gnawing holes in the plastic bottle; a fruit juice can cut in half can be slipped over both ends of the water bottle.

Figure 10–6   Aquariums make nice cages, and the wire mesh top prevents the hamster from escaping. *Courtesy of Carloyn Miller.*

When constructing a cage, consideration should be given to the ease of cleaning and in making sure that it is escape-proof. Cages with wood bottoms are harder to clean, and the wood will soak up urine and smell. If wood is used, the wood should be coated with linseed oil. Wood bottoms should be thick enough to prevent the hamster from gnawing through and escaping.

Cages can be constructed easily using galvanized metal for the sides and a wood floor. A small hole can be drilled in one side and a water bottle hung on the outside with the drinking tube extended into the cage through the hole. Sheet metal screws are used to hold the floor in place and allow the floor to be easily removed and replaced.

The cage should be large enough so an exercise wheel can be added to the cage. Exercise wheels are available in plastic or metal. Hamsters will gnaw on the plastic ones and may soon destroy them; plastic exercise wheels do have the advantage of making very little noise. Remember that the hamster is nocturnal and will run an exercise wheel all night long. A squeaky, metal exercise wheel may quickly become a nuisance. *See Figure 10–7.*

Figure 10–7   Metal exercise wheels are available in most pet stores. *Courtesy of PhotoDisc.*

The hamster cage should have plenty of clean, fresh bedding. Bedding can be paper confetti, paper strips, wood chips, wood shavings, ground corn cob, hay, or straw. Ground corn cob or wood shavings usually work the best. They absorb most of the urine. Cedar should be avoided because of concern about fumes that can cause respiratory problems. *See Figure 10–8.*

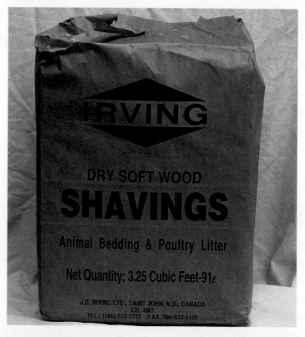

Figure 10–8   Wood shavings made from pine make ideal bedding for all small animals. Large bags like this 50-pound bag are available at feed stores and grain elevators at economical prices. *Photo by Dean Warren.*

FEEDING

The easiest way to feed hamsters is to purchase commercially prepared pellets specifically designed for small animals. These pellets are especially formulated and contain all of the vitamins and minerals necessary. The pellets are hard, and in order for the animal to consume them, they must gnaw on the pellet. This also aids in wearing down the continuously growing front incisors of these rodents. *See Figures 10–9 and 10–10.*

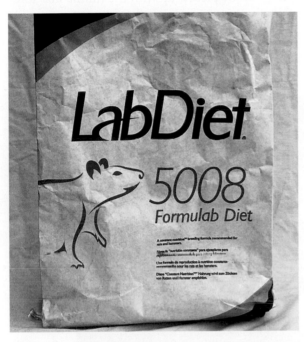

Figure 10–9 Purina Mills, Inc., markets Formulab chow for small animals. *Photo by Dean Warren.*

Figure 10–10 Purina Formulab pellets are oval shaped. They are ½ to 1 inch long, ½ inch wide, and ⅜ inch thick. *Photo by Dean Warren.*

In mixing a ration, it is important that the animal receive a wide range of foods to ensure a nutritionally balanced ration. The diet should contain seeds such as corn, millet, wheat, oats, sorghum, and rape seeds. Carrots, pieces of potato, fresh clover, or alfalfa hay should be supplemented into the diet, as well as dried peas, beans, and nuts. Dry dog pellets or biscuits can be fed also. These contain trace minerals and vitamins and are also helpful in grinding the teeth down. *See Figure 10–11.*

It is important that the diet be consistent. One must not change diets suddenly or feed greens and fruits in large amounts because this may cause diarrhea. Any soft-type food not eaten by the hamster should be removed from the cage before it becomes spoiled. Consumption of spoiled food can throw the animal off its feed and cause diarrhea or illness. Soft foods may also get lodged in the pouch of a hamster and become spoiled, causing infection and illness.

HANDLING

As mentioned earlier, hamsters are nocturnal animals and should not be disturbed while sleeping. If a sleeping hamster is poked or touched, it may awaken quickly and its reaction is to bite in self-defense. Occasionally, a

MICE — Adult mice will eat 4 to 5 grams of pelleted ration daily. Some of the larger strains may eat as much as 8 grams per day per animal. Feed should be available.	
Crude protein not less than	23.0%
Crude fat not less than	6.5%
Crude fiber not more than	4.0%
Ash not more than	8.0%
Added minerals not more than	2.5%
Ingredients: Ground yellow corn, soybean meal, ground wheat, fish meal, wheat middlings, animal fat preserved with	

Figure 10–11 The back of a Formulab bag shows the analysis and ingredients, plus feeding instructions for all small animals.

hamster may need to be awakened during the day to clean its cage. To awaken a sleeping hamster, one should approach it carefully, tap lightly on the cage, and speak to it. Hamsters learn to recognize their handler's voice, and speaking to it will calm it. The handler should attempt to stroke or pet it and see how it reacts. It may allow itself to be picked up; however, if it retreats to a corner or stands on its rear legs with its paws upraised, one must be very careful because this is the hamster's defensive posture. The handler should continue to talk to the animal and, if it leaves the corner, try to pet or stroke it. When the hamster allows it, it can be picked up with the thumb and forefinger placed right behind the front quarters. Hamsters are quite strong and very dexterous in their attempts to escape. They can twist their bodies and use their rear feet to free themselves. *See Figure 10–12.*

Figure 10–12   A hamster should be picked up firmly between the thumb and forefinger. Holding it in this manner usually prevents it from wriggling free or from turning and biting. *Courtesy of Stacey Riggert.*

The more a hamster is handled, the tamer it will become. The hamster will soon learn the voices of its handlers and come readily to the edge of the cage to be picked up.

A way to get a hamster familiar with members of the family is to place it on a table with the family members seated around the table, letting the hamster roam the table freely, talking to it, and attempting to pet or stroke it. The hamster will soon become familiar with the voices of the family members, and within a short time, the hamster will become a gentle, friendly pet.

Young hamsters should not be handled until their eyes open; this will be at about 14 days of age. They quickly get used to human smell and will allow themselves to be picked up without any fear. Care must be taken when handling young hamsters because they can move so quickly and are hard to hold without hurting them. They may jump from the handler's grasp in an effort to escape and fall to the floor, which could result in serious injury.

Children need to be watched carefully when they handle small hamsters because they tend to want to hold them too tightly.

The hamster's eyesight is not especially good; this is another reason for approaching a hamster carefully. Hearing and sense of smell are its most acute senses.

If they evade the handler's hands, squeal, or make a chattering noise with their teeth, they want to be left alone. Remember to move slowly and be patient. Most people get bitten by a hamster because they poke the hamster when it is asleep. Hamsters may also bite the end of a finger because they think they are being offered a treat.

DISEASES AND AILMENTS

Hamsters are generally healthy animals, and there are only a few diseases that may cause problems. The most important disease is hamster enteritis, commonly called **wet tail.** The disease is also sometimes referred to as *proliferative enteritis* and *regional enteritis.* Among the suspected causes are the bacteria *Lawsonia intracellularis* and *Campylobacter jejuni.* Poor diet, poor sanitation, and stress may also be involved. Many times, the disease is associated with general neglect and poor care. Affected animals should be taken to a veterinarian so an exact cause can be determined and treatment started as soon as possible. Treatment results are often poor.

Symptoms of wet tail include wetness around the tail and rear area of the animal, caused by a runny diarrhea. Affected animals go off their feed, have rough hair coats, become weak, and are feeble. **Rectal prolapse,** seen as a red protrusion from the anus, may occur. High death rates often occur within 1 to 2 days after the clinical signs.

The bedding of the infected animal should be changed, and the cage, food bowls, and water container should be disinfected. Infected animals should be kept separate, and all animals should be closely observed.

Tysser's disease, a digestive illness caused by the bacteria *Clostridium piliformis* (formally *Bacillus piliformis*), can have many of the same signs as wet tail. These signs may include loss of appetite, dehydration, watery diarrhea, and sudden death. A veterinarian can diagnose the illness and begin proper treatment. The illness usually has a high mortality rate.

Common diarrhea, caused by incorrect diet, should not be confused with wet tail or other serious ailments. Feeding an overabundance of green leafy materials, vegetables, or fruits may cause diarrhea. The soft foods should be removed and the dry grains and seeds increased in the hamster's diet; this should solve the problem.

Fleas and lice can infest hamsters, especially if cats, dogs, rats, and mice have access to them. If the presence of fleas or lice is suspected, the cage must be cleaned regularly and flea powder sprinkled in the bedding; a flea powder recommended for cats should be used. Flea powder used for dogs is too strong; hamsters and cats lick their fur and will get an overdose of the medication if dog powder is used.

Notoedres notoedres species of mites are found in hamsters; they spread by direct contact from one animal to another. This family of mites burrows within the skin, and the entire life cycle is spent on the host animal.

Demodex criceti and *Demodex aurati* are two species of **demodectic mites** that may occasionally be found on hamsters. These mites are usually considered normal inhabitants on the skin. In large numbers, they can cause the skin disease called **demodicosis** or demodectic mange.

Lymphocytic choriomeningitis (LCM) is **meningitis** that normally affects wild mice. The causative agent is the lymphocytic choriomeningitis virus (LCMV). Common symptoms are loss of weight, drowsiness, ruffled coat, and lethargic movement. This type of meningitis usually is not fatal to the animal but causes slower growth than normal. It is usually spread by mice, and the disease can spread to humans, causing flu-like symptoms.

"Humans can develop LCMV infection from exposure to urine, droppings, saliva, or nesting material of infected rodents. LVMV infection can also occur when these materials are inhaled or directly introduced into broken skin or into the nose, eyes, or mouth, and possibly by a bite from an infected animal.

"Adults with normal immune systems can be infected with LCMV without symptoms, or they may develop a mild illness with symptoms that include the following: fever, lack of appetite, muscle aches, headache, chills, nausea, and vomiting. Some people may have meningitis (inflammation of the brain covering) approximately 7 to 15 days after the start of fever. People with weakened immune systems may have more severe or fatal illness when infected with LCMV.

"Women who become infected with LCMV during pregnancy may have a spontaneous abortion, or their baby may have severe birth defects, including congenital hydrocephalus (fluid on the brain), chorioretinitis (inflammation of the eye), blindness, or mental retardation.

"To avoid contact with wild rodents, pregnant women who reside in a household with a wild rodent infestation should have the infestation controlled promptly by a professional pest control company or another member of the household. To avoid contact with pet rodents, remove the pet rodents from the house or keep pet rodents in a separate part of the home. Pregnant women should ask another family member or friend to clean the cage and care for the pet or arrange for temporary adoption of the pet by a responsible person. Pregnant women should avoid any room where a rodent resides."[1]

REPRODUCTION

Young litters of hamsters should be separated at 5 or 6 weeks of age, not only to prevent fighting, but also to prevent unwanted matings. Hamsters may become sexually mature around 6 weeks of age. Like other animals, the females should not be mated at too early an age because small, weak litters may result and the life span of the female may be shortened. *See Figures 10–13* and *10–14.*

Hamsters can reproduce rapidly; females come in heat every 4 days and, if bred, produce a litter in 16 days. Females should be 10 to 12 weeks old when mated and of good size (approximately 3½ ounces); small females should not be mated until later. Increased activity is usually a sign of heat in the female hamster. The heat period usually occurs in the evening hours between 5 and 11 o' clock. During the hot summer days, they may come in heat earlier in the day.

When mating hamsters, the female should always be placed in the male's cage; this usually results in less fighting. They should be observed carefully.

Figure 10–13   The genital area of a female hamster. *Photo by Isabelle Francais.*

Figure 10–14   The genital area of a male hamster. *Photo by Isabelle Francais.*

If the female is not in heat and not receptive to the male, fighting will soon occur, and the female should be removed immediately. If the female is in heat and receptive, she will usually accept the male within a couple of minutes. The male will sniff the female and lick her fur, checking to see whether she is receptive. If the female is receptive, she will go very rigid and arch her back with her tail raised. Actual mating occurs very quickly.

After mating, the male will rest for a few moments, clean himself, and then pursue the female again. They may mate several times in a 15- to 20-minute period.

When the female is satisfied, she will turn on the male and bite him. At this time, the female should be removed from the male's cage and placed back in her cage. If the female is not removed from the male's cage, she will attack the male and fierce battles will ensue. The female may even kill the male.

One must be very careful in attempting to separate a pair of fighting hamsters; in the fury of the fight, they may bite. One should use a brush, a piece of cardboard, or anything available to slip between the two animals and separate them.

When the female is introduced into the male's cage and is observed rubbing her rear on the floor of the cage, this is a sign that she is not receptive and should be removed immediately. When she rubs her rear on the floor, she leaves an offensive-smelling fluid that is intended to distract the male. If the female is not removed, fighting will soon occur.

The gestation period is 16 days and seldom varies. Two or three days before the female is due, her cage should be cleaned and fresh bedding provided. The mother and her young should not be handled until the young are about 14 days old. Litters consist of 2 to 16 young, with 6 to 8 being average. The young are born without fur, are blind, are totally helpless, and weigh about ¼ to ⅛ ounce. Attempts to handle the newborn or disturb the female and her litter may result in the mother biting and killing the babies because of nervousness and fright.

The young grow rapidly; at about 10 days, they begin to move about the cage. The mother will often be seen picking them up and carrying them back to the nest. The eyes of the young hamsters open at 14 days. Their appetite will increase, and they will begin consuming large amounts of food.

The young should be weaned at about 24 to 28 days. If left longer, the mother may bite them in an attempt to drive them away. When separating the young, the males should be placed in one cage and the females in another; this will prevent inbreeding and unwanted matings. Good homes should be found for the young hamsters because they will soon start fighting if left together. Fighting is common when too many hamsters are kept in too small of a cage; torn ears, torn faces, punctured eyes, severe injuries, death, and cannibalism may result.

SUMMARY

The Golden hamster, *Mesocricetus auratus,* is the most important of the hamsters. This species has been most often used in research and is the species most commonly found in pet stores.

The Golden hamster is native to the desert areas of Syria and is sometimes referred to as the Syrian Golden hamster. It was discovered in 1930 by Professor I. Aharoni of the Department of Zoology, Hebrew University, in Jerusalem.

Another species of hamster that can be found in pet shops is the small desert or dwarf hamster, *Phodopus sungorus.*

Hamsters are nocturnal animals; they sleep during the day and are active at night. In the wild, hamsters travel long distances at night foraging for food, which they carry back to their burrows in cheek pouches.

Hamsters are solitary animals. The young can usually be kept together until they are about 5 weeks old; after that they will begin to fight and may need to be separated.

Hamster cages are available on the market or can be constructed. Cages need to be large enough to allow plenty of room for exercise and should be gnaw-proof.

The easiest way to feed hamsters is to purchase commercially prepared pellets specifically designed for small animals. These pellets are especially formulated and contain all of the vitamins and minerals necessary. The pellets are hard, and in order for the animal to consume them, they must gnaw on the pellet. This also aids in wearing down the continuously growing front incisors of these rodents.

Hamsters can reproduce rapidly; females come in heat every 4 days and, if bred, produce a litter in 16 days. This litter usually consists of 2 to 16 young, with 6 to 8 being average. Young litters should be separated at 5 or 6 weeks of age, not only to prevent fighting, but also to prevent unwanted matings. Hamsters may become sexually mature around 6 weeks of age.

Hamsters are generally healthy animals, and there are only a few diseases that may cause problems. The most important disease is hamster enteritis, commonly called wet tail. The exact cause is not known, but several different causes such as bacteria, viruses, diet, and poor sanitation may be involved. Many times, the disease is associated with general neglect and poor care.

DISCUSSION QUESTIONS

1. What is the history of the common hamster?

2. What are common types of hamsters found in pet shops?

3. What foods can be fed to hamsters?

4. How should a hamster be handled?

5. List the common diseases and ailments of hamsters. What are the symptoms? How can we control each of them?

6. What common mating method is used with hamsters?

SUGGESTED ACTIVITIES

1. With the approval of the instructor, bring hamsters to the classroom. The class can observe the animals. Correct methods of handling can be demonstrated to the class.

2. With the approval of the instructor, bring a hamster or pair of hamsters to the classroom. Leave the cage and animals for a few days. Class members can then become familiar with all aspects of hamster care and management.

3. With the approval of the instructor, have a local veterinarian come to the classroom to discuss common diseases and ailments of hamsters. The veterinarian can also discuss treatments and prevention of hamster diseases and ailments.

ADDITIONAL RESOURCES

1. American Animal Hospital Association (AAHA)
   http://www.healthypet.com/Library/_main.aspx

2. Animal Health Center
   http://www.caringtogether.com/index.html

3. Animals in Science
   http://www.ahc.umn.edu/rar/MNAALA/index.html

4. Exotic Pets
   http://exoticpets.about.com

5. ASPCA (The American Society for the Prevention of Cruelty to Animals)
   http://www.aspca.org/site/PageServer?pagename=pets_hamstercare

6. Pet Web Site
   http://www.petwebsite.com/hamsters/hamster_health.htm

7. Animal Hospitals—USA
   http://www.animalhospitals-usa.com

8. Hamster Land
http://www.hamsterland.com

9. Simple Pet Care
http://www.simplepetcare.com

10. Diseases of Hamsters
http://www.afip.org/vetpath/POLA/99/1999-POLA-Mesocrictetus.htm

END NOTE

1. Center for Disease Control and Prevention, 1600 Clifton Road, N.E. Atlanta, GA 30333. http://www.cdc.gov/ncidod/dvrd/spb/mnpages/dispages/lcmv/qa.htm

CHAPTER 11

Gerbils

OBJECTIVES

After reading this chapter, one should be able to:

🐾 describe the characteristics of gerbils.

🐾 list practices for general care of gerbils.

🐾 describe the common diseases and ailments of gerbils.

TERMS TO KNOW

agouti

camouflage

monogamous

mutations

obesity

red nose

selective breeding

Tyzzer's disease

CLASSIFICATION

Kingdom	—	Animalia
Phylum	—	Chordata
Class	—	Mammalia
Order	—	Rodentia
Suborder	—	Myomorpha
Family	—	Muridae
Subfamily	—	Gerbillinae
Genus	—	*Meriones*
Subgenus	—	*Pallasiomys*
Species	—	*M. unguiculatus*

HISTORY

There are four subgenera and 14 species of common gerbils. The best known of these is the Mongolian desert mouse or Mongolian gerbil, *Meriones unguiculatus*.

Japanese scientists were the first to breed Mongolian gerbils in captivity; a Tokyo research institute received 20 pairs in 1935. In 1954, 11 pairs were sent to Dr. Victor Schwentker in the United States. They were found to be easy to work with, gentle, and active during the day; had no special food or housing requirements; drank little water; were virtually odorless; and seldom would bite. Because of these qualities, their popularity grew rapidly, and today gerbils are common household pets.

Gerbils, also called sand rats, include a large number of species of burrowing rodents forming the subfamily Gerbillinae, of the family Muridae. There are about 100 different species of these rat-like rodents, which are found in the dry, desert regions of Africa and Asia.

The Mongolian gerbil is found in parts of China, in the former Soviet Union, and throughout most of Mongolia. In the wild, they live in colonies and dig burrows in sandy soils for shelter. Mongolian gerbils are active both day and night. Their diet consists of seeds, leaves, roots, stems, and occasionally insects.

GROUPS AND BREEDS

The Mongolian gerbil reaches a size of 6 to 8 inches long from nose to tail; the body is about 3 to 4 inches long. A mature gerbil weighs 3 to 4 ounces.

The body is short and thick and has a hunched appearance when squatting on its hind legs. The forelegs are very short, with the forepaws being very similar to hands and used to hold food. The hind legs and feet look very similar to those of a kangaroo and are large and furry with long toes and claws. The large feet enable the gerbil to stand firmly in the sand of its desert environment. The gerbil walks on all fours but, when stopped, will stand up on its hind legs to observe the immediate area. In the wild, gerbils use their strong hind legs to leap and jump, but those raised in captivity rarely jump, except in an effort to escape from the cage. *See Figure 11–1.*

Figure 11–1   A normal-colored gerbil. *Photo by Isabelle Francais.*

The tail is covered with fur, has a bushy tip, and is used for support when standing and acts as a rudder and stabilizer when jumping or leaping. The skin covering the tail is loose and, if handled roughly, can become torn and pulled from the tailbone.

The head is broad and wide, and the eyes are large, usually jet black, and protrude slightly. The ears are small, rounded, and covered with soft fur; its sense of hearing is very acute because of the enlarged cavities in the skull that amplify the slightest sounds.

The natural color of the Mongolian gerbil is a reddish brown to dark brown. The fur is composed of dark guard hairs that give the overall coat a darker appearance. The underside fur is usually lighter brown, cream, or even white. This natural color pattern is referred to as an **agouti** color and serves as a **camouflage** against the sand and rock of its environment. The light underside fur serves to reflect heat from the desert sand.

Through **mutations** (a sudden change in an inherited trait) and **selective breeding,** gerbils are found today in several different color patterns. Black, white, black and white, white-spotted agouti, charcoal and white, and orange and white are common color patterns. The red-eyed white gerbil is not considered a true albino because some of the fur on the back and tail turns light brown as the animal matures. *See Figures 11–2 and 11–3.*

Figure 11–2 A black gerbil. *Courtesy of Stacey Riggert.*

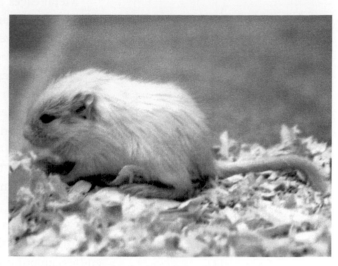

Figure 11–3 A white gerbil. *Courtesy of Stacey Riggert.*

CHARACTERISTICS

Both male and female gerbils have a scent gland on their stomach that looks like a bald patch or strip; this patch is more visible on mature males. The adults leave their scent by sliding their stomach along an object or area.

Gerbils are very quiet animals. A shrill squeak is sometimes made as an alarm and during mating by the female. Young gerbils can be heard making squeaking sounds, but these sounds diminish as they get older.

Another method of communication is a drumming sound made by standing upright and pounding with the rear feet; this drumming sound is also used as an alarm and by the male during mating.

Gerbils are short-lived animals. The life expectancy is usually 2 years but may be up to 4 years.

HOUSING AND EQUIPMENT

Most of the housing and equipment available for hamsters is also suited for gerbils. Being more active, gerbils should have more cage space than hamsters. A breeding pair should have approximately 150 square inches of floor space. Several gerbils kept in a cage together should have approximately 36 square inches per animal. If cage space is limited, cannibalism may occur; in the case of a breeding pair, they may not mate.

Like hamsters, gerbils are very adept at finding ways to escape. Gerbils can jump, so it is important that the cage be covered. A wire mesh cover can be purchased or constructed to prevent animals from escaping.

Several types of cages available on the market are suitable for gerbils; they all have advantages and disadvantages. Plastic cages are lightweight, inexpensive, and easy to clean. They are not, however, chew-proof, and a gerbil will gnaw at areas of the plastic that are not protected.

Metal cages are chew-proof but generally are more expensive, heavier, and harder to handle. Metal that is not protected with paint will soon rust. Gerbil urine is very concentrated and acidic and will quickly eat away paint and cause bare metal to rust.

If wood cages are used, they need to be made of heavy materials to prevent the gerbil from gnawing its way out. Wood will soak up urine and is harder to clean, especially if cleaning is delayed for long periods.

The common fish aquarium makes an excellent cage for gerbils; its greatest advantage is visibility. Gerbils are very curious and will watch people and come up to the glass when humans get close. Aquariums also have the advantage of being easy to clean, although children may find them heavy and hard to handle.

Several different types of bedding materials are available. Wood chips or shavings are the most desirable. They soak up urine, and the gerbil will spend endless amounts of time gnawing and chewing on the larger pieces. The bedding material should be dust free. Gerbils dig and kick in their bedding, and the fine materials can get in their eyes, causing irritation, and can also get in the nose and lungs. Cedar shavings should be avoided because they may cause respiratory problems in small animals.

If a large cage is available, an area with sand would be a nice addition; the gerbils will spend time digging and playing in the sand. An attractive-looking cage can be set up using sand and rocks that will simulate a gerbil's natural environment.

Cotton and wool should be avoided as bedding and nesting materials because these can be ingested by the animal and cause blockage of the digestive system.

Gerbils, being very active, like to be provided with toys to keep them busy. Remember that gerbils will gnaw on anything soft, so toys should be made of a solid material.

Cardboard tubes from rolls of toilet paper and paper towels make excellent tunnels for gerbils but will soon be destroyed as the gerbils gnaw and chew on them.

Care should be taken in providing gerbils with an exercise wheel; the tail can easily get caught in the spokes. The skin on the tail is loose, and as the gerbil fights to free itself, the skin may break loose from the tail. Metal exercise wheels should especially be avoided. Some plastic wheels are made solid without spokes; these are safe for gerbils but will probably be gnawed on.

Glass or ceramic cups and toys used in fish aquariums, toys used for small birds and tin cans provide gerbils with hours of play and exercise. Be careful to avoid any items with sharp or pointed edges that could cause injury.

FEEDING

The easiest way to feed gerbils is using commercial pellets made especially for small rodents. These will be complete balanced rations, with all of the required vitamins and minerals added.

If preferred, rations can be prepared by mixing different types of foodstuffs. Common grains used are corn, oats, wheat, and barley. Mixing these grains can be a good base for a balanced diet. Gerbils eat only the heart out of the corn kernel, and just the kernel from oat seeds. Waste can be prevented by feeding breakfast foods made from corn flakes and rolled oats.

Linseed, millet, canary seed, rape seed, and hemp seed found in mixtures for parakeets and other birds add variety to the diet. Although gerbils love sunflower seeds, they should be fed carefully because they contain a high percentage of fat and overconsumption may lead to **obesity.**

Green foods such as cabbage, carrots, turnips, and beets also add variety to the diet and are usually rich in minerals. Green foods should be washed to remove any dirt, chemical residues, and insects. These materials should be fed sparingly because they can cause diarrhea and upset stomachs.

Vitamin and mineral supplements are available that can be added to the food or water; however, unless recommended by a veterinarian, these supplements are not needed.

Gerbils consume very little water; however, a supply should always be available. The easiest way to provide water is by way of the glass drip-feed or automatic drinking bottle; several types are available. These should be hung on the outside of the cage with the drinking spout extending into the cage. Plastic bottles and the rubber stoppers on glass bottles will soon be chewed if hung on the inside of the cage. If water bottles have to be hung on the inside, one should provide a metal protective covering for the bottle. *See Figures 11–4 and 11–5.*

Figure 11–4 Vacuum-type water bottles are commonly available in 8-, 16-, and 32-ounce sizes. The 8-ounce size is ideal for hamsters, gerbils, rats, and mice. Pictured here are the 8- and 32-ounce sizes. *Photo by Dean Warren.*

Figure 11–5 Feed bowls are available in a wide range of sizes and made from several different materials. Ceramic bowls like the ones pictured here are heavy, so they won't tip over, and durable so small animals can't chew them. *Photo by Dean Warren.*

HANDLING

Gerbils are gentle and curious animals, and little skill is needed to tame one. Seldom will a gerbil bite, and then only if it is mistreated or provoked. Moving slowly and gently usually gets a gerbil to feel comfortable so that it can be picked up. Many times, they will readily climb into an owner's hand.

The easiest way to pick up a gerbil is to place your hand over the gerbil's back and encircle his body with your thumb and forefinger. While holding the gerbil, one should stroke the head and back; the animal should be calm before any attempt is made to pick it up. One should be careful not to squeeze the

Figure 11–6   Gerbils should never be picked up by the tail. The skin can break and pull off, leaving the tailbone exposed.

animal too tightly, as it will struggle to escape. Many times, a gerbil will nibble on a handler's finger, not to inflict a bite, but to let the handler know that it has been held long enough and wants down. After a gerbil has become accustomed to being handled, it will normally climb right into the owner's hand.

A gerbil should never be picked up by the tail because the skin can tear and pull loose, leaving the tailbone exposed. *See Figure 11-6.* This is extremely painful to the animal. The exposed portion of the tail will normally die and fall off. If the exposed tail does not fall off or if it should become infected, the tail will need to be amputated by a veterinarian.

Young gerbils are especially hard to handle. They are small, nervous, and very quick. Be careful in handling them, because they may jump from the hand in an attempt to escape and fall to the floor; this can result in injury.

Tame gerbils readily learn to accept food from the handler's fingers. Most will readily come forward to take treats from the fingers. If they hesitate, be patient and they will soon get over any fear.

DISEASES AND AILMENTS

Gerbils are very hardy and, if well cared for, will seldom be affected by disease.

Respiratory infections are the most common ailment of gerbils; common symptoms are loss of appetite, constant sneezing, and runny eyes and nose. The animal may also appear lethargic or inactive. Respiratory infections are usually caused by temperature changes, overcrowded conditions, and drafty or damp environments. If correcting these problems does not lead to rapid improvement, consult a veterinarian about treatment.

Nasal dermatitis, also referred to as **red nose,** is a common condition that causes hair loss and red, swollen areas of the skin around the nose and muzzle. *Staphylococcus* bacteria are usually the cause, although other bacterial organisms or trauma may be involved. Overcrowding, high humidity, coarse bedding, and other environmental conditions may also be contributing factors. One should watch the animal carefully, and if its health doesn't rapidly improve, the animal should be taken to a veterinarian.

The bacteria *Clostridium piliforme,* the cause of **Tyzzer's disease,** also affects gerbils and may cause high death rates. Affected animals have rough hair coats, diarrhea, lose weight, and appear listless and inactive; treatment is usually ineffective. This is a highly contagious disease. Sick animals should be separated from healthy ones. Accurate diagnosis and treatment should be done by a veterinarian.

Gerbils generally are not troubled by external parasites, but they can become infected with fleas, lice, and *Demodex* species of mange mites, which are transferred from other animals. General signs of these pests are dull fur, fur falling out, and unusual or continuous scratching by the animal. Animals suspected of having these parasites should receive a dusting from a flea powder formulated for cats or small animals. The animal should be dusted and the powder brushed into the fur; after 20 to 30 minutes, the powder should be brushed out of the fur. The dusting should be repeated in 24 hours. It is important that the cage be cleaned, disinfected, and dusted also.

Gerbils are occasionally affected by seizures. These attacks are usually observed during handling. In some cases, the animal appears to go into a trance, with the body remaining motionless or slightly trembling; these attacks last only a few seconds. Treatment is not needed because the animal will return to normal activity. These attacks may also be observed when cleaning the animal's cage. Animals that show signs of seizures should not be used for mating because this condition appears to be genetic.

REPRODUCTION

A pair of gerbils should be introduced to each other as early as possible; older animals may not accept each other, and fights may occur when they are introduced. Putting the female in the male's cage usually results in less fighting, but the pair should be observed closely. If introduced at about 9 to 12 weeks of age, they will readily accept each other and mate shortly. However, it may be 5 or 6 months before a litter is produced, and many times, first-litter parents may neglect their young. Gerbils are primarily **monogamous,** and the pair bond may last throughout the life of the animals.

Young gerbils become sexually mature between 9 and 12 weeks of age. The female comes into heat every 4 to 6 days, and mating usually takes place during the evening hours. Actual mating is very brief but occurs several times. During events leading up to actual mating, the male will sniff the female and make drumming noises with his rear feet. High-pitched squeaks may be heard from the pair. The gestation period varies between 24 and 26 days. *See Figures 11–7 and 11–8.*

Newborn gerbils are born naked, blind, and completely helpless. Litters consist of 1 to 10 young, with 3 or 4 being the average. Their ears are folded shut, they do not have any teeth, and they rely entirely on mother's milk for nourishment. Newborn gerbils are about 1 inch long. During the first few

Figure 11–7   The genital area of a female gerbil. *Photo by Isabelle Francais.*

Figure 11–8   The genital area of a male gerbil. *Photo by Isabelle Francais.*

days after birth, high-pitched squeaks can be heard coming from the nest; these squeaking noises diminish as the young get older. At about 5 or 6 days of age, the body will become covered with a light covering of fur.

Young gerbils are very tough. They take a lot of bumping about from the mother and often will be kicked from the nest. Later, the mother will gather them up and return them to the nest. The male may also help gather the young up and many times will hover over them, keeping them warm and protected. Gerbils do not seem to mind early handling of their young, and the handler can assist in returning lost young to their nest.

Young gerbils grow very rapidly and begin to crawl out of the nest even before their eyes are open. At about 5 days of age, the ears unfold; at about 14 days of age, the eyes open. Their incisors have developed by this time, and they will start gnawing on solid food and drinking water. They are very active and will jump and run about the cage and engage in fights with their littermates.

SUMMARY

There are four subgenera and 14 species of common gerbils. The best known of these is the Mongolian desert mouse or Mongolian gerbil, *Mericones unguiculatus.*

The Mongolian gerbil is found in parts of China, the former Soviet Union, and throughout most of Mongolia. In the wild, they live in colonies and dig burrows in sandy soils for shelter. Mongolian gerbils are active both day and night. Their diet consists of seeds, leaves, roots, stems, and occasional insects.

Most of the housing and equipment available for hamsters is also suited for gerbils; however, being more active, gerbils should have more cage space than hamsters. Like hamsters, gerbils are very adept at finding ways to escape. Gerbils can jump, and it is important that the cage be covered.

The easiest way to feed gerbils is using commercial pellets made especially for small animals. They are complete balanced rations, with all of the required vitamins and minerals added.

Young gerbils become sexually mature at between 9 and 12 weeks of age. The female comes into heat every 4 to 6 days; mating usually takes place during the evening hours. The gestation period varies between 24 and 26 days.

Gerbils are very hardy and, if well cared for, will seldom be affected by disease.

DISCUSSION QUESTIONS

1. What are the general characteristics of gerbils?

2. How do hamsters and gerbils differ?

3. What types of food are usually fed to gerbils?

4. How should one pick up and handle a gerbil?

5. What are the common diseases and ailments of gerbils? What are their symptoms and how should they be treated?

SUGGESTED ACTIVITIES

1. With the approval of the instructor, bring gerbils to the classroom. The class can observe the animals. Correct methods of handling can be demonstrated.

2. With the approval of the instructor, bring a gerbil or pair of gerbils to the classroom. Leave the cage and animals for a few days. Your classmates can then become familiar with all aspects of gerbil care and management.

3. With the approval of the instructor, have a local veterinarian come to the classroom to discuss the common diseases and ailments of gerbils. The veterinarian can also discuss the prevention and treatment of gerbil diseases and ailments.

ADDITIONAL RESOURCES

1. The National Gerbil Society
   http://www.gerbils.com.uk

2. Gerbil Care
   http://www.aspca.org

3. American Animal Hospital Association
   http://www.healthypet.com

4. Exotic Pets
   http://www.exoticpets.about.com

5. ABC Gerbils
   http://www.Abcgerbils.com

6. The American Gerbil Society
   http://www.agsgerbils.org

Rats

CLASSIFICATION

Kingdom	—	Animalia
Phylum	—	Chordata
Class	—	Mammalia
Order	—	Rodentia
Suborder	—	Myomorpha
Family	—	Muridae
Subfamily	—	Murinae
Genus	—	*Rattus*
Species	—	*R. norvegicus*
	—	*R. rattus*

OBJECTIVES

After reading this chapter, one should be able to:

🐾 detail the history of the brown and black rat.

🐾 describe the characteristics of common rats.

🐾 list the common methods of feeding and caring for rats.

🐾 describe common diseases and ailments of rats.

TERMS TO KNOW

albino	carriers	phenol
Black Plague	gregarious	pyrethrin
caped	inbreeding	rodents

HISTORY

Rats belong to the order of gnawing mammals called **rodents** (Rodentia). Together with mice, they compose the subfamily Murinae of the family Muridae. True rats form the genus *Rattus*. Rats are larger than mice and have more rows of scales on their tail.

The genus *Rattus* contains from 78 to 570 species, depending on which classification is used. Only two species, the black rat *(Rattus rattus)* and the brown rat *(Rattus norvegicus),* have been domesticated and used for research and as pets.

The black rat is sometimes referred to as the roof rat, climbing rat, or gray rat. It is believed to have originally come from Southern Asia. This species was well established in Europe during the 1200s and was the major reservoir for the **Black Plague** that killed more than one-fourth of the population during the 1300s.

The black rat reached North America during the 1500s and today is found along both coasts and throughout the southeastern states.

The brown rat is sometimes referred to as the Norway rat, barn rat, sewer rat, or wharf rat. It originated in Eastern Asia and Northern China; it found its way to Europe and North America during the 1500s. Today, the brown rat is well established throughout the United States.

Rats have spread to all parts of the world. Their success is attributed to their ability to adapt to many different habitats, environments, and food sources, as well as their ability to reproduce rapidly.

Rats in the wild cause tremendous damage in the United States every year. They destroy eggs, fruits, stored grain, and vegetables, and they attack other animals. Rats cause damage to buildings and cause fires by chewing through insulation of electrical wires. Rats are **carriers** of disease and have attacked human beings, especially babies and young children.

With all their destructiveness and bad characteristics, the black and brown rats, in the form of the white **albino** laboratory rats, have been of major importance in medical, biological, and psychological research. Laboratory rats have been used extensively in developing drugs, studying diseases, learning about nutrition, studying aging, and many other studies. Because they are intelligent and have shown an ability to learn, they have also been used in many behavioral studies.

MAJOR COLOR GROUPS

The Black Rat

The black rat has a head and body length of 7 to 8 inches and weighs 4 to 12 ounces. The tail is longer than the head and body, and the ears are about half as long as the head. The color is usually black or dark gray with a brown or gray-white underside. The black rat is an excellent climber and jumper. *See Figure 12–1.*

The Brown Rat

The brown rat is larger than the black rat; its body and head are 7 to 10 inches long, and the tail is always shorter than the head and body. The brown rat is thicker and more robust than the black rat and weighs 7 to 17 ounces. Its ears are shorter and more rounded, and the fur is a dark to gray-brown on the back, with lighter colors of brown and gray on the underside. The brown rat is not a climber and prefers to live in tunnels, burrows, sewers, and basements. *See Figure 12–2.*

Varieties

Because of mutations and selective breeding, both the black and brown rat can be found in several colors. The white laboratory rat is a descendant of albino strains. Creams, fawns, and light-gray varieties are common today; another common variety is the hooded rat. It has a colored head and dorsal stripe on a white body; the colors are brown, black, fawn, and cream. The **caped** variety has a colored head but lacks the colored dorsal stripe. *See Figures 12–3 through 12–8.*

CHARACTERISTICS

Rats are agile climbers, especially the black rat. They use their scaly tail to help maintain balance. Rats can jump up to 2 feet. They are also excellent swimmers and can stay submerged under water for 2 to 3 minutes.

Figure 12–1 A black rat *(Rattus rattus). Courtesy of Stacey Riggert.*

Figure 12–2 A brown rat *(Rattus norvegicus). Courtesy of Stacey Riggert.*

Figure 12–3 A rat with hooded markings. *Courtesy of Stacey Riggert.*

Figure 12–4 Caped markings.

Figure 12–5 Variegated markings.

Figure 12–6 Dalmatian markings.

Figure 12–7 Masked markings.

Figure 12–8 A rat with Berkshire markings. *Courtesy of Brian Yacur.*

Rats are very curious. They will climb a cage or stand on their rear legs in an attempt to get a better view of their surroundings, and they will soon learn their owner's voice and will come running when food is offered.

Rats are surprisingly clean and odor-free; they will groom and clean themselves and each other. If the bedding is changed regularly, very little odor is noticed.

Domesticated rats seldom bite; young rats bite only if they are being handled roughly or being abused. Adult rats will bite if surprised or scared. Handlers should go slowly and be patient when handling or feeding adults, especially if the hand is put into the cage. Females with young can be protective and will bite their handler if they feel the litter is being threatened.

Rats can become very upset and angry; when upset, they stiffen their body and arch their back. Every hair on the body will stand on end, giving it an imposing appearance. The tail will wag back and forth, and the animal will shake. Mature males will usually fight when put together, and fierce battles can result. Extreme care must be taken when handling rats that are upset or that are fighting.

Rats are **gregarious.** They do best when kept with other rats. Animals that have been raised together usually get along. Care should be taken, however, when introducing a strange rat into an occupied cage because the stranger will not be accepted immediately, or maybe not at all, and fighting will occur.

Rats are primarily nocturnal. Domestic rats that are kept as pets are active all day and do not mind being bothered; however, care should be taken when arousing a sleeping rat.

Rats in the wild will rarely live more than a year because of the large number of predators; domestic rats may live up to 3 years.

HOUSING AND EQUIPMENT

Two cubic feet per rat is the recommended cage size for each rat. Cages that are 12 inches wide by 24 inches long and 12 inches high would be the minimum size for a single rat. Larger cages are recommended. Cages that are 24 inches wide by 48 inches long and 24 inches high would be adequate for eight rats.

Remember that rats are rodents and will gnaw on soft materials. Wire mesh cages with solid bottoms or pull-out trays work well and are easy to clean. These cages are available with doors at the end or with tops that hinge open.

Cages can be constructed from wood and wire mesh; materials should be thick enough to withstand gnawing of the rats. Wood will soak up urine and will be harder to clean, however.

Glass aquariums work very well for rats. They are easy to clean and, with a lid, are escape-proof; a weight may need to be applied to the lid to keep the rats from escaping.

Water bottles should be hung on the outside of the cage with the spout extending into the cage. If the water bottle has to be hung on the inside, one must make sure it is protected with a metal shield. Feed bowls should be made of glass or ceramic materials; these are gnaw-proof, easy to clean, and will not rust like metal.

Rats are very active, and if a large cage is available, it should be equipped with ladders and ropes for the rats to get exercise. All rats also enjoy playing on exercise wheels.

A number of different bedding materials are sold for use in rodent cages. Wood shavings, ground corn cob, pelleted sawdust, newspaper or straw, and CareFresh® (a fluffy recycled paper product) are some examples. There is concern that cedar and pine wood shavings commonly sold for small animal bedding contain a chemical called **phenol** that can be toxic or cause airway damage. For this reason, some rodent keepers recommend using other types of bedding. Other materials can be provided for the female to build a nest, such as paper towels, paper napkins, or toilet tissue. The female will shred these into a nest.

FEEDING

The easiest way to feed rats is to use commercially prepared pellets formulated for small animals; these contain all of the nutrients that are necessary. Commercial rodent diets can be supplemented with small amounts of dry dog food, crackers, cereal, fruits, or vegetables. Overfeeding a ration of fruits and vegetables may cause diarrhea. In general, such treats should make up less than 10% of the daily diet.

Vitamins and mineral supplements are also available at most pet stores. These supplements are available in liquid form and can be added to the water or given directly to the rat. There is a risk when choosing to add supplements to the drinking water; the rat may ingest too big a dose if it is thirsty, or it may find the taste objectionable. If the latter occurs, the rat may refuse to drink and become dehydrated as well as not ingesting the vitamins. Commercially formulated rodent pellets contain all the vitamins and minerals the rat requires. Further supplementation should not be necessary if the pellets are fresh (less than 6 months old).

Rats will consume an ounce of water or more per day. Water should not be given in an open dish because rats will generally make a mess.

Figure 12–9 Young rats are easily tamed and can be picked up and handled very easily.

HANDLING

Young rats can easily be picked up by grasping them around the body just behind the front legs. *See Figure 12–9.* They may also be picked up by the tail; the tail should be grasped at the base close to the body and the rat lifted. *See Figure 12–10.*

When handling older rats, be careful and talk so that they will relax. How the handler approaches the animal depends a great deal on how tame it is. If the animal is not tame, one must go slowly, get its attention, grasp it by the base of the tail, and lift. If further restraint is needed, the handler should take the other hand and grasp the animal around the body just behind the front legs and restrict the movement of its head with the thumb and forefinger.

A rat should never be grasped by the tip of the tail. In an effort to free itself, it will twist and squirm. Its tail or the skin on the tail can break loose and be pulled off, leaving the tailbone exposed. The rat is also very dexterous and can turn and climb up its tail. Leather gloves should be used if unsure of the situation.

Figure 12–10 An adult rat that may not be completely tamed can be picked up by the tail if it is grasped up close to the body. Rats that are tame can be picked up by grasping them around the body.

DISEASES AND AILMENTS

Rats in the wild have been recognized as carriers of many types of diseases and parasites; however, rats that are kept in captivity are fairly free of diseases and ailments.

Respiratory disease caused by the organism *Mycoplasma pulmonis* is one of the more common diseases. Signs include nasal discharge, snuffling, rattled breathing, rubbing the eyes and nose, tilted head, loss of coordination, and circling.

Secondary invaders such as *Streptococcus pneumoniae, Pasteurella, Bordetella, Mycoplasma,* and *Pseudomonas* may cause pneumonia-like symptoms in animals. Increased respiratory difficulties and sudden death may also occur.

Treatment of respiratory diseases with antibiotics is often helpful, but complete isolation or elimination of affected animals may be the only way to eliminate the problem.

Good sanitation and ventilation are extremely important in controlling respiratory disease. The buildup of ammonia in unsanitary cages exacerbates the symptoms.

Polyplax spinulosa is a type of louse that may infest rats and causes loss of hair and intense itching. *Dermanyssus, Notoedres muris, Radfordia ensifera,* and *Ornithonyssus bacoti* are species of mites that may affect rats; these mites may cause hair loss and skin irritation. In more serious infestations, small fluid-filled lesions, swellings, and inflammation of the skin may be evident.

External parasites such as lice and mites may be controlled by using a shampoo containing **pyrethrin.** Pyrethrin is a natural product and should not be confused with other over-the-counter products containing the synthetic permethrin. Newer parasite control products available from a veterinarian reduce the number of treatments needed as well as the stress to the animal caused by administering the treatment.

Figure 12–11 Genital area of a male rat. *Courtesy of Dean Warren.*

Figure 12–12 Genital area of a female rat. *Courtesy of Dean Warren.*

REPRODUCTION

Rats may become sexually mature at 6 weeks of age. Litters should be weaned and separated at 4 weeks to prevent **inbreeding** and unwanted litters. Young females bred at too young of an age usually have small litters, and the babies are small and weak. Breeding at an early age may also shorten the productive life of the female. *See Figures 12–11* and *12–12.*

Breeding should be delayed until the rats are 16 weeks old. At this age, the female is strong enough and mature enough to raise a litter.

The heat cycle occurs every 4 days and lasts about 12 hours. When the female is in heat, the male will sniff the female and attempt to mount her. The actual act takes place very quickly and will be repeated several times.

Several different methods may be used in breeding rats; the easiest way is to put one male in with one to three females. This depends on the size of the cage and how many offspring are wanted.

The gestation period is 21 to 24 days. As the delivery day approaches, the female will busy herself shoving bedding up in a corner to make a nest.

The young are born pink, naked, blind, deaf, and completely helpless. There will be up to 16 in a litter. The young can be heard making bird-like peeps. In a couple of days, soft fur will become visible; if the young rat is going to be two-colored, the darker color will be more visible. At about 4 days, the ears open. Young rats will start to explore the cage at about 10 days of age. Teeth will begin to develop at 10 days, and they will begin to nibble on hard foods. The water bottle should be hung low enough so that they can start drinking water. Their eyes will open at about 14 days.

The male can be left in the cage or removed after mating. He will play no active part in the care of the young but will not harm them either. The male is usually very tolerant of the young, allowing them to crawl or lie all over him. The female comes in heat very soon after giving birth, and the male should be removed to prevent another litter. Both parents will be very protective of the young, and care must be taken when feeding and caring for rats that have litters.

The cage should be cleaned a couple of days before the litter is expected. The mother and litter should not be disturbed until the young start exploring from the nest. If the cage gets especially messy, the litter may need to be moved so that the cage can be cleaned.

The young will be fairly self-sufficient at 3 weeks of age, and should be weaned and separated at 4 weeks of age. At 4 weeks of age, young rats are very curious, friendly, and easy to handle. They can be tamed at this time and will soon learn to eat from the handler's fingers.

If two females occupy a cage and both have litters, they will share responsibilities for nursing and caring for the young. If the litter is large, the female may divide the litter into two or three groups and then spread her time among the groups. Nursing a large litter can be very taxing on the female.

SUMMARY

Rats belong to the order of gnawing mammals called rodents (Rodentia). Together with mice, they compose the subfamily Murinae of the family Muridae; true rats form the genus *Rattus*. Rats are larger than mice and have more rows of scales on their tail.

The genus *Rattus* contains 78 to 570 species, depending on which classification is used. Only two species, the black rat *(Rattus rattus)* and the brown rat *(Rattus norvegicus),* have been domesticated and used for research and as pets.

Because of mutations and selective breeding, both the black and brown rat can be found in several colors. The white laboratory rat is a descendant of albino strains of the brown and black rat. Creams, fawns, and light-gray varieties are common today. Another common variety is the hooded rat; it has a colored head and dorsal stripe on a white body. The caped variety has a colored head but lacks the colored dorsal stripe; colors are brown, black, fawn, and cream.

Rats are surprisingly clean and odor-free. They will groom and clean themselves and each other. If the bedding is changed regularly, very little odor is noticed.

Domesticated rats seldom bite; young rats bite only if they are being handled roughly or abused.

The minimum cage size for rats is 2 cubic feet. Good cages are built of materials that rats can't gnaw through and are furnished with wheels or other toys.

Rats can be fed commercially prepared pellets formulated for small animals; these pellets contain all of the nutrients that are necessary.

Young rats may become sexually mature as early as 6 weeks of age. The heat cycle of the female rat occurs every 4 days and lasts about 12 hours; the gestation period will be 21 to 24 days. The young are born blind, naked, deaf, and completely helpless; there can be up to 16 in a litter.

Rats in the wild have been recognized as carriers of many types of diseases and parasites; however, rats that are kept in captivity are fairly free of diseases and ailments.

DISCUSSION QUESTIONS

1. What are the two species of rat that have been domesticated?

2. Describe the common types of rats used for research and as pets.

3. How should one handle pet rats?

4. What are the common foods for pet rats?

5. What are the common diseases and ailments of pet rats?

SUGGESTED ACTIVITIES

1. With the approval of the instructor, bring rats to the classroom. The class can observe the animals. Correct methods of handling can be demonstrated.

2. With the approval of the instructor, bring a rat or pair of rats to the classroom. Leave the cage and animals for a few days. Class members can then become familiar with all aspects of rat care and management.

3. With the approval of the instructor, have a local veterinarian come to the classroom to discuss the common diseases and ailments of rats. The veterinarian can also discuss the prevention and treatment of rat diseases and ailments.

4. With the approval of the instructor, bring rats and cages to the classroom. Conduct feeding trials using different types of foods (e.g., candy bars, cakes, and cookies).

5. If available and with the approval of the instructor, invite laboratory workers in to speak on the research they do using rats.

6. Write to research companies for information on research using rats. If Internet access is available, one may find information on companies doing research with rats.

7. With the approval of the instructor, construct a maze for rats. Set up different situations and use different treats to see if the rats react differently and learn some mazes more quickly than others.

ADDITIONAL RESOURCES

1. Fancy Rats
   http://www.altpet.net/rodents/rats/rf32.html

2. The National Fancy Rat Society
   http://www.cableol.net/nfrs

3. Rats
   http://www.webcom.com/lstead/rodents/rats.html

CHAPTER 13

Mice

CLASSIFICATION

Kingdom — Animalia
Phylum — Chordata
Class — Mammalia
Order — Rodentia
Suborder — Myomorpha
Family — Muridae
Subfamily — Murinae
Genus — *Mus*
Species — *M. musculus*

OBJECTIVES

After reading this chapter, one should be able to:

- detail the history of the common house mouse.
- describe the clan or colony structure of the house mouse.
- list practices for the general care of pet mice.
- list common feeds for mice.
- describe the common methods of handling mice.
- describe common diseases and ailments of pet mice.

TERMS TO KNOW

clan	piebald
colony structure	subordinate

HISTORY

The common mouse is the most familiar and most widely distributed rodent in the world; there are four subgenera and 36 species. The best known species is the common house mouse *(M. musculus);* it dwells wherever humans live.

The word *mouse* comes from an old Sanskrit word meaning "thief"; Sanskrit is an ancient language of Asia, where scientists believe the house mouse originated.[1] The house mouse was originally a wild field animal. Grain and other foodstuffs stored in human settlements offered an attractive source of food for mice, which adapt quickly to new environments.

From Asia, the house mouse spread throughout the world. The common house mouse is one of the world's most successful mammals; they multiply extremely fast and are able to adapt to almost any condition. Their reputation has been as a dirty animal associated with carrying diseases and as a very destructive animal capable of causing tremendous economic losses of stored grains and other property.

Subsequently, mice have gained a place in history. The ancient Egyptians and Romans had a phrase that described the mouse's reproductive ability; they used to say, "its raining mice," or, "mice are made of raindrops." The Greeks and the people of India believed that mice were lightning bolts born of thunderstorms. In some places, mice were kept in temples and worshiped as sacred animals. They were also viewed as an instrument of punishment. People would present sacrificial offerings to stop or prevent a mouse plague.

More than 4,000 years ago, the Cretans built a temple in Tenedos, Pontus, where they fed and worshiped mice. In the Cretan victory over Pontus, according to legend, mice helped by chewing through the leather straps of the shields of the Pontic soldiers so that they were unable to defend themselves.

Paintings on ancient bowls and other clay artifacts tell us that Egyptians also kept mice about 4,000 years ago. The first white and spotted mice appeared about 4,000 years ago in Greece, Egypt, and China. White mice in particular were considered sacred; people used them to predict the future, as lucky charms, or to keep wild mice away. Waltzing mice originated in China 2,000 to 3,000 years ago. The Japanese bred white and colored mice systematically 3,000 years ago. In the 1800s, people in Europe began to exchange "fancy mice." Fancy mice have different coat colors and patterns. *See Figure 13–1.*

In ancient Rome and during the Middle Ages, mice were used as medicine against all kinds of diseases. They were dried, pulverized, sliced, marinated in oils, and then used as a compress or taken internally. Their blood was a favored ingredient of drugs and tonics. Mice were supposedly good for flesh wounds, snake bites, warts, bladder irregularities, diabetes, enlarged thyroid glands, diseases of the eye, and loss of hair.

Even though mice were used for medicinal purposes, they were also targets for extermination. All kinds of methods were recommended for getting rid of them, except for the use of cats. During the Middle Ages, cats were not recognized as a natural predator of the mouse nor were they considered pets. Cats were hated as accomplices of sorcerers and witches and suffered extreme cruelty.

In more modern times, the house mouse, especially the albino strains of *M. musculus,* has been widely used for medical and biological research, particularly in the study of heredity.

Figure 13–1 A white mouse. *Courtesy of Stacey Riggert.*

By careful breeding over the years, today's mice are gentler and less timid than their ancestors. Pet mice are relatively free of disease, and if handled frequently, they show little tendency to bite or escape. Today, these curious and interesting mammals are available in many different colors and color combinations.

CHARACTERISTICS

Mice have a pointed nose and a split upper lip. The tail is usually hairless with visible scale rings. Mice have four toes on their forefeet and five toes on their hindfeet.

The house mouse is approximately 2½ to 3½ inches long excluding the tail; the tail is generally the same length as the body. This small mammal will weigh ½ to 1 ounce. The house mouse has a small head and a long, narrow snout. Several long, thin whiskers are used by the animal to help find its way about in dark, tight areas. The fur of the house mouse is grayish-brown on the animal's back and lighter colored to white on the underside. *See Figure 13–2.*

The eyes of the house mouse are fairly large, round, and black; they cannot see very well. Because of the large, spherical shape of the eye, they have an almost perfect visual field, but because of the position of the eyes on the side of the head and the shape of the eye, they are incapable of detailed vision. Mice can detect movement but may be unable to determine what the movement is. Cats, in their pursuit of a mouse, move ever so slowly and may remain motionless for long periods.

Mice have fairly large ears and possess a highly developed sense of hearing. They perceive sounds in very high frequency ranges, up to 100,000 Hz.

Smell is the most highly developed sense that mice possess. Mice use smell for locating food, identifying family and colony members, finding their way around in darkness, and identifying enemies.

Figure 13–2   A brown house mouse. *Courtesy of PhotoDisc.*

Mice are primarily nocturnal in habit and usually seek hiding during the day, but pet mice, feeling safe around humans, may come out during the day. Pet mice can recognize by smell those who feed and handle them and will feel at ease in their presence.

Mice are gregarious and prefer the company of other mice. The house mouse lives in a **clan** or **colony structure.** This colony structure gives the colony more protection against enemies and makes it easier for the colony members to find and store large quantities of food. Females in the colony share duties of raising the young, providing warmth for the young, and nursing and feeding the young.

Members of the colony leave urine markings as they explore new areas and search for food. Mice have glands in the soles of their feet that leave secretions that other members of the colony can follow. These scent markings also mark the territory of a colony and serve as a warning to members of other colonies. Members of colonies, upon coming in contact with markings or scent from another colony, will not venture into the territory.

The mouse colony is ruled by a lead or head buck; all other males in the colony are **subordinate** to him. The lead buck is the only male in the colony allowed to mate with the females. The subordinate males are kept from mating the females by the lead buck, and they also are rejected by the females.

Subordinate males may challenge the lead buck for the position by attacking him; subordinate males attempt to bite and fight. During the fight, the males attack each other with their front feet and continuously try to bite each other. If one animal in the fight wants to stop, he will make a squeaking noise and standing on his rear legs, he may press his forefeet against his body, which signifies defeat and submission.

Should the lead buck be defeated in a fight, all males in the colony will bite him, and he will move to the bottom of the social structure. Many times, the lead buck will be an older male and will be defeated by a younger, stronger male. The defeated male may also be bitten by all the females in the colony. The defeated male may essentially be an outcast and go off alone, refuse to eat, and eventually die.

Today, mice can be found in a wide variety of body colors, combinations of colors, and with red or black eyes. There are approximately 50 different varieties of fancy mice. These are divided into five basic groups:

1. Self colors or mice of one color
2. Tans, mice of any recognized color with tan belly color (*See Figure 13–3.*)
3. **Piebald** or pied-marked mice—mice with spots, patches, or broken patterns (*See Figures 13–4* and *13–5.*)
4. Satins of any color or markings, but with satin coat
5. Any other variety that does not fit into any other group (*See Figure 13–6.*)

Males tend to produce more odor than females and leave more scent markers; however, males may learn tricks faster and usually are easier to tame and handle.

Figure 13–3   A mouse with a tan or fox pattern.

Figure 13–4   A mouse with a broken color pattern.

Figure 13–5   A mouse with a variegated pattern.

Figure 13–6   A mouse with a Dutch color pattern.

HOUSING AND EQUIPMENT

Homes for pet mice can be purchased from pet shops, or they can be constructed from wood and wire screen. A cage for a pair of mice should have at least 72 square inches of floor space and a height of 8 inches. Larger-size cages would allow for small nest boxes, exercise wheels, and toys to be added to the cage. Also, in the case of a breeding pair, a larger cage would be needed for growing litters. Mice will stay healthier and be more active if they have plenty of space to move around.

Many of the cages and the equipment used for other types of small animals can be used for mice. Cages that have plastic bottoms and metal bars will work for mice, but the bars need to be close together to prevent mice from escaping. The bars should have about 5/16 inch between them. Cages of this type are easy to clean and disinfect because the plastic bottom can be easily removed.

Aquariums and terrariums also make excellent cages for mice. Small aquariums with dimensions of 6 inches by 12 inches by 8 inches will work for a pair of mice; however, larger aquariums will allow the mice to have more space for exercising. Exercise wheels and nest boxes can be easily added to larger aquariums.

Cages can be constructed from wood and wire screen. Wood cages, however, are harder to clean. The urine will soak into the wood and make cleaning difficult. Mice are fairly clean and odor free, but mouse urine is very strong. Treating the wood with linseed oil or other nontoxic preservative will make wooden cages a little easier to clean and make them last longer.

All cages for mice must have some type of cover or lid because mice are excellent climbers. Most of the cages with metal bars have hinged covers. Covers or lids for aquariums can be purchased at most pet shops or can easily be constructed with wood and wire screen.

In a classroom situation, mice may run for hiding in the nest boxes because they are wary of all the noise and activity, but as soon as the activity dies down in the late afternoon, they will venture out to feed and play.

The same bedding materials used in rat cages may be used for mice. Although cedar and pine chips and shavings are commonly used, there is concern that they may be hazardous to the health of the mice. Avoid all bedding materials that have a lot of dust in them. The dust can cause breathing and respiratory problems for the mice.

Bedding should be changed at least once a week or more often if the cage begins to smell. Care should be taken when cleaning the cages containing female mice with newly born litters so that the female does not get overly excited or nervous.

Newspaper, paper towels, and toilet paper also make suitable bedding material. Mice will chew and shred these materials to make comfortable nests.

Mice usually urinate in the corners of their cages. Placing cat litter or baking soda in the corners of the cage will help minimize the mousy odor. Although this minimizes the smell, it does not eliminate the need for cleaning and disinfecting the cages.

Mice do not drink very much water, but water should be available at all times. The common vacuum-type water bottles that are available at most pet shops are ideal. Water should not be placed in small bowls or plates because they will soon be contaminated with urine and feces.

FEEDING

Mice in the wild will eat almost anything, but their main diet consists of grain and seeds. Either a formulated commercial feed or a mixture of grain and seeds can be used.

Commercial feeds come in a hard pellet form and contain all necessary nutrients. The hard pellet is also beneficial in keeping the continuously growing front incisors worn down.

A ration can be purchased or prepared using corn, oats, and wheat, and then adding small amounts of millet, barley, and buckwheat. Mice love sunflower seeds, which should be added sparingly because they are high in fat content. Small amounts of oily seeds are good for adding shine to the hair coat. Commercially prepared feeds are recommended because the amounts and ingredients are chosen to ensure that the diet is complete and balanced.

In feeding a ration, various greens and vegetables can be added. Fresh dandelions, carrots, apples, dates, and raisins can be added. Clean, fresh hay can supplement the diet and will also serve as a nesting material. It is important that these foodstuffs be free of any chemicals, contaminants, or molds that can affect the animal's health.

Mice will not overeat. Care should be taken so that waste is minimized and soft foods are not allowed to spoil. Too many treats may cause upset stomachs and may lead to diarrhea.

HANDLING

Mice can provide hours of entertainment as they climb ropes, climb ladders, and run on exercise wheels. When purchased at a young age, mice will tame very quickly; they will take food from fingers and eventually climb onto the owner's hand.

Small children need to realize that mice are not toys; they are delicate animals that can easily be crushed by excited children.

A mouse should not be picked up by the end of the tail or nape of the neck because this causes discomfort and pain. If necessary, a mouse can be grasped by the base of the tail and lifted. Another method of getting a mouse out of a cage until it gets used to being handled is to scoop it up in a cup. A mouse will usually go into a cup because of its curious nature. *See Figure 13–7.*

Figure 13–7 One way of picking up a mouse is to use a cup and scoop the mouse up. Hold the cup down and the mouse will usually go into the cup.

Figure 13–8 Mice can also be picked up by lifting the base of the tail. This is probably the easiest way to lift a mouse.

Figure 13–9 Mice that are tame can be picked up by "cupping" in the hands. Mice that are familiar with their handler and not afraid will usually climb up into the hands.

A mouse brought home should be left alone for a day or two. After this period of adjustment, taming can begin. Treats can be offered from the fingertips. Later, food can be put in the palm of the hand; soon, the mouse will come onto the hand and eat the food. An attempt can then be made to pet and stroke the mouse; a mouse likes to have its head and ears stroked. *See Figures 13–8 and 13–9.*

DISEASES AND AILMENTS

If mice are given proper diet, housing, and sufficient exercise, they will usually be free of disease. Problems that usually arise are nutrient deficiencies or obesity because of improper diets. Mice that do not receive sufficient exercise may become lethargic and lazy; their overall life span will also be shortened.

Several organisms can cause respiratory diseases in mice: *Mycoplasma pulmonis, Sendai virus, Pasteurella pneumotropica, Pseudomonas aeruginosa, Klebsiella pneumoniae,* and *Bordetella bronchiseptica.*

Symptoms may appear as squeaking or rattling breath, running nose, and watering eyes. The fur loses condition and the animals sit around in a huddled position with their backs arched; they are also listless and inactive. To reduce the risk of contracting respiratory diseases, new mice should be quarantined for 2 weeks before being housed in the same room as other mice in the home or facility. During this quarantine period, the new mice should be handled as little as possible, and caretakers need to wash their hands and change clothes before handling other mice. If the new mice show no symptoms after 2 weeks, it is unlikely that they are infected and they can be released from quarantine. Sick mice must remain in quarantine or be euthanized. Keeping cages clean, avoiding the use of dusty beddings or ones containing phenol, and locating cages in draft-free, well-ventilated areas may help reduce the incidence of respiratory disease.

Salmonella and related bacteria are responsible for many of the serious infectious diseases. Unclean cage conditions, contaminated feed or water, and flies transmit disease agents.

Mice are commonly infected with a variety of mites *(Myobia musculi, Mycocoptes musculinus, Myocoptes rombutsi, Radfordia affinis, Ornithonyssus bacoti)* that can cause severe irritation and inflammation due to self-trauma and hair loss. Topical insecticides have been used to try to eliminate mites. It is important to remember that products labeled for use in other species may be toxic to mice or might need to be used in reduced dosages. It is wise to seek advice from a veterinarian familiar with rodent health before using any insecticide on a mouse.

Mice can be infested with pinworm *(Syphacia obvelata, Aspiculuris tetruptera)* and tapeworm *(Hymenolepis nana* and *Hymenolepis diminuta).* These parasites rarely cause clinical diseases.

The protozoans *Spironucleus muris* and *Giardia muris* are capable of causing high death rates in mice between 3 and 5 weeks of age. Affected animals become lethargic, have hunched posture, do not gain weight, develop sticky stools, and have gas- and fluid-filled intestines because of the presence of large numbers of parasites.

REPRODUCTION

Female mice become sexually mature and have their first heat period as early as 4 weeks of age; males become mature as early as 5 weeks of age. A nonpregnant female comes in heat every 3 to 6 days. The heat period lasts approximately 12 hours, depending on the age. Young females usually have short cycles, and older females have longer cycles.

When a female is in heat, the male mouse continuously follows her around in an attempt to mate. When the female is ready, she remains still and moves her tail to one side. The actual mating takes only a few seconds and is repeated several times. *See Figures 13–10 and 13–11.*

The gestation period is normally 21 days but may vary by 1 to 3 days.

As gestation progresses, the female seeks out a nesting area that will provide security, darkness, and warmth. Newspaper, paper towels, straw, or other materials should be provided for a nest.

As with other animals, the female will tear open the placental membranes, bite off the umbilical cord, and clean each newborn. The female will eat the placenta and afterbirth and prepare for the next delivery. The actual birth of the litter will take about 15 to 30 minutes, depending on the size of the litter.

A mouse giving birth should never be disturbed; she may become nervous and panic. In her panic, she may kill and eat the newborn. The new litter should not be touched for at least a week.

A litter normally consists of 5 to 10 but may vary from 1 to 20. The young are extremely small, weighing about 1 to 2 grams and measuring about ½ inch long. The newborns will be completely helpless, hairless, and blind; they will depend on their mother's milk for the first 2 weeks. Hair will start to develop at 2 or 3 days, and they will have all their hair at 8 to 10 days. They will start to move around the nest at about 5 days, but they will not venture out until about 10 days of age.

Figure 13–10 The genital area of a female mouse. *Courtesy of Dean Warren.*

Figure 13–11 The genital area of a male mouse. *Courtesy of Dean Warren.*

During the first 8 to 10 days, the mother mouse will lick away the urine and soft feces of the newborn. This licking and massaging is important to stimulate the digestive and excretory systems of the newborns.

The incisors of the newborn become visible at about 10 days of age. Their eyes open at about 14 days, at which time they become more active. They begin to explore their surroundings and start to nibble on solid foods.

The owner may want to start handling the young when they reach 14 days of age; the young soon become tame through continued handling. Care needs to be taken because they may become frightened and leap from the hands in an attempt to escape.

The young should be separated by sex at about 3 weeks of age to prevent unwanted matings. Mice can reproduce rapidly, and if the owner is not keeping a close watch, the numbers can get out of control quickly.

SUMMARY

The common mouse is the most familiar and widely distributed rodent in the world; there are four subgenus groups and 36 species. The best known species is the common house mouse *(M. musculus).*

By careful breeding over the years, today's pet mice are gentler and less timid than their ancestors. These mice are relatively free of disease, and if handled frequently, they show little tendency to bite or escape. These curious and interesting mammals are available in many different colors and color combinations.

Mice are gregarious and prefer the company of other mice. The house mouse lives in a clan or colony structure. This gives mice more protection against enemies, and it is easier for colony members to find and store large quantities of food. Females in the colony share duties of raising young, providing warmth, nursing, and feeding the young.

Suitable homes for pet mice can be purchased from pet shops or can be constructed from wood and wire screen. A cage for a pair of mice should have at least 72 square inches of floor space and a height of 8 inches. Larger-size cages would allow for small nest boxes, exercise wheels, and toys to be added to the cage; also, in the case of a breeding pair, a larger cage would be needed for growing litters. Mice will stay healthier and be more active if they have plenty of space to move around.

Mice in the wild eat almost anything, but their main diet consists of grain and seeds. When feeding a pet mouse, a formulated commercial feed is recommended.

Mice can make enjoyable pets and provide hours of entertainment as they climb ropes, climb ladders, and run on exercise wheels. When purchased at a young age, they usually tame very quickly. As they become familiar with their owner, they will take bits of food from their owner's fingers and will eventually climb onto the owner's hand.

If mice are given proper diet, adequate housing, and sufficient exercise, they will usually be free of disease. Problems that usually arise are nutrient deficiencies or obesity because of improper diets. Mice that do not receive sufficient exercise will become lethargic and lazy, and their overall life span will be shortened.

Female mice become sexually mature as early as 4 weeks of age. The heat cycle occurs every 3 to 6 days and lasts approximately 12 hours. The gestation period is normally 21 days but may vary by 1 to 3 days. A litter consists of 5 to 10 but can vary from 1 to 20.

DISCUSSION QUESTIONS

1. What is the history of the domesticated mouse?
2. Mice were used in ancient Rome for medicinal purposes. What were they used to treat?
3. What are the characteristics of the common house mouse?
4. What is a clan or colony structure? Describe the advantages of such a structure.
5. What is the main diet of mice in the wild?
6. How should a pet mouse be handled?
7. What are the common foods for pet mice?
8. What are the common diseases and ailments of pet mice? What are the symptoms and treatments for them?

SUGGESTED ACTIVITIES

1. With the approval of the instructor, bring a pair of mice to the classroom. Classmates can observe the habits of these small animals and the class can discuss their observations.
2. With the approval of the instructor, bring in mice and cages or an aquarium to house them. Class members can conduct feeding trials with mice. Use various types of food. Class members can keep records and observe the results.
3. With the approval of the instructor and if materials are available, build a maze for mice. Class members can conduct various tests to see how quickly the mice learn the maze.

ADDITIONAL RESOURCES

1. American Fancy Rat and Mouse Association
   http://www.afrma.org
2. The Rat and Mouse Club of America
   http://www.rmca.org
3. Exotic Pets
   http://www.exoticpets.about.com

END NOTE

1. Adapted from *The World Book Encyclopedia.* © 1993 World Book, Inc. By permission of the publisher.

CHAPTER **14**

Guinea Pigs

OBJECTIVES

After reading this chapter, one should be able to:

- detail the history of the guinea pig.
- list and describe the seven varieties of guinea pigs.
- list practices for maintaining the good health of guinea pigs.
- describe how to pick up and handle a guinea pig.
- describe the common diseases and ailments of the guinea pig.

TERMS TO KNOW

abscess	malocclusion	rosettes
cavy	mane	ticking
crest	oral mucosa	vivariums
kinked	peripheral vision	

HISTORY

The guinea pig is more accurately called a **cavy;** there are seven species of cavy. The exact history of the domesticated cavy *C. porcellus* is not known, but it is believed to be derived from *C. aperea, C. tschudii,* or *C. fulgida.* The domesticated guinea pig has been bred for meat production in South America for at least 3,000 years. During the period of the Inca Empire, from about 1200 to 1532 AD, highly selective breeding produced a variety of strains, differentiated by color pattern and meat flavor. The range of domestication extended from Northwestern Venezuela to Central Chile. Subsequent to the Spanish conquest, there was a disruption of breeding and a contraction of range. The guinea pig is still widely kept as a source of food by the native people of Ecuador, Peru, and Bolivia. The guinea pig was brought to Europe in the sixteenth century and, since the mid-1800s, has been used by laboratories around the world for research on pathology, nutrition, genetics, toxicology, and development of serums. It also makes an ideal pet.[1]

CLASSIFICATION

Kingdom	—	Animalia
Phylum	—	Chordata
Class	—	Mammalia
Order	—	Rodentia
Suborder	—	Hystricomorpha
Family	—	Caviidae
Genus	—	*Cavia*
Species	—	*C. porcellus*

The guinea pig is neither a pig nor does it come from Guinea; the actual reason for the name is not known, but there are several explanations. The species name *porcellus (C. porcellus)* for the domestic guinea pig means "little pig" in Latin. The only similarities to a pig are the low grunts, the squeals they make when they are hungry, and their fat little bellies.

After obtaining some of the animals, Dutch and English traders stopped off in Guinea on their way back to Europe. The first guinea pigs were sold in England for one English coin called a *guinea.* Another explanation could be that guinea comes from the word *guine* in Portuguese, which means "far away and unknown lands."

Guinea pigs have been instrumental in research on nutrition, tuberculosis, and other infectious diseases, as well as in the discovery and production of serum against diphtheria. They were also used by Louis Pasteur in his research on rabies. Guinea pigs' contribution to medical science has saved countless human lives.

MAJOR VARIETIES

There are seven common varieties of guinea pigs: the Abyssinian, American, Peruvian, Satin, Silkie, Teddy, and the White Crested.

The Abyssinian

The Abyssinian has a rough, wiry hair coat. The hair is made up of swirls or cowlicks called **rosettes.** The more rosettes on the animal, the more desirable it is. The Abyssinian is found in all colors and color combinations. *See Figure 14–1.*

Figure 14–1   Abyssinian guinea pig. *Courtesy of Isabelle Francais.*

The American

The American is the most common of the guinea pig varieties. The hair of the American is short, very glossy, and fine in texture. This short-haired guinea pig can be found in a wide variety of colors and color combinations; its short hair makes it easy to care for. *See Figure 14–2.*

Figure 14–2   American guinea pig. *Courtesy of Isabelle Francais.*

The Peruvian

The Peruvian is the long-haired variety. Length, evenness, and balance of the hair coat are the deciding features when judging this variety. The Peruvian can be found in many of the same colors and color combinations as the short-haired, American variety. The hair coat takes a great deal of time and effort to keep clean and may reach a length of 20 inches. Because the guinea pig does not have a tail, the Peruvian has the appearance of an animated mop, and it may be difficult to tell which end is which. *See Figure 14–3.*

Figure 14–3   Peruvian guinea pig. *Courtesy of Dean Warren.*

The Satin

The luxurious, shiny fur of the Satin is its major characteristic. The coat is fine, dense, and soft and has a sheen. Satins are found in the same colors and color combinations as the other guinea pig varieties. *See Figure 14–4.*

Figure 14–4   Satin guinea pig. *Courtesy of Isabelle Francais.*

The Silkie

The Silkie, sometimes referred to as the Sheltie, is another long-haired variety, but unlike the Peruvian, there is no long frontal sweep over the head. The Silkie has a **mane** that sweeps back from the head, between the ears, and back over the back and down the sides. The Silkie is also found in the same colors and color combinations as the other varieties. *See Figure 14–5.*

Figure 14–5   Silkie or Sheltie guinea pig. *Courtesy of Isabelle Francais.*

The Teddy

The Teddy variety has short, kinky hair. The fur is short, resilient, and thick and lies close to the body; the whiskers are also **kinked.** The Teddy can be found in the same colors and color combinations as the other guinea pig varieties. *See Figure 14–6.*

Figure 14–6   Teddy guinea pig. *Courtesy of Isabelle Francais.*

The White Crested

The White Crested guinea pig has short hair and resembles the American short-hair variety except for the **crest,** which is a rosette and radiates evenly with a clearly defined center from the forehead. The White Crested is primarily found in the self, solid, and agouti colors; the crest is, of course, white. *See Figure 14–7.*

Figure 14–7   White Crested guinea pig.

CHARACTERISTICS

Adult guinea pigs reach a length of 8 to 10 inches and weigh 1 to 2 pounds. The males are larger and have larger, more rounded heads. The body of the guinea pig is short, stocky in build, with short legs; there are four toes on the forefeet and three toes on the hindfeet. They have short, sharp claws.

The nose of the guinea pig is short, blunt, and rounded. The mouth is small, and the upper lip, like that of all members of the rodent family, is split. They have long whiskers on the sides of the upper lip; these whiskers serve the same function as the whiskers on a cat, allowing the animal to make its way in dim light. The ears are short and have very little fur on them.

A guinea pig has very sensitive hearing; it can detect frequencies beyond the range of the human ear. The upper level of the range of the human ear is around 15,000 to 20,000 Hz. The guinea pig, however, can hear frequencies up to 30,000 Hz. Guinea pigs communicate with a series of various sounds; many are high-pitched squeaks or whistles. When threatened, the fur on their neck stands up, and they chatter their teeth in an attempt to scare off an enemy.

Their sense of smell is also highly developed; they learn to recognize human scent and can identify individual humans. They recognize their owner and whistle when the owner comes into the room.

The eyes of the guinea pig are located on the sides of the head; this gives them excellent **peripheral vision** that allows them to see in a wide circle or a large surrounding area. This allows them to spot approaching enemies from all directions.

Guinea pigs in the wild live in colonies or clans; there usually is 1 dominant male and 5 to 10 females in a colony. Two adult males cannot exist peacefully in a colony; younger and subordinate males usually exist on the edges of the colony. As younger males become independent of the colony, they try to lure females away from existing colonies and start their own.

Colors and color combinations of guinea pigs are agouti, selfs, solid, and marked. Agouti is described as having a ground or undercolor, determining the base color, with dark **ticking** over the entire animal except the belly, which should be devoid of ticking. Selfs or self-colored animals have the same color of hair over the entire body. Solid or solid-colored animals have color uniformly over the entire animal. This color uniformity may be obtained by the intermingling of different-colored hair shafts (Brindles, Roans) or agouti-colored (one base color, one tip color) hairs over the entire body. The solids should be devoid of markings and shadings. Animals with marked coloring have a basic color broken up by orderly placement on a white background.

HOUSING AND EQUIPMENT

Cages are available at most pet shops. The typical cage has a plastic bottom that is 3 or 4 inches deep and has vertical wire bars for the upper part; the plastic bottom can be easily removed for cleaning. The plastic bottom trays should be at least 3 or 4 inches deep, with deeper trays preferred to prevent the animals from scattering bedding or shavings out of the cage. Cages for a single guinea pig should be at least 12 inches by 24 inches; smaller cages will not allow the animal to have sufficient room to exercise.

Cages do not need covers or tops; guinea pigs usually will not climb out of cages. Larger aquariums or **vivariums** will also work; although they are somewhat heavy, they are easy to clean. Cages with vertical wire bars and glass aquariums allow the guinea pig to see out; they like to see what is going on in their surroundings.

Square or rectangular cages can be made of wood for very little cost. Wood cages are not as suitable because the wood will soak up urine and is hard to clean; eventually, these cages will become a permanent source of odors.

Plastic tubs normally used to store engine or machinery parts make ideal houses and can usually be purchased from hardware stores. These tubs should be at least 6 to 8 inches deep. *See Figure 14–8.*

Figure 14–8 Plastic tubs that are designed for machine parts can be used to house guinea pigs. This black plastic tub was purchased at a local hardware store. *Courtesy of Dean Warren.*

If more than one animal is going to be housed, each adult should have a floor space of at least 180 square inches.

Guinea pigs can be kept outside during the summer once the temperature moves above 50°F. Small houses similar to a dog house can be constructed to protect the animals from the elements, and a run can be constructed from boards and chicken wire. Each animal should have 3 square feet of run. Portable-type houses and runs can be moved as the grass gets eaten and tracked down. The runs should have covers for protection from dogs, cats, and wild animals. Runs also need to be anchored down to prevent predators from digging under the wire and attacking the guinea pigs and to prevent the guinea pigs from escaping.

Food bowls should be made of heavy glass or earthenware to prevent the bowl from being knocked over. Guinea pigs drink a lot of water, which should be provided in a vacuum-type water bottle. Guinea pigs should not be provided water in a bowl, which can be tipped over or contaminated with feces and urine.

FEEDING

Guinea pigs are vegetarians. Their large front incisor teeth are very sharp and are used to bite food. They have eight molars in both the upper and lower jaws that are used to grind food. All of their teeth grow continuously; therefore, they must receive solid foods to keep their teeth worn down.

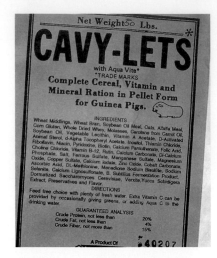

Figure 14–9 This is a label from a bag of guinea pig pellets. Note that the feed contains ascorbic acid (vitamin C). *Courtesy of Dean Warren.*

Guinea pig pellets are available that contain all of the nutrients needed. These pellets are hard and serve to wear the teeth down.

Guinea pigs, unlike many other animals, cannot synthesize vitamin C in their body and must be supplied with it in their diet. A lack of vitamin C causes scurvy, which results in dehydration, poor appetite, diarrhea, rough hair coats, lethargy, and weight loss. Small pinpoint hemorrhages on the gums and joint swellings with lameness also occur. The animals have severely reduced resistance to infectious agents causing respiratory disease. Most commercially prepared pellets have vitamin C added; however, it breaks down rapidly, and opened bags of pellets should be used within 30 days. *See Figure 14–9.*

Pellets that have high alfalfa content and dry alfalfa hay are suitable for young, growing guinea pigs, and for pregnant females. Clover and grass hay pellets and dry clover and grass hays that have lower protein content are more desirable for grown animals. It is important that hay not be moldy, musty, or dusty; moldy hay can be fatal. The pellets and dry hay also helps to wear the animal's teeth down.

Romaine lettuce, green leaf lettuce, kale, carrots, apples, pears, turnips, beets, and cucumbers are also favorite foods of guinea pigs. It is important that these foodstuffs be rinsed to remove any residue of chemical pesticides. Iceberg lettuce should be avoided because it is less nutritious and may lead to diarrhea. Wheat, corn, and oats are high-protein foods that can be added to the diet. Fresh foods left in the cage can mold and spoil; consuming moldy or spoiled foods can result in upset stomachs, diarrhea, or even death. Uneaten fresh foods should be removed from the cage each evening.

Fresh water should always be available for guinea pigs. Vacuum-type bottles work very well. Guinea pigs will drink when their mouth is full of food, and this food can get pushed up inside the tube. The bottle and tube need to be cleaned out frequently to prevent growth of bacteria.

CARE AND HANDLING

Figure 14–10 The guinea pig should be brushed daily to remove old, dead hair.

A guinea pig should be brushed daily to remove old, dead hair. A guinea pig with rosettes and long hair will need to have the hair combed out in addition to brushing (*See Figure 14–10*). Long-haired animals may also need the hair clipped around the rear to prevent it from becoming matted with feces. Long-haired animals exhibited in shows have the rear hair tied up in bows and wrapped with cloth. Well-groomed animals not only look nice, but also will be more healthy. Long-haired guinea pigs may need their long hair trimmed occasionally to keep it from dragging in the bedding and becoming dirty and matted.

Guinea pigs that are exhibited in shows may also need to be given an occasional bath. Guinea pigs are not especially fond of water. When bathing guinea pigs, a mild shampoo similar to what would be used on cats should be used; baby shampoo also works. Care must be taken not to get the guinea pig chilled; it needs to be dried and kept warm during and after the bath.

The toenails of guinea pigs grow constantly and will need to be trimmed. These can be trimmed with regular toenail clippers that can be found in the pet shop. Care must be taken when trimming the nails so that you do not cut into the pink or quick of the nail, because this will cause pain to the animal and bleeding. Be especially careful with animals that have dark-colored toe-nails because the quick cannot be seen. *See Figure 14–11.*

Guinea pigs are usually friendly and easy to handle; however, they may be timid and hide when approached. A guinea pig brought home should be allowed to explore its new home for several hours before handling. If the animal seems unusually frightened, leave it alone for a day or two.

If the handler offers treats of fresh greens, carrots, or apple, the guinea pig will get over this shyness and venture forth. If the animal remains shy, the treat should be dropped in the cage. By continuing to offer treats, the animal should soon take treats from the hand. Sometimes picking the animal up and petting it will help it get over its shyness.

Figure 14–11 The toenails of a guinea pig need to be checked regularly and trimmed if needed.

A guinea pig should be grasped firmly around the front shoulders with one hand and have its rear supported with the other hand. This will prevent the guinea pig from struggling. Guinea pigs can injure themselves if allowed to struggle while being picked up with one hand. Once picked up, it should be cradled in the palm and forearm and held close to the body. In this position, the animal will feel safe and secure. The handler must not allow the animal to fall; guinea pigs are not agile and can injure themselves if allowed to fall on a hard surface. *See Figures 14–12* and *14–13.*

Figure 14–12 When picking up a guinea pig, place one hand in front of the animal and with the other hand behind the animal, scoop it up. In this manner, the animal's rear is supported. *Courtesy of Dean Warren.*

Figure 14–13 When holding a guinea pig, make sure its rear is supported. A guinea pig can injure its back if it is not held properly. *Courtesy of Dean Warren.*

DISEASES AND AILMENTS

Guinea pigs are usually very healthy animals, especially if they are well cared for. Care must be taken to eliminate any major changes in temperature, drafts, and humidity. All conditions that might cause stress to the animals must be eliminated.

Respiratory diseases are the major problems of guinea pigs. *Bordetella bronchiseptica, Streptococcus pneumoniae, Pseudomonas aeriginosa, Klebsiella pneumoniae,* and *Streptococcus pyogenes* are types of bacteria responsible for respiratory problems.

Animals with respiratory diseases are usually lethargic and listless, have discharges from the nose, and sneeze. Guinea pigs do not appear to have a great deal of strength or resolve to fight off these diseases. Animals showing signs of respiratory diseases should be taken to a veterinarian and treatment begun immediately, or death may occur.

Guinea pigs may occasionally have large swellings on the neck. These swellings are a result of enlarged **abscessed** lymph nodes caused by the organisms *Streptococcus zooepidemicus* and *Streptococcus moniliformis.*

The organism usually enters through cuts or abrasions of the **oral mucosa** or upper respiratory tract. Affected animals should be isolated because the drainage from abscesses can infect other animals in the colony. They then should be taken to a veterinarian.

Pregnant females may develop toxemia in late pregnancy. Affected females will be lethargic, go off their feed, refuse to eat, and have difficulty breathing. Females that are fat and overweight appear to be most affected. A high-quality diet is essential during the late stages of pregnancy.

The mites *Chirodiscoides caviae* and *Trixacarus caviae,* and the biting lice *Gyropus ovalis* and *Gliricola porcelli* are the most common external parasites. *Chirodiscoides* mites usually cause severe lesions, and hair loss may occur over the entire body. Insecticide preparations used on cats are generally effective.

If guinea pigs do not receive a diet containing solid foods, their teeth can grow to the point that they cannot bite or even close their mouths. If their teeth overgrow to this point, they will need to be clipped or the animal will starve. A veterinarian or experienced breeder can do this. **Malocclusion** is a condition in which the upper and lower teeth do not come together properly; the condition is probably an inherited trait, and animals showing this condition should not be used in breeding. The condition can possibly be corrected by a veterinarian, depending on the severity. *See Figure 14–14.*

Male guinea pigs should be checked occasionally to make sure the penis is free of shavings and hair, which will occasionally mat around the penis and render the male unable to mate.

Figure 14–14 The teeth of guinea pigs need to be checked occasionally to see if they come together properly and are wearing down normally.

REPRODUCTION

Guinea pigs breed throughout the year, with increased births in late spring and summer. Females (sows) can produce up to five litters in a year, with the average litter size being two to four, although litters of six are not uncommon. It is important that the first breeding of a female guinea pig occur before she is 7 months of age. If a female is bred for the first time after this age, there is a high probability that her pelvic bones will not be able to separate to allow for the delivery of the babies. If this occurs, a cesarean section is the only way to save the sow and babies.

The heat cycle in females averages about 16 days. Females come into heat within 24 hours after giving birth and can subsequently be bred if a male is present.

The heat period normally lasts 24 hours; the male (boar) chases the female around the cage until she is ready to mate. She then flattens herself and raises her hindquarters. The actual mating lasts only a few seconds but may be repeated several times.

About 4 to 5 weeks into the pregnancy, the female becomes large and looks similar to a water balloon. The young weigh about half the weight of the female. As the delivery date approaches, the young can be seen moving about in the womb. The female guinea pig does not attempt to make a nest.

The gestation period ranges from 56 to 74 days, with an average of 68 days. The young will weigh 1½ to 4½ ounces. Depending on the size of the litter, the actual birth will take 10 to 30 minutes. During the delivery, the mother is usually in a hunched position and receives the newborn under her. She will immediately tear off the enveloping membrane, eat it, and then lick the newborn's nose, eyes, and mouth to remove any fluid.

The mother is usually very clean and will eat and clean up the placenta. If the birth takes place at night or early morning, the mother will have all traces of birth cleaned up by daylight; except for some soiled bedding, it would be hard to recognize that a birth took place.

The young are fully developed at birth and will be running around and consuming solid food within 24 hours. Their eyes are open, and they are fully furred. It is hard for the novice owner to believe that they are newborn.

In the wild, guinea pigs have very little defense against their enemies except to be able to run and hide. The young, being fully developed at birth, are able to run and hide shortly after birth.

The young normally nurse for 3 weeks. The female guinea pig has only two teats located far back on her belly. While nursing, the mother stands and hunches over the young. If there are more than two in the litter, the others wait their turn. Occasionally, guinea pigs may lose their mothers and need to be hand fed. A substitute for mother's milk can be prepared using a combination of goat's milk, Karo syrup, and powdered milk. Normally a mixture containing 8 to 10 ounces of goat's milk, 2 tablespoons of powdered milk, and a teaspoon of Karo syrup will be accepted by the guinea pigs. One can experiment with the mixture and see what the young guinea pigs will accept. One may also want to consult with a veterinarian about substitute formulas for orphaned guinea pigs. *See Figure 14–15.*

Figure 14–15 Occasionally guinea pigs and other small animals may lose their mothers and need to be hand fed. A substitute for mother's milk can be prepared using goat's milk, Karo syrup, and powdered milk. *Photo by Dean Warren.*

To prepare the formula, heat the goat's milk in a small pan, then add the powered milk and the syrup and stir the mixture. Put the mixture into a small bottle. Small bottles for feeding small animals are available at most pet supply stores. Be careful that the mixture is not to hot for the small animals.

Never force the milk substitute into the young animal's mouth. If the young guinea pig does not swallow the mixture, it may be inhaled into its lungs. This can cause the death of the young guinea pig. Positioning the baby in a natural "belly-down" position, rather than on its back, during feeding will also help prevent aspiration of the formula into the lungs.

The young can be separated from the mother at 4 weeks of age. Sexual maturity in females occurs at 2 months of age and in males at 3 months, although it is not uncommon for them to become sexually mature earlier. To prevent unwanted matings, the litter should be separated by sex at 4 weeks of age.

Breeding of guinea pigs is usually done by either pair breeding or in colonies. In colony breeding, one male is put with 2 to 20 females. Remember that each adult animal should have approximately 180 square inches of floor space; an adult female with a litter should have about 2 square feet of space. *See Figures 14–16* and *14–17.*

Figure 14–16   The genital area of a female guinea pig. *Courtesy of Dean Warren.*

Figure 14–17   The genital area of a male guinea pig. *Courtesy of Dean Warren.*

SUMMARY

The guinea pig is more accurately called a cavy; there are seven species of cavy. The exact history of the domesticated cavy *C. porcellus* is not known, but it is believed to be derived from *C. aperea, C. tschudii,* or *C. fulgida.*

There are seven common varieties of guinea pig: the Abyssinian, America, Peruvian, Satin, Silkie, Teddy, and White Crested.

Cages are available at most pet shops. The typical cage has a plastic bottom that is 3 or 4 inches deep and has vertical wire bars for the upper part.

Guinea pigs are herbivores. Their large front incisor teeth are very sharp and used to bite food. Guinea pig pellets are also available that contain all of the nutrients the animal needs; these pellets are hard and serve to wear the teeth down. Guinea pigs, unlike many other animals, cannot synthesize vitamin C in their body and must be supplied with it in their diet.

When picking up a guinea pig, grasp it firmly around the front shoulders with one hand and support its rear with the other hand; this will prevent the guinea pig from struggling. A guinea pig can injure itself if allowed to struggle.

Guinea pigs are usually very healthy animals, especially if well cared for. Care must be taken to eliminate any major changes in temperature, drafts, and humidity.

Guinea pigs breed throughout the year, with increased births in late spring and summer. Females produce up to five litters in a year, with the average litter size being two to four, although litters of six are not uncommon.

The heat cycle in females averages about 16 days and lasts for 24 hours. The gestation period ranges from 56 to 74 days, with an average of 68 days. The young are fully developed at birth and will be running around and consuming solid food within 24 hours. The young normally nurse for 3 weeks.

DISCUSSION QUESTIONS

1. Where were guinea pigs originally found?

2. Where does the name *guinea pig* come from?

3. What is the more correct name for the guinea pig?

4. What are the seven varieties of guinea pig? What are the characteristics of each type?

5. How should a guinea pig be picked up?

6. What is the importance of vitamin C in the guinea pig diet?

7. What are the common foodstuffs for guinea pigs?

8. What are the common diseases and ailments of guinea pigs? What are the causes, symptoms, treatments, and control methods?

SUGGESTED ACTIVITIES

1. With the approval of the instructor, bring guinea pigs of various varieties to class and talk about the characteristics of the varieties. Class members can observe the guinea pigs.

2. With the approval of the instructor, invite a local veterinarian into the classroom to talk about the diseases and ailments of guinea pigs.

ADDITIONAL RESOURCES

1. Guinea Pig Care
   http://www.aspca.org/

2. Guinea Pigs: Dr. Sue's Pocket Pets
   http://www.virtual-markets.net/vme/DrSue/guinea.html

3. Diseases of Guinea Pigs
   http://netvet.wustl.edu/species/guinea/gpigs.txt

4. Mancini, Julie R. 2006. *Guinea Pigs*. T.F.H. Publications, Inc., One TFH Plaza, Third and Union Avenues, Neptune City, NJ 07753.

5. Pavia, Audry. 2005. *Guinea Pig*, 2nd Edition. Howell Book House, Wiley Publishing, Inc., Hoboken, NJ.

ENDNOTE

1. Nowak, R. M., and J. L. Paradise. 1991. *Walker's Mammals of the World*, 5th Edition. Baltimore/London: Johns Hopkins University Press, p. 808. Reprinted by permission.

CHAPTER **15**

Chinchillas

OBJECTIVES

After reading this chapter, one should be able to:

- detail the history of the chinchilla.
- describe the two species of chinchilla.
- list practices for care of chinchillas.
- describe the methods of handling a chinchilla.
- describe the common diseases and ailments of the chinchilla.

CLASSIFICATION

Kingdom	—	Animalia
Phylum	—	Chordata
Class	—	Mammalia
Order	—	Rodentia
Suborder	—	Hystricomorpha
Family	—	Chinchillidae
Genus	—	*Chinchilla*
Species	—	*C. laniger*
		C. brevicaudata

TERMS TO KNOW

grotzen	pathogenic organisms	progeny
impaction	polygamous	pyometra
metritis	prime fur	trophozoites
otitis	priming line	veil

HISTORY

Chinchillas belong to the order Rodentia and are descended from Paramys of the Paleocene and Eocene epochs, although the actual connection is obscure. In *Modern Chinchilla Fur Farming* by Willis D. Parker, reference is made to Megamys, a chinchilla-like animal that existed millions of years ago. Megamys is described as being the size of an ox or larger. Because of the abundant vegetation that existed millions of years ago, it is believed the predecessor of the modern, small-sized chinchilla was probably a much larger animal.[1]

A thousand years ago, the Incas used chinchillas as a source of furs. They used the skins for garments and blankets and even made thread from the fur. The softness and distinctiveness of the fur was valued highly and provided them with a badge of distinction. Subsequently, chinchillas were used by a tribe of Indians known as Chinchas; the animal was named for them and means "Little Chincha." Chinchilla fur reached Europe during the 1500s but did not become popular until the 1700s. Not until the 1800s did a real volume of pelts begin satisfying a market hungry for this prized fur.

South American pelts were shipped to Richard Gloek of Leipzig, the "Chinchilla King." He handled 300,000 pelts in 1901, making Leipzig the fur capital of the world and chinchilla the toast of European royalty. About the same time, South American pelts reached the New York fur markets.

Chinchillas nearly became extinct in the wild state but were revived as a ranch-raised animal. Greedy trappers invaded regions of South America and killed the little animals indiscriminately to feed the fur markets.

South American governments realized that destruction of this resource must be stopped, so they placed an export duty on all skins leaving the country. Unfortunately, this only encouraged smuggling. Later, the governments established regulations to prohibit killing the animals and encouraged trapping for the purpose of raising chinchillas in captivity. This was the beginning of fur ranches supervised by the governments.

A few private ranches were set up, and the industry as we know it today was started by Mathias F. Chapman, a mining engineer with Anaconda Copper Company. Chapman applied to the Chilean government for permission to capture several animals and transport them to the United States. After 3 years, he succeeded in trapping only 11 quality animals.

Because the chinchilla's native habitat was approximately 12,000 feet above sea level, Chapman gave the animals time to acclimate themselves before boarding a ship for California. His chinchilla cages were lowered down the mountains gradually during a 12-month period, cooled with blocks of ice when necessary, and shaded from the direct sunlight. His careful handling paid off, and all 11 chinchillas survived the trip down the mountains.

The animals required special conditions and care aboard the ship, too.[2] A constant supply of ice was needed to keep the specially constructed cages and passengers cool.

Over the years, Chapman experimented with both housing and diet. As the herd adjusted and thrived, livestock was finally offered for sale. Most people purchasing chinchillas at that time did so on a group basis, with members buying shares in a pair of animals, hoping to sell lots of livestock in the future.[3]

The first organized offering of chinchilla skins, approximately 10,000, took place at the New York Auction Company on June 21, 1954.[4] Today, Poland, Hungary, Slovakia, and Croatia supply almost 50 percent of the annual 200,000 chinchilla pelts produced worldwide.[5] Chinchillas were quite expensive, especially in the early years. As the numbers continued to increase, some animals were offered as pets.[6]

MAJOR COLOR GROUPS

The Standard

The most common color of chinchillas is the "standard" or blue-gray color; however, they are available in other colors. Common mutant colors include white, beige, and black. *See Figure 15–1.*

Figure 15–1   A "standard" or gray-colored chinchilla. *Courtesy of Brian Yacur and Guilderland Animal Hospital.*

Many breeders have been reluctant to breed the mutant colors. It takes 120 to 150 pelts to make a full-length coat and would take considerable time before the mutant strains could be produced in large enough quantities to supply a market. However, with the increase in the pet industry, the demand for mutant colors continues to increase.

The White

The first important mutant was a white male born in 1955. It was not a true albino because it had black eyes. The male was used in further crossings, and the offspring are known as Wilson Whites. Mating whites with standards usually produces 50 percent whites and 50 percent standards in the first generation. Further matings may produce silver-white and platinum-white animals. *See Figure 15–2.*

Figure 15–2   A white chinchilla. *Courtesy of Michael Gilroy.*

The Beige

Another mutant was a beige female born in 1955 on an Oregon ranch. Matings between beige bucks and standard females give light beige (pearl-colored) and medium- to dark-beige shades (pastel colored). Mating beige with white produces cream-colored **progeny** that are called Rose or Apricot. *See Figure 15–3.*

Figure 15–3   A beige-colored chinchilla. *Courtesy of Brian Yacur and Guilderland Animal Hospital.*

The Black

Black mutations called Black Velvet were developed in 1956 by an American breeder. The fur on these animals has a black undercoat, a very narrow gray-white band, a jet-black **veil** (tips of the fur), and a very high density. Mated with clear-colored standards, it produces the most beautiful mutation ever seen in chinchillas, the Blue Black Velvet. Matings of black males with females from other mutations have produced Sapphire Velvets, Pastel Velvets, and Brown Velvets.[7] *See Figure 15–4.*

Figure 15–4   A black-colored chinchilla. *Courtesy of Stacey Riggert.*

The Sullivan Violet

One of the newest mutations to be raised in the United States is called the Sullivan Violet, an animal with a clear white belly and a lavender (pinkish-violet) **grotzen** (the center back strip of a fur pelt). The first violets were born on a ranch in Rhodesia (now Zimbabwe) Africa in the mid-1960s. The animals were put up for sale and are now housed on the Loyd Sullivan ranch in California.[8]

CHARACTERISTICS

The native habitat of the chinchilla is the barren areas of the Andes Mountains at elevations up to 20,000 feet. Chinchillas shelter in crevices and holes among the rocks; they are primarily nocturnal. To eat, they sit upright and hold food in their forepaws; the diet consists of any available vegetation.

Chinchillas are classified into two species: *C. brevicaudata* and *C. laniger.* *C. brevicaudata* is native to Peru, Bolivia, and Northwestern Argentina. The native range of *C. laniger* is Northern Chile.

C. brevicaudata has a thick neck and shoulders and is heavily furred with light-gray coarse hair that is often tinged with a yellowish cast. Its ears are shorter than those of *C. laniger,* and its nose is flatter, giving the animal a stocky appearance. *C. laniger* has a narrow neck and shoulders; the fur is very silky and medium to dark gray, with a bright bluish cast.[9]

Today, most of the chinchillas raised domestically are of the *C. laniger* species or a close relative designated the "Costina" type. The *C. brevicaudata* is a larger animal, but its fur is not as desirable as the crisp blue fur of the *C. laniger.* Some attempts were made to cross-breed the two species; however, the male offspring of the crossings were found to be sterile.

Chinchillas resemble small rabbits with short ears and a short, bushy tail much like that of a squirrel. Chinchillas range from 9 to 15 inches long, have a tail of 3 to 10 inches, and weigh from 1 to 2 pounds.

Their thick, shiny fur is about 1 inch long; the softness of the fur is due to the fewer number of guard hairs as compared with those of fur-bearing animals. Chinchillas usually shed about every 3 months. New growth starts at the neck and progresses toward the rear of the animal. Where the new hair is coming up through the old hair a darker, a distinct line known as the **priming line** can be seen. When the priming line reaches the tail, it is time to take the pelt. The hair coat remains **prime fur** for several weeks, and then the process begins again.

The head of the chinchilla is broad, with large ears and eyes; they have very long, stiff whiskers. Both the short forefeet and the narrow hindfeet have four digits, with stiff bristles surrounding soft, weak claws; the pads of the feet are large and rubbery. Females have one pair of mammary glands on the lower abdomen and two pair higher up near the chest.

Chinchillas make very little noise but do make a barking sound similar to a squeak toy. When annoyed, they can be heard making this sound.

Chinchillas are very clean, almost odor-free animals. The feces are normally dry and produce very little odor; the urine does have an unpleasant odor if allowed to accumulate.

HOUSING AND EQUIPMENT

Chinchillas are nocturnal; when planning housing for them, one should consider a location where it will be fairly quiet during the day. This is important to the successful raising of chinchillas. The area should be fairly dry, have adequate ventilation, and not get too hot. As temperatures reach 80°F, chinchillas become uncomfortable; at 90°F, they are subject to heat prostration.

Cages used for other small animals can be used to house chinchillas. A cage of approximately 24 inches wide by 24 inches long and 14 inches high is adequate for a single chinchilla; however, cages that are taller are recommended. Cages that are 24 inches high or even 36 inches high allow for ladders and shelves to be included, which will give the animal room to move about and get more exercise.

Cages should be made of metal and wire; any plastic used will soon be chewed through. Cages can be constructed of wood and wire mesh screen; however, the wood must be on the outside of the wire mesh or it will be destroyed.

The size of openings in the wire mesh is very important, especially if the owner plans to breed the chinchillas and produce young ones. Small chinchillas can crawl through openings about an inch square. They are very active shortly after birth, and the owner needs to take precautions not to lose any young.

Chinchillas raised as littermates can usually be kept together in a cage; others should be introduced together slowly. This can be done by placing the cages containing two chinchillas close together to allow them to get used to each other. After a while the two can be put together in a neutral cage. Chinchillas introduced together into a cage should be watched closely in case they do not get along.

Polygamous breeding cages are available for owners who want to breed more than one pair of animals. *See Figure 15–5.* This type of cage usually has three separate cages connected together with holes in the back of the cages that lead to a tunnel that connects all three cages. Females are fitted with a plastic collar that prevents them from leaving their cage and entering the tunnel. *See Figure 15–6.* Two females and one male are placed in the cages, and the male can use the tunnel to go from one cage to the other. If one of the females allows him to stay in her cage, a third female can be added to the cage where the male was put first. Additional cages can be added to house up to 20 females. Allowing the male to run the tunnel keeps him in good condition.

Cages usually have wire mesh or solid bottoms. With the wire bottoms, newspaper can be placed below the cage to catch feces and litter. Solid-bottom cages usually have a pan that can be removed for cleaning. This pan is usually filled with some type of absorbent material; wood shavings are probably best to use.

Wood shavings make good, inexpensive bedding material. Ground corn cobs are also inexpensive and absorbent. Hay and straw can be used, but these materials can become moldy and may stain the animal's fur.

An owner may want to put a wooden box or a large can in the cage to give the animal a place to hide and sleep. If a tin can is used, the ends should be cut out and the can flattened on one side so it cannot roll around.

A small wooden block should be placed in the cage to give the animal something to chew on and to play with. Remember that the chinchilla is a rodent with continuously growing teeth, and it will need something to chew on to wear the teeth down.

Chinchillas need regular baths to rid their fur of excess moisture and oil; they do not bathe in water but in finely ground powder. In their natural setting, chinchillas bathe in volcanic ash of the Andes Mountains. A very similar powder can be purchased from pet stores for this purpose. A small rectangular

Figure 15–6 Females are fitted with a plastic or aluminum collar that prevents them from entering the tunnel. *Courtesy of Isabelle Francais.*

Figure 15–5 Polygamous breeding cages are developed so that one male can breed with three or more females. A tunnel at the upper back of the cage connects to three or more cages. The male can move through the tunnel and enter any of the female's cages. *Courtesy of Michael Gilroy.*

pan with about 2 or 3 inches of powder can be used; the pan needs to be large enough for the animal to roll around in. The powder may be used several times. *See Figure 15–7.*

The animal should receive a bath about twice a week. During hot, humid days, daily baths may be needed. The pan with the powder can be placed in the cage and left for about 5 minutes; that should be sufficient time for the animal to finish a bath. Chinchillas should not be given a bath in water.

Cages should be fitted with metal, creep-type feeders; these are the easiest for the animal to use. Heavy ceramic bowls can also be used. These will not tip over, and the animal cannot destroy them.

Cages should be fitted with vacuum-type water bottles. Chinchillas do not drink very much water, but a clean, fresh supply should always be available.

FEEDING

Various pelleted foods are available today for chinchillas. These pellets consist primarily of ground alfalfa hay, corn gluten, corn tailings, wheat germ, and bran. These foods usually have all of the vitamins and minerals necessary. If pelleted

Figure 15–7   A chinchilla taking a dust bath. *Courtesy of Michael Gilroy.*

chinchilla feed is not available, pelleted guinea pig feed will be sufficient. The pelleted foods may be supplemented with fresh, well-dried alfalfa or timothy hay. The key is to make sure it is not moldy or dusty and does not contain weeds.

The diet can be supplemented with grass, carrots, and celery. Pears and apple slices are readily consumed by chinchillas. Limit the amounts of greens and other supplements because these can upset the digestive process and lead to serious problems. As special treats, chinchillas love raisins. The animals will jump and run around their cages on hearing a bag of raisins being opened. Owners should avoid wet grass and hays that may spoil and cause digestive problems and diarrhea.

Clean, fresh water should be available to chinchillas on a pelleted diet, or they will become dehydrated.

HANDLING

When a chinchilla is first brought home, it should be allowed to get used to its new surroundings. When approaching the cage, one should go quietly and speak to the animal softly. Chinchillas dislike loud noises and quick, sudden movements; they may become frightened and keep to the corner of the cage. Being nocturnal animals, they may not appreciate being approached during the day. Even a very tame chinchilla may raise up on its rear legs and bark. They are able to shoot a stream of urine at their intruder.

Several methods are used to handle chinchillas. Ranchers will grasp the animal's tail close to the body and pick the animal up. Ranchers are very careful not to grab the animal in any way that may cause it to lose some of its fur. An animal with a patch of fur missing would mean a pelt that is worthless; it would take several weeks or months for the fur to grow back.

Chinchillas are able to release their fur as a defense mechanism. If the animal is grabbed by a predatory animal, it will leave the predator with a

Figure 15–8 A chinchilla can be picked up by its tail if the tail is grasped close to the body.

Figure 15–9 Chinchillas that are tame can be picked up by cupping the hands around the animal. Place one hand behind the animal and one hand in front and cup it in the hands.

mouthful of fur. The release of fur in chinchillas is called *fur slip*. If the chinchilla owner handles the animal roughly, the same thing can happen.

The recommended and gentlest method of picking up a chinchilla is to cup both hands and place one hand in front and one hand behind the animal. The handler should then cup the animal in the hands. This needs to be done fairly quickly, or the animal will jump over the handler's hands and head back to a safe corner. The animal should be held close to the handler's body so that it feels safe and secure. For a chinchilla that is somewhat frightened, the handler may want to hold on to the tail also. *See Figures 15–8 and 15–9.*

DISEASES AND AILMENTS

Chinchillas are very healthy if they receive proper nutrition; have a clean, dry, draft-free environment; and are free of any stressful conditions. However, there are diseases and ailments that can affect these animals.

Pseudomonas aeruginosa is a bacterium that may be found in unclean drinking water or is associated with unclean cage conditions. Loss of appetite, loss of weight, diarrhea or constipation, inflammation of the eyes and ears, pneumonia, inflammation of the intestines, inflammation of the uterus in females, abortion, and infertility are signs of infection. A veterinarian can evaluate the signs and begin proper treatment. Proper cleaning and sanitation are important in preventing infection.

Swelling and redness around the eyes, sensitivity to light, watering, and pus formation are common symptoms of eye inflammation (conjunctivitis). The eyes should be washed gently and frequently with a eyewash solution; several ophthalmic medications are available. Advanced cases should receive the attention of a veterinarian. Cages and feed containers should be washed and disinfected; infected animals should be isolated.

Twisting and lowering of the head, hanging the head to one side, and running around in a circle are common signs of an inner ear infection (**otitis**). The ear should be cleaned and ear drop medications applied. These are available from your veterinarian. The animal should be kept warm and free of drafts.

Listlessness, failure to eat, and difficulty breathing are signs of pneumonia. Other signs may include sneezing, fever, weight loss, a lack of energy, and a thick yellowish discharge from the nose. The ailment is usually associated with cold, damp conditions that lower the animal's resistance. A veterinarian can diagnose the condition and begin proper treatment.

"Loose or watery stools occur in chinchillas with intestinal disorders caused by nutritional, bacterial, protozoal, parasitic, or stress-induced illnesses. Animals with these disorders may die quickly without signs or, in long-term cases, show a range of signs including lack of energy, loss of appetite, rough hair coat, staining near the anus, hunched posture, listlessness, dull eyes, dehydration, weight loss, stomach pain, excessive gas, fever, or lower-than-normal body temperature, and diarrhea or constipation. The feces may look bloody or include mucus. Straining to have a bowel movement can lead to the lower part of the large intestine slipping out of place. Sudden changes in diet and the giving of inappropriate antibiotics can affect the bacteria that normally live in your chinchilla's digestive tract, causing digestive problems."[10] A veterinarian needs to evaluate the signs and begin proper treatment.

Reproductive disorders in females may include infertility, absorption or resorption of fetuses, retained or mummified fetuses, difficulty giving birth, **metritis, pyometra,** lack of milk, and mastitis. Metritis is inflammation of the uterus. Pyometra is an accumulation of pus in the uterus, which usually follows a case of metritis. Vaginal discharges in the form of watery, white pus or pus having an unpleasant odor, increased urination, loss of weight, and fur chewing at the flank and abdomen are signs that should alert one to possible problems. Infections in the uterus can spread to other parts of the body, can be caused by retained placentas, and can be spread by males in a polygamous breeding system. If any of these signs are observed, the animal should be taken to a veterinarian for evaluation, diagnosis, and treatment.

Pathogenic organisms and their toxins can be absorbed into and distributed throughout the body by the circulatory system. Animals appear listless, go off their feed, and lose weight.

Listeria monocytogenes is a pathogenic organism that infects chinchillas. Affected animals go off their feed, lose weight, develop paralysis in the rear quarters, and go into a coma. Many animals may die without any clinical signs, and mortality is high. The infection is usually associated with unsanitary conditions and contaminated food and water.

Giardia species is a common intestinal parasite found among chinchillas. Cysts and **trophozoites** can be seen in the feces. Transmission is from feces to mouth. These cysts and trophozoites can also be seen in intestinal scrapings from chinchillas that have died after a period of intermittent watery diarrhea or **impaction.** Mortality may be high in such outbreaks, and often, no other cause of death can be found.

Trichophyton mentagrophytes is the cause of dermatophytosis or ringworm in chinchillas. Affected animals usually have a loss of hair and reddening around the nose, eyes, and base of the tail. Affected animals should be isolated and treated. Ringworm is contagious to other animals and to humans.

Impaction is a condition in which the cecum and large intestines become tightly packed with feces; many times, impaction follows diarrhea. The condition is usually caused by moldy or spoiled feed, poor nutrition, stress, insufficient water intake, pressure on the intestines from pregnancy, and infectious agents. Common symptoms include the reduction or absence of feces, lack of appetite, rough hair coat, and an animal humped up with its chin between the front feet and resting on the cage floor. The animal may show no desire to move about and may appear glassy-eyed. Several remedies may be successful in relieving impaction; if the chinchilla is still eating, a dropper full of mineral oil can be administered daily or the dietary fiber can be increased. Otherwise, a veterinarian needs to be consulted to determine the cause and appropriate action.

Fur chewing is a common problem of chinchillas; the animals will chew the fur on the flanks, leaving patches of short fur. Poor nutrition, small cages, drafty conditions, boredom, and high humidity have been suggested as causes.

Malocclusionis is a dental deformity in which the molars and incisor teeth are overgrown or do not meet properly. Affected animals fail to eat, lose weight, and may salivate or slobber; death may result from starvation. The condition is caused by abnormal growth or inappropriate diet; it may be a result of genetics. If diagnosed early, a veterinarian may be able to trim or file the teeth so the animal can eat more normally.

"Thiamine (vitamin B) is required for normal carbohydrate metabolism and protein synthesis. Deficiency of this vitamin causes damage to peripheral motor nerves that is often reversible when thiamine is restored to the diet. Affected chinchillas may show neurological signs such as trembling, circling, convulsions, or paralysis. Your veterinarian can treat this deficiency with injections of thiamine or B-complex vitamins. Sources of natural thiamine include leafy vegetables, high-quality hay, and wheat germ meal, or supplements can be added to the diet (1 milligram of thiamine per kilogram of feed) if needed."[11]

"A dietary imbalance in the ratio of calcium to phosphorus or phosphorus deficiency may result in severe muscle spasms in young or pregnant chinchillas. Muscles of the hind limbs, forelimbs, and face are affected. Treatment with calcium gluconate, given by a veterinarian, is recommended. Prevention is accomplished by feeding well-balanced, nutritionally complete diet formulated for chinchillas."[12]

Mastitis may appear in nursing mothers; swelling, redness, and hardness of the mammary glands are common symptoms. Pain may occur while the young are trying to nurse, and the mother will often not allow nursing. The condition is caused by infection following biting of the nipples by the young, weaning the young too early, unsanitary conditions, or incomplete nursing. The owner should consult with a veterinarian to try to determine the cause, and begin treatment as necessary. The cages and nest box should be cleaned and disinfected thoroughly. The owner may have to supplement milk to the young.

REPRODUCTION

The estrous cycle in female chinchillas varies from 28 to 35 days. A young female chinchilla first comes in heat at about 5 months of age but should not be bred until about 8 months.

Figure 15–10 The genital area of a female chinchilla. *Courtesy of Michael Gilroy.*

Figure 15–11 The genital area of a male chinchilla. *Courtesy of Michael Gilroy.*

A female in heat shows general signs of restlessness, frequently urinates, and scratches the bedding. Males become aroused when around females in heat. *See Figures 15–10 and 15–11.*

In pair mating of chinchillas, the male should be placed near the female's cage for a while so that they can become familiar; the male is then placed in the female's cage. A box should be placed in the cage so that the male can find an area of safety in case the female is unwilling to mate and prefers to fight. The pair should be watched closely; if the female continues to fight, the male should be removed from the cage. If the female rejects the male, she will usually stand on her rear legs, aim, and shoot urine at the male. Fighting almost always occurs, but the pair should settle down after a while. Mating usually takes place in the evenings. A copulatory plug forms in the female's vagina after mating. This plug is produced from the male ejaculate. The plug is about 3 cm long and white to yellowish in color. The plug will shrink and drop out a few hours after mating.

The male and female can be left together after mating. The male may even remain in the cage during birth, but the female comes into heat again within 24 hours. Unless a second mating is desired, the male should be removed; he may be returned to the cage after a couple of days.

In polygamous matings, the male is left in the tunnel and can visit the cages of several females at any time. The opening to the tunnel can be closed to prevent a female from being rebred right after giving birth.

With all mating methods, the owner should always observe the chinchillas and be aware of overly aggressive animals; serious fighting may ensue and death can result.

The gestation period for chinchillas is 111 days. Any stress can cause premature birth, abortion, or resorption of the fetuses. Pregnant females should be handled very carefully, if at all. Chinchillas usually do not need any assistance at birth; birth usually occurs during the early morning hours.

Figure 15–12   An adult female with a young chinchilla.
Photo by Isabelle Francais.

Emergency Formula for Orphan Chinchillas

- ⅓ ounce (10 g) ground chinchilla pellets (make sure these contain calcium)
- ⅓ ounce (10 g) oat cereal
- ¹⁄₃₀ ounce (1 mL) dextrose (available from your veterinarian)
- ⅙ ounce (5 mL) Pedialyte
- 10 mg vitamin C (available in tablet or liquid form from your pharmacy)
- Bottled water, enough to make a liquid when ingredients are blended

1. Blend thoroughly, and warm the formula to body temperature (do not make it hot).
2. Use an eyedropper to feed, and feed as much as the kit will eat on its own. Feed slowly, and do not force feed.
3. Make sure the mixture is prepared fresh daily and is fed at body temperature.

Figure 15–13   Emergency formula for orphan chinchillas. *Reprinted with permission from the Chinchilla Handbook, by Sharon L. Vanderlip, DVM, 2006, Barron's Educational Series, Inc. 250 Wireless Boulevard, Hauppauge, NY 11788.*

When delivery is near, the female stretches, rears up, and emits low, moaning sounds. During a normal birth, the female pulls out the young, removes the placental sac, cleans fluid from around the young's nose and mouth, and dries it off.

Litters vary from one to three, but the average is two per litter. The young are born fully furred, eyes open, with a full set of teeth, and have lots of energy. Most parents do not mind the young being handled, but handling should be put off until the young are dry and warm.

If a polygamous breeding system is used, the young should not be allowed into the cages of other females because they will probably be killed (*See Figure 15–12*).

The owner should observe the young carefully during the first few days to see if they are getting enough to eat. If the litter consists of more than two, the mother may not be able to provide enough milk. Babies that do not get enough to eat fail to grow; young chinchillas can be successfully fed using a dropper or a small bottle (*See Figure 15–13*). The young grow fast and are ready for weaning at about 6 weeks of age.

SUMMARY

Chinchillas belong to the order Rodentia and are descended from the Paramys of the Paleocene and Eocene epochs, although the actual connection is obscure. A thousand years ago, chinchillas were used by the Incas.

In the 1800s, chinchilla fur became highly prized in Europe. This demand became so great that the animals almost became extinct in the wild; they eventually recovered by being raised in ranches. M. F. Chapman is responsible for the ranch-raised chinchilla industry.

Chinchillas are classified into two species: *C. brevicaudata* and *C. laniger*. Because of its more desirable blue fur, the *C. laniger* species was more commonly raised and bred.

Chinchillas are nocturnal; when planning housing for them, one should consider a location that will be fairly quiet during the day.

Chinchillas are able to release their fur as a defense mechanism. If the animal is grabbed by a predatory animal, it will leave the predator with a mouthful of fur; if the chinchilla owner handles the animal roughly, the same thing can happen.

Various pelleted foods are available on the market for chinchillas. These pellets consist primarily of ground alfalfa hay, corn gluten, corn tailings, wheat germ, or bran. If pelleted chinchilla feed is not available, pelleted guinea pig feed is sufficient.

The estrus cycle in females varies from 28 to 35 days; the gestation period is 111 days. Litters vary from one to three; the average is two. The newborn are fully furred, their eyes are open, they have a full set of teeth, and they have lots of energy.

Chinchillas are very healthy if they receive proper nutrition; have a clean, dry, draft-free environment; and are free of any stressful conditions.

DISCUSSION QUESTIONS

1. Where did chinchillas originate?
2. What are the two species of chinchilla?
3. Who is responsible for the chinchilla industry in the United States?
4. What are the mutant colors of chinchilla?
5. What are common foods for chinchillas?
6. Describe a polygamous mating system.
7. List the common diseases and ailments of chinchillas. What are the causes, symptoms, and treatments?

SUGGESTED ACTIVITIES

1. With the approval of the instructor, invite local breeders or pet shop owners to bring chinchillas into the classroom to talk about the animals.
2. With the approval of the instructor, invite a local veterinarian into the classroom to discuss the diseases and ailments of chinchillas.

ADDITIONAL RESOURCES

1. Etc. Chinchilla Page
   http://www.etc-etc.com/chin.htm

2. Underhill Chinchillas
   http://members.aol.com/UHChins/UnderhillChinchillas/main.htm

3. Chinchillas: Dr. Sue's Pocket Pets
   http://www.virtual-markets.net/vme/DrSue/chinchil.html

4. Chinchilla Information Site
   http://members.aol.com/sirchin/chininfo.htm

5. About.com: Exotic Pets
   http://www.exoticpets.about.com

6. Dr. Jungle's Animal-World
   http://www.animal-world.com

7. Chin Care
   http://www.chincare.com

8. Chinchilla World
   http://www.chinchillaworld.com

9. Alderton, David. 2007.*Chinchillas.* T.F.H. Publications, Inc., One TFH Plaza, Third and Union Avenues, Neptune City, NJ 07753.

10. Vanderlip, Sharon L. DVM. 2006. *Chinchilla Handbook.* Barron's Educational Series, Inc., 250 Wireless Boulevard, Hauppauge, NY 11788.

11. Low Country Chinchillas
    http://www.huggablepets.com

END NOTES

1. Parker, W. D. 1975. *Modern Chinchilla Fur Farming.* Alhambra, CA: Borden Publishing Company.

2. Zeinert, K. 1986. *All About Chinchillas.* Neptune City, NJ: T.F.H. Publications, Inc., pp. 12–14.

3. Ibid.

4. Bowen, E. G., and R. W. Jenkins. 1969. *Chinchilla, History, Husbandry, Marketing.* Hackensack, NJ: Adler Printing Company, pp. 1–10.

5. Thissen, Simone DVM. VN Times. Volume 6, Number 5, Page 18, May 2006. VBD, Olympus House, Werrington Centre, Peterborough PE4 6NA.

6. Zeinert, K. *All About Chinchillas.*

7. Mosslacher, E. 1968. *Feeding and Caring for Chinchillas.* Neptune City, NJ: T.F.H. Publications, Inc., p. 116.

8. Zeinert, K. *All About Chinchillas.*

9. Ibid.

10. *The Merck/Merial Manual for Pet Health.* Copyright © 2007 by MERCK & CO., INC., Whitehouse Station, NJ. All rights reserved. Used with permission.

11. Ibid.

12. Ibid.

CHAPTER 16

Ferrets

CLASSIFICATION

Kingdom	— Animalia
Phylum	— Chordata
Class	— Mammalia
Order	— Carnivora
Family	— Mustelidae
Genus	— *Mustela*
Subgenus	— *Mustela*
Species	— *M. putorius*
	(European polecat)
Subspecies	— *M. putorius furo*
	(Domestic ferret)

OBJECTIVES

After reading this chapter, one should be able to:

- detail the history of the domestic ferret.
- list and describe the various color patterns of ferrets.
- list the general care practices for ferrets.
- list the common feeds for ferrets.
- describe the correct method for handling a ferret.
- describe the common diseases and ailments of ferrets.

TERMS TO KNOW

aplastic anemia	ferreting	kits
bib	hobs	mitt
estrogen	jills	ovulation

HISTORY

The domestic ferret belongs to the family Mustelidae and the genus *Mustela*, which also includes weasels, ermines, mink, and polecats. There are four subgenus of *Mustela;* all members of this genus have long, limber, slender bodies. The tail is long and is usually about half the length of the head and body. They have short legs and small, rounded ears. They have rather long, oval-shaped heads and a pointed snout; the eyes are bright and clear.

The European polecat (*M. pulorius*) is a wild animal that is common to the open forest areas and meadows of Europe. It makes its den in hollow logs, crevices, and burrows made by other animals. It is nocturnal and capable of climbing. The European polecat has been tamed for centuries and used in **ferreting,** or hunting rabbits. The polecat will readily go into a burrow and chase the rabbits out. Polecats have also been used to drive rats from their burrows and tunnels. The domestic ferret (*M. pulorius furo*) is believed to be a descendant of the European polecat (*M. pulorius*).

Ferrets have been in the United States for more than 300 years; they were used in the 1800s for rodent control. The "ferret meister" would come with his ferrets to a farm or granary and release his ferrets. These working ferrets ran into holes and hiding places of rodents and the rodents ran out. The people outside would wait with shovels and terrier-type dogs and kill as many rodents as they could. In the 1800s, ferrets were often kept on small farms and at feed mills for rodent control. The normal range of a ferret is about 200 yards from the place it considers home. Ferrets traveling through rat tunnels leave trace odors that trigger fear in rats and mice, causing them to flee.

Ferrets have also been used successfully to help wire airplanes in hard-to-reach places. Because ferrets love to enter anything that looks like a tunnel, it is a simple matter to attach a wire and have them run down a tunnel and return. Ferrets have also been used in scientific research because they catch the same colds as humans.[1]

People have recently found that ferrets make wonderful pets. No wild colonies of the domestic ferrets exist. After many generations of domestication, they are unable to survive in the wild.

The possession of ferrets as pets is legal in 48 states. They are not legal in California and Hawaii. Permits are required for possession as pets or for breeding in Kentucky, New York, and Rhode Island.

Some local governments or cities have adopted regulations prohibiting ownership of ferrets. Among them are New York City, New York; Dallas/Fort Worth, Texas; and Washington, D.C.

One needs to check local, county, and state laws and/or ordinances to make sure ferrets can legally be kept as pets in their area of residence. Laws and ordinances are constantly changing and being updated, so it is important to check on the latest status of laws and/or ordinances regarding ferrets.

All ferrets sold in Massachusetts must be spayed or neutered and have initial rabies and distemper vaccinations. Only licensed breeders or research institutions are allowed to possess whole or unaltered ferrets. Other states may have similar laws.

The only wild ferret in the United States is the black-footed ferret (*M. nigripes)*, which is found in small numbers in Arizona. The black-footed ferret is very similar in size and build to the domestic ferret but is lighter in color and has very dark feet and legs. It has a broad black stripe across the eyes, and the tail is black-tipped. The ears are also larger, more rounded, and higher on the head. The black-footed ferret is on the endangered species list.

MAJOR COLOR GROUPS

Several colors are now being bred: sable (the most common), red-eyed white, silver mitt, sterling silver, white-footed sable, butterscotch, white-footed butterscotch, and the rare cinnamon. *See Figure 16–1.*

Figure 16–1   A common sable ferret. Note the mask across the face that is characteristic of the sable ferret. *Photo by Isabelle Francais.*

The Common Sable

The sable ferret ranges from light to dark, depending on the shade of both the underfur and guard hairs; the underfur ranges from white to beige. The guard hairs are longer and are usually black. A well-marked sable ferret should have a definite mask or hood pattern over the face. The preferred nose color is black.

The White

The red-eyed white is often referred to as an albino and has the characteristic red eyes of an albino. The fur of the albino ranges from white to yellowish-white; the nose of the albino is pink. *See Figure 16–2.* Black-eyed white ferrets are white or yellowish-white with black eyes.

Figure 16–2   A white ferret. This is a red-eyed white ferret. There are black-eyed white ferrets, but they are not as readily available. *Photo by Isabelle Francais.*

The Silver Mitt

A silver mitt ferret has underfur of white or off-white and guard hairs of black and white. This combination of white and black gives the ferret a silvery appearance. The feet of the silver **mitt** (feet) are white. The silver mitt has a white **bib** (the patch of white below the chin and on the throat) and eyes that appear black but are often a deep burgundy. *See Figure 16–3.*

Figure 16–3   A silver mitt ferret. The underfur of the silver mitt ferret is white. The guard hairs are black and white. *Photo by Isabelle Francais.*

The Sterling Silver

A sterling silver ferret is very similar to a silver mitt but has more white guard hairs that give the animal a lighter appearance. *See Figure 16–4.*

Figure 16–4   A sterling silver ferret. The sterling silver ferret is similar to the silver mitt, but there are more black guard hairs, giving the ferret a darker shading. *Photo by Isabelle Francais.*

The White-Footed Sable

White-footed sables are marked like the regular sable ferret but have four white feet and a white bib.

The Butterscotch

The underfur of the butterscotch ferret is the same as the sable, but the guard hairs, legs, mask, and hood coloring are butterscotch instead of black. *See Figure 16–5.*

The white-footed butterscotch ferrets have the same markings and color combinations, except for having four white feet and a white bib.

Figure 16–5   A butterscotch ferret. Butterscotch ferrets will have white to beige-colored underfur and butterscotch guard hairs. *Photo by Isabelle Francais.*

The Cinnamon

The cinnamon ferret has underfur that is white or off-white and guard hairs that are a rich red-brown or cinnamon color.

CHARACTERISTICS

The domestic ferret, like the other members of the genus *Mustela*, has a body that is elongated, lean, slender, and muscular. The legs are short, and the feet have five toes with claws. The fur of ferrets is dense and very soft; they change coats completely twice a year.

Ferrets have oval-shaped heads and pointed snouts. The head of the male usually is broader and less pointed than that of the female.

Male ferrets (**hobs**) are 16 to 20 inches long and weigh 3 to 5 pounds. Females (**jills**) are a little smaller, usually 12 to 14 inches long, and weigh 1½ to 3 pounds.

There is little difference between a male and female ferret as a pet. Males are usually about twice the size of the females. Both develop a strong musky odor as they reach maturity; the odor is strongest in the males. The odor originates partly from anal glands that can be surgically removed. In males, however, this procedure alone does not eliminate the odor; castration is also necessary. Glands in the skin also cause a musky scent. With occasional baths, the male ferret that has been descented and castrated has very little odor. If a pet owner is not interested in breeding, a spayed and descented female would make the best pet.

Ferrets that have been descented, spayed, or neutered may not have the same definite color pattern around the eyes. These patterns usually come back in a paler color following shedding.

Ferrets do not have a very well-developed sense of sight; being primarily nocturnal animals, they do not see well in bright light. They do, however, have highly developed senses of hearing, smell, and touch. When put in new surroundings, they use their sense of smell to check out the area thoroughly. Ferrets learn their owner's voice and usually come to the door of a cage when the owner approaches.

Ferrets that are well cared for live from 8 to 11 years. Normal body temperature is 101.8°F; the heart rate is between 180 and 250 beats per minute, and respiration is between 30 and 40 breaths per minute.

Ferrets have a total of 34 teeth. On each side, there are three incisors on top and three on the bottom, one canine teeth on the top and bottom, three premolars on top and three on the bottom, and one molar on top and two on the bottom.

HOUSING AND EQUIPMENT

The type of housing depends on the owner's situation. A single ferret should have a cage at least 12 inches wide, 24 inches long, and 10 inches high. This amount of space will sufficiently house a ferret and provide space for a litter, feeding, and sleeping area; however, a ferret is very active and, if kept in this size of cage, should be allowed out frequently for exercise.

Cages that are 24 inches wide, 24 inches long, and 14 inches high are more desirable. A cage approximately 2 feet wide and 4 feet long is appropriate for a pair of ferrets; this allows plenty of room for a sleeping or nesting box, a litter box, and space for exercising. Cages of this size can easily be constructed from wood and wire screen. *See Figure 16–6.*

Figure 16–6   This cage is constructed from plywood and wire screen. *Courtesy of USDA.*

Glass aquariums are not desirable because they do not allow air to circulate and can become overly warm in the summer.

Whatever is used to house a ferret, two things should be kept in mind:

1. If there is a way to escape from a cage, a ferret will find it. All materials should be of good quality, and latches on the doors should lock securely.

2. Ferrets enjoy watching activity around them. Cages should be constructed to allow ferrets to see out and observe what's going on; they want to be part of the action.

Food bowls should be made of earthenware. These are heavy and almost impossible for the ferret to upset. A constant supply of fresh water should be available; the larger vacuum-type bottles used for rabbits and guinea pigs are ideal.

Ferrets are very playful, and almost anything makes a suitable toy. Toys that are available for cats will usually work well for ferrets. Ferrets, however, will be harder on the toys than cats. Ferrets will chew on the toys and are capable of breaking small pieces off the toy, and then the ferret can swallow the pieces, causing digestive tract blockage problems. Soft rubber and sponge toys should be avoided. Plastic pipes that they can crawl through and plastic milk jugs with holes cut in them provide ferrets with a place to explore and play.

FEEDING

Ferrets are easy to feed. They are carnivores. They have a quick metabolism; food material passes through their digestive system quickly. They will eat frequently and should have food available to them at all times. Ferrets will eat ¼ to ½ cup of food daily. Ferrets are not prone to overeating and usually will not become obese. Food for ferrets should be high in animal protein, relatively high in fat, and low in fiber. An appropriate food for ferrets should contain approximately 35 percent protein, and 20 percent or more in fat and less than 3 percent fiber. Commercial diets are now being formulated for ferrets; however, one needs to check the label carefully because not all foods are formulated the same. Dry foods are recommended because they will help keep the ferret's teeth and gums in good condition, cost less than moist foods, and are easier to feed.

If formulated ferret foods are not available in your area, then dry foods developed for kittens will work. Foods formulated for kittens will have a higher protein amount than foods formulated for mature cats. *See Figure 16–7.*

When feeding dry foods, it is important to remember to make sure the animal has plenty of water available because no water is contained in the dry product. *See Figure 16–8.*

Young ferrets reach 90 percent of their adult size in about 14 weeks. Large amounts of food are consumed during this time, so young, growing ferrets should have food available at all times.

If an animal is receiving a well-balanced diet, it should not be necessary to provide additional vitamins or mineral supplements.

Older ferrets may need vegetable oil added to their diet to aid in digestion and bowel movements and help maintain a healthy coat; about ½ teaspoon per day will be sufficient. Check with a veterinarian for his or her recommendation.

Ferrets love treats. Treats should be given in small amounts and should be meat based. Cooked meat scraps (chicken, turkey, beef), cracklings, and jerky can be used as treats. Feed treats in limited amounts so they do not interfere

Figure 16–7 Ferrets can be fed dry kitten food. A large, heavy bowl prevents the ferret from tipping the bowl over. *Courtesy of Carolyn Miller.*

Figure 16–8 This ferret is being kept in a wire cage typical of the type that can be purchased at pet shops and discount stores. The cage is equipped with a vacuum-type water bottle so that the ferret has a supply of water at all times. *Photo by Isabelle Francais.*

Figure 16-9 Ferrets can be picked up by the nape of the neck. Their bodies will become limp and allow one to pick them up in the same manner that their mother did when they were young. *Photo by Isabelle Francais.*

with the ferret's regular diet. Although ferrets love raisins, dates, licorice, and ice cream, these and other sweet, sugary treats should not be given to ferrets as they can lead to tooth decay and interfere with the function of the pancreas. Dairy products can cause diarrhea.

Bones should never be used as treats; slivers may be broken off and eaten by the animal, causing internal injury.

HANDLING AND TRAINING

A ferret should be allowed to get used to its new home for a couple of days before the handler begins training. Young ferrets, although they may be used to humans, need some time to get used to new surroundings. When approaching a ferret, one should speak to it in a soft, gentle voice before attempting to pick it up.

When ferrets are young, their mother will pick them up by the skin on the nape of the neck; handlers can pick up ferrets in a similar manner. *See Figure 16-9.* One should grasp the ferret firmly around the body behind the forelegs. The ferret will go limp and allow itself to be picked up. Placing the other hand under the rear of the animal supports its body. *See Figures 16-10 and 16-11.*

Ferrets have very tough skin and, while playing with their littermates, will play rough and bite one another. This playful bite may be painful to a human, and the animal should be disciplined. A firm "no" each time it bites should be sufficient for the young ferret to realize this is not approved behavior. If the ferret bites while one holds it, the ferret should not be put down because it will learn to use this behavior when it wants down.

Figure 16-10 Ferrets are easy to pick up and handle. Grasp the animal around the body just behind the front legs. The animal will usually go limp and allow itself to be picked up. *Photo by Isabelle Francais.*

Figure 16-11 After picking the ferret up, continue to hold the animal with one hand under the front legs and support the rear of the animal with the other hand. *Photo by Isabelle Francais.*

Figure 16–12 Two items necessary in training ferrets are Linatone® and BitterApple®. *Courtesy of Carolyn Miller.*

Figure 16–13 Brushing a ferret's fur will help to remove loose hair and stimulate the skin. *Photo by Isabelle Francais.*

Three products that may be helpful in training ferrets are Ferretone®, Linaton®, and BitterApple®; these items are available at pet shops. Ferrets love Ferretone® and Linatone® and they can be used as a treat and reward for good behavior or in association with training. Ferretone® and Linatone® will help keep the animal's coat in good condition. Ferretone® and Linatone® should be limited to three to five drops per day because they contain vitamin A and can be harmful if fed in excess. *See Figure 16–12.*

BitterApple® works in the opposite manner. Ferrets dislike it and will not touch anything to which BitterApple® has been applied. It should be used on electrical and telephone cords until the animal learns not to chew them.

Ferrets enjoy being stroked and talked to. The owner's voice and hands are the best training tools for ferrets. *See Figure 16–13.*

Ferrets prefer to relieve themselves in a corner and may therefore be litter trained. A litter box should be placed in a corner of the cage with some of the ferret's feces in it. Place the litter box away from where the ferret sleeps and eats. The ferret should soon learn to use their litter box. Ferrets will usually relieve themselves shortly after waking up, so some time should be allowed for this before letting it out of the cage. If the ferret is allowed to run free in a room, the ferret will usually go to a corner to relieve itself. When the ferret is playing outside the cage, it is a good idea to have a litter box in that room. If you see your ferret backing up into a corner, pick the animal up and place it in the litter box. Ferrets are not as dependable as cats and accidents will happen. A litter box placed in the corner should have high sides towards the walls and low sides in the front. The high sides will prevent accidents occurring against the wall, and the low sides allow the animal to enter the box.

Ferrets allowed to run free should be fitted with a collar and a tiny bell. *See Figure 16–14.* If they should get outside, they will be easier to locate, and

Figure 16–14 Ferrets allowed to run free should be fitted with a collar and a tiny bell. *Photo by Isabelle Francais.*

Figure 16–15 Here is a ferret being exercised on a leash. *Photo by Isabelle Francais.*

the bell will signal to others that is a pet and not a wild animal. Ferrets can also be taken outside for exercise on a harness and leash. The owner can be sure the pet is not going to wander away. *See Figure 16–15.*

DISEASES AND AILMENTS

Ferrets are susceptible to canine distemper. This highly contagious viral disease is fatal to ferrets; vaccination is important. Young **kits** will receive immunity from their mother's colostrum if she was properly vaccinated. This immunity will last about 9 weeks. A series of vaccinations to the young kits should then be given at 2-week intervals; a booster should then be given annually. Discharges from the eyes and nose, difficulty breathing, and diarrhea are early signs of the disease. The feet of the animal may become swollen and the pads of the feet thick and dry. Once an animal has canine distemper, there is very little hope of saving it.

Rabies among ferrets is rare because the animals are usually confined to cages; however, there is a risk for a ferret escaping to the outdoors and becoming exposed to a rabid animal. An owner may choose to have the pet vaccinated as a precaution. Some states require ferrets to be vaccinated for rabies.

"*Helicobacter mustelae* bacteria are found in the stomach and in the small intestines of most, if not all, ferrets after weaning. It can cause chronic, persistent inflammation of the stomach (common in ferrets over 4 years of age) and cause ulcers similar to those found in humans. Gastric lymphoma, a stomach cancer, may occur in chronic cases. Signs include lack of appetite, vomiting, teeth clenching or grinding, diarrhea, stools stained black by blood, increased salivation, and abdominal pain. Lethargy, weight loss, and dehydration can also occur. An accurate diagnosis requires a biopsy and tests on the tissue collected, but often a suspected diagnosis can be made based on the history and the signs of the ferret. Treatment is with multiple antibiotics and usually lasts for 21 days.

"*Lawsonia intracellularis* bacteria can cause a bowel disease characterized by excessive growth of intestinal tissue, especially in younger ferrets. Acute cases are often associated with stress. Signs include diarrhea, weight loss, and rectal prolapse, in which the rectum protrudes out of the anus and can be damaged or prevent defecation. Treatment is with antibiotics for 14 to 21 days."[2]

Ferrets are very susceptible to botulism caused by the toxin produced by the bacterium *Clostridium botulinum*. Clinical signs include breathing difficulty and paralysis; however, death may occur without clinical signs. The disease is usually seen in large colony operations and is caused by contaminated food.

Ear mites of the *Otodectes cynotis* variety commonly affect dogs and cats and can also infect ferrets. Common signs of infection are discomfort, scratching, head shaking, and a buildup of debris in the external ear canal. A veterinarian can determine if mites are present in the ear canal and prescribe treatment. *See Figure 16–16.*

The mite *Sarcoptes scabei* may infect ferrets, causing hair loss and skin lesions commonly referred to as mange. Animals with mange will scratch the infected area in an effort to relieve the itching. A veterinarian can take a skin scraping and identify the mites under a microscope and then recommend treatment. *See Figure 16–17.*

Figure 16–16 The ears of a ferret can be cleaned with cotton-tipped applicators and peroxide. Dip the cotton tip into the peroxide and squeeze out any excess. Carefully clean out any wax or dirt from the outer ear. Make sure the ear is dry. *Photo by Isabelle Francais.*

Figure 16–17 Proper grooming and cleaning can help keep a ferret healthy. Ferrets may need to be bathed occasionally. Wet the ferret thoroughly, apply a ferret shampoo or cat shampoo, rinse the ferret to make sure no soap remains in the fur, and then dry the ferret completely and place it in a warm area. *Photo by Isabelle Francais.*

Figure 16–18 The nails of a ferret should be kept trimmed. Clippers used to cut the nails of cats can be used, or regular clippers used by humans will work. Be careful not to cut into the blood vein that runs down the inside of the nail. *Photo by Isabelle Francais.*

Fleas may infect ferrets. This is especially true in households that also have cats and dogs. Large, untreated flea-infested areas can cause a decrease in the red blood cells and cause weakness in the animal. A veterinarian can diagnose and prescribe treatment for fleas.

Physical injuries are common with ferrets. Doors, rocking chairs, and refrigerator and freezer motors may be hazardous to these small, agile animals. Houses where ferrets are allowed to run loose should be "ferret-proofed" so that physical injuries can be eliminated. Their nails should be kept trimmed to avoid injury. *See Figure 16–18.*

REPRODUCTION

Female ferrets come into heat for the first time at about 6 to 12 months of age. Under natural conditions, ferrets are seasonal breeders; the female comes in heat the spring following birth. The males are also sexually inactive during the winter months because of a decline in the production of testosterone. This yearly cycle of seasonal activity is regulated by the length of the day. As spring days lengthen, sexual activity starts to increase. If mating does take place before the male is seasonally ready, the sperm that is produced will usually be sterile, and pregnancy will not occur. Ferrets kept indoors under artificial light may come in season at anytime during the year. *See Figures 16–19 and 16–20.*

When the female comes in heat, the vulva will show signs of swelling. A thin, watery fluid is secreted to cause wetting to the area of the female; there is an obvious increase in odor, even if the female has been descented. A female ferret in heat usually shows very little change in behavior.

As spring approaches, there is an increase in the level of testosterone production. This increase in testosterone production causes the male to become sexually active. As the level of testosterone increases further, a production of fertile sperm occurs.

Figure 16–19 The genital area of a female ferret.

Figure 16–20 The genital area of a male ferret.

The female ferret should be put with the male for breeding when she comes in heat. When the female is first put into the male's cage, she may seem disinterested and go about exploring; this is normal behavior. The male will become very aggressive and grab the female by the nape of the neck with his teeth. It is not uncommon for the male to throw the female back and forth and drag her around the cage. Fur may be pulled from the neck of the female, and the skin may be broken, causing bleeding. When he has a firm hold, the female will become limp and passive and mating can then occur.

After successful mating, the two may remain locked together for up to 3 hours. After they separate, the female may be removed from the cage, although it is not necessary because the two will get along together. If they are left together, additional mating will probably occur during the next few hours. After mating, **ovulation** occurs. If the male was fertile, pregnancy will result.

The pregnant female can go about normal activity until the young are due. The average gestation period is approximately 6 weeks. The female should be allowed to eat as much as she wants; however, she should not be allowed to put on excess fat. Your veterinarian can advise you on proper diet and whether a supplement is needed for your pregnant female. Her activity may decline somewhat, and she will be probably sleep more often.

If the male was left in with the female, he should be taken from the cage about 2 weeks before the arrival of the litter. A bed or nest box should be provided for the female to deliver her young. A clean towel, an old shirt, or a piece of old sheet can be added for bedding.

Female ferrets usually have no problems giving birth and probably will not need any assistance. Sometimes, first-time mothers may need some assistance because they may be nervous and not know exactly what is happening.

Most mothers are very protective of their young, and they should not be handled unless absolutely necessary. As each young kit is born, the sac that surrounds the newborn is broken open by the mother and cleaned away from the nose of the young kit so that it can breathe; if this is not done, the animal will suffocate. Some assistance may be necessary at this point. A clean, damp washcloth can be used to gently wipe the membrane from the young kit's face.

Occasionally, a first-time mother will have nothing to do with the young; she may even eat each new kit as it is born. Unless another female is available to foster the kits, the litter will be lost. Second litters will usually be well cared for, however.

If everything is going well, the newborns will arrive at intervals of 5 to 10 minutes. If the female is nervous or upset it may take her longer to deliver.

The average litter has 6 to 8 kits but may vary from 1 to 14. The young are pink and covered with an almost invisible, thin, white fuzz. They are born with their eyes closed and ears sealed shut. Each kit weighs about ½ to ¾ of an ounce and is about 1½ inches long, with a tail about ½ inch

Figure 16–21 One can start handling young kits at two weeks of age. Handling at this early age is important so that the young kit can become familiar with humans. *Photo be Isabelle Francais.*

long. The young will cry and be very noisy at first but will quiet down after a few days.

After delivery, they should be left alone so that the female remains calm. The progress of the litter can be checked every 3 or 4 days during the next 2 weeks. Always remember to approach the female and her litter slowly and with caution.

The young litter will grow very fast, and with the approval of the female, the owner may start to handle the young kits at 2 weeks of age. *See Figures 16–21* and *16–22.* Handling at an early age is important so that the young kits become familiar with humans, even though their eyes are not open yet. At 4 weeks of age, the young start to crawl around the nest and may even venture out. Their eyes should start to open at this time. The young can start to receive dry kitten or ferret food soaked with water. At 6 weeks, the young kits can be weaned from their mother.

Figure 16–22 A female with her litter. *Photo by Isabelle Francais.*

When a female ferret comes in heat, she remains in heat if she is not bred. The breeding cycle is controlled by induced ovulation, meaning the eggs are not released into the oviduct for fertilization until mating has taken place. The continued production of **estrogen** during the heat period can lead to **aplastic anemia.** The seriousness of this ailment is the female's loss of weight and the deterioration of her general condition. Signs of aplastic anemia during the continued heat period include listlessness, depression, and apathy along with pale mucous membranes, dehydration, and hind-end weakness. If the animal is not bred and she remains in heat, she could die.

One should consult a veterinarian about the proper course of action. The female will need to be bred, spayed, or brought out of heat artificially.

One method of ending heat in a jill is by administering hormones. Ovulation induced by hormones may result in a false pregnancy. During a false pregnancy, the abdomen and mammary glands of the female may enlarge, thus showing the signs of a true pregnancy.

SUMMARY

The domestic ferret belongs to the family Mustelidae and the genus *Mustela,* which also includes weasels, ermines, mink, and polecats. All members of this genus have long, limber, slender bodies. The tail is usually about half the length of the head and body. They have short legs and small, rounded ears. They have long, oval-shaped heads and pointed snouts; the eyes are bright and clear.

The domestic ferret *(M. pulorius furo)* is believed to be a descendant of the European polecat *(M. pulorius);* the two are very similar in appearance.

Several colors are now being bred: sable (the most common), red-eyed white, silver mitt, sterling silver, white-footed sable, butterscotch, white-footed butterscotch, and the rare cinnamon.

The type of housing used depends on the owner's situation. Two things should be kept in mind: (1) If there is a way to escape a cage, a ferret will find it. All materials should be of good quality, and all latches on the doors should lock securely. (2) Ferrets enjoy watching the activity around them. Cages should be constructed to allow ferrets to see out; they want to be part of the action.

Ferrets are easy to feed; the major part of their diet consists of a high-quality commercial dry kitten or ferret food. It is important that the food contain approximately 35 percent animal protein.

Ferrets are highly susceptible to canine distemper. This contagious viral disease is usually fatal to ferrets. Ferrets should receive regular vaccinations for canine distemper.

Female ferrets come into heat for the first time at about 6 to 12 months of age, usually in the spring following birth. The gestation period is approximately 6 weeks. The average litter has 6 to 8 kits but may vary from 1 to 14.

When a female ferret comes in heat, she remains in heat if not bred. The breeding cycle is controlled by induced ovulation, meaning the eggs are not released into the oviduct for fertilization until mating has taken place. A female that is allowed to remain in heat for a long period may die from uterine and reproductive tract infection and from aplastic anemia.

DISCUSSION QUESTIONS

1. What is the ancestor of the domestic ferret?

2. What are the characteristics of the ferret?

3. What are the eight color variations of ferrets?

4. How should ferrets be handled?

5. What type of housing and equipment are necessary for ferrets?

6. What are ferrets commonly fed?

7. What happens if a female ferret is not bred when she comes into heat? Why does this happen?

8. List the common diseases and ailments of ferrets. What are the causes, symptoms, and treatments?

SUGGESTED ACTIVITIES

1. With the approval of the instructor, invite local breeders, pet shop owners, or class members who have ferrets to bring the animals to the classroom and talk about the characteristics of the ferrets. Class members can observe the animals and their habits.

2. With the approval of the instructor, invite a local veterinarian into the classroom to talk about the diseases and ailments of ferrets.

3. If access to the Internet is available, research and report to the class on the black-footed ferret *(M. nigripes).*

ADDITIONAL RESOURCES

1. Ferret.com
   http://www.ferret.com

2. The American Ferret Association, Inc.
   http://www.ferret.org

3. Ferret Central
   http://www.ferretcentral.org

4. Pet Education.com
   http://www.peteducation.com

5. The Ferret Owner's Manual
   http://www.thechipster.com/fert-man.html

6. Answers.com
   http://www.answers.com/ferrets?cat=health

7. McKimmey, Vickie. 2007. *Ferrets.* New Jersey: T.F.H. Publications.

8. Morton, E. Lynn "Fox". 2000. Ferrets, A Complete Pet Owner's Manual. New York: Barron's Educational Series, Inc.

END NOTES

1. Morton, C., and F. Morton. 1985. *Ferrets, A Complete Pet Owner's Manual.* New York: Barron's Educational Series, Inc.

2. *The Merck/Merial Manual For Pet Health.* Copyright © 2007 by MERCK & CO., INC., Whitehouse Station, NJ, USA. All rights reserved. Used with Permission.

Amphibians

OBJECTIVES

After reading this chapter, one should be able to:

- describe the characteristics of amphibians.
- compare the three orders of amphibians.
- describe and construct an aquatic, semiaquatic, and land habitat aquarium.
- describe the methods used in handling and feeding amphibians.

TERMS TO KNOW

amphibians	metamorphosis	osmosis
amphiumas	newts	sirens
cloaca	olm	spermatophore

HISTORY

E vidence indicates that the lobe-finned fish were ancestors of amphibians. Lobe-finned fish were abundant during the Devonian period, 345 to 405 million years ago. This period of time was believed to be a period of changing weather conditions during which the weather varied from periods of rain and flood conditions to dry conditions. When ponds and flooded areas dried up, lobe-finned fish were forced to move to another area with water. Their strong lobe-fins allowed them to do this. These fish also possessed a type of lung developed as an outgrowth of the pharynx. Over time, the lobe-fins developed into legs, and these fish were able to utilize atmospheric oxygen. The skull and tooth structure of lobe-finned fish is very similar to those of the earliest known amphibians of the late Devonian period.

CHARACTERISTICS

Amphibians live the larval part of their lives in water and their adult lives partially or completely on land. They are cold-blooded and do not have scales.

Approximately 4,000 species of amphibians exist, belonging in three different orders: Gymnophiona, which is made up of worm-like amphibians called caecilians; Caudata, which is made up of newts and salamanders; and Salientia, which is the frogs and toads.

Amphibians have thin, moist skin that serves a variety of purposes. They can breathe through their skin by the process of **osmosis.** During osmosis, oxygen is absorbed through the skin from the surrounding air into the bloodstream. To keep their thin skin from drying out, amphibians must live in moist environments. Amphibians also absorb water through their skin and lose water from the skin by evaporation. Amphibians that live in drier areas have thicker skin, which enables them to conserve moisture.

The skin of most amphibians is covered with fluid-secreting glands that produce a slimy mucus. This mucus serves several purposes. It helps conserve moisture, prevents too much water from being absorbed into the body when the animal is in the water, and makes the animal slippery, which aids in its defense. Some amphibians have toxin-producing glands. The toxic substance produced by these glands serves as a means of defense against predators; some species of frogs and toads produce a toxin that can be deadly to predators and humans.

The tongues of amphibians may vary considerably. Some do not have tongues, whereas others have very long tongues with sticky tips that they can stick out very quickly to capture small insects. Amphibians do not have teeth; they crush their prey with their jaws and swallow it whole. Amphibians are capable of consuming large numbers of insects and prey; they feed primarily on spiders, insects, and insect larvae. Larger species can also feed on mice and small rodents.

CLASSIFICATION—NEWTS AND SALAMANDERS

Kingdom — Animalia
Phylum — Chordata
Class — Amphibia
Order — Caudata (Newts and Salamanders)
Family — 9 different families
Genus — 60 genus groups
Species — 358 species

MAJOR FAMILIES—NEWTS AND SALAMANDERS

The Family Hynobiidae

The family Hynobiidae is referred to as the Asiatic salamanders because they are found in Northern and Eastern Asia; this family is made up of nine genus groups consisting of 33 different species. The characteristics of this family vary considerably; they range in size from about 4 to 8 inches.

The Family Cryptobranchidae

The family Cryptobranchidae is referred to as giant salamanders. The family is made up of two genus groups and consists of three different species.

The Hellbender

The Hellbender, *Cryptobranchus alleganiensis,* which is native to the Central and Eastern United States, may reach a length of 1½ feet. The Hellbender lives in fast-moving mountain streams. The Japanese giant salamander, *Megalobatrachus japonicus,* may reach a length of 5 feet and weigh more than 200 pounds. *See Figure 17–1.*

Figure 17–1 Hellbender.

The Family Ambystomatidae

The family Ambystomatidae, or mole salamander, is made up of four genus groups and consists of 35 different species. Mole salamanders burrow in decaying, damp soil, where they live most of their lives. They are small, although some species may reach a length of 10 inches. They are more commonly seen during the breeding season, when they move to water to mate and lay eggs. Mole salamanders have thick bodies and smooth skin with bright, colorful markings; they are found in North America.

The Spotted Salamander

The spotted salamander, *Ambystoma maculatum,* is a member of the family Ambystomatidae, which are commonly called mole salamanders. The spotted salamander is blue-black with yellow spots. *See Figure 17–2.*

Figure 17–2 Spotted salamander.

The Tiger Salamander

The tiger salamander, *Ambystoma tigrinum,* is another member of the family Ambystomatidae. It is an olive-green color with yellow spots. *See Figure 17–3.*

Figure 17–3 Tiger salamander.

The Family Dicamptodontidae

The family Dicamptodontidae is more commonly referred to as Pacific mole salamanders. This family is very similar to mole salamanders but is found along the Pacific coast of North America.

The Family Salamandridae

The family Salamandridae is made up of the more commonly known fire salamanders, brook salamanders, and **newts.** The family is made up of 14 genus groups and consists of 53 different species. Fire salamanders are brightly colored and are found in Europe. They have black and yellow or black and orange coloration. They have glands in their skin that produce an irritating fluid. This fluid is especially irritating to the mucous membranes of the mouth; many predators avoid this animal. Fire salamanders return to

Figure 17–4 Fire salamander.

Figure 17–5 California newt.

Figure 17–6 Eastern newt.

Figure 17–7 Congo eel.

Figure 17–8 Mud puppy or water dog.

the water only to breed. Three species of brook salamanders are found in isolated, highland areas of Europe.

The Fire Salamander

The fire salamander, *Salamandra salamandra*, belongs to the family Salamandridae. It is black with yellow spots. *See Figure 17–4.*

The California Newt and the Eastern Newt

Two species of newts are found in the United States: the Californian newt, *Taricha torosa*, and the Eastern newt, *Notophthalmus viridescens*.

Newts spend a good part of their time in water; they need the moisture to keep their thin, soft skin from drying out. They are able to replace lost or damaged parts of their bodies by regeneration and periodically shed their skin. *See Figures 17–5* and *17–6.*

The Family Amphiumidae

The family Amphiumidae is made up of one genus group and consists of three different species of **amphiumas.**

The Congo Eel

Commonly called Congo eels, amphiumas have long, round bodies with tapered heads. They have two pairs of tiny, weak limbs with one, two, or three toes on each limb. They range from 18 to 46 inches in length; all three species are found in muddy waters and swamps of the Southern United States. Amphiumas usually spend the day submerged in the mud or among stands of water weeds and hunt for food at night. *See Figure 17–7.*

The Family Proteidae

The family Proteidae is made up of water dogs, mud puppies, and the **olm.** There are two genus groups and six different species.

The Mud Puppy

Water dogs and mud puppies are very similar in appearance; they may reach a length of up to 18 inches. They have rather broad, flattened heads, with bodies that taper down to a short, flattened tail. They breathe by means of external, bushy gills found on both sides of the head. The mud puppy, *Necturus maculosus*, is fairly common in the Eastern and Central United States. Four species of water dog are found in the Eastern and Southern United States. The olm, *Proteus anguinus*, is a dull, white animal with red, feathery, external gills. It is about a foot long and is found in underground lakes, rivers, and streams in Austria, Italy, and Yugoslavia. The olm is basically blind. *See Figure 17–8.*

The Family Plethodontidae

The family Plethodontidae is commonly referred to as lungless salamanders. There are 24 genus groups consisting of 209 species; this is the largest group of salamanders. They are mostly found in caves, under rocks, and in other damp habitats throughout the United States.

The Red Salamander

This is a lungless salamander. Lungless salamanders breathe through their skin and throats; the oxygen is absorbed directly into blood vessels in the skin and throat. *See Figure 17–9.*

Figure 17–9   Red salamander.

The Family Sirenidae

The family Sirenidae is made up of two genus groups and three species of **sirens.**

The Siren

Sirens are eel-like amphibians that live in freshwater areas of the Southern and Central United States. They have feathery, external gill structures. Sirens have small, weak front legs and no rear legs. The greater siren, *Siren lacertina,* may grow to a length of 29 inches. The lesser siren, *Siren intermedia,* grows anywhere from 7 to 27 inches in length. The third member of the family, the dwarf siren, *Pseudobranchus striatus,* reaches a length of only 4 to 8 inches. *See Figure 17–10.*

Figure 17–10   The greater Siren, *Siren lacertina.*

CLASSIFICATION—FROGS AND TOADS

Kingdom — Animalia
Phylum — Chordata
Class — Amphibia
Order — Salientia (Frogs and Toads)
Family — 20 different families
Genus — 234 genus groups
Species — 3494 different species

MAJOR FAMILIES—FROGS AND TOADS

The Family Leiopelmatidae

The family Leiopelmatidae is made up of two genus groups and three species. Two species are commonly referred to as New Zealand frogs because they are found in New Zealand. The third species is the 2-inch-long, tailed frog *Ascaphus truei* that is found in the Western United States. These species of frogs do not have tails but have retained two tail-moving muscles from their tadpole days. The muscles extend about ¼ inch. This extension allows the male to fertilize the female internally; this is the only species to use this method of fertilization.

The Family Discoglossidae

The family Discoglossidae is made up of five genus groups and 14 different species. They are commonly referred to as disk-tongued toads. They are found in Europe, Western Asia, the Philippines, Borneo, and other areas of the Far East. These toads do not extend their tongues to catch their prey but catch them with their mouths.

The Family Pipidae

The family Pipidae is made up of four genus groups and 26 different species. In this family are the clawed and Surinam toads that are found in Africa and parts of Central and South America. These toads do not have tongues but use their front toes to dig for and pick up food. Some references refer to the clawed and Surinam toads as "frogs" because they are aquatic animals. This species has fairly flattened bodies, with small eyes on the top of their heads that aid them in looking up from ponds or streams. They also have large, webbed rear feet. The Surinam toad of Mexico and Central America is unique in that the female carries her eggs on her back; they are placed there by the male during fertilization.

The African Clawed Frog

The African clawed frog, *Xenopus laevis,* is an aquatic frog that seldom leaves the water. Its forelimbs are very weak and do not support the frog's body while on land. *See Figure 17–11.*

Figure 17–11   African clawed frog. *Photo by Isabelle Francais.*

The Family Rhinophrynidae

The family Rhinophrynidae is made up of only one genus group and consists of one species. The burrowing toad, *Rhinophrynus dorsalis,* is found in Central America. The species has spade-like structures on the first toe of its hind feet that enable it to dig tunnels. It is about 3 to 4 inches long and feeds on termites and other soil insects.

The Family Pelobatidae

The family Pelobatidae is made up of 10 genus groups and consists of 88 different species. In this family are the spade-foot frogs, horned toads, and parsley frogs that are found in North America, Europe, Indonesia, the Philippines, and parts of Asia.

The Spade-Foot Frog

The spade-foot frogs have developed spade-shaped structures on their hind feet that enable them to burrow into the soil. They are nocturnal and usually spend the day in their damp tunnels. The horned toads get their name from the horn structures that are found on their head. Parsley frogs closely resemble other frogs in their general shape. Their name is derived from the spotted, green patterns on their backs. They live among rocks or dig tunnels. Their hind feet are not as well developed for digging as the spade-foot or horned toads. *See Figure 17–12.*

Figure 17–12   Spade-foot frog. *Photo by Isabelle Francais.*

The Malaysian Horned Frog

The Malaysian horned frog, *Megophrys monticola,* is native to Southeast Asia, Indonesia, and the Philippines. This is a ground-dwelling frog that lives among the dead leaves on the forest floor. *See Figure 17–13.*

Figure 17–13   Malaysian horned frog. *Photo by Isabelle Francais.*

The Family Ranidae

The family Ranidae are the true frogs. There are 40 genus groups and 611 different species. True frogs can be found in almost all parts of the world and range in size from ½ inch up to 14 inches in length. Frogs have slender bodies and long, muscular legs that enable them to be excellent jumpers and swimmers; a few species are adapted to tree life.

The Green Frog

Included in the family Ranidae is the American bullfrog, *Rana catesbeiana,* and the green frog, *R. clamitans,* both of which are common in the United States. *See Figure 17–14.*

Figure 17–14   American bull frog. *Photo by Isabelle Francais.*

Also belonging to this family is the goliath frog, *Gigantorana goliath,* the largest member. It may reach a length of 16 inches from snout to rear, and with its legs stretched out, it may exceed 30 inches in length.

The Family Dendrobatidae

The family Dendrobatidae is made up of four genus groups and consists of 116 different species. These frogs are commonly referred to as poison-arrow frogs. This family is found in the tropical areas of South and Central America. Poison-arrow frogs are brightly colored and are noted for the extremely potent poisons that are secreted from their skins. The bright colors serve as a warning to would-be predators. The poison from one species, the Kokoi

Poison-Arrow Frog, *Phyllobates bicolor,* found in Colombia is used by the local Choco Indians to tip the points of their arrows. The Kokoi Poison-Arrow Frog is bright red with black spots. They are very small in size. The male poison-arrow frog usually carries the fertilized eggs on his back until they hatch; then he carries the tadpoles to the water, where they continue to develop.

The Family Rhacophoridae

The family Rhacophoridae is made up of 10 genus groups and 184 different species. This family is commonly referred to as Old-World tree frogs, which have adapted themselves to life in the trees. They have flattened bodies and disks on their feet that aid them in clinging to tree limbs. About one-third of the species are referred to as flying frogs because they are able to stretch their bodies and limbs and glide from tree to tree. The female tree frog lays her eggs in moist, foamy masses on branches that overhang ponds and streams. The male mounts the female from the rear and fertilizes the eggs as the female lays them. The foamy mass then hardens. After the tadpoles have developed, the foamy mass breaks open and the liquid within the foamy mass runs down to the water below, carrying the tadpoles with it. The Old-World tree frogs are found in tropical areas of Africa and Asia.

The Family Hyperoliidae

The family Hyperoliidae is made up of 23 genus groups and consists of 292 different species. This family is referred to as sedge frogs and bush frogs. They are found in habitats throughout Africa. They resemble tree frogs in that they have sucker disks on the toes of their feet. Most species, however, are found around swamps and ponds. The female deposits eggs in a jelly mass on reeds or limbs above the water. The developed larval stage drops to the water, where they continue to grow. These frogs can change their color as a result of temperature changes, humidity changes, or nervous stimuli.

The Family Sooglossidae

The family Sooglossidae is made up of two genus groups and three different species. These species are found on the islands of Mahe and Silhouette, which are part of the Seychelles Islands. The small, 1-inch-long Seychelles frog, *Sooglossus seychellensis,* is the best known of this group. Because there are no still-water breeding ponds available, the female lays her eggs in gelatinous masses on the ground. When the eggs hatch, the larval stage of the tadpoles make their way to the back of the male, where they are attached by means of a sticky substance secreted by the male; here they develop until they are able to fend for themselves.

The Family Microhylidae

The family Microhylidae is made up of 61 genus groups and consists of 281 species; they are referred to as narrow-mouthed frogs. These frogs have small, tapered heads, narrow mouths, and pointed snouts. They dig and burrow into the soil. They are found in Indonesia, Northern Australia, South America, Africa, southern parts of North America, and Asia.

The Family Pseudidae

The family Pseudidae is made up of two genus groups and consists of four different species. This family is commonly referred to as shrinking frogs because they grow smaller as they mature. They live in shallow lakes and seldom leave the water. One species lives in shallow ponds that evaporate during the summer. They bury themselves in the mud and lie dormant until the next rain.

The Family Bufonidae

The family Bufonidae is made up of 25 genus groups and 339 different species. This family represents the true toads, which are found in most parts of the world. Toads have short, thick bodies and short legs. Although toads can jump, their legs are too short to allow them to jump the distances that frogs are capable of jumping. The toad's skin is dry and has a rough appearance; true toads have skin glands that produce a poison that can be irritating to the mucous membranes of predatory animals. Toads eat a wide variety of foods, but insects make up the main part of their diet. During the winter months, toads hibernate under rocks, under logs, or in underground holes. They emerge in the spring and head for lakes and ponds to breed. Toads do not make noises like frogs and use their sense of smell to locate the opposite sex for mating.

The American Toad

The American toad, *Bufo terrestria,* is a common toad of the eastern parts of the United States. *See Figure 17–15.*

Figure 17–15   American toad. *Photo by Isabelle Francais.*

The Marine Toad

The marine or cane toad, *Bufo marinus,* is native to extreme Southern Texas and to Northern South America. It is a large toad that grows to 9 inches. The marine toad is poisonous to dogs, cats, and other possible predators. *See Figure 17–16.*

Figure 17–16   Marine or cane toad. *Photo by Isabelle Francais.*

The Family Brachycephalidae

The family Brachycephalidae is made up of only one genus group and consists of two different species. These are commonly called gold frogs and are found in the southern parts of Brazil. The frogs are about 1 inch long and are brilliant orange in color. A feature of the frog is a bony, cross-shaped structure on the back of the animal.

The Family Hylidae

The family Hylidae is made up of 37 genus groups and 637 different species. These are the true tree frogs and resemble the Old-World tree frogs in the flattened shape of their bodies and the adhesive disks on their toes. These tree frogs live in South and Central America; species can also be found in North America, Central Asia, and Europe. *See Figure 17–17.*

Figure 17–17 Tree frogs have long, slender limbs and adhesive disks on the ends of their toes. *Photo by Isabelle Francais.*

The Red-Eyed Tree Frog

The red-eyed tree frog, *Agalychnis callidryas,* is an arboreal frog of Central America. Its eyes are red with vertical pupils. Its primary color is green. The color markings on its sides are yellow and blue, and the toes are red. It also has opposable thumbs, which allow it to take a strong grip on leaves and branches. *See Figure 17–18.*

Figure 17–18 Red-eyed tree frog. *Photo by Isabelle Francais.*

The White's Tree Frog

The White's tree frog, *Litoria caerulea,* is native to Australia. This small, bloated-appearing, light-green frog is commonly found around houses and breeds in wet grassland areas. *See Figure 17–19.*

Figure 17–19 White's tree frog. *Photo by Isabelle Francais.*

The Green Tree Frog

The green tree frog, *Hyla cinerea,* is found in the Mississippi Valley and Southeastern United States. It is yellow-green with a yellow stripe running from the lower jaw back along its side. *See Figure 17–20.*

Figure 17–20 Green tree frog. *Photo by Isabelle Francais.*

The Family Leptodactylidae

The family Leptodactylidae is made up of 51 genus groups and consists of 722 different species. This is the largest family in the order Anura. Because of the great number of species, there is a large variety of colors, characteristics, and behavior within this family. Most of these frogs are found in tropical areas of South and Central America. One of the most common members of this family is the South American bullfrog, *Leptodactylus pentadactylus.*

The Ornate Horned Frog

The ornate horned frog, *Ceratophrys ornata,* belongs to the family Leptodactylidae. It is a carnivorous frog that lives among the leaves and debris on the rainforest floors of South America. Its primary color is a dark green with brown markings above and yellow below. The legs are yellow with green and brown markings. *See Figure 17–21.*

Figure 17–21   Ornate horned frog. *Photo by Isabelle Francais.*

The Colombian Horned Frog

The Colombian horned frog, *Ceratophrys calcarata,* also belongs to the family Leptodactylidae. It is primarily light brown with dark-brown markings. *See Figure 17–22.*

Figure 17–22   Colombian horned frog. *Photo by Isabelle Francais.*

The Family Rhinodermatidae

The family Rhinodermatidae is made up of one genus group and consists of two different species. This group is commonly referred to as mouth-brooders. They are unique because of their parental rearing habits. The female deposits her eggs on the ground; when the larvae start to move within the egg, the male gathers them up in his mouth and holds them there as they develop. The male retains the eggs in his mouth for about 3 weeks; during that time, the tadpoles hatch. When they are able to survive on their own, the male spits them out.

The Family Heleophrynidae

The family Heleophrynidae is made up of one genus group and consists of four different species. This family of frogs is commonly called ghost frogs. These small frogs live in fast-flowing mountain streams in Africa. The tadpoles have a large suction disk around the mouth that is used to help them cling to rocks. The skin on the belly of the adults is white and so thin that their internal organs and muscles can be seen.

The Family Myobatrachidae

The family Myobatrachidae is made up of 20 genus groups and 100 different species. This family of frogs is found in Australia, New Zealand, and Tasmania. The family varies a great deal in appearance and habits, but their feet are usually less than half webbed, and their toes may have small disks. Included in this family are the pseudophryne frogs, which are among the smallest frogs.

The Family Centrolenidae

The family Centrolenidae is made up of two genus groups and consists of 64 different species. These are small, bright-green frogs that have transparent skin on their bellies; their internal organs and muscles are visible. Because of their visible internal structure, these tree-dwelling frogs of Central America, Northern South America, and Southern Brazil are commonly referred to as glass frogs.

CLASSIFICATION—CAECILIANS

Kingdom	—	Animalia
Phylum	—	Chordata
Class	—	Amphibia
Order	—	Gymnophiona (Caecilians)
Family	—	Caecilliidae
Genus	—	16 genus groups
Species	—	50 different species

MAJOR FAMILY—CAECILIANS

The family Caecilliidae is made up of 16 genus groups and consists of about 50 different species. The caecilians are seldom seen, and not very much is known about their habits. References differ as to the actual number of families, genus groups, and species that make up the order. Caecilians are worm-like in appearance; they do not have limbs and are usually without sight. They burrow into the decayed material on top of the soil and into the soil. They vary in length from 3 to 59 inches and have tentacle-like structures on their heads that enable them to find their way in their tunnels. Caecilians are found in tropical areas of Southern Asia, Africa, and South Africa. *See Figure 17–23.*

Figure 17–23 The family Caecilliidae consists of about 50 species of worm-like amphibians called caecilians.

HOUSING AND EQUIPMENT

In discussing housing and equipment, one must realize the vast numbers of different species and the large variation in environments. The information in this text is intended as a general guide only. One should duplicate as closely as possible the animal's natural environment.

The best container for most types of amphibians is the common rectangular aquarium. Consideration for the size of the animal is very important when deciding what size aquarium to use.

Three specific types of habitats need to be considered for amphibians: (1) aquatic habitat for tadpoles and completely aquatic species, (2) semi-aquatic habitat for species that live partially in the water and partially on land, and (3) woodland habitat for species that live almost entirely on land.

When preparing an aquarium for tadpoles or aquatic species, it can be set up like a true aquarium with filtration and aeration systems; no land area is needed. One should prepare the aquarium by placing a 1-inch layer of rock or gravel in the bottom; then a 3-inch layer of sand can be added. The sand should be sloped gradually so that one end is shallower. Water, to a depth of 1 to 2 inches on the shallow end, should be added; tadpoles will use the shallow end to rest, and this is where they are fed. Other species may prefer a different habitat, so it is important to research the species that is being put into the aquarium. Depending on the species, a source of heat may be needed.

Dechlorinated water should be used for the aquarium. Chlorinated water can be treated with decholorinating tablets or allowed to sit in open containers for 2 days to allow the chlorine to evaporate.

Plants, rocks, and tree branches should be added; they provide hiding places. As tadpoles develop legs, they will crawl up onto the rocks and branches.

If the aquarium does not have a filtration and aeration system, the water will need to be changed every 2 or 3 days; a buildup of excrement and the loss of oxygen will kill aquatic animals.

Semiaquatic aquariums are set up for species that spend part of their lives in water and part on land. As tadpoles develop into toads and frogs, they require semiaquatic habitats.

To set up this type of habitat, an aquarium needs to be divided in half. A piece of glass or Plexiglas can be cemented to the bottom to divide it into two watertight halves; the divider should be about one-third the height of the aquarium. This again depends on the species.

To set up the land habitat half, a 1-inch layer of small rocks or gravel should be placed in the bottom of the aquarium. A 1-inch layer of charcoal is placed over this; then, a 2- to 3-inch layer of sand or sandy loam is placed over the charcoal. On top of this, a layer of topsoil is added so that it is level with the divider. Finally, leaves, branches, bark, and some plants are added to give the aquarium a natural look. The water in the semiaquatic habitat should be a dechlorinated. Water is added to the aquatic side of the aquarium until it is level with the land habitat. This allows the animals to move freely between the water habitat and the land habitat. The water should be changed every 2 or 3 days if a filtration and aeration system is not used. Plants and soil should be sprinkled daily with dechlorinated water so that the habitat does not dry out, and a desirable, damp environment is maintained.

The aquarium needs to have a cover to keep the animals from escaping; a piece of window screen or hardware cloth can be used. Temperature of the aquarium should be maintained at 60 to 80°F, depending on the species. A small light bulb can be added if heat is needed. Salamanders prefer temperatures around 60 to 70°F, whereas frogs and toads should be kept in an environment around 70 to 80°F. All amphibians require full-spectrum lighting. Full-spectrum bulbs are available from pet stores. These bulbs can emit full spectrum light for only a limited time and should be replaced every 4 to 6 months, even if they have not burned out.

In preparing a woodland aquarium, one should proceed just as when preparing the land half of the semiaquatic aquarium. A shallow bowl of water should be provided; amphibians absorb the water through their skin. Make sure the bowl is level with the soil so that the animals can crawl in and out freely. Remember that plants and soil should also be kept moist. A small light bulb can be added if the temperature drops too low.

During the winter months, most amphibians from northern climates will begin to burrow into the soil to hibernate. They will begin to hibernate when the temperatures reach 40°F or below. Amphibians that are being kept indoors will not need to go through hibernation unless one is interested in breeding them. If one is interested in breeding amphibians, then the aquarium should be placed in a cool place. One must make sure the aquarium is not in an area where it will receive a hard freeze; basements and cellars are ideal locations. Those amphibians that are not adept at digging will hibernate in cracks and crevices in rocks or logs. Some will bury as deep as they can in the leaves and grass. Not all amphibians hibernate in the soil; aquatic amphibians will hibernate at the bottom of ponds, lakes, and rivers.

FEEDING

Most amphibians are insect eaters. Depending on the size and species, they also consume earthworms, small birds, and newborn rodents. Tadpoles and aquatic species feed on vegetation, aquatic plant life, and algae; they also feed on dead animals.

In an aquarium habitat, tadpoles can be fed algae or tadpole powder, which is available in pet stores. Tadpoles can also be fed pieces of boiled lettuce, various floating or sinking flake fish food products, and the fry (hatchlings) of fish. One needs to be careful and not put more feed in the water than the tadpoles can consume or the water will become cloudy and suffocate them.

As the tadpoles change to adults, they need to be fed insects. Bait and pet stores are sources of crickets, mealworms, wax worms, earthworms, and nightcrawlers for tadpoles. Flies, crickets, grasshoppers, and other insects can be collected with a butterfly net along roadways and in open fields. Care should be taken to ensure insects are not collected from areas where insecticides have been used.

Salamanders prefer to be fed earthworms. Crickets, wax worms, and other insects can also be fed to salamanders.

HANDLING

Frogs and toads are best picked up by grasping them around the body just in front of the rear legs. They should be held firmly enough to prevent escape but gently enough to prevent injury. They should not be held by the rear legs because they can injure themselves as they struggle to escape; small frogs and toads can be cupped in the hand. *See Figure 17–24.* It is best for handlers to

Figure 17–24 Frogs and toads can easily be picked up by carefully grasping them around the body. *Photo by Isabelle Francais.*

wear latex or vinyl gloves to avoid transmission of disease or damage to the amphibians' skin. Soap residue from the handler could also be absorbed by the animal if gloves are not worn.

When amphibians are transported, they should be placed in a moist cloth bag to prevent excess moisture loss.

Salamanders have soft, tender skin, and before picking them up, a person should run water over the hands; this prevents injury to the skin of the salamander. One should gently grasp the salamander around the body; its tail should not be grabbed because the tail may be released in one's hand. Some salamanders can release their tail as a defense mechanism; the tail usually grows back.

The skin glands on some frogs and toads produce toxic substances. After handling frogs and toads, one should wash his or her hands and be careful not to touch the face and eyes before washing. Again, it is important to know the species one is handling.

DISEASES AND AILMENTS

Red Leg

Red leg is a disease of frogs caused by the bacterium *Aeromonas hydrophila*, which is found in standing water. It is primarily a disease of frogs kept in laboratory or aquarium situations. Normal, healthy frogs living in a natural environment are usually not affected. The added stress of capture and captivity in aquariums seems to be a contributing factor in causing the disease. Red blotches appear on the skin, especially on the inside of the thighs. Antibiotics given by stomach tube may be effective.

Fungus Infection

Fungus infection may affect newts and salamanders. Their skin can develop abrasions from handling or crawling over rocks; these abrasions may be attacked by fungus. Some of the fungus treatments used for fish are effective.

REPRODUCTION

Male frogs and toads call females to breeding sites with distinctive calls that each species makes. Most species mate in the water; the male grasps the female from the back, and as she releases the eggs, he fertilizes them. The eggs then usually attach to vegetation in the water.

Eggs hatch into the larval form of tadpoles. The newly hatched tadpoles lack legs, eyelids, lungs, and true jaws; they have a long tail with a high fin. Tadpoles breathe through external gills and feed on vegetation.

Tadpoles go through a series of physical changes called **metamorphosis.** During these changes, the tadpole develops rear legs, then front legs; eyelids and jaws also develop. The tail is absorbed, and internal lungs develop. Tadpoles gradually develop into the adult form over a period of a few weeks to 3 years, depending on the species. *See Figure 17–25.*

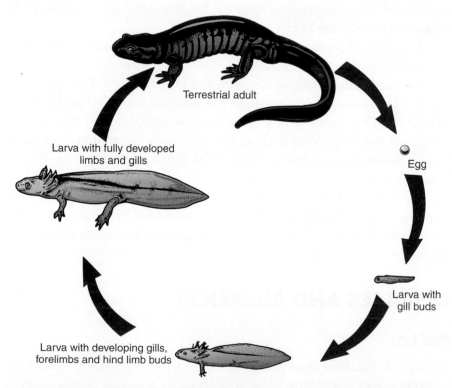

Terrestrial adult

Egg

Larva with fully developed limbs and gills

Larva with gill buds

Larva with developing gills, forelimbs and hind limb buds

Figure 17–25 The life cycle of an amphibious salamander consists of stages referred to as larval stages, during which the small salamanders have external gills. As they develop, some species lose the external gills, whereas other species retain them and remain completely aquatic. Terrestrial salamanders are usually miniature adults and do not have external gills.

Most salamanders and newts breed during the spring; they do not make any noise, so males and females must find each other by smell. Breeding in most cases occurs in the water; in some cases, the male deposits sperm on the eggs as the female releases eggs into the water. In some species, the male deposits a packet of sperm on the bottom of the pond; this packet of sperm is called a **spermatophore**. The female rubs her **cloaca** (the cavity where the intestinal and genitourinary tract empties) on the packet and picks up the sperm; fertilization takes place internally. The eggs are then released into the water. The larval forms that hatch resemble their parents in appearance and have bushy external gills. Those species that move onto land lose their external gills.

SUMMARY

Amphibians live the larval part of their lives in water and their adult lives partially or completely on land.

Approximately 4,000 species of amphibians exist, belonging in three different orders: Gymnophiona, which is made up of worm-like amphibians called caecilians; Caudata, which is made up of newts and salamanders; and Salientia, which is the frogs and toads.

There is great diversity among the species; however, most of them have thin, moist skin and live in moist environments. Dryness and excessive heat are enemies of most amphibians.

The skin of amphibians is covered with fluid-secreting glands that produce a slimy mucus. This mucus serves to conserve moisture and makes the animal slippery, which aids in its defense. Some amphibians also have skin glands that produce toxic substances. Amphibians absorb water and, sometimes, oxygen through the skin. It is important to wear gloves during handling to protect the health of the animal and the handler.

Most amphibians feed on spiders, insects, insect larvae, and other small prey. Larger species can also feed on newborn mice (pinkies) and other small rodents.

There is a wide variation in the habits and environments of amphibians. Three specific types of habitat need to be considered: (1) water or aquatic habitat for tadpoles and completely aquatic species, (2) semiaquatic habitat for species that live partially in water and partially on land, and (3) woodland habitat for species that live almost entirely on land.

If interested in keeping amphibians in captivity, it is important that one duplicates as closely as possible the animal's natural environment, including lighting, temperature, and humidity.

Amphibians return to water to reproduce; most species lay eggs that cling to vegetation in the water. The eggs hatch into the larval form of tadpole; the tadpoles go through a series of changes called metamorphosis. During this metamorphosis, the tadpoles develop into the adult form.

DISCUSSION QUESTIONS

1. What is an amphibian?
2. What are the three different orders of amphibians?
3. What are caecilians?
4. What are the characteristics of a newt?
5. What are the characteristics of a salamander?
6. What are the differences between a frog and a toad?
7. Describe how to set up an aquatic, semiaquatic, and land habitat in an aquarium.

SUGGESTED ACTIVITIES

1. Visit the local pet shop and make a list of the different species of amphibians that are available. Using the local library, compile as much information as you can about the species.
2. Set up aquatic, semiaquatic, and woodland habitats in aquariums. Obtain some amphibians from pet shops or capture them in local ponds to place in the aquariums; observe them and their habits.

ADDITIONAL RESOURCES

1. Amphibian Species Identification Guide
   http://www.npwrc.usgs.gov/narcam/idguide/specieid.htm
2. Living Amphibians
   http://phylogeny.arizona.edu/tree/eukaryotes/animals/chordata/living_amphibians.html
3. Checklist of Florida Amphibians and Reptiles
   http://www.flmnh.ufl.edu/natsci/herpetology/fl-guide/flaherps.htm
4. Florida International University, General Biology 11, Lecture Notes
   http://www.jochemnet.de/fiu/BSC1011/BSC1011_18/index.htm
5. Zoo Lab: A Website for Animal Biology
   http://bioweb.uwlax.edu/zoolab/index.htm
6. University of Michigan Museum of Zoology, Animal Diversity Web
   http://animaldiversity.ummz.umich.edu/site/index.html
7. About.COM: Exotic Pets
   http://exoticpets.about.com
8. Tree of Life Web Projects
   http://tolweb.org/tree/phylogeny.html

CHAPTER 18

Reptiles

TERMS TO KNOW

arboreal	hemipenes	probing
brille	lamellae	scutes
brood	melanin	terrapins
carapace	oviparous	terrarium
casque	ovoviviparous	tympanum
chromatophores	plastron	vivarium
crepuscular	popping	viviparous
dimorphism	preferred optimal temperature zone (POTZ)	
ectotherms		

CHARACTERISTICS

Reptiles are cold-blooded vertebrates that possess lungs and breathe air. They also have a bony skeleton, scales, or horny plates covering the body, and a heart that has two auricles and one ventricle.

There are approximately 6,500 species of reptiles belonging in four different orders: Testudines (called Chelonia in some references), which are made up of turtles, tortoises, and terrapins; Serpentes, which includes snakes, pythons, and boas; Squamata, which includes iguanas and lizards; and Crocodilia, which includes crocodiles, alligators, caimans, and gharials.

Fossil remains of reptiles date back to the Carboniferous period 340 million years ago. Other fossil remains dating back 315 million years are of small, agile, lizard-like animals. The evolution of reptiles continued through the Permian and into the Triassic periods. During the Triassic period, reptiles became the dominant group of terrestrial vertebrates inhabiting the earth; they continued their dominance for about 160 million years. The reptiles reached their greatest number during the Cretaceous period; then many became extinct. Today, only four orders of the fifteen that existed during the Cretaceous period have survived.

Present-day reptiles have three body types. The first have long bodies and clearly defined tails like the crocodiles and lizards; the second have long bodies that taper into tails, like the snakes; and the third have short, thick bodies encased in shells like the turtles and tortoises.

The limbs of crocodiles and lizards are paired and attached to the body at right angles. This enables the animal to lift its body up off the ground when moving. The limbs of crocodiles and alligators are strong and powerful; the limbs of most lizards are weak, with some having very short, stump-like limbs.

Snake-like reptiles do not have limbs, and their movement is a result of undulating movements of the body. The scales on the underside of the body project outward as the muscles are contracted and relaxed; these scales exert pressure on the surface and move the animal forward.

All turtles and tortoises have limbs that enable them to walk on land. These reptiles have modified limbs that enable them to swim as well. Some turtles and tortoises spend most of their time in water, with their limbs primarily adapted for swimming, and move about on land with creeping and crawling motions.

The skin of reptiles has a horny surface layer. In lizards and snakes, this layer forms a hard, continuous covering of scales. This outer layer is shed, allowing for growth. Crocodiles, alligators, and some lizards have bony, dermal scales covered by a horny epidermal layer that must also be shed to allow for body growth.

Turtles and tortoises do not molt their thick epidermal skin; a new epidermal scale is formed beneath the old one as they grow. This allows for growth of the animal. The growth of the animal and the formation of new epidermal scales is a result of the amount of food available. These epidermal scales, or **scutes**, form rings that can be used to estimate the age of the animal; however, this is method of counting the rings may not give an accurate indication of the animal's age because more rings may be formed during a good year.

The scutes are divided into three types: the marginal scutes are located around the outer edge of the shell, the central scutes are located down the center, and the costal scutes are located between the marginal and central scutes. The hard, tough layer of scales prevents moisture from dissipating from the body. In turtles and tortoises, this horny epidermal layer forms an exoskeleton that has two parts: the upper part is called the **carapace** and forms a part of the vertebrate and ribs of the animal; the lower portion is called the **plastron.** *See Figures 18–1* and *18–2.*

Snakes cannot close their eyes; instead of eyelids, they have a transparent layer, or **brille,** that permanently covers the eye. During the shedding process, this layer over the eyes is shed and replaced with a new covering. Nocturnal

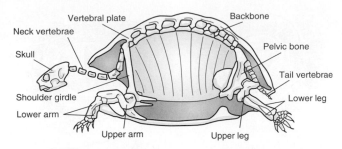

Figure 18-1 The parts of a turtle's skeleton.

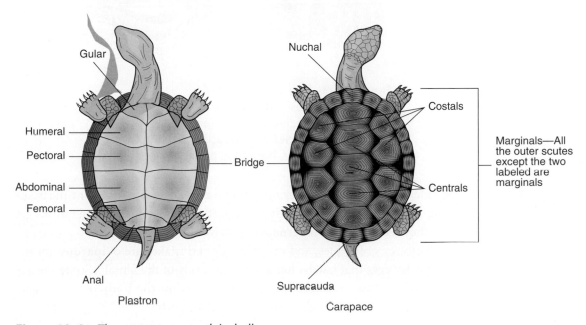

Figure 18-2 The scutes on a turtle's shell.

species have vertical pupils that open very wide in dim light and close to a small slit or a pinhole in bright light; diurnal species have round pupils. Burrowing species have an opaque scale covering the eye for protection and are almost blind. Some arboreal (see Glossary) snakes have horizontal pupils, large protruding eyes, and tapered snouts, which give them binocular vision. Some lizard species also do not have eyelids. *See Figure 18–3.*

Lizards and snakes have teeth that are fused to the jaw bones; some snakes also have teeth fused to the palate bones. In some snakes and in two species of lizards, the teeth are connected to poison glands. Crocodile teeth are set in sockets similar to those of mammals. Turtles and tortoises do not have teeth; their jaws form sharp crushing plates. The front part of the jaws forms a horny beak.

The tongues of reptiles vary considerably. Some have short, fleshy tongues that have little movement, whereas others have tongues that are long, slender, and forked, such as some lizards and all snakes.

The lungs of most lizards are the same size, but in some lizards, the left lung is reduced in size, with the right lung being very long and lying between internal organs. All reptiles have a three-chambered heart, which has two auricles and one ventricle. The three-chambered heart is associated with a greatly reduced metabolic rate. This low metabolic rate means that not much heat is produced within the body; therefore, they must function at the same temperature as their surroundings. Most reptiles are found in tropical areas.

Figure 18–3   The eyes of snakes.
(a) Diurnal species, round pupils.
(b) Nocturnal species, vertical pupils.
(c) Burrowing species, opaque scale covering the eye.
(d) Arboreal species, horizontal pupils.

Outside the tropical areas, they must spend a great amount of their time basking in the sun to obtain the body warmth they need. When temperatures get cool, they seek places to hibernate to survive the cold weather.

Tortoises, crocodiles, and many lizards and snakes are **oviparous,** meaning they lay eggs that hatch after leaving the body of the female. Some species are **ovoviviparous,** meaning the eggs hatch within the body of the female, with the live young emerging from the female's body.

Hibernation

Reptiles that live in temperate zones (area between the tropical zones and polar circles) must hibernate during the winter; winter temperatures are too cold for them to remain active. The length of the hibernation period varies depending on how long the cold weather lasts. In the wild, reptiles seek out dens or tunnels; burrow under rocks, logs, and leaves; or burrow into the soil to find a suitable place to hibernate. If they are to survive, they must be below the frost line. During hibernation, the reptile's body temperature drops to a point where the body's systems barely function. Respiration and heartbeat are greatly reduced, and oxygen is absorbed through the skin. Hibernation is associated with bringing the reptile into breeding condition; temperate-zone reptiles that are not allowed to hibernate do not breed.

Many reptiles must go into hibernation for breeding. The temperature change stimulates ovulation and production of sperm. If one isn't interested in breeding reptiles, then a hibernation period is not necessary.

Plans should be made for temperate-zone reptiles in captivity to go through a period of hibernation. As the time nears, one should reduce the amount of food and gradually reduce the temperature in the vivarium. Temperature for the hibernating reptile should be between 39 and 50°F, depending on the species. The length of the hibernation period varies from 4 to 12 weeks, again depending on the species and its original environment. The vivarium can be put in the basement where the temperature is lower. To bring the animal out of hibernation, one can simply reverse the procedure, gradually increasing the temperature until normal temperatures are reached. Most reptiles coming out of hibernation do not eat for a few days.

CLASSIFICATION—TURTLES, TORTOISES, AND TERRAPINS

Kingdom — Animalia
Phylum — Chordata
Class — Reptilia
Order — Testudines or Chelonia (turtles, tortoises, and terrapins)
Suborder — (1) Pleurodira (side-necked turtles)
Suborder — (2) Cryptodira (hidden-necked turtles and tortoises)
Family — 12 families

MAJOR FAMILIES—TURTLES, TORTOISES, AND TERRAPINS

The Family Chelidae

The family Chelidae consists of the long, side-necked turtles of South America, Australia, and Papua New Guinea. These turtles have very long necks, and only a few can withdraw them into their shells; those that can withdraw their necks do it in a sideways manner so that the head and neck are on the side of the shell when retracted. There are nine genus groups consisting of 37 species in this family. *See Figure 18–4.*

Figure 18–4 A member of the family Chelidae, which consists of long-necked turtles. *Photo by Isabelle Francais.*

The Family Pelomedusidae

The family Pelomedusidae consists of side-necked turtles found in tropical South America and Africa. There are five genus groups and 24 species in this family. *See Figure 18–5.*

Side-necked turtle

Figure 18–5 (a) Most common turtles are hidden-necked turtles and withdraw their head and legs back into the shell. (b) Side-necked turtles bring their heads backward and to the side as pictured.

The Family Kinosternidae

The family Kinosternidae is made up of the mud and musk turtles. These are small-sized turtles that are found in North America, Mexico, and Central America. They are aquatic turtles that live in rivers, swamps, and ponds, where they feed on insects, worms, small fish, and plants. *See Figures 18–6* and *18–7.*

Figure 18–6 Common mud turtle. *Photo by Isabelle Francais.*

The Mud and Musk Turtles

Mud turtles have flattened shells and are able to completely retract their head, tail, and webbed feet into their shells. Musk turtles get their name from the strong, musky-smelling liquid they produce from glands between their upper and lower shells. Their lower shell does not hinge like that of the mud turtles.

Figure 18–7 Common musk turtle of the genus *Sternotherus. Photo by Isabelle Francais.*

The Family Chelydridae

The family Chelydridae consists of two genus groups and two species. *Chelydridae serpentina* is the common snapping turtle. *Macroclemys temminckii* is the alligator snapping turtle.

The Common Snapping Turtle

The common snapping turtle, *Chelydridae serpentina,* is the most widespread turtle in North America. It can reach a length of over 12 inches and weigh up to 60 pounds. Snapping turtles feed on fish, frogs, small ducks, geese, and small mammals. They usually live in ponds and swamps but will venture out to find food. They are very aggressive and, if threatened, will open their mouth and snap their powerful jaws. Snapping turtles are eaten and are becoming scarce. *See Figure 18–8.*

Figure 18–8 Common snapping turtle. *Photo by Isabelle Francais.*

The Alligator Snapping Turtle

The alligator snapping turtle, *Macroclemys temminckii,* is found in the Mississippi River system including the Missouri River and rivers that drain into the Gulf of Mexico. This is the largest of all freshwater turtles, reaching a total length of up to 5 feet and weighing 200 pounds. It usually lies motionless on the bottom of ponds and slow-moving streams, where it waits for its prey to swim by. Attached to its tongue is a red, worm-like growth that lures small fish in close so they are captured in the turtle's powerful jaws (*See Figure 18–9*).

Figure 18–9 The alligator snapping turtle is the largest freshwater turtle. They can reach a weight of up to 200 pounds. *Photo by Isabelle Francais.*

The Family Platysternidae

The family Platysternidae consists of one genus group and one species.

The Big-Headed Turtle

The big-headed turtle, *Platysternon megacephalum,* is native to streams in China, Thailand, and Burma; its head is huge and out of proportion with the rest of its body. It feeds primarily on shellfish, which it crushes with its powerful jaws, but will also feed on fish and small animals. They have slightly webbed toes with strong claws, and they have a long tail that is almost as long as their shell. This turtle lives in rivers and fast-moving streams. Because of their strong claws, a wide sharp beak, a long tail, and adaptations to their carapace and plastron, they are able to climb over rocks and bushes and climb up trees found along the river's edge. *See Figure 18–10.*

Figure 18–10 Big-headed turtle, *Platysternon megacephalum. Photo by Isabelle Francais.*

The Family Emydidae

The family Emydidae, with approximately 30 different genus groups and 85 different species, is the largest family within the order Testudine/Chelonia. Commonly called pond and river turtles, they are found in parts of Europe, North Africa, Asia, South America, Central America, and North America. These turtles are primarily found in freshwater habitats but live in brackish habitats or on land. Those that live in freshwater spend warm, sunny days basking on logs and rocks.

Turtles of the Emydidae family vary from about 4 inches up to 2 feet in length. The bog turtle, *Clemmys muhlenbergi,* at 4 inches in size, may be the smallest of all turtles and is found in shallow bogs along the East Coast of the United States. The largest Emydidae turtle is the batagur, *Batagur baska,* which reaches a length of 2 feet and weighs 50 to 60 pounds. The batagur is found in Sumatra, Burma, and Thailand.

The Emydidae turtles are noted for their bright colors and distinct patterns. This group of turtles makes up most of the turtles that are traded as pets and used for research.

The Red-Eared Slider

The red-eared slider, *Trachemys scripta elegans,* is one of the most popular turtles kept as pets. It is found from the Southwestern United States eastward to Kentucky and Indiana. Red-eared turtles have a characteristic red streak behind the eye; they may reach an adult length of up to 12 inches. *See Figure 18–11.*

Closely related to the red-eared turtle is the yellow-bellied turtle, *Trachemys scripta scripta.* This turtle is found in the Southeastern United States. They are usually dark green with a yellow plastron and a broad yellow crescent-shaped marking behind the eyes. They are 9 to 12 inches long.

Figure 18–11   Red-eared turtle. *Photo by Isabelle Francais.*

The Red-Bellied Turtle

Red-bellied turtles, *Pseudemys nelsoni* and *Pseudemys rubriventris,* have reddish-colored plastrons. *Pseudemys nelsoni* are found in Florida, and *Pseudemys rubriventris* are found in areas along the East Coast of the United States. Adult red-bellied turtles are usually 9 to 15 inches long. *See Figure 18–12.*

Figure 18–12   Red-bellied turtle. *Photo by Isabelle Francais.*

The Coastal Plain Turtle

Pseudemys floridana floridana, or Coastal Plain turtle, is found along the Southeastern Coast of the United States. It is brown with yellow stripes and a yellow plastron; it reaches a length of 15 inches. *Pseudemys floridana peninsularis,* or peninsula turtle, is a common turtle found throughout the peninsula of Florida; this turtle has a greenish tint to its yellow plastron.

The Missouri Slider

The Missouri slider, *Pseudemys floridana hoyi,* is found from Alabama north to Southern Illinois and in parts of Kansas, Oklahoma, and Texas. It is somewhat smaller than its southeastern relatives. Its plastron is yellow or orange. There are no markings on the plastron of the *floridana* species of turtles.

The Concinna Species

The *concinna* species of turtles are similar to the *floridana* except for the dark markings along the plastral seams; these markings appear as loops and whorls. Adults of this species range from 9 to 16 inches in length. Among the *concinna* species are the river cooter, *Pseudemys concinna concinna,* found in Southeast Virginia, North and South Carolina, and the eastern edge of Alabama; *Pseudemys concinna hieroglyphica,* or the hieroglyphic turtle, found in Texas, Oklahoma, Kansas, and Southern Illinois; *Pseudemys concinna mobiliensis,* or the mobile turtle, found along the Gulf Coast from Florida to Texas; and *Pseudemys concinna texana,* found in Southern Texas and northern areas of Mexico.

Blanding's Turtle

Blanding's turtle, *Emydoidae blandingi,* is found in the Great Lakes area of the United States. This semiaquatic turtle is found in ponds, in small streams, and around lakes. The color of the Blanding's turtle varies from black, brown, to olive with yellow spots and markings. The neck is speckled, and the lower jaw, chin, and throat are a bright yellow. This turtle is usually 7 or 8 inches long. It is carnivorous, feeding on worms, small fish, crustaceans, and insects. *See Figure 18–13.*

Figure 18–13   Blanding's turtle. *Photo by Isabelle Francais.*

The European Pond Turtle

The European pond turtle, *Emys orbicularis,* is a semiaquatic turtle found in North Africa and most of Europe. This is a small turtle; adults usually measure 4 to 5 inches long. The color of the European pond turtle varies, but the carapace (upper shell) is usually black with yellow spots, specks, or short lines. The head, neck, and limbs are also covered with yellow markings. This small turtle is carnivorous, feeding on worms, frogs, mollusks, and small fish.

Painted Turtles

The painted turtles of the genus *Chrysemys* are another group of brightly colored turtles. They are different from the genus *Pseudemys* because they have smooth shells, lack the serrations to the rear shell margins, and are somewhat smaller. There is one species of painted turtle, *Chrysemys picta,* and four subspecies.

The Southern Painted Turtle

The southern painted turtle, *Chrysemys picta dorsalis,* is probably the most familiar of the painted turtles. This turtle is dark in color with yellow stripes along its head and an orange-yellow stripe running down the center of its carapace. The southern painted turtle is found in Louisiana and surrounding states.

The Eastern Painted Turtle

The eastern painted turtle, *Chrysemys picta picta,* is found along the Eastern Coast of the United States and as far north as Nova Scotia. It has a dark-olive to dark-brown upper shell and a yellow plastron. Two characteristic yellow marks are found behind the eyes, and there are yellow stripes along the chin and neck; these stripes turn into orange as they run down the neck.

The Midland Painted Turtle

The midland painted turtle, *Chrysemys picta marginata,* is found in the area around the Great Lakes, from Canada South to Northern Alabama. It is very similar in appearance to the eastern painted turtle but lacks the yellow margins on the carapace scutes, and there is a dark marking along the midline of the plastron.

The Western Painted Turtle

The western painted turtle, *Chrysemys picta belli,* is the largest of the painted turtles; it may reach a length of 10 inches. The carapace is green with yellow lines; the head and limbs have yellow stripes. The plastron is pink or red with a center marking that branches out along the seams. This turtle is found from Kansas and Missouri westward and from Southern Canada to Northern Mexico. *See Figure 18–14.*

Figure 18–14   Western painted turtle. *Photo by Isabelle Francais.*

Box Turtles

Box turtles belong to the genus group *Terrapene.* They are easily recognized by their domed shells and a hinged plastron. They can withdraw their head, tail, and limbs, and the hinged plastron closes, completely sealing the animal from outside predators. *See Figure 18–15.*

Figure 18–15   The common or eastern box turtle. This picture shows the hinged plastron that can close and completely seal the animal from outside predators. *Photo by Isabelle Francais.*

The Common Box Turtle

The common, or eastern, box turtle, *Terrapene carolina carolina*, is usually found from Illinois and Mississippi eastward. They are brown to black with yellow or orange markings; the plastron is dark brown to yellow. Males have bright-red eyes. The eastern box turtle is usually found in open woodland areas. They may be found near water but rarely enter water over a few inches deep. Adults are usually 4 to 5 inches long. Box turtles are long-lived and may easily live 40 to 50 years. *See Figure 18–16.*

Figure 18–16   The common, or eastern, box turtle. *Photo by Isabelle Francais.*

The Florida Box Turtle

The Florida box turtle, *Terrapene carolina bauri*, is found throughout the peninsula of Florida. The carapace of the Florida box turtle is dark brown to black with a yellow marking down the center. The scutes of the carapace have yellow streaks radiating outward; the plastron usually lacks any color. *See Figure 18–17.*

Figure 18–17   Florida box turtle. *Photo by Isabelle Francais.*

The Gulf Coast Box Turtle

The Gulf Coast box turtle, *Terrapene carolina major*, is found along the Gulf Coast from the Florida panhandle to Texas. This is the largest box turtle, reaching a length of 7 to 8 inches. It is usually a dark color and, in many cases, lacks the markings common to other subspecies. *See Figure 18–18.*

Figure 18–18   Gulf Coast box turtle. *Photo by Isabelle Francais.*

Map Turtles

Map and sawback turtles belong to the genus group *Graptemys*; these are deepwater, diving turtles. Adults are wary and hard to capture. They are recognized by their striped heads and limbs, serrated rear shell margins, and center scutes that have raised or pointed projections. Females may be twice as large as males, and develop huge heads with large jaws. *See Figure 18–19.*

Figure 18–19   Common map turtle, *Graptemys geographica*. *Photo by Isabelle Francais.*

Barbour's Map Turtle

Barbour's map turtle, *Graptemys barbouri*, is found in the Chipola, Apala-chicola, and Flint rivers of the Florida panhandle and in

Southwestern Georgia. Females usually reach 10 inches in length and have large, toad-like heads; males reach only 5 inches in length and have small, narrow heads. The center scutes have raised projections. The carapace is usually olive green with yellow, wavy lines on the coastal scutes and inverted L-shaped markings on the marginal scutes. There are yellow-green stripes on the head, neck, limbs, and tail.

The Alabama Map Turtle

The Alabama map turtle, *Graptemys pulchra,* is found in most of the rivers that empty into the Gulf of Mexico. The females reach lengths of up to 11 inches, and the males reach about 4 inches in length. The carapace is olive with distinctive yellow, curved markings on each of the marginal scutes; the plastron is yellow. The head marking is very distinctive and has a single pale green or white marking between the eyes that tapers toward the nose; there are also large markings behind the eyes. There are several yellow lines on the neck and limbs. *See Figure 18–20.*

Figure 18–20 The Alabama map turtle. *Photo by Isabelle Francais.*

The Black-Knobbed Sawback Turtle

The black-knobbed sawback turtle, *Graptemys nigrinoda,* is found in several rivers of Alabama and the Tombigbee River of Mississippi. The central scutes have raised projections that are blunt, rounded, and knob-like. The carapace is gray-brown, and there are dark-edged circles on the coastal scutes. Females of this species reach 8 inches in length, and males reach about 4 inches in length.

The Yellow-Blotched Sawback Turtle

The yellow-blotched sawback turtle, *Graptemys flavimaculata,* is a smaller sawback turtle found in the Pascagoula River system of Mississippi. The females reach a length of only 7 inches, with the males being much smaller. There are large yellow blotches on the coastal scutes; the rear margin scutes have large serrations.

The Ringed Sawback Turtle

The ringed sawback turtle, *Graptemys oculifera,* is very similar to the yellow-blotched sawback except the markings form circles on the coastal scutes and incomplete circles on the marginal scutes. The ringed sawback turtle is found in the Pearl River of Louisiana and Mississippi.

The Mississippi Map Turtle

The Mississippi map turtle, *Graptemys kohni,* is a common species found in pet stores. This species is found in most of the lower Mississippi River drainage system. It can be recognized by its distinctive eye; the iris appears as a white ring. There is also a distinctive crescent-shaped marking behind the eyes.

The Spotted Turtle

The spotted turtle and the wood turtles belong to the genus *Clemmys.* The spotted turtle, *Clemmys guttata,* is common in the Eastern United States. This is a small turtle that reaches about 4 inches in length. These turtles are easily recognized by the yellow spots found on the black carapace; young turtles do not have spots; they develop as the turtle matures. Yellow spots are also found on the head, neck, and limbs.

The Wood Turtle

The wood turtle, *Clemmys insculpta,* is found in the Eastern United States, Southern Ontario, and Northern Wisconsin. It is easily recognized by its rough carapace and orange throat and limbs. It is primarily terrestrial and usually reaches 6 to 7 inches in length.

Terrapins

The genus *Malaclemys* is commonly referred to as **terrapins.** The term *terrapin* is applied to turtles found in freshwater and brackish water that are considered excellent eating; these turtles were very popular as a gourmet food before World War I.

The northern terrapin, *Malaclemys terrapin terrapin,* is found from the Cape Cod area to Cape Hatteras. The southern terrapin, *Malaclemys terrapin centrata,* is found from the Cape Hatteras area south to Florida. The Florida East Coast terrapin, *Malaclemys terrapin tequesta,* is found along the East Coast of Florida.

Other species include the diamond terrapin, *Malaclemys terrapin rhizophorarum,* which is found in the lower Gulf Coast area of Florida and the Florida Keys. The ornate diamondback, *Malaclemys terrapin macrospilota,* is found along the Gulf Coast of Florida. The Mississippi diamondback, *Malaclemys terrapin pileata,* is found along the Gulf Coast from Florida to Louisiana. The Texas diamondback, *Malaclemys terrapin littoralis,* is found from Mississippi to Texas.

Females of terrapin species are larger than the males and range in size from 6 to 7 inches; the males are usually 4 to 5 inches long. The Mississippi diamondback is the largest species.

The Family Trionychidae

The family Trionychidae consists of the soft-shelled turtles. They are commonly found in freshwater habitats of North America, Africa, Asia, Indonesia, and the Philippines. There are six genus groups consisting of 22 species. They have flat, flexible shells that lack scutes; soft-shelled turtles have three claws on each foot. Three species are found in the United States; the largest is *Trionyx ferox,* the Florida soft-shelled turtle. Females are approximately 12 to 15 inches long, and the males are somewhat smaller. They are found throughout most of Florida. The spiny soft-shelled turtle, *Trionyx spinifer,* is the most widespread species. This species has cone-shaped projections along the front margin of the carapace. There are six subspecies of spiny soft-shelled turtles. The smooth or spineless soft-shelled turtle, *Trionyx muticus,* is common in the Central United States; two subspecies are recognized. *See Figure 18–21.*

Soft-shelled turtles settle on the bottom of streams, ponds, and lakes. Most have snorkel-shaped snouts that enable them to breathe while they remain beneath the water's surface. Most species are carnivorous and feed on crustaceans, mollusks, insects, worms, frogs, and fish.

Soft-shelled turtles are able to exchange oxygen through their shell and skin. This allows them to remain submerged longer than other turtles. Soft-shelled turtles can often be seen sunning themselves on the banks of streams, ponds, and lakes.

Figure 18–21   Soft-shelled turtle. *Photo by Isabelle Francais.*

The Family Testudinidae

The tortoises make up the family Testudinidae. There are 10 genus groups consisting of 41 different species. Tortoises are terrestrial and are primarily herbivorous, some are omnivores. They vary in size from 6 inches to more than 50 inches in length.

The Egyptian tortoise, *Testudo kleinmanni,* is the smallest species, with the Aldabra tortoise, *Aldabrachelys elephantina,* being the largest. The Aldabra tortoise is found on the Seychelles, Mascarene, and Aldabra islands. The most famous tortoises are the large island tortoises, *Chelonoidis elephantopus,* found on the Galapagos Islands. *See Figure 18–22.*

Figure 18–22   Large island tortoise, *Chelonoidis elephantopus. Photo by Isabelle Francais.*

The Gopher Tortoise

Gopher tortoises of the genus group *Gopherus* are found in North America. The desert tortoise, *Gopherus agassizi,* is found in the Southwestern United States. Adults are about 11 inches long. The desert tortoise has a uniform brown carapace and wanders over desert areas during the day and retires to burrows at night. It digs short burrows at the base of bushes to escape from the cool night temperatures. *See Figure 18–23.*

Figure 18–23   This picture shows two adult and two young gopher tortoises. *Photo by Isabelle Francais.*

The Texas Tortoise

The Texas tortoise, or Berlandier's tortoise, *Gopherus berlandieri,* is found in Southern Texas. This tortoise is about 6 to 7 inches long and dark brown. The color usually becomes lighter as the tortoise ages. The Texas tortoise's front limbs are not as well developed for digging as those of the desert tortoise. They may dig shallow trenches to protect themselves from the night cold. The Texas tortoise feeds on grass, cactus pods, flowers, and fruits.

The Florida Gopher Tortoise

The Southeastern United States is home for the Florida gopher tortoise, *Gopherus polyphemus.* This tortoise is 9 to 10 inches long and has a uniform brown color with some light mottling; the young are bright yellow-orange. This species is well adapted to digging and may make burrows up to 30 feet long and 12 feet deep. The Florida gopher tortoise feeds on grass and leaves. Although no laws prevent the keeping of these tortoises, it is illegal to buy and sell them in Florida. *See Figure 18–24.*

Figure 18–24 Florida gopher tortoise and plated lizard.
Photo by Isabelle Francais.

The Family Dermochelyidae

The family Dermochelyidae consists of one genus group and one species, the leatherback sea turtle or leathery turtle, *Dermochelys coriacea.* The leatherback sea turtle is the largest species of turtle in the world. The average carapace length is about 61 inches, and leatherbacks average about 800 pounds. Some leatherbacks have been reported at 8 or 9 feet in length and weighing more than 1,000 pounds. The leatherback is primarily black with numerous white spots. The carapace of the leatherback is without scutes and is somewhat flexible.

The Family Cheloniidae

The family Cheloniidae consists of four genus groups and six species of hard-shelled sea turtles. The hawksbill turtle, *Eretonochelys imbricata;* the loggerhead turtle, *Caretta caretta;* and the green sea turtle, *Chelonia mydas,* are found in the Atlantic, Indian, and Pacific Oceans. The flatback turtle, *Chelonia depressa,* has its nesting area only on the Northern Coast of Australia. There are two species of Ridley sea turtle. The Pacific (or olive) Ridley sea turtle, *Lepidochelys olivacea,* is found in the tropical areas of the Pacific, Indian, and Atlantic Oceans. The Kemp's (or Atlantic) Ridley sea turtle, *Lepidochelys kempi,* is found in the Gulf of Mexico and the North Atlantic.

Both the leatherback and the hard-shelled sea turtles have been hunted for their meat and eggs; today, many species are threatened or endangered.

The Family Dermatemyidae

The family Dermatemyidae consists of one genus group containing one species, the Central American river turtle, *Dermatemys mawi.* This species of turtle is found in Belize, Guatemala, and Southern Mexico. The carapace of both the male and female of the species reaches a length of about 18 inches. The shell scutes of this turtle are very thin. The turtle spends most of its time in the water. It floats in the water and does not climb on logs or riverbanks to bask in the sun.

The meat of the Central American river turtle is considered excellent; therefore, the turtle has been hunted extensively for human consumption.

The Family Carettochelyidae

The family Carettochelyidae consists of one genus group and one species, the Papua New Guinea plateless river turtle, *Carettochelys insculpta.* This species is found in the remote areas of Papua New Guinea and Northern Australia. Its carapace reaches a length of about 18 inches. The upper surface of the turtle is gray, and the underside is white. Its carapace does not have scutes but consists of a layer of skin. The tail is covered with a series of crescent-shaped scales.

CLASSIFICATION—SNAKES, PYTHONS, AND BOAS

Kingdom	—	Animalia
Phylum	—	Chordata
Class	—	Reptilia
Order	—	Serpentes (snakes, pythons, and boas)
Family	—	11 families, 2,400 species

MAJOR FAMILIES—SNAKES, PYTHONS, AND BOAS

The Family Leptotyphlopidae

The family Leptotyphlopidae is commonly referred to as thread snakes. They are among the smallest in the world and range from 3 to 16 inches in length. There are 78 species of thread snakes found in arid areas of Africa, Arabia, North America, Central America, and South America. The western blind snake, *Leptotyphlops humilis,* and the Texas blind snake, *Leptotyphlops dulcis,* are found in the Southwestern United States. Their eyes are very small, and they spend most of their time in tunnels. They feed on insects, primarily termites and ants. They are light brown to pink.

The Family Typhlopidae

The family Typhlopidae is made up of 180 different species commonly referred to as blind snakes. They inhabit moist, tropical areas, where they burrow under leaves, around roots, and in decaying plant material. These snakes have small eyes, and their scales are smooth, which aids in their movement through tunnels. They differ from thread snakes in that they have teeth on the upper jaw only, whereas thread snakes have teeth on the lower jaw only.

The Family Anomalepidae

The family Anomalepidae consists of about 20 species, commonly referred to as dawn blind snakes. They range in size from 5 to 6 inches in length. They also have small eyes and burrow among the leaves and mats of tropical Central and South America. They differ from the first two families because they show no evidence of pelvic limb bones, which the families Leptotyphlopidae and Typhlopidae possess.

The Family Acrochordidae

The family Acrochordidae consists of two species, the Javan wart snake (or elephant's trunk snake), *Acrochordis javanicus,* and the Indian wart snake (or Asian file snake), *Chersydrus granulatus.*

The Wart Snake

Wart snakes are aquatic and found in streams, canals, and rivers in Asia and Australia; they feed primarily on fish. Wart snakes can reach a length of 6 feet.

The Family Aniliidae

The family Aniliidae consists of three genus groups and 10 species. They are commonly referred to as pipe snakes. All species except one is found in Asia and Indonesia; the one exception is found in South America.

Pipe snakes get their name from the shape of their bodies, which is almost the same circumference from head to tail.

The Coral Pipe Snake

The coral pipe snake (or fake coral snake), *Anilius scytale,* of South America is ringed with brilliant scarlet and black that resembles the poisonous coral snakes.

The Family Uropeltidae

The family Uropeltidae is made up of eight genus groups containing 44 different species. They are commonly referred to as shieldtail snakes. These snakes have an enlarged, thick tail with spines on the upper surface. This adaptation may serve to help the snake move through its tunnel as well as serving as a plug for the tunnel. They have brightly colored red, orange, or yellow markings, and they feed primarily on earthworms.

The Family Xenopeltidae

The family Xenopeltidae is made up of one genus group containing only one species, the sunbeam snake (or sunbeam python), *Xenopeltis unicolor.*

The Sunbeam Snake

The sunbeam snake grows to about 3 feet in length. It burrows under leaves and decaying matter and is primarily active at night. It is brown, and its smooth scales shimmer in the sunlight, giving rise to its name.

The Family Boidae

The family Boidae contains the pythons and boas; there are 27 genus groups containing about 88 species. Included in this family are the largest species of snakes, the reticulated python, *Python reticulatus,* and the anaconda, *Eunectes murinus.* Both may reach a length of up to 33 feet and weigh more than 400 pounds. The family Boidae is divided into several subfamilies.

The subfamily Pythoninae includes about 24 species found in Africa, India, Southeast Asia, and Australia. Members of this subfamily are egg layers. The female may build a nest; after the eggs are laid she coils around the eggs and protects them until they hatch.

The Reticulated Python

The reticulated python is found in Southeast Asia and the Philippines. It has a network pattern of brown, buff, yellow, and reddish-brown. These large pythons feed on deer, pigs, and birds. The reticulated python, *Python reticulatus,* may reach a length of more than 33 feet. *See Figure 18–25.*

Figure 18–25 Reticulated python. *Photo by Isabelle Francais.*

The India Python

The India python, *Python molurus,* looks very similar to the reticulated python but is smaller; the colors are similar. There is a distinctive arrow-shaped marking on the head of the India python. This python is found in India, Sri Lanka, and the East Indies. India pythons that are light in color are said to tame more readily than the darker-colored ones. *See Figure 18–26.*

Figure 18–26 India python. *Photo by Isabelle Francais.*

The Burmese Python

The Burmese python, *P. m. bivittatus,* is a subspecies and is darker in appearance. Also included in this subspecies is the albino Burmese python. *See Figures 18–27* and *18–28.*

Figure 18–27   Burmese python. *Photo by Isabelle Francais.*

Figure 18–28   Albino Burmese python. *Photo by Isabelle Francais.*

The Ball Python

The ball python, *Python regius,* is a small species found in West Africa; it reaches a length of 6 to 7 feet. This python is dark brown with light-brown markings. The color helps it blend into the dry forest and bushlands where it is found. It protects itself from predators by coiling up into a tight ball; the ball python is sometimes referred to as the royal python. *See Figure 18–29.*

Figure 18–29   Ball python. *Photo by Isabelle Francais.*

The Rock Python

The rock python, *Python sebae,* is a large python found in Africa that reaches a length of more than 20 feet. It is dark brown with light-brown to yellow markings; it has a dark, arrow-shaped marking on its head. This python feeds on antelope, gazelle, and other mammals found on the grassy plains.

The Green Tree Python

The green tree python, *Chondropython viridis,* is a tree-dwelling python of Northeast Australia, Papua New Guinea, and the Solomon Islands. It is bright green with small blue, red, and white markings; its underside is yellow or white. The eggs of the tree python are laid on the ground, where they hatch into yellow or red young; the young soon take to the trees. These pythons feed on roosting birds that they seize with extremely sharp teeth. They usually rest during the day, curled up on the limbs of trees. *See Figure 18–30.*

Figure 18–30   Green tree python. *Photo by Isabelle Francais.*

The Carpet Python

The carpet python, *Morelia spilotes variegata,* and the diamond python, *Morelia spilotes spilotes,* are found in Australia, Papua New Guinea, and other Pacific islands. These pythons reach 6 to 10 feet in length. The diamond python is strictly an **arboreal** (tree-dwelling) species, but the carpet python feeds both in trees and on the ground; they feed on birds, mammals, and lizards. The diamond python is light brown with light-brown to yellow diamond-shaped markings; the carpet python is similarly marked. *See Figure 18–31.*

Figure 18–31   Carpet python. *Photo by Isabelle Francais.*

Boelen's Python

Boelen's python, *Liasis boeleni,* is found in the high mountain jungles of several Pacific islands. It is black with yellow markings.

The Amethystine Python

The amethystine python, *Liasis amethystinus,* is a long, slender python reaching lengths of 25 feet. It is found in Australia and Papua New Guinea. *See Figure 18–32.*

Figure 18–32   Amethystine python. *Photo by Isabelle Francais.*

The Black-Headed Python

The black-headed python, *Aspidites melanocephalus,* is different from other snakes because it feeds on other snakes, although it also eats other warm-blooded animals. Its head, neck, and throat are black in contrast to the rest of its body, which is light brown.

The Blood Python

The blood python, *Python curtus,* is found in Asia. This small python reaches a length of only 4 to 5 feet. They vary in color from yellow, orange, to blood-red. *See Figure 18–33.*

The subfamily Boinae is made up of about 39 species. Boas are very similar to pythons except that they bear live young instead of laying eggs. Most boas live in Central and South America, although two species are found in North America.

Figure 18–33   Blood python. *Photo by Isabelle Francais.*

The Anaconda

The anaconda, *Eunectes murinus,* is a semiaquatic boa found in the river basins of the Amazon and Orinoco Rivers of South America. This large boa may reach a length of up to 33 feet. It is grayish-brown with large black markings. Anacondas are excellent swimmers; their nostrils are located on the top of their snout, which allows them to breathe easily while in the water. They feed in trees, on land, and in the water. Included in their diet are birds, peccaries, caiman, tapirs, capybara, fish, and turtles. Anacondas do not adapt well to captivity.

The Emerald Tree Boa

The emerald (or green) tree boa, *Corallus caninus,* is similar to the green tree python in appearance. It lives in trees and feeds primarily on birds. It grows to 6 to 7 feet in length and is bright green above with a yellow underside. The young vary in color from rusty red to blue-green. *See Figure 18–34.*

Figure 18–34 Emerald (or green) tree boa. *Photo by Isabelle Francais.*

The Boa Constrictor

The boa constrictor, *Boa constrictor,* is one of the most popular of the common boas. They are popular in the pet trade because they tame quickly and breed well in captivity. Their color varies from a light-brown, yellow, or orange background with dark bars and markings. They grow up to 18 feet in length but average 12 to 14 feet. They can be found in trees and on the ground and are excellent swimmers. Boa constrictors are found throughout much of South America and north into Central America and Mexico. They feed on all types of small animals, birds, eggs, and fish.

The Red-Tailed Boa

The red-tailed boa, *B. constrictor constrictor,* is a subspecies of the boa constrictor. They are sold in the pet trade as Mexican red-tailed boas, Peruvian red-tailed boas, or Columbian red-tailed boas. The red-tailed boas are found in Mexico and Central and South America. *See Figure 18–35.*

Figure 18–35 Red-tailed boa. *Photo by Isabelle Francais.*

The Yellow Anaconda

The yellow anaconda, *Eunectes notaeus,* is a smaller anaconda that grows up to 15 feet in length. The young are bright yellow with black markings; the colors fade somewhat as the animal ages.

The Rainbow Boa

The rainbow boa, *Epicrates cenchris,* is one of the most attractive species because of the blue-green iridescence of its scales. Its color ranges from light brown to bright reddish-orange. They grow to about 8 feet in length and feed at night on rodents and birds. This boa is found throughout the Caribbean region, Central America, and Northern South America. *See Figure 18–36.*

Figure 18–36 A rainbow boa, one of the most attractive species because of its blue-green iridescent scales. *Photo by Isabelle Francais.*

The Central American Dwarf Boa

The Central American dwarf boa, *Ungaliophis panamensis,* is a small boa that reaches a length of only 30 inches. It is found in the rainforests of Central and South America. It is light brown with dark oblong spots, and it feeds on lizards, frogs, small birds, and rodents.

The Rubber Boa

Two boas are found in North America. The rubber boa, *Charina bottae,* is found in the Western United States; its olive-colored body scales have a rubbery feel. This small boa reaches a length of 6 feet, although it usually averages about 30 inches. It is harmless and usually rolls up into a ball when threatened. It feeds on lizards, rodents, and small birds. The rubber boa may burrow in search of food. *See Figure 18–37.*

Figure 18–37   *Rubber boa,* Charina bottae. *Photo by Isabelle Francais.*

The Rosy Boa

The rosy boa, *Licanura trivirgata,* is found in the drier areas of the Southwestern United States. Its background color varies from pink to gray, and it has three black stripes that run the length of its body. These boas are nocturnal and spend the day hiding among rocks and logs. *See Figure 18–38.*

Figure 18–38   Rosy boa. *Photo by Isabelle Francais.*

The Family Colubridae

The family Colubridae is the largest family of snakes; there are approximately 300 genus groups consisting of 1,562 species. They can be found on all five continents and as far north as the Arctic Circle and south to the tip of South America. Only the more common genus groups are covered here. There are 20 subfamilies. Eight subfamilies have grooved fangs at the rear of their jaws that can inject venom into their prey. These snakes are referred to as rear-fanged snakes. Most of these snakes are small and cannot inflict injury to humans. Two species, however, can grow to over 5 feet. The boomslang, *Dispholidus typus,* and the twig snake, *Thelotornis kirtlandii,* both found in Africa, have been known to inflict serious injury and even death to humans. Species of rear-fanged snakes are found in Southeast Asia, New Guinea, and Northern Australia. Because rear-fanged snakes may be dangerous to handle, they are not discussed further here.

Common Water Snakes

The subfamily Natricinae consists of the common water snakes and garter snakes found in North America. This group ranges in length from 24 inches to more than 6 feet; they all give birth to live young. Water snakes are semiaquatic and are seldom found far from water; they spend a lot of their time basking in the sun. Most water snakes, although harmless, are very aggressive. If provoked or cornered, they will strike. To keep water snakes in captivity, the snakes must have an aquatic environment; cleaning and maintaining this aquatic environment can be difficult and time consuming. Water snakes feed on fish, frogs, salamanders, and toads. These snakes are gray-brown with darker cross-bands. One common species is the northern water snake, *Natrix sipedon.* There are about 18 different subspecies of northern water snakes. Three other common species in North America include the red-bellied water snake, *Natrix erythrogaster;* the green water snake, *Natrix cyclopion;* and the brown water snake, *Natrix taxispilota. See Figure 18–39.*

Figure 18–39   Brown water snake. *Photo by Isabelle Francais.*

The Common Garter Snake

The common garter snake, *Thamnophis sirtalis,* and about 15 subspecies of garter snakes are the most widespread snakes in North America. Almost all youngsters have encountered garter snakes in the outdoors, where the snakes can be found among leaves, under logs, and under rocks. Young garter snakes feed on earthworms, mealworms, and other insects; larger garter snakes feed on rodents, fish and frogs. Garter snakes adjust readily to captivity. Common species are the eastern garter snake, *Thamnophis s. sirtalis;* the checkered garter snake, *Thamnophis marcianus;* and the black-necked garter snake, *Thamnophis cyrtopsis.* Most species of garter snakes average 2 to 3 feet in length. The smallest species is the Butler's garter snake, *Thamnophis butleri,* which reaches a length of only 18 inches at maturity. The giant garter snake, *Thamnophis couchi gigas,* may grow to over 4 feet. *See Figure 18–40.*

Most species of garter snakes are easily recognized by their vivid stripes running the length of their body on a dark background. Some species have a checkerboard pattern (checkered garter snake), whereas others may be solid black.

Figure 18–40   Common garter snake. *Courtesy of PhotoDisc.*

Ribbon Snakes

Ribbon snakes are closely related to garter snakes; they are semiaquatic and more slender. The eastern ribbon snake, *Thamnophis sauritus sauritis,* is black or dark brown with yellow stripes running the length of the body. They can usually be found near water and can be seen basking in the sun on branches and logs. They feed on small fish, frogs, and salamanders.

Racers

The genus group *Coluber* is made up of the racers; they are common in North America and are found from coast to coast. They are among the fastest moving snakes, reaching speeds of 3.5 miles per hour. Species found in the Western United States are usually brown or olive with a lighter underside. Those found in the East are usually black or blue-black. Racers may be difficult to keep in captivity; they are aggressive and have been known to chase those who encounter them. Many times when encountered, they will stand their ground and move their head back and forth while flicking their tongue. In captivity, angry racers kept in glass aquariums may strike at the glass as people approach it. If provided with a hiding place, they may stay in hiding and not venture out. Racers do not like to be handled. Racers feed on earthworms, insects, frogs, salamanders, and rodents; racers in the wild swallow their food live. The young are hatched from eggs.

Whipsnakes

Whipsnakes are closely related to racers. The coachwhips, *Masticophis flagellum,* may reach a length of 8 feet; whipsnakes need dry environments.

Rat Snakes

Rat snakes make up the genus *Elaphe.* There are more than 40 species of rat snakes. Several species are brightly colored and boldly marked. Rat snakes are constrictors; they feed almost entirely on mice and rats.

The Black Rat Snake

The black rat snake, *Elaphe obsoleta obsoleta,* is black with white on the chin and throat. It may reach lengths of 4 to 8 feet and is usually found in wooded and rocky hillsides. *See Figure 18–41.*

Figure 18–41   Black rat snake. *Photo by Isabelle Francais.*

The Corn Snake

The corn snake, *Elaphe guttata,* is found from New Jersey to Florida and across to Utah. It is yellow or gray with black-edged, red blotches. This snake, also referred to as a red rat snake, may reach a length of 6 feet. *See Figure 18–42.*

Figure 18–42   Corn snake. *Photo by Isabelle Francais.*

The Fox Snake

The fox snake, *Elaphe vulpina,* is found in the central states from Wisconsin to Missouri. This snake receives its name from the fox-like odor it releases when captured. It is yellowish to pale brown above with dark blotches; its underside is yellow with black, checkered markings.

The fox snake reaches a length of about 4 feet and feeds on rodents and birds.

The Yellow Rat Snake

The yellow rat snake, *Elaphe obsoleta quadrivittata,* is a bright-yellow snake with four black stripes running the length of its body. It is found in the southern coastal states of North and South Carolina, Georgia, and Florida. *See Figure 18–43.*

Figure 18–43   Yellow rat snake. *Photo by Isabelle Francais.*

The Everglades Rat Snake

A subspecies, the Everglades rat snake, *Elaphe obsoleta rossalleni,* is found in the Everglades. It is orange or pinkish with black stripes. *See Figure 18–44.*

Figure 18–44   Everglades rat snake. *Photo by Isabelle Francais.*

The Gray Rat Snake

Another rat snake found in the Southern Mississippi drainage area is the gray rat snake, *E. o. spiloides;* it is gray with darker blotches. *See Figure 18–45.*

Figure 18–45   Gray rat snake. *Photo by Isabelle Francais.*

The Common King Snake

The genus *Lampropeltis* consists of the king snakes and milk snakes. They are constrictors and feed primarily on small animals and birds; however, they have been known to feed on other snakes. King snakes are immune to the venom of venomous snakes.

The common king snake, *Lampropeltis getulus,* is found over the southern third of the United States. It is also referred to as a chain snake because its black body has cream-colored markings that form a chain-like pattern. *See Figure 18–46.*

Figure 18–46   Common king snake. *Photo by Isabelle Francais.*

The Scarlet King Snake

There are several subspecies of common king snake. The scarlet king snake, *L. doliata,* has a scarlet red background with black and yellow bands down the length of its body; it is often confused with the venomous coral snake. *See Figure 18–47.*

Figure 18–47   Scarlet king snake. *Photo by Isabelle Francais.*

The Sonora Mountain King Snake

One of the most beautiful snakes is the Sonora Mountain king snake, *L. pyromelano.* This snake is black with white markings down the length of its body. The speckled king snake, *L. getulus holbrooki,* is black with yellow specks all over the body. Most king snakes adapt to captivity and can be fed mice and small rats.

The Eastern Milk Snake

The eastern milk snake, *L. triangulum,* is found in the Northeast United States and South to the Gulf Coast and west to Texas. In the East, they are usually a dark brown with black-edged, gray chain bands running the length of the body. In southern areas, they are usually found with yellow and red colorings and can be mistaken for the venomous coral snake. Milk snakes are fairly small, reaching lengths of only 30 inches. Milk snakes usually do not do well in captivity and must be given some branches or bark to hide under. *See Figure 18–48.*

Figure 18–48 Southern milk snake. *Photo by Isabelle Francais.*

The Ring-Neck Snake

The ring-neck snakes belong to the genus group *Diadophis.* Ring-necked snakes are small snakes that inhabit forested areas; they are usually less than 2 feet in length. They hide under leaves and logs, where they feed on insects and earthworms. The common ring-neck snake, *Diadophis punctatus,* is black with a red or orange underside. A narrow ring is located behind the head.

The Pine Snake

The genus *Pituophis* consists of pine snakes and bull snakes. The pine snake, *Pituophis melanoleucus melanoleucus,* is a fairly large snake reaching lengths of about 5 feet. It is white with black markings and is found in pine forest areas of the East Coast, where it feeds primarily on rodents. *See Figure 18–49.*

Figure 18–49 Pine snake. *Photo by Isabelle Francais.*

The Bull Snake

The bull snake, *P. m. sayi,* is an even larger snake, reaching lengths of over 8 feet; it is yellow-brown with dark-brown to black patches along the sides and back. The underside is pale yellow with dark spots. When the snake is annoyed it will take an aggressive posture and make a hissing noise. They feed on mice and most adapt readily to captivity; however, some may retain their aggressive behavior. *See Figure 18–50.*

Figure 18–50 Bull snake. *Photo by Isabelle Francais.*

The Rough and Smooth Green Snake

The genus group *Opheodrys* consists of the rough green snake, *O. aestivus,* and the smooth green snake, *O. vernalis.* The smooth green snake is small, reaching a length of only 30 inches. The rough green snake is somewhat larger, reaching a length of up to 3 feet. They are both bright green and feed primarily on insects.

The final two families of snakes are the Elapidae and Viperinae. These are not discussed here because they are venomous snakes and should be kept only by those with considerable experience.

CLASSIFICATION—IGUANAS AND LIZARDS

Kingdom — Animalia
Phylum — Chordata
Class — Reptilia
Order — Squamata (iguanas and lizards)
Family — 16 families, 3,750 species

MAJOR FAMILIES—IGUANAS AND LIZARDS

The Family Gekkonidae

The family Gekkonidae are commonly referred to as geckos. The family includes 85 genus groups consisting of approximately 800 species. Geckos are found in tropical and semitropical regions throughout the world. They have four limbs with five toes on each limb; the toes of most geckos have adhesive pads called **lamellae.** These pads give the gecko the ability to climb up almost any surface, including the glass walls of an aquarium. Most geckos are nocturnal and have vocal cords, allowing them to communicate in the darkness.

The geckos that are nocturnal have large eyes; the pupil changes depending on the light intensity. The pupil closes to a narrow, vertical slit or to four small pinhole openings when the light intensity increases. Geckos, with the exception of the leopard gecko, do not have eyelids, but have a transparent covering over the eye. The eye covering (eye cap) is replaced each time the animal sheds its skin. *See Figures 18–51* and *18–52.*

Most geckos are insect eaters. Others feed on the nectar of plants, small mammals, and birds. Some geckos eat their skin after shedding.

Figure 18–51 The toes of most geckos have adhesive pads called lamellae. These pads enable geckos to climb almost any surface.

Figure 18–52 Most geckos are nocturnal and have large eyes with vertical pupils that close to a small slit or to four small pinhole openings when the light intensity increases.

The Tokay Gecko

The tokay gecko, *Gekko gecko,* is one of the largest geckos; it reaches a length of about 12 inches. The tokay gecko is found in the rainforests of Southern Asia, India, and Papua New Guinea. The body is a grayish-blue with small orange spots. They have large yellow eyes with slit pupils and feed on insects and small rodents. In captivity, they feed on insects and baby mice and are especially fond of wax worms. They obtain water by licking it from plants and logs. Tokay geckos get their name from the "to-kay," "to-kay" sound they make. Tokay geckos have powerful jaws and will bite. *See Figure 18–53.*

Figure 18–53 This is a tokay gecko. Notice the grayish-blue body color with orange spots. *Photo by Isabelle Francais.*

The Leopard Gecko

The leopard gecko, *Eublepharus macularus,* is a medium-sized gecko that reaches a length of about 8 inches. It is found in dry areas from Iraq to Northern India. This gecko has claws instead of the adhesive lamellae and has a movable eyelid. Its body is a gray-buff color with dark spots and blotches. The skin has wart-like projections all over its body. It feeds on insects and small rodents; water is licked from plants, logs, and rocks. *See Figure 18–54.*

Figure 18–54 Leopard gecko. *Photo by Isabelle Francais.*

The House Gecko

The house gecko, *Hemidactylus frenatus,* is commonly referred to as a chitchat gecko. It is found around human dwellings in Southeast Asia and on many islands in the Pacific Ocean. It reaches a length of about 6 inches and has slit pupils and adhesive lamellae; the lamellae do not reach the tip of the toes. House geckos feed on insects, spiders, and small rodents. Water is licked from plants, logs, and rocks. They are sandy-brown with darker spots and blotches.

The Gold Dust Madagascar Day Gecko

The gold dust Madagascar day gecko, *Phelsuma laticauda,* is found on the island of Madagascar and several surrounding islands. It is diurnal. It reaches a length of about 5 inches, and its eyes are large, round, and black and do not have slits like those of the nocturnal geckos. The day gecko feeds on insects, spiders, and fruit nectar; water is usually licked from plants. It is light yellow-green with a dusting of yellow spots. It has small blue blotches above the eyes and two orange stripes; one running across the nose and one between the eyes. *See Figure 18–55.*

Figure 18–55 Gold dust Madagascar day gecko. *Photo by Isabelle Francais.*

The Large Madagascar Gecko

The large Madagascar gecko, *Phelsuma madagascariensis*, grows to 10 inches in length. It is found on the island of Madagascar and neighboring islands. This gecko is light blue-green below with the upper body a darker green. Numerous pinkish-orange blotches and spots appear on its back and on the head. This gecko is diurnal and has large, round, black eyes. It is found in wooded areas, where it feeds on spiders, insects, and fruit nectar. Water is obtained from the leaves of plants. *See Figure 18–56.*

Figure 18–56   Large Madagascar gecko. *Photo by Isabelle Francais.*

Kuhl's Flying Gecko

Kuhl's flying gecko, *Ptychozoon kuhli*, has large, webbed feet and a flap of skin along each side of the body. These two adaptations allow the gecko to "glide" from branch to branch. This gecko is native to Southeastern Asia. It is brownish-gray with darker blotches and markings, and it grows to a length of about 6 inches. It is nocturnal, has slit pupils, and feeds on insects; water is obtained from leaves.

The Banded Gecko

The banded gecko, *Coleonyx variegatus*, is a small, thin-skinned gecko found in the arid areas of the Southwestern United States. This gecko reaches a length of only about 3 inches. It is light yellow with dark bands across its back and tail, and feeds on small insects.

The Mediterranean Gecko

One of the gecko species that has been introduced into the United States is the Mediterranean gecko, *Hemidactylus turcicus*. It is about 3 inches long and can now be found in the southern states, usually around homes; it feeds on insects. This gecko is also referred to as the Turkish gecko.

The Family Pygopodidae

The family Pygopodidae is made up of eight genus groups consisting of 31 species, and is commonly referred to as snake lizards; most species are found in Australia. They are limbless lizards that move like snakes. Most species feed on insects and rodents. Little is known about their habits, and very few are kept in captivity.

The Family Xantusiidae

The family Xantusiidae is made up of two genus groups consisting of four species; this family is commonly referred to as night lizards. They are nocturnal and are found in arid, rocky areas.

The Yucca Night Lizard

The yucca night lizard (or desert night lizard), *Xantusia vigilis,* is a small lizard found in the Southwestern United States and Mexico. It feeds on insects and spiders.

The Island Night Lizard

The island night lizard, *Klauberina riversiana,* is found on the islands off the southern coast of California; these lizards grow to about 8 inches in length. This species is primarily nocturnal but may be seen during the day. They feed on leaves, blossoms, and seeds of plants. Night lizards give birth to live young, usually one to eight after a gestation period of about 90 days. Night lizards are fairly secretive, and if kept in captivity, they should be provided with areas to hide.

The Family Dibamidae

The family Dibamidae is made up of one genus group consisting of four species; the family is commonly referred to as blind lizards. Little is known about these lizards. Three species are found in Asia and a fourth, the Mexican blind lizard, *Anelytropsis papillosus,* is found in Eastern Mexico.

The Mexican Blind Lizard

The Mexican blind lizard is flesh-colored to purplish-brown. It lacks limbs and burrows under leaf litter, in the soil, and around rotten logs. The eyes are very small.

The Family Iguanidae

The family Iguanidae is made up of 55 genus groups consisting of more than 650 different species; most species are found in the Americas, with a range from Southern Canada to the tip of South America. A few are found on the islands of Fiji and Tonga in the Pacific Ocean, the wetlands of Madagascar, and the Galapagos Islands. Members are active during the day and most lay eggs, although a few give birth to live young. There is a great variation in body size and body shape among the members of this family.

The genus *Anolis* includes the anoles and related species that make up about 50 percent of the Iguanidae family. They are primarily found in South America and the West Indies island group.

The Green Anole

This species is found in the Southern United States. The green anole or American anole, *Anolis carolinensis,* is found in trees, in shrubs, and in and around houses; it is also commonly sold in pet shops. They have adhesive lamellae. The green anole is able to change color from various grays to brown, depending on its mood, surroundings, temperature, stress, and breeding status. The male is usually a little larger than the female and displays or extends a bright-pink to red dewlap during sexual activity or to scare predators. The green anole reaches a size of about 6 to 8 inches in length. It feeds on insects and obtains water from vegetation.

The Cuban Brown Anole

Another species found in Southern Florida is the Cuban brown anole, *Anolis sagrei sagrei.* This anole is brown with light-brown to gray markings. The male displays a reddish-orange dewlap during sexual activity or when threatened. It is 6 to 7 inches long and is found on and around trees and shrubs. *See Figure 18–57.*

Figure 18–57 Cuban brown anole. *Photo by Isabelle Francais.*

The Knight Anole

The Knight anole, *Anolis equestric,* is the largest of the anoles. This species, found in Cuba, reaches a length of about 18 inches. It is green with white markings; the eyes have a blue circle around them. The males extend a pink-colored dewlap during sexual activity or when threatened. It feeds on insects but is large enough to feed on mice and small birds. Water is usually obtained from vegetation. *See Figure 18–58.*

Figure 18–58 Knight anole. *Photo by Isabelle Francais.*

The Common Green Iguana

The genus *Iguana* consists of three similar species. The common green iguana, *Iguana iguana,* is popular in the pet trade. Found in Central and South America, this arboreal lizard reaches a length of about 6½ feet. Young iguanas are about 8 inches long when they hatch from eggs. The young are omnivorous, feeding on insects, earthworms, and leaves. The adults are herbivorous and feed on fruits, leaves, blossoms, and buds. In captivity they can be fed collard greens, turnip greens, mustard greens, dandelion greens, water cress, green beans, parsnips, acorn squash and butternut squash, along with some other vegetables and small amounts of fruit. The common iguana is green and has a crest running from the neck to the tail. Two similar species are the

Iguana iguana delicatissima and the *Iguana iguana rhinolopha*. The *Iguana iguana delicatissima* lacks the large, round scale under the ear and has large scales along the side of the lower jaw. *Iguana iguana rhinolopha* has the large scale below the ear but differs because it has several erect spines on its snout. Adult male iguanas are often aggressive. In addition, the large size of this lizard makes housing it difficult as an adult. Many iguanas end up in rescue shelters as a result. *See Figure 18–59.*

Figure 18–59   Common (or green) iguana. *Photo by Isabelle Francais.*

The Galapagos Islands Marine Iguana

The genus *Amblyrhynchus* consists of one species, the Galapagos Islands marine iguana, *Amblyrhynchus cristatus.* This marine species lives near seawater and feeds only on certain types of seaweed. Because of its special diet, it is found only in a few zoos. It is dark green to black and grows to about 4½ feet in length.

The Galapagos Land Lizard

The genus *Conolophus* contains one specie, the Galapagos land lizard, *Conolophus subcristatus.* This land-dwelling iguana of the Galapagos Islands feeds on bark, flowers, fruits, and insects. This species grows to about 4 feet.

The Rhinocerios Iguana

The genus *Cyclura* consists of about five species. The rhinocerios iguana, *Cyclura cornuta,* is found in Haiti and the Dominican Republic. It has three blunt horns on its snout. It is dark brown to dull gray and grows to more than 5 feet in length. It is primarily a vegetarian but also eats rodents, birds, and other small mammals.

Turks Island and Bahama Iguanas

Two other species are the Turks Island iguana, *Cyclura carinata,* and the Bahama iguana, *Cyclura baeolopha,* found on the islands of the Bahamas and Jamaica. They are smaller but similar in appearance to the rhinocerios iguana.

The Spiny-Tailed Iguanas

The genus *Ctenosaura* contains species commonly referred to as spiny-tailed iguanas. The common spiny-tail iguana, *Ctenosaura hemilopha,* is found from Central Mexico to the border of the United States. It is primarily terrestrial, is gray, and has a short, spiny tail. Authorities disagree on the number of different species in this genus group.

The Desert Iguana

The genus group *Dipsosaurus* contains three species. The desert iguana, crested iguana, or northern crested iguana, *Dipsosaurus dorsalis,* is found in the Southwestern United States and Northern Mexico. It grows to about 18 inches in length. It is light brown with shadings of gray and brown; black lines and white spots may also appear on the body. It feeds on insects and the flowers and buds of plants.

The Helmeted Iguana

The genus *Corythophanes* is represented by the helmeted iguana, *Corythophanes cristatus,* found in Central America. Not much information exists about these iguanas. They have a large crest or "helmet" on the neck and are arboreal.

Spiny Lizards

The genus *Sceloporus* consists of the largest number of species. These lizards are referred to as spiny lizards, fence lizards, and swifts. They are found from Canada to Southern South America and the Caribbean; most species reach a length of about 10 inches. They have overlapping scales with a ridge that ends in a sharp point or spine. They are primarily terrestrial but can be found on tree stumps and among rocks.

The Eastern Fence Lizard

The eastern fence lizard, *Sceloporus undulatus,* is found along the East Coast of the United States from New Jersey and across the Southern United States. It is commonly referred to as a fence lizard because it can be found on fences basking in the sun, or as a pine woods spiny lizard because it can sometimes be found in pine woods habitats. Fence lizards have sharp claws that enable them to climb almost any rough surface. They can move very quickly. Males have bright blue-green markings on their stomachs and throats. Fence lizards feed primarily on insects.

The Rock Crevice Spiny Lizard

The rock crevice spiny lizard, *Sceloporus poinsetti,* lives among the rocks and boulder habitats of New Mexico and Texas. This lizard has a distinguishing black collar around its neck and black bands around its tail; insects are its primary food source. It gives birth to live young.

The Clark's Spiny Lizard

The Clark's spiny lizard, *Sceloporus clarki,* is found in the Western United States; it is one of the larger spiny lizards. It is light brown to gray-brown with small, darker markings along the back and head.

The Emerald Swift

The emerald swift, *Sceloporus malachitus,* is found in Mexico and Northern Central America. This is a bright-green lizard; the male has patches of blue-green on its stomach and throat. The emerald spiny lizard feeds on insects and gives birth to live young.

The Curly-Tailed Lizard

The genus group *Leiocephalus* is represented by the curly-tailed lizard, *Leiocephalus carinatus.* This lizard is found on several Caribbean islands and has found its way to Southern Florida. It is a dark-brown lizard with lighter stripes and markings on its body; the tail has bands and ends in a curl.

The Keel-Tailed Lizard

The genus *Tropidurus* is made up of keel-tailed lizards. Two species from South America are *Tropidurus semitaeniatus* and *Tropidurus torquatus.* They have a raised, spiny keel that runs from the lower back down the top of the tail. They primarily feed on insects and spiders but also eat small rodents and plant material.

Basilisks

The genus *Basiliscus* consists of lizards commonly referred to as basilisks. They are found from Southern Mexico, throughout Central America, and into Northern South America. These tree-dwelling lizards can reach lengths of up to 3 feet, with the major portion of their length being a long tail. Males have a large crest or **casque** along their backs and tails that looks like a sail or fin. Basilisks are primarily carnivorous, feeding on insects, small mammals, and birds. Basilisks live near water and are powerful swimmers. Because of the surface tension of water, these lizards have the ability to run short distances on the surface of water. They have long rear toes with fringed scales on the outer edges of the toes. By holding its body almost upright and using its tail as a counterweight, it moves its feet rapidly and runs across the surface of the water. Species that may be found in pet stores are the common basilisk, *Basiliscus basiliscus;* the double-crested basilisk, *Basiliscus plumifrons;* and the banded basilisk, *Basiliscus vittatus.* The common basilisk is brown with darker stripes on the back. It has a white throat and lower jaw, with a white stripe running from each eye back across the head and down the back. Another white stripe runs from the white throat patch back across the front shoulders. The double-crested basilisk is green with orange eyes. *See Figures 18–60 and 18–61.*

Figure 18–60 This is a female double-crested basilisk. The double-crested basilisk is also referred to as the crested basilisk, plumed basilisk, and leafy-tree basilisk. *Photo by Isabelle Francais.*

Figure 18–61 This is a male double-crested basilisk. Notice the casque, a helmet-like ridge on the head of the male. *Photo by Isabelle Francais.*

The Family Agamidae

The family Agamidae is made up of over 30 genus groups consisting of approximately 300 species; this family is referred to as chisel-teeth lizards. Their teeth are fused at the base on the ridge of the upper and lower jaws, and the front teeth are long and chisel-like. These lizards are found in Africa, the Middle East, South and Central Asia, and Australia. Very few are available in the pet trade; several species live in special habitats that would be hard to duplicate.

The genus group *Physignathus* is commonly referred to as water lizards or water dragons; they are similar in appearance to iguanas.

Water Dragons

The Asian water dragon, Chinese water dragon, or Oriental water dragon, *Physignathus cocincinus,* is commonly found in pet stores. They are native to tropical forest areas of Southeast Asia. They reach a length of about 30 inches, two-thirds of it being the tail. These lizards are green with light bands around the body and tail. There may be light blue-green patches on the throat. They feed on insects, earthworms, fish, and small rodents. Their diet is more carnivorous than that of the common iguana. *See Figure 18–62.*

Figure 18–62 Young water dragon. *Photo by Isabelle Francais.*

The water lizard, *Physignathus lesueurii,* of Australia and Papua New Guinea grows to more than 3 feet in length. It is olive-brown with darker bands around the tail. A dark stripe extends from the eye back toward the shoulders; the stomach is red. It is carnivorous and feeds on insects, small fish, small birds, and rodents. *See Figure 18–63.*

Figure 18–63 Asian water dragon. *Photo by Isabelle Francais.*

The Bearded Dragon

The genus *Amphibolurus* contains the bearded lizard, *Amphibolurus barbatus;* it is found in Australia and Papua New Guinea. This lizard inhabits the grassland area and is usually found around tree stumps and among rocks. The scales on its neck are elongated into spines that can be raised when the lizard is threatened. Its primary diet is insects, but it will eat vegetative material. This lizard is brown with lighter spots along the sides. Light bands encircle the tail, the throat and stomach area is light brown, and a light-brown stripe extends from the eye backward. The bearded lizard reaches a length of 18 to 22 inches. *See Figure 18–64.*

Figure 18–64 Bearded dragon. *Photo by Isabelle Francais.*

The Sail-Tailed Water Lizard

The sail-tailed water lizard (or soasoa), *Hydrosaurus amboinensis,* is found on the Indonesian island of Sulawesi. This is the largest member of the Agamidae family, reaching a length of more than 3 feet. This lizard has a large crest that runs from head to tail; the crest on the tail is especially large. The sail-tailed water lizard is brown and is a tree-dwelling species that takes to the water when alarmed; it is an excellent swimmer. Its diet consists of a large variety of items, including insects, earthworms, fish, birds, small rodents, and fruit.

Spiny-Tailed Water Lizard

The genus *Uromastyx* is made up of the spiny-tailed lizards, which have short, thick, spiny tails. They are found in hot, dry habitats; in captivity, this habitat must be duplicated if they are to do well. Representatives of this group are the black spiny-tailed lizard, *Uromastyx acanthinurus;* the Egyptian spiny-tailed lizard, *Uromastyx aegyptius;* and the Hardwick's spiny-tailed lizard, *Uromastyx hardwicki.* Spiny-tailed lizards change skin color during the day; in the morning, they are usually darker to help absorb the sun's rays; they become lighter as temperatures rise.

The Common House Agama

The common house agama (or common agama), *Agama agama,* is found throughout Africa. Most of them are very colorful, and several species change color during the day. In the morning, they have dark-brown coloration, and as they absorb the sun's rays, they become lighter. The male of the Namibian rock agama, *Agama planiceps,* is dark blue to purple with a red or orange head and tail; the females are usually brown or gray.

Flying Lizards

Flying lizards of the genus *Draco* have folds of skin that they extend and use to glide among the trees. Several species are known; most are native to Southeast Asia and the East Indies, where they feed on ants. Most are small and delicate and do not live long in captivity.

The Family Cordylidae

The family Cordylidae consists of sungazers, girdle-tails, and plated lizards. This family is made up of 10 genus groups and consists of about 50 species. They are inhabitants of the African deserts, where they are usually found among rocks; they feed primarily on insects and spiders.

Sungazers

The sungazers get their name because they lie with the front portion of the body raised and appear to be staring into the sun. Their scales are sharply pointed. One or two large young are born in February to March.

The Giant Girdle-Tailed Lizard

The giant girdle-tailed lizard, *Cordylus giganteus,* is the largest of the family, growing to about 18 inches. They give birth to live young.

Plated Lizards

Plated lizards also belong to the Cordylidae family. The giant plated lizard, *Gerrhosaurus validus,* is the largest member, reaching a length of more than 2 feet. It is found among the rocks and boulders of the African savanna. Its diet includes insects, spiders, rodents, and plant material. The desert plated lizard, *Angolosaurus skoogii,* is found in the sand dunes of the Namib Desert. It feeds on insects and, when disturbed, it buries itself in the sand. Plated lizards are an egg-laying species. *See Figure 18–65.*

Figure 18–65   Giant plated lizard. *Photo by Isabelle Francais.*

The Yellow-Throated Plated Lizard

Another species is the yellow-throated plated lizard, *Gerrhosaurus flavigularis.* It is found in South Africa and Madagascar. Plated lizards have a smaller lower jaw, and the throat area is covered with large scales. In some references, the plated lizards are placed in a separate family, Gerrhosauridae. *See Figure 18–66.*

Figure 18–66   Yellow-throated plated lizard. *Photo by Isabelle Francais.*

The Family Teiidae

The family Teiidae are commonly referred to as whiptails, race runners, and tegus. There are 39 genus groups consisting of 227 species. They are found from the northern parts of the United States down through Central America to Argentina and Chile.

The Six- and Seven-Lined Race Runner

The six-lined race runner, *Cnemidiphorus sexlineatus,* and the seven-lined race runner, *Cnemidiphorus inornatus,* are found in the Southern United States. The seven-lined race runner is brown with seven lighter lines running from the head down the back. The underside, throat area, legs, and tail are a light blue. Race runners live among the rocks in dry desert habitats. They feed on insects and spiders.

The Common Tegu

The common tegu, *Tupinambis teguixin,* grows to 4 feet in length. Tegus are heavy-bodied lizards that are native to South America, where they are found from the tropical rainforest to fairly dry, arid habitats. Tegus are terrestrial and dig tunnels; they are diurnal and return to their tunnels at night. They are dark brown to black with numerous white patches and markings over the entire body, limbs, and tail. They feed on snails, mice, small birds, fish, and eggs. Tegus may be difficult to keep in captivity because of their size, difficulty in getting them to eat, and difficulty maintaining proper temperature and humidity. *See Figure 18–67.*

Figure 18–67 Common tegu. *Photo by Isabelle Francais.*

The Golden Tegu

The golden tegu, *T. t. nigropunctatus,* is black with yellow markings. The golden tegu grows to 30 to 36 inches in length. They tend to be very aggressive, strong, and active. Over time and with proper handling they may become tame. They are not recommended for children or others who have no experience in keeping lizards. *See Figure 18–68.*

Figure 18–68 The golden tegu. *Photo by Isabelle Francais.*

The Jungle Runners

The jungle runners (sometimes referred to as rainbow runners), *Ameiva ameiva,* are very colorful lizards found from Central America to Northern South America; they have also been introduced into Florida. The color of this lizard varies considerably, but it usually has a brown background with green vertical stripes on the sides of the body. Blue stripes may appear on the stomach and underside of the tail as well. They may grow to 20 inches in length. They are diurnal and return to their tunnels at night. They feed on insects, small birds, and rodents. A smaller but similar lizard is the rainbow runner, *Cnemidiphorus lemniscatus.*

The Family Lacertidae

The family Lacertidae is referred to as the true lizards. Other common names are sand lizards and collared lizards. Most species have a horizontal fold between the throat and chest scales. There are 25 genus groups consisting of about 200 species.

The Green Lizard

The green lizard, *Lacerta viridis,* is a bright-green lizard found in Central and Southern Europe. It is a diurnal species that prefers the open woodland, fence rows, and hedge rows, and it feeds primarily on insects. A closely related species is the eyed lizard or ocellated green lizard, *Lacerta lepida,* found in Southern Europe. Both lizards are considered delicate and may be difficult to raise in captivity.

The Wall Lizard

The wall lizard, *Podarcis nuralis,* is a small lizard growing to about 7 inches. It may be found in a wide range of colors, from brown to green with specks and patterns of blues and yellows. Native to Central and Southern Europe, wall lizards are found on the walls of buildings, around ruins, and on rocky walls; a related species is the ruin lizard, *P. sicula.* Like the green lizards, wall lizards are considered delicate and not easily kept in captivity.

The Canary Island Lizard

The Canary Island lizard, *Gallotia galloti,* grows to about 18 inches. It is greenish-brown, and males may have blue spots on the shoulders and sides. This species is diurnal and is found around rocks. Their diet consists of insects, earthworms, snails, small rodents, fruits, and seeds.

The Pityuses Lizard

The Pityuses lizard, *Podarcis pityusensis,* is found on the islands of the Pityuses group in the Mediterranean. It is found among rocks, stone fences, stone piles, and buildings. It is diurnal and feeds on insects and spiders.

The Family Scincidae

The family Scincidae is made up of 85 genus groups consisting of more than 1,275 species; they are found in tropical areas throughout the world. Skinks are secretive, and very little is known about them. Most skinks are small, but the largest members may reach 2 feet in length. Skinks are usually terrestrial and burrow under leaves and debris on the forest floor. They are almost round in cross section, and they have indistinct necks and short limbs. The limbs of some species may be so short that they are unable to lift their bodies off the ground; some species may not have limbs.

The Five-Lined Skink

The five-lined skink, *Eumeces fasiatus,* is found in the Eastern United States. They are black when young, with white or yellow stripes running from the head to the tail, which is blue. As they mature, the body turns reddish-brown. Males may have a reddish tinge to their head. The five-lined skink is diurnal and feeds on insects and earthworms.

The Australian Blue-Tongued Skink

The Australian blue-tongued skink, *Tiliqua scincoides,* is native to Eastern and Northern Australia. Its habitat ranges from the rainforests to dry forest and scrubland areas. It has a large, broad, blue tongue that it extends when threatened. They have been bred in captivity and can produce up to 25 live young. Skinks become very tame in captivity and are considered excellent pets. They feed on snails, earthworms, and insects that they find among the ground cover and rotten vegetation. In captivity, they feed on baby mice, lean raw meat, and high-quality soaked dry cat food, along with small amounts of greens and vegetables. *See Figure 18–69.*

Figure 18–69   Australian blue-tongued skink. *Photo by Isabelle Francais.*

The Solomons Giant Skink

Other related species are *T. occipitalis, T. nigrolutea, T. gigas,* and *T. gerrardii.* The Solomons giant skink, *Cornucia zebrata,* is an arboreal skink that has a prehensile tail. It too has been bred in captivity and is commonly found in pet stores. They are **crepuscular** (active at twilight or before sunrise) to nocturnal and grow to 24 inches in length. This giant skink is native to the forests of San Cristobal Island on the edge of the Solomon Island chain. They feed primarily on plant material; in captivity, they feed on chopped fruit and vegetables. They give birth to live young. *See Figure 18–70.*

Figure 18–70   Solomons giant skink. *Photo by Isabelle Francais.*

The Family Anguidae

The family Anguidae is made up of glass lizards, slow worms, and alligator lizards. There are eight genus groups consisting of 75 species. The limbs of glass lizards are usually reduced in size and are basically useless, whereas the limbs of slow worms are absent.

The Eurasian Glass Lizard

The Eurasian glass lizard, *Ophisaurus apodus,* is the largest of this family, reaching a length of up to 5 feet. It is tan to bronze-brown with a yellowish head and underside. They are found in open woodland and rocky areas, where they hide among rocks and burrows that they have made in the loose soil. They feed on snails, earthworms, small rodents, and eggs. Eurasian glass lizards are native to Eastern Europe and parts of Asia.

The Slender Glass Lizard

The slender glass lizard, *O. attenuatus,* is native to North America. It is found in tall-grass prairie habitats, where it hides among grass and vegetation; it may also occupy the burrows of other animals. The slender glass lizard is primarily active during the day, but if the weather gets too hot, it becomes active at twilight and during the night.

The Northern and Southern Alligator Lizards

The northern alligator lizard, *Elgaria coerulea,* and the southern alligator lizard, *E. multicarinata,* are found in the Western United States from Southern Washington to Baja, California; both are about 12 inches in length. The southern alligator lizard lays eggs in burrows, whereas the northern alligator lizard gives birth to live young. The northern species prefers cooler, moister habitats. Both are primarily terrestrial but can climb trees and branches aided by a partially prehensile tail.

Arizona and Texas Alligator Lizards

Two other species are the Arizona alligator lizard, *E. kingi,* and the Texas alligator lizard, *E. liocephalus.* The Arizona species has black and white spots on its upper jaw. The Texas species is the largest of the alligator lizards, reaching a length of about 16 inches. Alligator lizards have the ability to drop off their tails as a decoy for predators as the lizard makes an escape. *See Figure 18–71.*

Figure 18–71   One of the alligator lizards, *genus* Elgaria. The exact species is unknown. *Photo by Isabelle Francais.*

The California Legless Lizard

The California legless lizard, *Aniella pulchra,* is found in California and Mexico. This limbless lizard is about 6 to 10 inches long. It burrows in the soil along stream banks, beaches, and sand dunes. It feeds on insects and earthworms. In appearance, it looks very much like a snake, but has a moveable eyelid.

The Family Xenosauridae

The family Xenosauridae contains two genus groups with only four species. They are commonly referred to as crocodile lizards. These lizards have large scales along the back. They are primarily terrestrial and live in damp habitats. Their main diet consists of fish, tadpoles, and amphibians. They give birth to live young. Three species of the genus *Xenosaurus* are found in the rainforests of Mexico and Guatemala. The fourth species, belonging to the genus *Shinisaurus,* is found in Southern China.

The Family Varanidae

The family Varanidae is made up of one genus group consisting of 31 species. They are commonly referred to as monitor lizards; most are found in Australia and Southeast Asia. Monitor lizards vary greatly in size from the 8-inch short-tailed pygmy monitor, *Varanus breicauda,* to the 10-foot giant Komodo dragon, *Varanus komodonsis.* All monitors have long necks, powerful limbs, strong claws, and powerful tails, and all are egg layers. Most monitors are large animals and can be dangerous; therefore, they are not recommended as pets.

The lace monitor, *V. varius;* the yellow monitor, *V. flavescens;* and the sand monitor (*see Figure 18–72*), *V. gouldii,* are species that may occasionally be found in pet shops. The savannah monitor, *V. exanthematicus,* is another monitor that is sometimes found in the pet trade, but it can reach an adult length of more than 6 feet and become hard to handle. The savannah monitor feeds on mice, rats, small birds, eggs, hamsters and gerbils. In captivity, it feeds on dog and cat food. *See Figure 18–73.*

Figure 18–72   Sand (or Gould's) monitor. *Photo by Isabelle Francais.*

Figure 18–73   Savannah monitor. *Photo by Isabelle Francais.*

The Family Chamaeleonidae

The family Chamaeleonidae consists of the true chameleons. The family is made up of four genus groups with 855 species. Most are found in Africa and Madagascar, with a few found in Europe and Asia. Chameleons have the ability to change color. The outer layer of skin doesn't contain any pigmented cells. Specialized pigmented cells called **chromatophores** lie in layers under the outer layer. The first layers of pigmented cells are red and yellow. Under the red and yellow layers are layers of blue and white pigment. Changes in light, temperature, and the emotional state of the chameleon cause the cells to enlarge or shrink, allowing various colors of light to enter or be reflected. This causes the changes in color. A chemical called **melanin** in the cells causes the cells to darken. Chameleons have tongues as long as 5½ inches, and the tip is covered with mucus. They can shoot their tongue out in a flash to catch insects.

Chameleons also have excellent eyesight. Their eyes can move independently of each other and can focus on different objects at the same time if necessary. Most are arboreal with opposing toes that allow them to grasp branches and limbs; most have prehensile tails.

The Common Chameleon

The common chameleon, *Chamaeleo chamaeleon,* grows to about 10 inches in length and is found along the coasts of North Africa and Southern Europe. It feeds on mealworms, flies, and spiders.

Meller's Chameleon

Meller's chameleon, *Chamaeleo melleri,* is one of the largest species, growing to up to 2 feet in length. Their natural color is green with yellow stripes and black spots. They have a single small horn growing from their forehead and a tongue that is up to 20 inches long that they flick out very quickly to capture their prey. They feed on insects and small birds. The Meller's chameleon is common to East Africa, Northern Mozambique, and Tanzania.

Veiled Chameleon

The veiled chameleon, *Chamaeleo calyptratus,* is one of the more commonly bred and available species of chameleons. They are relatively hardy, large, beautiful, and prolific. The veiled chameleon is native to Yemen and Saudi Arabia. One of the distinctive characteristics is the casque, a helmet-like ridge, on the top of its head. *See Figure 18–74.*

Figure 18–74   A veiled chameleon. Notice the crest on the head of the animal. *Photo by Isabelle Francais.*

Jackson's Chameleon

The males of some species, such as the Jackson's chameleon, *Chamaeleo jacksonii,* have projections or horns on their head. Several species, including the South American dwarf chameleon, give birth to live young. *See Figures 18–75 and 18–76.*

Figure 18–75   Female Jackson's chameleon. *Photo by Isabelle Francais.*

Figure 18–76   Male Jackson's chameleon. *Photo by Isabelle Francais.*

The Family Helodermatidae

The family Helodermatidae is made up of one genus group with two species; they are referred to as beaded lizards and are the only venomous lizards in the world. Beaded lizards are heavy-bodied, terrestrial lizards that inhabit dry areas. Their scales are small and beadlike and do not overlap. The venom glands are on the lower jaw, and when the lizard bites, the venom flows by capillary action upward into the wound; they do not inject the venom like a snake. The venom of beaded lizards is usually not fatal to humans but can cause swelling around the bite wound, severe pain, vomiting, and fainting. The beaded lizard has powerful jaws and can hold onto its prey for long periods.

The Gila Monster and Mexican Beaded Lizard

The Gila monster, *Heloderma suspectum,* of the Southwestern United States and Mexico, and the Mexican beaded lizard, *H. horridum,* of Mexico and Guatemala are the two species in the family Helodermatidae. These two species are the only species of venomous lizards. Both grow to about 24 inches in length. The Gila monster is black with yellow or white markings, and the Mexican beaded lizard is black with pinkish markings. They have small eyes, and their vision is not very good. They rely on their senses of hearing and smell to catch their prey. They feed on small mammals, birds, and eggs; beaded lizards are oviparous. One needs to check local laws because many states prohibit the keeping of these two species. *See Figure 18–77.*

Figure 18–77   Gila monster. *Photo by Isabelle Francais.*

The Family Lanthanotidae

The family Lanthanotidae contains only one genus group and one species—the Borean earless lizard, *Lanthanotus borneensis.* This nocturnal lizard lives in the lowland areas of Northern Borneo. They are semiaquatic and feed on earthworms and small prey they find around rivers and streams. The Bornean earless lizard lacks a **tympanum,** or external ear, but is capable of detecting sounds.

CLASSIFICATION—CROCODILES, ALLIGATORS, AND GHARIALS

Kingdom — Animalia
Phylum — Chordata
Class — Reptilia
Order — Crocodilia (crocodiles, alligators, and gharials)
Family — 3 families exist

MAJOR FAMILIES—CROCODILES, ALLIGATORS, AND GHARIALS

The order Crocodilia consists of the largest reptiles. Today's living crocodilians are survivors of the Mesozoic era, when giant land and marine reptiles ruled the world. Crocodilians look very similar to their relatives of that era, who flourished 230 to 70 million years ago.

Crocodilians are easily recognized, and all are semiaquatic. They can walk on land but can move more easily in the water, propelled by their powerful tails. General characteristics are the large head and long jaws that are lined with teeth. They are almost all brown, olive-green, or black, which helps them blend into their habitat. Their bodies are covered with heavy, rough skin that is made up of scales called scutes. Bony plates under the scutes help give them added protection. Crocodilians are cold-blooded and spend long periods basking in the sun.

The Family Alligatoridae

The family Alligatoridae is made up of four genus groups and consists of seven species. Alligators and caimans differ from crocodiles in the way their teeth fit into the jaws. In alligators and caimans, the fourth tooth of the lower jaw fits into sockets in the upper jaw, and when the mouth is closed, the tooth is not visible. In crocodiles, the fourth tooth is visible when the mouth is closed.

There are two species of alligators: the American alligator, *Alligator mississippiensis,* and the Chinese alligator, *A. sinensis.*

The American Alligator

The American alligator is found along the East Coast of the United States from North Carolina south and across the Gulf of Mexico to Texas. They are black and reach a length of about 14 feet. The American alligator was once on the endangered species list. They were hunted for their meat, for their skins, and as pets. Several years of protection have brought them back to where the public is allowed to hunt them. Small alligators feed on insects, snails, and small aquatic life. Adults feed on larger fish, small mammals, turtles, and birds. *See Figure 18–78.*

Figure 18–78   American alligator. *Photo by Isabelle Francais.*

Caimans

Caimans are found from Central America southward to the central parts of South America; they commonly find their way into the pet trade. Caimans are heavily protected by an armored underside made up of overlapping, bony plates. The largest member of the caiman family is the black caiman, *Melanosuchus niger,* which can reach lengths of 15 feet. The common caiman, *Caiman crocodilus,* and the large-snouted caiman, *C. latirostris,* are found in tropical marsh areas of South America. The common caiman has been widely hunted for its skin. Dwarf caimans make up the genus *Paleosuchus;* two species are the dwarf caiman, *P. palpebrosus,* and the smooth-fronted caiman, *P. trigonatus.* Caimans adapt readily to captive situations. They are tropical animals and must have temperatures of 78 to 85°F. They are carnivorous and, in the wild, feed on earthworms, snails, minnows, frogs, and small animals. In captivity, they feed on small pieces of meat sprinkled with vitamins and bone meal to provide adequate nutrition. *See Figure 18–79.*

Figure 18–79   Small caiman of the genus *Paleosuchus.*
Photo by Isabelle Francais.

The Family Crocodylidae

The family Crocodylidae is made up of three genus groups consisting of 14 species. They range in size from 4 feet to the giant crocodiles, which may reach 25 feet in length and weigh more than 2,000 pounds. One species, the American crocodile, *Crocodylus acutus,* is found in extreme Southern Florida, where it survives in brackish waters. Crocodiles are generally not considered part of the pet trade. *See Figure 18–80.*

Figure 18–80 American (or Florida) crocodile. *Photo by Isabelle Francais.*

The Family Gavialidae

The family Gavialidae contains two genus groups and two species: the false gharial, *Tomistoma schlegelii,* and the true gharial, *Gavialis gangeticus.* Gharials have very long, narrow snouts that are used for catching fish, which is their main diet. Gharials are seldom exported and are not considered part of the pet trade. *See Figure 18–81.*

Figure 18–81 False gharial. *Photo by Isabelle Francais.*

HOUSING AND EQUIPMENT

It would be impossible in the scope of this text to cover all of the different housing and equipment requirements necessary to keep reptiles in captivity. In the following material, some general guidelines are given as a starting place in providing housing and equipment for reptiles. These guidelines can be altered to meet the requirements of specific species.

Habitats for reptiles generally fall into one of four types:

1. Terrestrial habitats for reptiles that live on land

2. Semiaquatic habitats for reptiles that live on land or in trees but spend part of their time in the water

3. Aquatic habitats for reptiles that live primarily in the water

4. Arboreal habitats for reptiles that live in trees

Reptiles can be kept in a terrarium, a vivarium, or an aquarium. A **terrarium** is a cage used for keeping land or terrestrial animals. A **vivarium** is a cage in which the environment is duplicated as closely as possible to the original environment of the species. An aquarium is a water environment for aquatic animals. A combination of two of these would be needed for animals that are semiaquatic. *Vivarium* is the term that is used in this discussion. *See Figure 18–82.*

Figure 18–82 A very nice, well-decorated vivarium for land turtles. A vivarium does not need to be this elaborate.

It is difficult to give general guidelines for the size of a vivarium. It would depend on the size of the species, the number of animals, how active they are, and how large the species will be when they mature. Some of the larger species of reptiles are less active than the smaller species; therefore, they may not need a proportionally larger vivarium.

Following are some general guidelines. A land turtle needs a vivarium that is at least six times, in length and width, the length of the turtle's shell. If the turtle's shell is 6 inches long, the vivarium should be approximately 36 inches long by 36 inches wide. If rocks, logs, or decorations are to be added, the vivarium should be larger. If additional turtles are going to be added, figure one-third more area needed for each additional turtle. A vivarium approximately 48 inches long and 48 inches wide would be needed for two turtles. For aquatic turtles, an area five times as long and three times as wide as the turtle's shell should be provided. Platforms made from wood or rocks should be provided so that the turtles can climb out of the water and bask in the rays of a heat lamp. The size of the basking area will influence the overall size of the vivarium. The depth of the water in the vivarium should be at least 1 foot. The height of the sides should be sufficient to prevent the turtle from climbing out and escaping.

The substratum in the bottom of a land turtle's vivarium can be 4 to 6 inches of soil, sand, dry leaves, or wood mulch. These materials allow for normal digging and hiding, but are difficult to clean. Lining the cage with newspaper, then providing a plastic box for hiding that contains organic bedding solves this problem. Misting the hide box materials with water also provides an area of increased humidity. Depending on the species, branches, large rocks, and logs can be added. All land turtles should be provided with a shallow container of water. The depth of the water in the container should never be more than the height to the front edge of the turtle's carapace. The sides of the container should be gradually sloping so that the turtle can easily get out of the water. No material or substratum is needed in the bottom of an aquatic turtle habitat.

Small lizards, such as geckos and anoles, that are 5 to 6 inches long can be kept in vivariums a minimum of 12 inches wide, 12 inches long, and 16 inches high. The vivarium should also contain rocks, plants, tree limbs, or branches to give these active lizards things to climb on. Some species are not social and should be housed individually except during breeding.

A ratio of length to width to height of 1:2:2 is sometimes used in determining the size of a vivarium for active tree- or wall-climbing lizards. In planning for terrestrial lizards, a ratio of 2:1:1 is sometimes used. Larger geckos and more active lizards need a larger vivarium. A vivarium 60 inches long, 24 inches wide, and 24 inches high may be needed to keep several spiny lizards. Common iguanas need a vivarium at least as large as 80 inches by 60 inches by 60 inches.

The most commonly used material for vivariums is glass aquariums; these are easily obtained from pet stores and discount stores.

Figure 18–83 The most commonly used material for vivariums is glass aquariums. The substratum can completely cover the bottom, be partitioned into dry and wet areas, or a water bowl can be used to provide a wet area.

They should be fitted with an appropriate cover to keep the animals from escaping. Covers can be purchased or constructed from wood and wire mesh screen. Plexiglas can also be used to construct vivariums, with holes drilled to provide ventilation. Glass and Plexiglas are definitely needed for aquatic, semiaquatic, or rainforest habitats. Wood and wire mesh screen can be used to construct cages for dry habitats. In providing vivariums for reptiles, the environment should be duplicated as close as possible to the original; however, consideration should also be given to the ease with which the vivarium can be cleaned and maintained. Very elaborate vivariums can be used but are not always necessary. *See Figure 18–83.*

Water in aquatic and semiaquatic vivariums should be changed every 3 to 7 days to prevent the buildup of bacteria and harmful wastes. An alternative to changing the water is to install a circulation pump and filter system, along with an aeration system, to help maintain water quality. *See Figure 18–84.*

Reptiles are **ectotherms,** meaning they cannot generate their own body temperature and take on the temperature of their environment. Because reptiles are unable to generate their own body heat, a heat source must be provided. An incandescent light with a reflector shield makes a good light as well as a good heat source. The wattage of the bulb used depends on the amount of heat needed and the distance the light is placed above the vivarium. Incandescent bulbs, ceramic heating bulbs, and under-tank heating pads can be used to create a gradient of temperatures within the cage so that the animal can bask or move to a cooler area as needed. Reptiles bask until their body temperatures reach an optimal level, then they seek shaded areas or areas of lower temperatures. A thermometer and thermostat in the vivarium is important so the temperature can be monitored and maintained in the **preferred optimal temperature zone (POTZ)** for the species. The temperature

Figure 18–84 Water in semiaquatic and aquatic systems must be changed every 3 to 7 days, or a circulation pump and filter system, along with an aeration system, needs to be used.

under the heat lamp may need to be as high as 100°F. To duplicate normal habitats, a timer may need to be added to a second heat lamp so that the lamp is turned off at night, thereby maintaining higher temperatures during the day and cooler temperatures at night. Relative humidity should be maintained in the normal range for the species and can be monitored with a hygrometer. *See Figure 18–85.*

Sunlight is important not only as a heat source, but also the skin absorbs the ultraviolet (UV) rays that are needed in calcium metabolism, formation of pigment, and vitamin D synthesis. Inadequate lighting can lead to metabolic bone disease and other ailments. A full-spectrum UVB fluorescent lamp is recommended. A UVB heat lamp will also provide the needed UV rays as well as heat. The lamps should not be placed on top of a piece of glass because the glass will filter out the UV rays. These lamps emit full-spectrum light for only a limited time and should be replaced every few months. In warm months, it is advisable to house the reptile outside during the day. Outside caging should be predator-proof and placed in an area that is not in full sun so it does not get too hot.

Vivariums for aquatic and semiaquatic species need a method of heating the water, which should be maintained at a temperature mimicking the natural environment of the species. Aquarium heaters can be used. To prevent reptiles from damaging the heating unit, it should be placed between a couple of rocks.

If additional heating is needed, a heating cable or a heating mat can be used. Heating cables run through the substratum in the bottom of the vivarium. A piece of wire mesh screen placed over the cables will prevent reptiles

Figure 18–85 Reptiles seek areas where the temperature feels good to them. A thermometer should be used to monitor temperatures.

from digging up the cables and possibly causing burns. Placing heat cable or mats under the bottom of the cage and varying the thickness of the substrata can increase safety. Heat rocks are not recommended, because many animals have been burned by them.

With the use of more than one heat source, the animals can move to the area they find most desirable. Remember that reptiles lie or bask in the warm areas until their body temperature reaches an optimal level, then they seek cooler areas.

Reptiles that originate in tropical rainforests may need to have sprinkler or misting systems to duplicate the daily rainfall of their original habitat; these sprinklers need to be on a timer. Heat and humidity requirements for this type of habitat will differ considerably from that of other types of habitats.

Before leaving this discussion, it is important to talk about electricity and its safe use.

1. Always purchase electrical equipment that has been UL approved.

2. Always read the instructions before installing and using a piece of electrical equipment.

3. Make sure all heat lamps are placed so that they will not be sprayed or splashed with water and so animals cannot come into direct contact with them.

4. Water heaters should be approved for underwater use.

5. Do not open the vivarium, attempt to handle animals, or make any changes to the vivarium until the electricity is turned off.

6. All electricity should flow through a switch box with appropriate fuses or circuit breakers.

7. All circuits should be equipped with Ground Fault Interrupters (GFIs); the GFIs will disconnect the power when a dangerous condition exists.

FEEDING

Turtles

Turtle should be fed every day. All aquatic turtles are primarily carnivorous or omnivorous. In captivity, aquatic turtles can be fed commercially prepared turtle pellets, leafy lettuce such as romaine, raw fish, small goldfish or minnows, and earthworms.

Most land turtles are primarily herbivores, feeding on grasses and plants. They should be given pieces of alfalfa, clover or timothy hay, leafy dark greens such as kale and mustard greens, carrots, sweet potatoes, green beans, peas, radishes, strawberries, and cantaloupe. A variety of foods should be fed. Omnivorous land turtles can be offered earthworms, wax worms, soaked dry low-fat dog food, or cooked chicken in small amounts, as well as the same foods offered to herbivorous land turtles.

Snakes

All snakes are carnivorous. Small snakes in captivity readily feed on small fish, earthworms, mealworms, crickets, and other insects. Medium-sized snakes feed on baby mice and rats. If available, they may also feed on frogs and toads. Big snakes feed on full-grown mice, rats, baby chicks, guinea pigs, chickens, and rabbits. In feeding snakes live food such as rats and mice, one should observe the snake closely. If the snake does not eat the rodent within a few minutes, the rodent should be removed from the snake's vivarium. A rat or mouse left in with a snake can bite and injure the snake. Feeding killed prey is preferred for safety. Frozen rats and mice can be purchased for feeding snakes.

If several types of food have been offered to a snake and it has not eaten in 4 weeks or more, one should consult with a veterinarian. The snake may need to be force-fed. Snakes that are about to shed do not eat. When the eyes cloud over, a snake is getting close to shedding.

Lizards

Feeding lizards in captivity is fairly easy. Most are insect eaters. Lizards feed on crickets, as well as mealworms and earthworms, which are readily available from pet shops. Iguanas and larger lizards eat mealworms, earthworms, lettuce, flower blossoms, fresh fruit, vegetables, ground meat, and dog or cat food (*See Figure 18–86*). Always research the proper diet for your specific species.

Figure 18–86 Iguanas will eat several different types of fruit. This iguana is being fed a piece of watermelon. *Photo by Isabelle Francais.*

HANDLING

Turtles

Most turtles found in pet shops can easily be handled. Small turtles are picked up by the shell with the fingers and thumb and placed in the hand. Larger turtles may necessitate use of both hands, placing the fingers under the lower shell (plastron) and the thumbs on the top shell (carapace). One should not be too alarmed if a hissing sound is heard because this is usually a rush of air being expelled from the lungs as the turtle withdraws its head and limbs. Turtles that are going to be transported can be placed in a cardboard box.

Snakes

Snakes that have not been handled will be frightened, and their reaction is to bite in self-defense. In handling a snake the first few times, a pair of leather

Figure 18–87 Small snakes usually coil up around the hand when being lifted. *Photo by Isabelle Francais.*

gloves should be worn. The bite of a small snake is not very painful, but sharp teeth can tear the skin as the hand is pulled back in a reflex action. Once the snake has been handled a few times and does not struggle or attempt to bite, the gloves will not be needed.

When picking up a small snake, one should grasp it firmly just behind the head. A small snake usually coils up around the hand and allows itself to be lifted. Larger snakes may need their body supported and lifted with the other hand. The snake should be held close to the handler's body. *See Figure 18–87.*

When transporting a snake, one should place it in a cloth bag; a pillowcase works well. A pillowcase is porous and allows the snake to have air. If the weather is cool, the bag can be placed in a Styrofoam box or cooler. Holes must be punched or drilled into the box to allow air circulation. If necessary, a hot-water bottle can be placed in the box to provide warmth. The snake must not be allowed to get chilled.

Lizards

The temperament of lizards may vary considerably. Some lizards may allow themselves to be handled, whereas others may remain wary and aggressive forever. Lizards that are obtained at a young age are probably the easiest to work with. Small lizards can be picked up by grasping them with the thumb and forefinger behind the head and holding the animal around the body with the rest of the hand. Larger lizards can be held in the same way, but the other hand should be used to support the rear of the lizard. The large iguanas, tegus, and monitors can be picked up using one hand to grasp the animal around the neck and the other hand to support the mid- and rear section of the animal. The tail can be held close to the body with the arm and elbow. *See Figures 18–88* and *18–89.*

The bite of small lizards most likely will not cause any pain or injury; they are usually not strong enough to even break the skin. Larger lizards, however, are capable of causing serious bites; some have powerful jaws and may clamp down and not let go. The handler should wear gloves to pick up animals that may possibly bite. Some lizards are capable of inflicting serious scratches from sharp and powerful claws. Any bites and scratches should be cleaned with soap and water and treated by a doctor.

Figure 18–88 Small lizards can easily be picked up with and held in one hand. *Photo by Isabelle Francais.*

Figure 18–89 Some lizards are capable of causing serious bites as well as inflicting serious scratches with their sharp and powerful claws. *Photo by Isabelle Francais.*

One should never pick a lizard up by the tail; lizards are capable of dropping or losing their tails as a defense mechanism. The predator is left holding the tail while the lizard heads for cover. The lizard will grow or regenerate another tail, but the new tail usually lacks the coloration and pattern of the original.

DISEASES AND AILMENTS

When obtaining a reptile, one should make sure the reptile is healthy. Captive-bred animals are usually healthier. Many reptiles are imported and have been subjected to stressful conditions, extreme changes in temperature, unsanitary conditions, and lack of food or water and inhumane handling. The reptile should be observed carefully, making sure it has been eating or is eating. An animal that has been eating and is well nourished will be filled out, and the ribs, vertebrae, and pelvic bones will be covered and not show. The animal's eyes should be open, and the animal should be alert to its surroundings. Its mouth should be closed; any frothing or foaming at the mouth, around the nose, or running at the eyes indicates an unhealthy animal. If the animal shows bite wounds or marks of other injuries, these should be healed and scarred over. One should avoid purchasing animals with open wounds, wounds that are seeping, or wounds that do not appear to be healing. Turtles that are sick usually drag the rear of their shell on the ground instead of carrying the shell horizontally off the ground. Aquatic turtles that are sick may lean to one side when in the water. *See Figure 18–90.* One should look carefully for parasites; reptiles can be infected with ticks and mites. Animals that appear to be infested should be avoided. Also, one should check to see what the policy of the pet shop is on returning sick animals or if they have a replacement policy for animals that may die after leaving the shop.

Figure 18–90 A healthy land turtle walks with its shell parallel to the ground, whereas a sick turtle drags the rear of its shell on the ground. An aquatic turtle should swim level in the water. If it swims at an angle, it may be sick.

Reptiles are surprisingly free of diseases and ailments. If one starts out with a healthy, active reptile, chances are very good that one can avoid disease problems in the future. It is important to keep the vivarium clean and sanitary. The major cause of ailments is poor diet or substandard housing. Feces should be removed and any leftover food materials should be removed daily before they spoil. One must make sure the reptiles have a supply of fresh water. It is vital to have proper heating, lighting, and humidity.

Bacterial Diseases

Salmonella

Salmonella is a bacteria that may appear normally in the digestive tract of a reptile. A normal population of *Salmonella* does not cause any problems, but when populations get out of control, prolonged diarrhea and anorexia may occur; this diarrhea is often watery, green, and foul smelling. *Salmonella* bacteria can be transmitted from reptiles to humans; children are especially susceptible. It is important to thoroughly wash the hands after handling reptiles and cleaning vivariums. If *Salmonella* infection is suspected, one should consult a veterinarian.

Aeromonas and Pseudomonas

Aeromonas and *Pseudomonas* are two types of bacteria that cause digestive infections, which can lead to poor nutrition and stress. Animals with digestive infections have diarrhea and are anorexic. If not treated, further problems in the form of respiratory ailments and pneumonia can develop. Symptoms of respiratory ailments are difficulty breathing, blocked nostrils, and nasal discharge; bubbles may be seen around the nostrils when the animal exhales. The animal may hold its mouth open in an attempt to breathe easier. Treatment includes making sure the animal has a clean, dry, warm vivarium and the use of antibiotics.

Pathogenic Amoebas

Pathogenic amoebas, *Entamoeba invadens,* may invade the digestive system and cause watery, slimy diarrhea and anorexia. Treatment by a veterinarian is recommended.

Mouth Rot

Mouth rot is a fungal infection with open wounds or sores on the mouths of reptiles. Reptiles that rub their mouths and snouts up against the side of the vivarium or on decorations in the vivarium are likely to become infected. Mouth rot needs to be treated by a veterinarian.

Shedding

Snakes and lizards in good health shed their skin several times during the year; snakes normally shed their skin in one piece. The first sign that a snake is getting close to shedding is that the covering over the eye becomes cloudy. The shedding starts at the snout. The snake rubs the snout on a rock or piece of wood to get the shedding process started. Lizards normally shed their skin in patches. The tokay gecko pulls the skin off in one piece and then eats it. This is normal. The process of shedding should not take more than a couple of days. A reptile that does not shed properly or completely may have a nutritional, environmental, or health problem. Moisture is usually needed for the skin to shed. Snakes may lie in their water container prior to shedding.

Turtles also shed their skin. Small pieces of old skin come off constantly; however, if there is a continuous shedding or peeling of the skin from the same area, a bacteria or fungus may be causing a problem. One should make sure the vivarium is not too damp or wet and provide the turtle with drier conditions. One must make sure the turtle is eating and has a good nutritional diet. A visit to the veterinarian may be in order so that the turtle can be treated.

External Parasites

Mites and Ticks

Mites and ticks are external parasites that often attach themselves to reptiles. One should not attempt to pull ticks out of the skin by their body; this may break the mouthparts off the tick and leave them in the skin, which can lead to an infection. The tick can be removed using a pair of tweezers to grasp it by the head. Individual mites are so small that they may not be noticed, but when an animal becomes heavily infested, the mites can be seen crawling around. Heavy infestations of mites can cause anemia, anorexia, depression, stress, listlessness, lack of appetite, and even death. If mites are visible or are suspected, a veterinarian should be consulted. Several methods are recommended for controlling mites, and a veterinarian can prescribe the best method. Both the reptile and its environment must be treated. Before new reptiles are introduced into a vivarium, they should be put in a quarantine vivarium and observed for ticks and mites.

Internal Parasites

Roundworms and Tapeworms

Roundworms and tapeworms are internal parasites that can occasionally become a problem. Reptiles caught in the wild almost always carry these worms. Symptoms are anemia, listlessness, loss of appetite, anorexia, and possibly death. One should check with a veterinarian about diagnosis and treatment.

Vitamin Deficiencies

Vitamin deficiencies often occur in captive reptiles and are usually related to improper diet and housing. Untreated nutritional deficiencies may lead to deformed shells on turtles, respiratory diseases, and pneumonia. Proper use of heat lamps, UVB fluorescent lamps, and vitamin and mineral supplements should prevent deficiencies.

REPRODUCTION

Turtles

Turtles are oviparous (lay eggs that hatch later). The female deposits eggs in a nest cavity that she digs in the soil. The eggs are covered and then left unattended. The eggs of turtles vary considerably in size, shape, shell texture, and number laid. Number laid may vary from 1 to 3 to as many as 200. Some species lay eggs with soft leathery shells, whereas others lay hard-shelled eggs. The incubation period also varies considerably; the eggs of soft-shelled turtles hatch in about 30 days, whereas the eggs of some land tortoises may take up to 18 months. Incubation time depends on the temperature of the soil. When the young hatch, they are on their own, relying on inherited instincts for survival.

It is almost impossible to tell the sex of young turtles; as they age, some differences begin to appear. On some species, the plastron of the male becomes somewhat concave. This may aid in the male's ability to mount the female; the tail of the male becomes thicker and longer than that of the female. The cloacal opening is located further down the tail on the male. The cloaca of reptiles, amphibians, and birds is a cavity into which both the intestinal and genitourinary tracts empty. *See Figure 18–91.*

During mating, the male climbs onto the carapace of the female. The claws on the front limbs are used to grasp onto the front edge of the female's carapace. The large domed shell of the box turtle presents some problems for the male during mating; it is impossible for the male to hold onto the female's carapace. To complete copulation, the male falls backward off the female's carapace.

Snakes

Snakes may be oviparous, ovoviviparous, or viviparous (see Glossary). Most are oviparous, laying their eggs in shallow holes covered with a thin layer of soil. Some snakes lay their eggs under or around rocks, in hollow logs, or around the stumps of trees. It is important that the eggs not get too hot during the day or too cool at night; temperature is important in the development of the embryos. Snakes that are ovoviviparous retain the eggs within their bodies until they hatch and then give birth to the young. The young receive no nourishment from the female, only the nourishment contained in the egg. The garter snake is an example of an ovoviviparous species. In a few species, the young are retained in a preplacental sac that allows for some exchange of oxygen and nutrients. This is referred to as **viviparous.** Pit vipers are an example of a viviparous species.

Most egg-laying snakes leave the eggs unattended after laying them. Pythons, however, coil themselves around their eggs and brood them until

Note the concave plastron

Figure 18–91 Determining the sex of turtles. The plastron of the male may be slightly concave. The tail of the male is thicker and longer than that of the female. The cloacal opening is located further down the tail on the male.

they hatch. Incubation periods vary with different species. Egg-laying snakes lay eggs from 30 to 85 days after mating, with the eggs incubating for 40 to 90 days. The gestation period for live-bearing snakes varies from 90 to 150 days. The number of eggs varies from 1 to as many as 60. Live young varies from 2 to as many as 100. The young free themselves from the eggs with the aid of a sharp egg tooth on the tip of their snout and slice open a hole in the egg through which they can emerge. Most young are on their own as soon as they hatch. Pythons may stay near the mother for 2 or 3 days before they move off on their own. Some species of rattlesnakes may stay with their young and defend them for several days, but most snakes show no interest in their young; some may even eat their own young.

The sex of snakes may be difficult to determine; very little difference exists between males and females. Females usually have a shorter tail. Many species of snakes will have spurs near the base of the tail; these are what remain of the legs of their ancestors. These spurs are usually longer in males than females. Females may also be heavier and longer than males. Accurate sex determination can be done by "**popping**" juvenile snakes or "**probing**" adult snakes. Both procedures should only be done by those who are adequately trained to prevent pain and injury to the snake.

Male snakes have a double penis, or **hemipenes.** The hemipenes often contains recurved spurs that aid the male in holding onto the female's cloaca during mating. During mating, the male and female usually entwine themselves around each other. The male moves forward and inserts one or both hemipenes in the cloaca of the female. The female may store the sperm in the oviduct for several months before fertilizing eggs.

Lizards and Iguanas

Lizards and iguanas may be oviparous, ovoviviparous, or viviparous; most species are oviparous. The females lay eggs in holes they dig in the soil; they may dig several holes before finding one that meets their approval. The time necessary for young to hatch varies considerably. For the smaller anoles and geckos, hatching takes place in 30 days or less, but for the larger monitor lizards, it may take as many as 120 days. Incubation periods vary depending on environmental factors. Anoles usually lay a single egg, whereas some species of the iguana family lay as many as 50 soft-shelled eggs. Geckos differ from other lizards because they lay two hard-shelled eggs, which are attached by an adhesive substance to cracks in tree bark.

Most of the young lizards have a sharp egg tooth on the end of their snout that they use to slice open the egg. Most lizard females do not stay around after the eggs are laid, so the young are on their own as soon as they emerge.

Many species of lizards show definite differences between the males and females. This is referred to as **dimorphism.** Determining sex in species with definite dimorphism is fairly easy; however, this characteristic may not develop until the lizards reach sexual maturity. Some differences may be observed to help determine the sex of young lizards. The male hemipenes, when

Figure 18–92 Male geckos can be distinguished from females by the presence of anal or preanal pores.

Figure 18–93 Male agamas and most species of the family Iguanidae have thigh or femoral pores.

not being used for copulation, is withdrawn into the cloacal vent. This gives the male a larger or swollen appearance to the area of the cloaca and tail base; the female has a thinner tail base. Male anoles have two large scales behind the cloaca. Male geckos have anal or preanal pores forward of the cloaca. Male agamas and iguanas have femoral pores on the inside of the rear thighs. *See Figures 18–92* and *18–93.*

During the period prior to mating, males may take on bright body colors, display brightly colored dewlaps, or engage in a series of ritual head bobbing or body movements to entice a female into mating. If the male finds an interested female, he may grasp or bite the female by the neck in an effort to hold onto her. The male then moves his cloaca as close as possible to the cloaca of the female. When the two cloacas are together, one of the male's hemipenes is inserted into the female's cloaca. The sperm flows into the female's cloaca, where it then enters the oviducts. In some species, the sperm can be retained or stored for long periods before fertilization takes place. In some cases, several fertilizations of eggs may occur from only one mating.

In species that bear live young, the mother usually shows no mothering instincts; the young are left to fend for themselves as soon as they are born.

Crocodilians

The reproduction habits of crocodilians are similar to that of lizards, except that crocodilians usually mate in the water. All crocodilians lay hard-shelled eggs, which are usually deposited in nests. The nests may be cavities that are dug into the soil. The eggs are laid into the cavity and then covered with soil. Nests may also be built into mounds from soil and vegetation, with the eggs being deposited in the center of the mound.

Female crocodiles lay between 16 and 100 eggs, alligators and caimans lay between 20 and 30 eggs, and gharials lay about 40 eggs. The number of eggs laid depends on the age and size of the female. Crocodilian females stay within the vicinity of the nest to guard it against predators. After about 90 days, the young inside the eggs start making high-pitched sounds that alert the female. The female begins to dig the top away from the nest so the young can emerge as they hatch; some species even gather up the young in their mouth and carry them to the water.

Environment

To get reptiles to mate and reproduce in captivity, the environment that the species came from must be duplicated as closely as possible. This may involve a series of heat lamps, ultraviolet lights, water heaters, timers, thermostats, and spray equipment.

Reptilian females in captivity should be provided with a medium in which to lay their eggs. It is important to know the habits of the species so that proper facilities can be provided. Normally, a tray or box filled with damp sand is sufficient.

The eggs of reptile species that do not **brood** (stay with and incubate their eggs) can be removed after the female has completed laying her clutch of eggs and then artificially incubated.

To artificially incubate eggs, they should be placed in a shallow container that has been filled with an incubation medium; this can be sterile peat moss or vermiculite. It should be of enough thickness so that it covers the eggs; an equal weight of water should be added to the medium. The medium should be damp but not overly wet. The eggs should be placed in the medium so that they are covered from one-half to two-thirds. One should try to place the eggs in the same position they were in originally; unlike bird eggs, they should not be turned or rotated. *See Figure 18–94.*

A cover with several small holes in it should be placed over the container; the holes allow oxygen to get to the eggs. The container should be placed in an incubator at about 85°F and the eggs observed daily. Most reptile eggs have a soft, leathery shell; if there is too much moisture, a white, fluffy mold may grow on the eggs. This mold should be wiped off the eggs, and some moisture should be allowed to evaporate. If a small indentation appears on the eggs, this is because moisture is lacking, and more water should be added to the medium.

As the young hatch, they can be moved from the incubator and placed in a nursery cage. If proper conditions exist, the young should begin feeding shortly after hatching. Proper knowledge of the species is important so that proper food is provided for the young. Several different kinds of food may need to be tried before the young will feed.

Lizards usually feed on insects such as crickets, flies, mealworms, wax worms, and small worms. Snakes are carnivorous; small snakes feed on earthworms, mealworms, crickets, and other insects. Larger snakes feed on baby mice, baby rats, and frogs.

Figure 18–94 An incubator using heated water to provide warmth and humidity. The eggs are placed in peat moss or vermiculite and then placed in a small container. The eggs should be covered one-half to two-thirds. Place the container with the eggs on a rack inside the incubator. Temperature should be maintained at 85°F with a heater that has a thermostat.

SUMMARY

Reptiles are cold-blooded vertebrates that possess lungs and breathe atmospheric air. They also have a bony skeleton, scales, or horny plates covering the body, and a heart that has two auricles and one ventricle.

There are approximately 6,500 species of reptiles belonging to four different orders: Testudines (called Chelonia in some references), which are made up of the turtles, tortoises, and terrapins; Serpentes, which includes snakes, pythons, and boas; Squamata, which includes iguanas and lizards; and Crocodilia, which includes crocodiles, alligators, caimans, and gharials.

Present-day reptiles basically have three body types. The first have long bodies and clearly defined tails like the crocodiles and lizards. The second have long bodies that taper into tails like the snakes, and the third have short, thick bodies encased in shells like the turtles and tortoises.

The skin of reptiles has a horny surface layer. This surface layer must shed regularly to allow for the continued growth of the reptile. Snakes and some lizards cannot close their eyes. Instead of eyelids, they have a transparent layer, or brille, that permanently covers the eye. When the animal sheds its skin, this covering is also shed and is replaced by a new covering.

Habitats for reptiles generally fall into one of four types: terrestrial habitats for reptiles that live on land, semiaquatic habitats for reptiles that live on land or in trees but also spend part of their time in the water, aquatic habitats for reptiles that spend their time primarily in the water, and arboreal habitats for reptiles that live in trees. To be successful at raising reptiles in captivity, it is important to duplicate the animal's original habitat as closely as possible.

Reptiles are ectotherms; they are unable to generate their own body heat and take on the temperature of their environment. Because they cannot generate their own body heat, a heat source must be provided.

Reptiles are generally healthy animals. One should observe a reptile carefully when considering obtaining one. If the animal appears healthy, if it is eating, and if it does not appear to have any parasites, chances are good the animal will survive for a long time in captivity.

Turtles are oviparous. They lay eggs that hatch after an incubation period. Snakes may be oviparous, ovoviviparous, or viviparous. Snakes that are ovoviviparous retain the eggs within their bodies until they hatch, then give birth to the young. The young receive no nourishment from the female, only the nourishment contained in the egg. In a few species, the young are retained in a preplacenta sac that allows for an exchange of oxygen and nutrients; this is referred to as viviparous. Lizards and iguanas may be oviparous, ovoviviparous, or viviparous. Crocodilians are oviparous and usually mate in the water.

DISCUSSION QUESTIONS

1. What is a reptile?

2. What are the four orders of reptiles?

3. What are the characteristics of turtles, tortoises, and terrapins?

4. What are the characteristics of lizards?

5. What are the characteristics of snakes?

6. What are the differences between pythons and boas?

7. What are the characteristics of crocodiles, alligators, caimans, and gharials?

8. Why do reptiles shed their skin?

9. What are scutes? What are the different types of scutes?

10. Discuss the terms *oviparous, ovoviviparous,* and *viviparous.* What would be some advantages and disadvantages of each?

11. What are the four types of habitats that reptiles live in?

12. Discuss the types of housing and equipment that would provide the four habitats required by reptiles in captivity.

13. What is an ectotherm?

14. Discuss the types of foods needed to maintain reptiles in captivity.

15. What are the signs of good health to look for when obtaining a reptile?

16. What are some signs of ill health?

SUGGESTED ACTIVITIES

1. Visit local pet shops and make a list of the different species of reptiles that are available. Using the library, compile as much information as you can about the species.

2. Obtain some aquariums or use wood and wire mesh screen to construct a habitat for a reptile. Obtain a reptile from the pet shop. Place it in the habitat and observe its habits.

3. Research the different orders, families, genus groups, and species of reptiles. Why are they classified as they are?

ADDITIONAL RESOURCES

1. Amphibian Embryology Tutorial
   http://worms.zoology.wisc.edu/frogs/index.html

2. Florida Museum of Natural History
   http://www.flmnh.ufl.edu/natsci/herpetology/
   herpetology.htm#Top

3. San Diego Zoo.org
   http://www.sandiegozoo.org/animalbytes/a-reptiles.html

4. The EMY System
   http://emys.geo.orst.edu/default.html

5. The Reptile Database
   http://www.tigr.org/reptiles/search.php

6. Welcome to Who Zoo
   http://whozoo.org/index.htm

7. The Natural Source
   http://www.northern.edu/natsource/fauna1.htm

8. Zipcode Zoo
   http://zipcodezoo.com

9. LLL REPTILE and Supply Company, Inc.
   http://lllreptile.com/infolibrary/animal-care-sheets

10. The New Reptile Database
    http://www.reptile-database.org

CHAPTER 19

Birds

OBJECTIVES

After reading this chapter, one should be able to:

- describe the characteristics of birds.
- compare and contrast the different orders of birds discussed.
- describe methods used in handling and feeding birds.
- describe the various housing and equipment needs of birds in captivity.
- describe the various diseases and ailments of birds.

TERMS TO KNOW

aviary	flight feathers	powder-down feathers
cere	grit	preen
clutch	isthmus	scalloped feathers
contour feathers	lores	
coverts	lutinos	scapulars
crown	mandibles	sternum
down feathers	mantle	
filoplume feathers	papilla	

CHARACTERISTICS

Birds are two-legged, egg-laying, warm-blooded vertebrates with feathers and wings, and they are believed to have evolved from prehistoric reptiles. Evidence that would suggest that birds evolved from reptiles include:

1. Birds have scales on their legs and feet similar to those of reptiles.
2. Birds are egg-laying like many species of reptiles.

3. Birds have feathers, which are modified scales.

4. Both birds and reptiles have similar soft anatomy, including musculature system, brain, heart, and other organs.

5. Birds and reptiles show many skeletal similarities.

6. Bird-like fossils have been discovered in several countries.

The first bird is believed to be *Archaeopteryx;* the fossil of this bird-like reptile was found in the Bavarian area of Germany. *See Figure 1–4* in Chapter 1. *Archaeopteryx* lived approximately 150 million years ago during the Jurassic period and was about the size of a pigeon. Its forelimbs formed short wings that were covered with feathers. *Archaeopteryx,* unlike modern birds, had teeth, a long tail, and three claw-like digits on the wings. Modern birds appeared about 80 million years ago during the Cretaceous period; by the Cenozic Era, about 70 million years ago, they were very similar to birds today. *See Figure 19–1.*

Modern birds have thin skin that consists of two layers: the outer layer, or epidermis, and the inner layer, or dermis. From the dermis emerge the bird's feathers, and there are basically four different types. **Contour feathers** give the bird its outward form; those contour feathers that extend beyond the body are used in flight and are called **flight feathers.** There are both primary and secondary flight feathers. The small, soft feathers located under the contour feathers are referred to as **down feathers;** their primary function is to conserve body heat. Short, hair-like feathers called **filoplume feathers** are found over the birds' entire body. Piloplume feathers may play a role in sensory function and serve as insulation. Their sensory function, is possibly to send information about how feathers need to be adjusted to keep them in place during preening, display, and flight. These **filoplume feathers** appear as the hairs on chickens that have been plucked of their contour and down feathers. **Powder-down feathers** have tips that break down as they mature and release a talc-like powder that provides waterproofing and luster; these feathers are found on parrots and cockatoos. Fully mature feathers molt. The molting process usually occurs once a year and is a gradual process. Some species may also molt or partially molt before the breeding season.

The skeletal structure of a bird is similar to that of other vertebrate animals but has several unique features that enable the bird to fly. The **sternum** of birds forms a large, thin keel; this design provides a large foundation for the attachment of the bird's powerful flight muscles. Many of the bones that make up the vertebra and neck are fused together, and several of the rib bones are in turn fused to the vertebra. This provides a strong, rigid structure necessary for flight. *See Figure 19–2.*

The forelimbs of the bird have been modified into wings. Wing bones are hollow, lack marrow, and contain air sacs that are attached to the bird's respiratory system. Air sacs are also found in other parts of the bird's body; these make the bird lighter and provide additional oxygen needed during flight.

The skull bones of the bird are elongated toward the front of the head. In most birds, the upper beak of the bird is fused to the skull. In others, both the upper part of the beak and the lower beak (**mandibles**) are hinged. This allows the mouth to open wide and provides greater flexibility when feeding. There are several beak types depending on the feeding habits of the bird. Most of the birds covered in this text are seed eaters and have short, powerful beaks for breaking open seeds.

ANATOMY

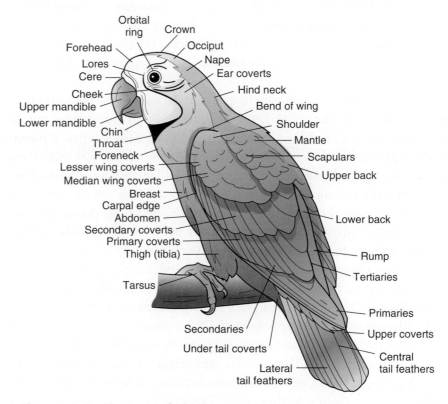

Figure 19–1 The parts of a bird.

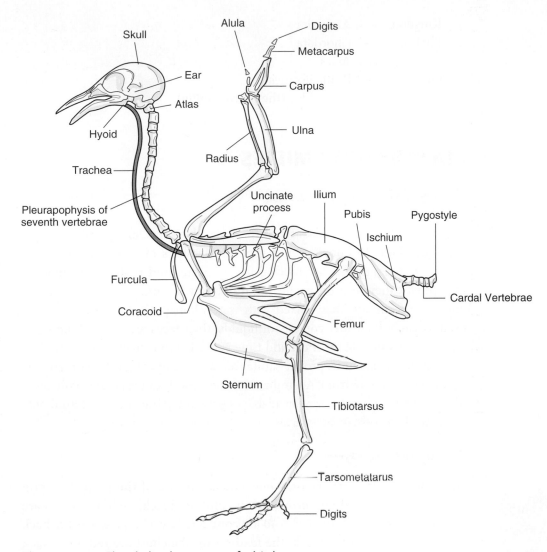

Figure 19–2 The skeletal structure of a bird.

CLASSIFICATION—MACAWS, CONURES, PARROTS, AND RELATED BIRDS

Modern birds are classified into 28 orders containing 170 families and approximately 9,000 species. This text, however, covers only three orders: the order Psittaciformes, which is made up of the parrots, cockatoos, macaws, lories, and parakeets; the order Piciformes, which is made up of the woodpeckers and related species (toucans); and the order Passeriformes, which is made up of the perching birds.

The order Psittaciformes consists of only one family, Psittacidae, and six subfamilies. The order Piciformes is made up of six families, with one family discussed in this text. The order Passeriformes is the largest order of birds, containing almost 60 percent of the bird species. This order is broken down into four suborders and 65 families. Of these 65 families, only 5 families have members kept in captivity and are discussed in this text.

Kingdom — Animalia
Phylum — Chordata
Class — Avies
Order — Psittaciformes
Family — Psittacidae (the parrot family)

MAJOR SUBFAMILIES

The Subfamily Loriinae

The subfamily Loriinae consists of the lories, lorikeets, and similar species; there are about 15 known genus groups. Members of this subfamily have brush-like tongues that are used to lap up nectar and pollen from flowers. In captivity, they feed on nectar mixtures, which are usually obtained in a powdered form and then mixed with water; honey or glucose can also be used. Because their diet consists of liquids, their feces are also soft and of a liquid consistency. More effort and time is needed to provide proper feeding and care for these birds; proper sanitation is very important. Most members of this family come from Australia, New Guinea, Indonesia, and Polynesia. Many are rare and not always available in pet stores. They may be found only in the wild, in zoos, or in aviaries.

The Rainbow Lory

The rainbow lory, *Trichoglossus haematodus,* is one of the more common species. The forehead and **crown** of the bird are black. Yellow patches are found on each side of the neck and are connected by a band across the back of the neck. The chin is black; the feathers on the chest are red with black margins; and the back, wings, and belly are green. The thighs are yellow with green margins. The bill is orange, and the feet and legs are gray. *See Figure 19–3.*

The Subfamily Cacatuinae

The subfamily Cacatuinae is commonly referred to as cockatoos, and there are six genus groups. Members of this subfamily are easily recognized by the crest or tuft of feathers on the top of the head. Cockatoos are found in Australia and on islands of the South Pacific. Cockatoos and related species make excellent pets. Most are beautiful birds, tame easily, are considered intelligent, and learn to mimic words and sounds. Most cockatoos breed in captivity, although this may present a challenge.

The Palm Cockatoo

The genus *Probosciger* consists of one species—the palm cockatoo, *Probosciger aterrimus.* This is a completely gray-black cockatoo that reaches a length of 30 inches. It has a large beak and long crest feathers. It feeds on seeds and nuts. This species is rare in captivity.

Figure 19–3 Rainbow lory. *Photo by Isabelle Francais.*

The Black Cockatoos

The genus *Calyptorhynchus* consists of four species. They range in size from 23 to 28 inches in length. Because the primary color is black, they are referred to as black cockatoos. In the wild they feed on grubs found in the bark and wood of trees, on seeds of the eucalyptus trees, seeds and blossoms of proteas, casuarinas and other small trees, insect larvae, and seeds of other plants. In captivity, they can be fed on seeds and nuts and pelleted food. Adult males have black beaks; the females have light-colored or white beaks. These large birds need large cages to be able to spread their wings and tails. Members of the genus group are the red-tailed black cockatoo, *C. magnificus;* the glossy black cockatoo, *C. lathami;* the white-tailed black cockatoo, *C. baudinii;* and the yellow-tailed black cockatoo, *C. funereus.*

The Gang-Gang Cockatoo

The genus group *Callocephalon* consists of a single species and two subspecies. *C. fimbriatum* is commonly referred to as the gang-gang cockatoo. It is about 14 inches long. The primary body color is gray, and the head is red or scarlet. The crest of red feathers is curly and appears to be in disarray. The gang-gang cockatoo is rare in captivity.

The Greater Sulphur-Crested Cockatoo

The genus group *Cacatua* consists of about 10 species and represents some of the most popular cockatoos. The greater sulphur-crested cockatoo, *C. galerita,* is the most common. *See Figure 19–4.* The bird is white with a bright-yellow crest. The feet and beak are black. Males have dark brown to black eyes, and the females have a reddish tinge to their black eyes. Surgical or DNA sexing is necessary for accurate sex determination. These birds become very tame and affectionate and like a lot of attention. There are five subspecies.

Figure 19–4 Greater sulphur-crested cockatoo. *Photo by Isabelle Francais.*

The Lesser Sulphur-Crested Cockatoo

The lesser sulphur-crested cockatoo, *C. sulphurea,* is similar to the greater sulphur-crested cockatoo except it is smaller in size. They are about 13 inches long. Five subspecies are recognized.

The Umbrella-Crested Cockatoo

The umbrella-crested cockatoo, *C. alba,* is also referred to as the white-crested or greater white-crested cockatoo. This bird is white with a slight tinge of yellow under the wings and tail.

The Moluccan Cockatoo

The Moluccan cockatoo, *C. moluccensis,* is about 20 inches long. The primary body color is a salmon-pink underlayer of feathers with white outer feathers. The crest is salmon-pink, and the beak is black. *See Figure 19–5.*

Figure 19–5 Moluccan cockatoo. *Photo by Isabelle Francais.*

The Philippine Cockatoo

The Philippine cockatoo, *C. haematuropygia*, is a small cockatoo that is about 13 inches long. It is reddish to pink with a short, white crest.

Leadbeater's Cockatoo

Leadbeater's cockatoo, *C. leadbeateri*, is about 15 inches long, and the primary body color is white. The underside of the wings, abdomen, neck, chin, and sides of the face are light pink. The crest feathers have bands of red and yellow with white tips. This cockatoo makes an excellent pet and is a good breeder in captivity. Two subspecies are recognized.

The Bare-Eyed Cockatoo

The bare-eyed cockatoo or Corella cockatoo, *C. sanguinea*, is considered one of the most intelligent of all birds. It is white with a short, white crest. Light-yellow feathers are found under the wings and tail, and a distinctive blue skin patch is found around and extends below the eye. The beak is white.

The Slender-Billed Cockatoo

The slender-billed cockatoo, *C. tenuirostris*, is very similar to the bare-eyed cockatoo except it has a very long, slender upper beak.

Ducorps' Cockatoo and Goffin's Cockatoo

Two other species are the Ducorps' cockatoo, *C. ducorps*, and Goffin's cockatoo, *C. goffini*. Both species are white with tinges of red and pink. They are about 11 inches long. *See Figure 19–6.*

The Rose-Breasted Cockatoo

The genus group *Eolophus* consists of one species and three subspecies. The rose-breasted cockatoo, *E. roseicapillus*, is another of the more popular species of cockatoo. The primary body color is gray, and the face, breast, and abdomen are light pink. The crest is also light pink. These birds are about 15 inches long. They may also be referred to as roseate cockatoos and Galah cockatoos. This species is considered mischievous and aggressive, but they do make excellent pets.

The Cockatiel

The genus *Nymphicus* consists of only one species, the cockatiel, *Nymphicus hollandicus*. Cockatiels are native to Australia but have become one of the most popular pet birds in the United States. These birds are about 12 inches long with the tail being about the same length as the body. They make very affectionate companions. Cockatiels are common in pet shops and are fairly reasonable in price. Although there is only one species, several color varieties are available. The gray or wild-colored cockatiel is the most readily available. White cockatiels are primarily white but have light-yellow and peach markings; their eyes are red. White cockatiels with black eyes are also available.

Figure 19–6 A Ducorps' cockatoo. *Photo by Isabelle Francais.*

Cockatiels with primarily yellow coloring are referred to as **lutinos.** Pearled cockatiels are light gray in color with white or yellow spots over the body and wings. Bordered cockatiels are similar to the pearled variety but have dark margins on the feathers. The pied cockatiels or harlequin variety have symmetrical pied markings. The cinnamon cockatiel is a brownish color with the feathers becoming yellow at the tips, giving the bird a cinnamon-colored appearance. *See Figure 19–7.*

Adult male cockatiels have bright-yellow faces and crests with a bright-orange patch on the cheeks. The underside of the adult male's tail feathers are a solid grey color with no barring. Females and immature birds have light-yellow faces and crest. The cheek patches are a duller shade of orange. The undersides of the female's tail feathers have a yellow barred pattern.

Cockatiels are considered ideal for beginning bird fanciers and for youngsters. They are easy to raise, breed readily in captivity, and make affectionate pets.

Figure 19–7 Cockatiel. *Photo by Isabelle Francais.*

The Subfamily Psittacinae

The subfamily Psittacinae consists of more than 330 species, including macaws, conures, parrots, parakeets, parrotlets, and other birds. All members of the subfamily have four toes; the two center toes face forward, and the two outside toes point backward. The bills of these birds are large and are used in climbing, clasping, and feeding and for defense. Their heads are large in comparison to the rest of their bodies; they have short necks and a large, fleshy tongue, and all have a hooked bill.

Macaws

Macaws are large, beautiful birds found from Mexico to Paraguay and Southern Brazil. They are hardy, long-lived birds that may live for 75 years or more. Macaws have large beaks, long tails, and bare areas around the eyes and face. The beak of the macaw is the largest of the parrot family. The power in the bird's beak is tremendous. They could easily cut off a human finger if they chose to do so; however, the birds are gentle and affectionate. They are considered highly intelligent, are very playful, and have an average ability to mimic the human voice. The macaw's natural voice is very loud and harsh. There are three genus groups of macaws.

The Hyacinth Macaw. The genus group *Anodorhynchus* consists of three species. The hyacinth macaw, *A. hyacinthus,* is one of the most beautiful macaws. The primary body color is a deep hyacinth blue. Yellow skin patches are present around the eyes and on the lower mandible. The beak and feet are black. The hyacinth macaw may reach a length of more than 34 inches. These birds are rare and are not readily available; prices may range from $6,500 to $12,000. *See Figure 19–8.* Two other species in this genus group are the glaucous macaw, *A. glaucus,* and the Lears macaw, *A. leari.* Both are smaller versions of the hyacinth macaw; however, they lack the bright, deep-blue sheen and are only about 30 inches long.

Figure 19–8 Hyacinth macaw. *Photo by Isabelle Francais.*

Figure 19–9 Blue and gold macaw. *Photo by Isabelle Francais.*

The Blue and Gold Macaw. The genus group *Ara* consists of about 13 different species. The blue and gold macaw, *A. ararauna,* is one of the more popular and commonly seen macaws. These macaws may reach a length of more than 30 inches. The neck and abdomen are a bright golden-yellow, and feathers on the crown are greenish-blue. Bright-blue feathers run from the back of the head down the back; the wing feathers and tail feathers are blue. The underside of the blue tail feathers are tinged with yellow. The blue and gold macaw has a white cheek patch with rows of tiny black feathers. There is a black collar or bib under the lower mandible; the beak and feet are black. This species is much lower in price than the hyacinth macaw but still commands prices from $900 to $1,400. The blue and gold macaw is considered more alert and more intelligent than other macaws. It is also very curious and mischievous. *See Figure 19–9.*

The Scarlet Macaw. The scarlet macaw, *A. macao,* is sometimes referred to as the red and yellow macaw. The primary body color is a bright red or scarlet. The wings are red, but bright-blue feathers and yellow feathers tinged with green are also found on the wings. The cheek patch is white with no rows of tiny feathers, and the upper mandible is white; the lower mandible is black. This macaw reaches a length of about 30 inches. It is one of the most commonly seen macaws but may command prices of more than $1,000.

The Yellow-Collared Macaw. The yellow-collared macaw (or yellow-caped macaw), *A. auricollis,* is a small parrot that reaches a length of about 16 inches. The primary color is green, with the head usually being a darker green than the body. A yellow collar on the lower back part of the neck is the distinguishing characteristic. The distinct yellow collar is broader at the back of the neck. There is a yellow patch of feathers above the nostrils. A few red and blue feathers are found on the wings. Smaller macaws like the yellow-caped macaw are sometimes referred to as dwarf macaws or mini-macaws. These smaller macaws are usually lower priced than the larger macaws. Yellow-caped macaws are considered good talkers. The severe macaw, *A. severa,* is another of the smaller macaws; it may reach a length of up to 20 inches. The primary color is green, with a patch of dark brown above the beak. The crown is blue; lines of dark feathers appear on a large, white, bare cheek patch. The beak and feet are black. There are a few blue wing feathers, and red appears on the underside of the wings and tail. The bird is considered a fairly good talker and makes an affectionate pet. A second species, *A. severa castaneifrons,* is a little longer.

Illiger's Macaw. Illiger's macaw, *A. maracana,* is also a small macaw; it reaches a length of about 15 inches. The primary color is a dull olive green. There is a red-orange patch of feathers above the nostrils; the crown and head are blue-green. Some red feathers appear on the abdomen and wings. The upper tail feathers show a reddish tinge, and a yellow tinge is found on the underside of the feathers. The beak is black.

The Red-Bellied Macaw. The red-bellied macaw, *A. manilata,* is another of the small macaws and reaches a length of about 15 inches. The primary color is a dull olive green. The forehead and crown are black, changing to blue-green on the nape of the neck. The long flight feathers are blue with black tips.

The upper side of the tail is tinged with red, and the underside is tinged with yellow. The beak is black, becoming lighter at the tip.

The Hahn's Macaw. The Hahn's macaw, *A. nobilis nobilis,* is the smallest of the macaws. It reaches a length of about 12 inches. The primary color is a dull green. The forehead and crown show a bluish tinge. Some red appears on the wings, and the flight feathers are tinged with black. The underside of the wings is tinged in red, and the underside of the tail is tinged in yellow. A bare, white skin area exists between the eyes and the beak. Two other subspecies are recognized: the noble macaw, *A. nobilis cumanensis,* and the long-winged macaw, *A. nobilis longipennis.*

The Red-Crowned Macaw. The red-crowned or red-fronted macaw, *A. rubrogenys,* is about 17 inches long. The primary color is a dull olive green. A bright red-orange patch appears on the forehead and crown. This bright red-orange also appears on the shoulders, **scapulars,** the small feathers that cover the base area, and the outer tips of the wing. The flight feathers are a dull blue; the underside of the tail is tinted gray. The upper side of the tail is tinged with blue and yellow. The feathers on the thighs are reddish-colored. There is a bare skin patch on the cheeks with lines of black feathers.

Other Macaws. Coulon's macaw, *A. couloni;* the great green macaw, *A. ambigua;* and Wagler's macaw, *A. caninde,* are three other species of rare macaws.

The Spix Macaw. The third genus group is *Cyanopsitta.* One species, the Spix macaw, *C. spixii,* is known. This is among the most rare species of macaws. Perhaps as few as 20 birds are known to exist in captivity; there may not be any left in the wild. The distinctive characteristic of this macaw is a head completely covered with feathers; there is no bare skin patch around the eyes or on the cheek like the other macaws. The primary color is a blue-green. The head is gray, and the beak is black. This macaw reaches a length of about 22 inches. Its tail is longer than that of other species of macaws.

Conures

Conures are about the size of the dwarf or mini-macaws. They are found in the wild from Mexico throughout Central America and most of South America. They differ from the macaws because they do not have a bare skin area around the eyes or cheek area; all macaws have a bare area to some extent. Conures are long and slender in shape and have large, tapered tails. They also have large heads and large, powerful beaks. Conures make very loud and harsh sounds. There are eight genus groups of conures; several are rare and unavailable in the pet market.

The Queen of Bavaria Conure. The genus group *Aratinga* consists of 18 species; 7 species are discussed here. The Queen of Bavaria conure, *A. guarouba,* is also referred to as the golden or yellow conure. The primary color is a bright golden-yellow; the long flight feathers are green. The beak and feet are pink, and it has orange eyes. It reaches a length of about 14 inches. This beautiful bird is rare and very expensive.

Figure 19–10 Sun conure. Note that the long flight feathers have been trimmed on this bird. *Photo by Isabelle Francais.*

The Sun Conure. The sun conure (or yellow conure), *A. solstitialis,* is considered by many to be the most beautiful and brilliantly colored of the conures. The primary color is a bright yellow-orange. The head shows more orange, especially around the eyes and in the cheeks. The wings are yellow with green and blue flight feathers. The eyes, beak, and feet are black. The sun conure is about 12 inches long. It is available in the pet market. *See Figure 19–10.*

The Janday Conure. The Janday conure, *A. jandaya,* is one of the more popular conures. The head is yellow. The neck, throat, and chest is yellow with a vivid mixture of red-orange. The wings are primarily green, but the longer flight feathers are blue. The tail feathers are an olive green shading to blue and ending with black tips.

The Petz' Conure. The Petz' conure, *A. canicularis,* is also known as the half-moon conure and orange-footed conure. The primary color is green; the upper parts are usually a darker green than the lower parts. The flight feathers are blue and green, and there is a bare yellow ring around the eye. There is an orange band on the forehead, and the crown is a dull blue. The Petz' conure becomes very tame and makes an excellent pet, except for its loud, harsh sound. There are two subspecies, *A. canicularis eburnirostrum* and *A. canicularis clarae.*

Wagler's Conure. The Wagler's conure (or red-fronted conure), *A. wagleri,* is a somewhat larger conure that reaches a length of about 14 inches. It is all green except for red on the forehead and crown. Some streaks of red appear on the throat; the beak is yellow.

The Cherry-Headed Conure. The cherry-headed (or red-headed) conure, *A. erythrogenys,* is about 13 inches long. The primary color is a bright green. The front part of the head is red; red covers the eye and cheek area. There are also red feathers on the shoulders, underside of the wings, and on the thighs. There is a large, yellow ring around the eyes. *See Figure 19–11.*

The Brown-Eared Conure. Eight subspecies of the *A. pertinax* are recognized. The common names of the various subspecies are derived from variations in color patterns. The primary color is a dull green, and they are not considered very attractive. The most common of the subspecies is the brown-eared conure, *A. pertinax ocularis.* It has brown on the sides of the face, cheeks, and throat. Other subspecies are the brown-throated conure, *A. p. aeruginosa;* the Margarita brown-throated conure, *A. p. margaritensis;* the Tortuga conure, *A. p. tortugensis;* the yellow-cheeked conure, *A. p. chrysophrys;* the Aruba conure, *A. p. arubensis;* the St. Thomas conure, *A. p. pertinax;* and the Bonaire conure, *A. p. xanthogenia.*

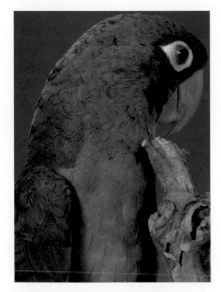

Figure 19–11 Cherry-headed (or red-headed) conure. *Photo by Isabelle Francais.*

The Cactus Conure. The cactus conure, *A. cactorum,* is a small bird that is about 11 inches long. The forehead and crown are brown; the throat and chest are brown to olive brown. Green tinges the cheeks, and the wings are green. Flight and tail feathers are gray-blue. There is a white ring around the eye; the beak is yellow.

The Golden-Crowned Conure. The golden-crowned conure, *A. aurea,* is about 10 inches long. There is a bright-orange patch above the nostrils and on the forehead followed by blue on the crown, changing to green on the head. The throat and chest are yellow, tinged with green. The upper parts of the wings and tail are green with yellow on the underside of the wings and tail; the beak is black.

The Nanday Conure. The genus *Nandayus* consists of one species—the Nanday conure, *N. nenday.* The primary color is green, and the head is black. The flight feathers and the tail feathers are a blue-black. It has bright-red thighs; the bill, legs, and feet are black.

The remaining genus groups of conures, *Leptosittaca, Cyanoliseus, Pyrrhura, Ognorhynchus, Microsittace,* and *Enicognathus,* are rare in captivity and are not discussed in this text.

Parrots

There are 19 species of what are commonly referred to as the true parrots.

The African Gray Parrot. The genus group *Psittacus* consists of only one species—the African gray parrot, *P. erithacus erithacus.* The African gray parrots are native to Africa, where they may be found from the Ivory Coast to Angola, and inland to Kenya and Tanzania; they are also found on islands near the Gold Coast. The primary color is gray. Various shadings of black and gray are found on the bird. Many of the feathers are tipped with white; the tail feathers are bright red. The beak is black, and the feet are a dusty black color. *See Figure 19–12.*

The African gray parrot is a beautiful bird, and many people consider it the best talker of all birds, with a voice most closely resembling a human's. They are also considered very alert and intelligent, and they make affectionate pets. The African gray parrot reaches a length of about 13 inches.

The Timneh Parrot. The Timneh parrot, *P. e. timneh,* is a smaller version of the African gray. It is a darker gray, and the tail is maroon. They are long-lived birds, with a normal life span of 20 to 25 years, but they may live for up to 70 years. They sell from $600 to more than $1,000.

The Yellow-Headed Amazon. The genus *Amazona* is the largest genus group, consisting of about 27 species. Most are considered loud and are not favorable pets; only the more common species are discussed. The Mexican double yellow head, *Amazona ochrocephala oratrix,* is also referred to as Levaillant's Amazon or the yellow-headed Amazon. *See Figure 19–13.* It is one of the most popular of the Amazon parrots. The primary color is a bright green. The adult has a yellow head and neck; red appears on the shoulders, wings, and near the base of the tail feathers. They reach a length of 14 inches. These parrots are considered excellent talkers.

The Yellow-Naped Amazon. The yellow-naped Amazon, *A. o. auropalliata,* is primarily green. There is a yellow patch above the upper beak and a yellow band on the nape of the neck. The parrot has a bare eye ring. The tail feathers are green with a reddish tinge near the base and a yellow tinge near the tips of the feathers; some red and blue appear on the flight feathers. The beak is gray, becoming darker at the tip. It is considered a good talker.

The Panama Amazon. The Panama Amazon, *A. o. panamensis,* is an excellent talker. It is very similar to the yellow-naped Amazon but does not have the

Figure 19–12 African gray parrots. *Photo by Isabelle Francais.*

Figure 19–13 Mexican double yellow head Amazon, also referred to as Levaillent's Amazon or the yellow-headed Amazon. *Photo by Isabelle Francais.*

yellow band on the nape of the neck. The beak is yellow with a dark tip. Both parrots are about 14 inches long.

The Yellow-Fronted Amazon. The yellow-fronted Amazon, *A. o. ochrocephala,* resembles the Panama Amazon. The yellow patch above the upper beak is larger and extends up into the crown. The eyes are orange with bare eye rings. It is considered a good talker. It is sometimes referred to as the Colombian Amazon or single yellow head.

The Green-Cheeked Amazon. The green-cheeked Amazon, *A. viridigenalis,* is also referred to as the Mexican red-headed parrot and the red-crowned Amazon. The primary color is green. The flight feathers are red and blue; the head is red, and the red descends down to the eyes. A blue-green strip runs from the eye backward and down the back of the head; the cheeks are green. The bill is a yellow-white color. This parrot is about 13 inches long. *See Figure 19–14.*

The Lilac-Crowned Amazon. The lilac-crowned Amazon, *A. finschii,* is also referred to as Finsch's Amazon. The primary color is a dull green. There is a band of bright purple or plum above the upper bill connecting the eyes; the color becomes lighter as it moves up into the crown and top of the head. The cheeks are a bright green; the feathers on the chest and abdomen have dark margins. The lilac-crowned Amazon is about 13 inches long. It is considered a fairly good talker and makes a good pet.

The Spectacled Amazon. The spectacled Amazon, *A. albifrons,* is a small parrot that reaches a length of about 11 inches. It is sometimes referred to as the red, white, and blue parrot because those three colors are found on the head. A red patch above the upper beak extends back to encircle the eyes; above this and extending up into the crown is white changing to blue. As with most of the Amazon parrots, the flight feathers are blue and green. The beak is yellow-white. Two subspecies are recognized: the lesser white-fronted Amazon, *A. a. nana,* and the Sonoran spectacled Amazon, *A. a. saltuensis.*

The Yellow-Lored Amazon. The yellow-lored Amazon, *A. xantholora,* is very similar to the spectacled Amazon except the patch above the bill and between the eyes is yellow. It has a black cheek patch, and there is a red ring around the eye.

The Yellow-Cheeked Amazon. The yellow-cheeked Amazon, *A. autumnalis,* is about 14 inches long. Like all Amazon parrots, the primary color is green. A bright-red patch above the upper beak connects the eyes; the forehead and crown are light purple to lilac, changing to light blue on the head and back down the back of the head. The cheeks are bright yellow. The beak is yellow-white with darker edges and tip. It is considered a good talker. Three subspecies are recognized: Salvin's Amazon, *A. a. salvini;* Diademed Amazon, *A. a. diadema;* and Lesson's Amazon, *A. a. lilacina.*

The Blue-Crowned Amazon. Four subspecies of *Amazona farinosa* are known. The blue-crowned Amazon, *A. farinosa guatemalae,* is considered the most colorful of the subspecies. The forehead and crown are a bright blue

Figure 19–14 Green-cheeked Amazon. *Photo by Isabelle Francais.*

fading to a lighter blue down the back of the head and nape of the neck. All four species are about 16 inches long.

The Blue-Fronted Amazon. The blue-fronted Amazon, *A. aestiva aestiva,* is primarily green. A blue patch above the upper bill extends into the forehead; the head has a mixture of blue and yellow feathers. The cheeks are yellow, and the flight feathers are red. The blue-fronted Amazon is considered an excellent talker. A subspecies, the yellow-winged Amazon, *A. a. xanthopteryx,* is very similar to the blue-fronted Amazon except that yellow appears on the wings, and there is a yellow band on the crown above the blue. The feathers on the neck, chest, and back have dark margins. *See Figure 19–15.*

The Orange-Winged Amazon. The orange-winged Amazon, *A. amazonica amazonica,* is similar to the blue-fronted Amazon except that the blue on the forehead is a brighter color, and there is a tinge of orange in the yellow cheeks. There is a yellow circle on the crown. A subspecies, the Tobago orange-winged Amazon, is very similar.

The Festive Amazon. The festive Amazon, *A. festiva festiva,* has a bright-red rump. When the wings are folded, the red does not show. A light-purple band is located above the upper bill, and the cheeks, throat, and crown are tinged with blue. A subspecies, Bodin's Amazon, *A. f. bodini,* is similar in appearance.

The San Domingo Amazon. The San Domingo Amazon, *A. ventralis,* is about 12 inches long. It is also referred to as Salle's or the white-headed Amazon because it has a white forehead.

The Red-Vented Parrot. There are eight species of the *Pionus* genus of parrots and about 28 subspecies. The red-vented parrot, *P. menstruus,* is about 11 inches long. It is also referred to as the blue-headed parrot. The primary color is green. The head is blue, and the blue extends downward onto the neck and upper part of the breast. The vent area and tail **coverts** are bright red. The red-vented parrot makes a quiet and affectionate pet but is not considered a good talker.

The White-Crowned Pionus. The white-crowned Pionus, *P. senilis,* is also referred to as the white-capped parrot or white-crowned parrot. It is about 10 inches long. White appears above the upper beak and extends upward onto the forehead and crown; white also appears on the throat area. The chest and abdomen are a dull blue, and bronze coloring appears on the green wings. There is a bare ring around the orange eyes. *See Figure 19–16.*

The Bronze-Winged Parrot. The bronze-winged parrot, *P. chalcopterus,* is about 11 inches long. The crown, cheeks, back of the neck, tail, and upper breast are dark blue-green. The wing feathers are bronze; specks of pink appear on the throat and breast. The vent area is red, and the bill is yellow-white.

The White-Bellied Caique. The *Pionites* genus of parrots is commonly referred to as caiques. Caiques are considered very intelligent and can be amusing comics. They are about 10 inches long. The white-bellied caique, *P. leucogaster leucogaster,* is the most common of the caiques. It is also referred

Figure 19–15 Blue-fronted Amazon. *Photo by Isabelle Francais.*

Figure 19–16 White-crowned (or white-capped) Pionus. *Photo by Isabelle Francais.*

to as the white-breasted caique. The entire head of the bird is orange-yellow down to and including the neck; the breast area is white. The wings and tail are green, and there is a white ring around the eyes. The thigh feathers of the bird are green; the yellow-thighed caique, *P. l. xanthomeria,* is similar to the white-bellied caique except that it has yellow thighs.

The Black-Headed Caique. The black-headed caique, *P. melanocephala melanocephala,* has a black forehead and crown. Blue feathers appear on both sides of the upper beak. Below the black cap, the feathers around the neck are yellow tinged with some white; the thigh feathers are yellow.

The Pallid Caique. The pallid caique, *P. melanocephala pallida,* is similar to the black-headed caique except the thigh color is an apricot orange.

Hawk-Headed Parrot. The genus *Deroptyus* contains two species, and they are referred to as hawk-headed parrots. The hawk-headed parrots are about 14 inches long and have a broad, square tail that is about 4 inches long. The forehead and crown are brown with streaks of light brown to white. The feathers from the top of the head and down the back of the neck are longer than normal; these feathers are maroon with bright-blue tips. When the bird is excited or threatened, these long feathers on the back of the head and neck become erect, presenting a threatening pose. The breast of the bird also has maroon feathers with blue tips. The wings and rump are green. The upper tail feathers are a mixture of blue, green, and brown. The base of the tail feathers on males is red. The female does not have the red on her tail feathers. The beak is black. The birds are considered affectionate and make excellent pets. When feeding hawk-headed parrots, it is necessary to add fruits to their diet of nuts and seeds. Two species are recognized: the Guiana hawk-headed parrot, *Deroptyus accipitrinus accipitrinus,* and the Brazilian hawk-headed parrot, *D. a. fuscifrons.*

The Red-Capped Parrot. There are six species of the genus *Pionopsitta.* They are small parrots about 7 to 9 inches long. Considered rare, they are not readily available in the United States. One species, the red-capped or pileated parrot, *Pionopsitta pileata,* is the most common. It is primarily green with red on the cheeks and wings. The primary flight feathers are blue.

The Brown-Necked Parrot. The genus group *Poicephalus* consists of nine species. They range in size from 8 to 13 inches. They are native to the African continent. The brown-necked parrot, *P. robustus fuscicollis,* is primarily green with red on the thighs, outer edges of the wings, and shoulders. The forehead of the male is brown, and the forehead of the female is red. The feathers on the head and neck are silver, tinged with brown. Two other subspecies are recognized. All are rare and not readily available in the pet market.

Jardine's Parrot. Jardine's parrot, *P. gulielmi gulielmi,* is also referred to as a red-headed or red-crowned parrot. It is primarily green with a red crown, thighs, and shoulders. There are dark, almost black feathers on the wings and tail. The upper bill is yellow-white with a black tip; the lower bill is black. The orange-crowned parrot, *P. g. fantiensis,* is very similar to the Jardine's parrot except the markings are orange.

The Yellow-Fronted Parrot. The yellow-fronted parrot, *P. flavifrons flavifrons,* is primarily green with bright yellow on the top of the head and facial area. The tail and flight feathers are a dark green tinged with bronze; the eyes are orange.

The Senegal Parrot. The Senegal parrot, *P. senegalus,* is one of the most common of the African parrots. The primary color is green; the head is gray, and they have yellow eyes. Immature Senegal parrots have gray eyes. The breast feathers are tinged with yellow. There are three subspecies of the Senegal parrot: the scarlet-bellied Senegal parrot, *P. s. versteri;* the orange-bellied Senegal parrot, *P. s. mesotypus;* and the Senegal parrot, *P. s. senegalus.* Hand-reared Senegal parrots can be affectionate and make good pets. *See Figure 19–17.*

Meyer's Parrot. Meyer's parrot, *P. meyeri,* is a small parrot native to Central and Eastern Sub-Saharan Africa. The primary color of this parrot is brown. Yellow appears on the wings; the lower breast and abdomen are green. Some blue appears on the rump and tail feathers; the eyes are red. The beak is black. There are six subspecies that are referred to as brown parrots.

The Orange-Bellied Parrot. The orange-bellied parrot, *P. rufiventris,* is also referred to as the Abyssinian or red-breasted parrot. The upper parts of this parrot are gray; the underparts are blue-green. The throat is tinged with orange changing to bright orange on the breast.

Grand Eclectus Parrot. The genus *Eclectus* consists of one species and 10 subspecies. These beautiful parrots are native to Australia, New Guinea, and the islands of the South Pacific. Eclectus parrots are about 12 to 14 inches long. Unlike most of the other parrots, the males and females show sexual dimorphism. The males are bright green with red patches on the sides of the breast. The upper side of the wings is green, and the underside is red; the upper side of the tail is green with blue and white tips, and the underside is a blue-black with yellow tips. The lower mandible is black, and the upper mandible is yellow-white with a pink-orange base. The female has a bright-red head and breast; the red changes to a bright blue down the neck. The lower breast and abdomen are blue. The upper part of the wings is maroon while the underside is a mix of blue, maroon, and green. The upper side of the tail is maroon, changing to orange at the tips. The beak is all black and is long. Eclectus parrots are considered very affectionate, make excellent pets, and are very good talkers; however, the female is generally more dominant and may show signs of aggressive behavior. There are 10 subspecies of the grand eclectus parrot, *Eclectus roratus.* Eclectus parrots command prices from $600 to more than $2,000. *See Figure 19–18.*

There are several species of the genus group *Tanygnathus.* Most are rare and are seldom seen in captivity. The great-billed or large-billed parrot, *T. megalorhynchos,* may be kept in some breeding aviaries but seldom reaches the pet market. The primary color is green, and the beak is a bright red.

Parakeets

The genus *Psittacula* is made up of the larger parakeets; the males all have red beaks.

Figure 19–17 Senegal parrot. *Photo by Isabelle Francais.*

Figure 19–18 Grand eclectus parrot. The female is the more brightly colored of the two. She has a bright-red head and breast. The red changes to a bright blue down the neck. The male is green with red patches on the sides of the breast. The male is the least colored of the two. *Photo by Isabelle Francais.*

Figure 19–19 Indian ringneck parakeet. This is a lutino or yellow-colored ringneck. *Photo by Isabelle Francais.*

Figure 19–20 Moustache parakeet. *Photo by Isabelle Francais.*

The Indian Ringneck Parakeet. The Indian ringneck parakeet, *P. krameri manillensis,* is one of the most beautiful of the larger parakeets. The birds are about 17 inches long with a long, tapering tail that makes up about half the body length. The primary color is pastel green. Males have a narrow, black band above the nostrils that connects the eyes; below the beak is a black band that goes down the throat and then spreads out and back, making a black band below the cheek. There is a light-blue band above the black, and a light-rose band below the blue. The tail and flight feathers show some yellow; some blue tinting may appear on the nape of the neck and down the back. The beak is red; an orange ring surrounds the eye. Females are a duller green and lack the black, blue, and rose-colored rings. Young ringnecks are dull in color, and it may take 2 to 3 years before the males develop their beautiful color markings. Indian ringneck parakeets make excellent pets and are good talkers. Ringnecks are also being bred in blue and yellow forms. Ringnecks are fairly common and may be found in the pet markets for $150 to $500. *See Figure 19–19.* An African subspecies, the African ringneck, *P. k. krameri,* is smaller and lighter in color.

The Alexander Ringneck Parakeet. The Alexander ringneck parakeet, *P. eupatria nidalensis,* is a larger version of the Indian ringneck. The species is about 20 inches long; several subspecies are recognized.

The Moustache Parakeet. The moustache parakeet, *P. alexandri fasciata,* is also referred to as the banded parakeet. It has a black band across the forehead connecting the eyes, and a black patch below the lower mandible that includes the chin and throat area; most of the body is green. The head is blue-gray, and the breast is reddish-brown with tinges of violet; the beak is red. Seven subspecies are recognized. *See Figure 19–20.*

The Derbyan Parakeet. The Derbyan parakeet, *P. derbiana,* is similar to the moustache parakeet but is about 20 inches long. The primary color is green. A black band across the forehead connects the eyes, and a black patch is on the chin and throat; the head is blue-gray. The wings are green with a yellow wing patch. The tail is 7 to 9 inches long. The upper mandible of the male is red with yellow edges and tips; the lower mandible is black with yellow edges. The beak of the female is black. The female is not as colorful as the male.

The Plum Head Parakeet. The plum head parakeet, *P. cyanocephala,* is a beautiful bird, with the head being a plum or light-purple color. A black band extends down from the lower mandible, curving backward and toward the nape of the neck. The primary body color is yellow-green. The wings are green with a maroon wing patch; the beak is yellow-white. Females have gray heads. The plum head parakeet is about 12 inches long. The slaty-headed parakeet, *P. himalayana himalayana,* resembles the female plum head parakeet.

The Malabar Parakeet. The Malabar (or blue-winged) parakeet, *P. columboides,* is another beautiful bird. The male is primarily gray. A black band completely circles the neck; a blue band is below the black band and changes to a light green. The wings and tail feathers are dark blue with a yellow tinge to the tips; yellow also appears on the vent and underside of the tail. Males have a red beak, and there is a tinge of green around the eyes. Females have a black head and do not have the collar.

The Barraband Parakeet.   The genus group *Polytelis* is made up of three species. The Barraband parakeet, *P. swainsonii,* is a very beautiful bird with a long, tapering tail. The bird reaches an overall length of about 16 inches. The forehead, crown, chin, and upper throat area are bright yellow. A bright-red patch appears on the lower throat area; the bill is yellow-white. The top of the head and down the back are green; the breast and abdomen are yellow-green. The wings are green with blue flight feathers; the tail is green with tinges of pink and black. The female does not have the red and yellow markings, and the overall colors are not as bright. They are considered very affectionate and make excellent pets.

The Rock Pebbler.   The rock pebbler, *P. anthopeplus,* is about 17 inches long. The primary color is olive yellow, with the upper parts being darker than the underparts. There is a green wing patch, and the flight feathers are dark blue; some of the shorter secondary flight feathers are red. The upper tail feathers are shades of dark blue, olive, and black; the underside of the tail is tinged with red. The beak is red. The female is more of an olive green with duller shades of blue and red.

The Princess Alexandra Parakeet.   The Princess Alexandra parakeet, *P. alexandrae,* is larger than the other two birds in this genus group. It is also referred to as the Princess of Wales, Princess parakeet, and Queen Alexandra's parakeet. The male is 18 inches long; the coloring is a mixture of soft pastels. The crown is blue, changing to bright green on the top of the head; this color changes to an olive green down the back. The wings and tail feathers are olive green; the throat is a reddish-orange. The breast and abdomen are a light yellowish-red. The female is duller and has shorter tail feathers.

The Crimson Rosella.   The genus group *Platycercus* contains nine species of parakeets; most are referred to as rosellas. Only four species are common in the United States. Rosellas are recognized by a black shoulder patch and **scalloped feathers** on the back, which are black with the outer edges being of a different color. The crimson rosella, *P. elegans elegans,* is about 14 inches long. The head, neck, and underparts are crimson red. The upper side of the tail is dark blue and green, and the underside is light blue; the cheek feathers are blue. The scalloped feathers on the back are black with green and red margins. The rump is red, and there is a violet-blue coloring to the wing feathers.

The Red Rosella.   The red rosella, *P. eximius eximius,* is smaller than the crimson rosella. The head, neck, and throat are bright red, which descends into the breast area in a V-shape; there is a white patch on the chin. The breast and abdomen are yellow. The scalloped feathers down the back are black with yellow margins. The flight feathers are dark blue; secondary flight feathers are lighter blue, and there is a red wing patch. The upper side of the tail feathers is greenish-blue, and the underside is light blue; the vent is red. The male is a very bright color with the patterns very distinct, whereas the female is duller in color with less distinct color patterns.

The Stanley (or Western) Rosella.   The Stanley (or Western) rosella, *P. icterotis icterotis,* is about 11 inches long. The head and underparts of the male are red; there is a yellow cheek and throat patch. The scalloped feathers on the back are black with green margins. The upper wing and tail feathers are blue, green, and black. The undersides of the wings and tail are green with a blue tinge.

The Blue (or Mealy) Rosella. The blue (or mealy) rosella, *P. adscitus adscitus,* is about 12 inches long. The head is yellow with tinges of white; there is a white chin and throat patch with a narrow, light-blue band below the white. The scalloped feathers on the back and wings are black with yellow margins; the flight feathers are blue. The breast is yellow tinged with blue; the vent is red. The females are duller in color.

The Elegant Parakeet. The genus group *Neophema* is made up of seven species commonly referred to as grass parakeets. They are about 8 to 9 inches long. The elegant parakeet, *N. elegans elegans,* is primarily olive yellow in color with yellow shading on the face, abdomen, and underside of the tail. The flight feathers are dark blue. Females are duller in color. The other six species are less known and not easily obtained in the United States.

The Quaker Parakeet. The genus *Myiopsitta* contains one species—the Quaker parakeet, *M. monachus.* The primary color is gray. A green area extends from the eye back to the nape of the neck and down to the shoulders and back. The wings and tail are green; the undersides of the wings and tail are gray. The vent and thighs are green; the gray feathers on the breast show light-gray margins. The bill is yellow-white. The Quaker parakeet is one of only a few parrot-like birds that builds a nest. In the wild, they live in colonies, and the nests may resemble large apartment complexes. The birds are about 12 inches long.

The Red Rump Parakeet. The genus group *Psephotus* is made up of five species; this group is fairly rare and not usually available in pet shops. The red rump parakeet, *P. haematonotus,* is primarily green with a red rump and yellow on the chest and lower abdomen. The females are duller in color. Although not one of the most attractive birds, the red rump parakeet makes an excellent pet. *See Figure 19–21.*

The Many-Color Parakeet. The many-color parakeet, *P. varius,* is about 11 inches long. The primary color is green. There is a yellow patch on the forehead, a red patch on the nape of the neck, and a yellow patch on the shoulder. The long flight feathers are blue, and there is a red patch on the green wings. The upper side of the tail is green, and the underside is yellow. Flecks of red appear on the green breast and the abdomen, and the vent area is red. The female is duller in appearance.

The Budgerigar. The genus *Melopsittacus* is made up of one species—the budgerigar, *M. undulatus.* The "budgie" is the most popular pet bird in the world. This small bird is about 7 inches long. Most people refer to the budgie as a parakeet. The budgerigar is native to Australia, where the aborigines used them for food. The term *budgerigar* in Aborigine means "good bird" or "good food." The wild budgerigar is primarily a yellowish-green. Brown lines are found on the head feathers and the feathers on the nape of the neck. These lines get wider on the back and wings; the underside is pale green. The throat is pale yellow with several small, blue spots and larger black markings; numerous color mutations are now available. The bill is olive gray and the legs are

Figure 19–21 Red rump parakeet. *Photo by Isabelle Francais.*

bluish-gray. Budgerigars are ground feeders and, in captivity, drop to the floor of cages to feed. They learn to talk if properly taught. It may be difficult to determine the sex of young birds. Adult males have a bright-blue **cere** (the fleshy portion of the beak that surrounds the nostrils), and on adult females, the cere is pink, light blue, or brown. Budgerigars are easy to tame, can be taught to talk, are easy to care for, make excellent companions, and are fairly inexpensive. Budgerigars normally live for about 4 to 6 years. *See Figure 19–22.*

Lovebirds

The genus group *Agapornis* are referred to as lovebirds and are native to Africa. There are nine species; three species are fairly common. Lovebirds are very hardy and long-lived. They make excellent pets and tame readily if obtained at a young age; older birds may become aggressive. Determining the sex of lovebirds is difficult.

The Peach-Faced Lovebird. The peach-faced lovebird (or rosy-faced lovebird), *A. roseicollis,* is the largest species of this group. The primary color is a soft, pastel green. Rose or peach color makes up the forehead and crown, encircles the eyes, and extends down to the chin, cheeks, and throat. The wings and the long, central tail feathers are dark green; the upper tail coverts are blue. The beak is yellow. *See Figure 19–23.*

The Black-Masked Lovebird. The black-masked lovebird, *A. personata,* is another very popular species. The head is black with a white ring around the eye. A yellow-orange patch appears on the throat; a faint yellow band encircles the neck. The rest of the body is primarily green. The wings and central tail feathers are a darker green; the underside of the tail is yellow-green. The beak is a bright red. The blue-masked lovebird is a mutation of the black-masked lovebird; the color pattern is the same except the green is replaced with blue, and the yellow-orange is replaced with an off-white to gray. The beak is light pink.

Fischer's Lovebird. Fischer's lovebird, *A. fischeri,* has similar markings to the black-masked lovebird. The head is bright orange-red; the beak is bright red, and there is a white ring around the eye. The rest of the body is primarily green. The rump, tail feathers, and flight feathers are tinged with blue.

The Nyassaland Lovebird. The Nyassaland lovebird, *A. lilianae,* resembles the Fischer's lovebird but is smaller in size; it is about 5 to 6 inches long. The Nyassaland is considered a more pleasant bird than the other lovebirds. Its voice is less harsh, and it is not as aggressive.

Parrotlets

The genus group *Forpus* consists of a group of small parrots referred to as parrotlets. There are seven different species in this genus group. Parrotlets are sexually dimorphic, meaning one can visually identify males from females. They are about 4¾ to 5½ inches in length. The Pacific parrotlet, *F. coelestis,* is probably the most well known. The Pacific parrotlet is primarily an olive

Figure 19–22 Budgerigar. *Photo by Isabelle Francais.*

Figure 19–23 Peach-faced (or rosy-faced) lovebird. *Photo by Isabelle Francais.*

green. The male has cobalt blue flight feathers and a blue streak that runs from the eyes to the back of the neck. The female lacks the blue in the wings, and the streak running from the eyes to the back of the neck is a lighter blue. Both have blue rumps, but the blue of the female is lighter. Their eyes are a dark brown, the beak is a grayish-white, and their legs are a pale pink. The personality of the parrotlets is similar to the lovebirds; however, they can be moody, jealous, territorial, and can become aggressive. Others in this genus group that are fairly common are the green-rumped parrotlet, *F. passerinus,* and the spectacled parrotlet, *F. conspicillatus.* These small parrots can be found in the pet trade from $100 to $400.

Hanging Parakeets

The genus group *Loriculus* is made up of a group of small, agile birds referred to as hanging parrots or hanging parakeets. Unlike other birds, they hang upside down when roosting; they have long claws that interlock to prevent them from falling. They are 5 to 6 inches long and have short tails. They feed on nuts and nectar. The blue-crowned hanging parakeet, *L. galgalus,* is the most common species. It is primarily green with a bright-blue crown and black beak. A red patch covers the rump and descends down into the upper tail coverts. The male has a red and yellow throat patch.

CLASSIFICATION—WOODPECKERS (TOUCANS)

The order Piciformes consists of woodpeckers and related birds; there are six families in the order. Most species in this order are strictly arboreal, rarely descending to the ground. They feed on fruits, vegetation, and insects. Only the toucans are kept as pets.

MAJOR FAMILY—WOOD PECKERS (TOUCANS)

The Family Ramphastitidae

Toucans belong to the family Ramphastitidae and are only occasionally available in pet shops. There are 40 species of toucans, the most common being the Toco toucan, *Ramphastos toco;* Swainson's toucan, *Ramphastos swainsonii;* and the channel-billed toucan, *Ramphastos vitellinus.* These are large birds that have about the same sized body as a macaw. The most distinctive feature of the toucan is its large bill, which can be almost as long as the bird's body.

The bill, despite its enormous size, is light in weight because of a honeycomb of fibers within the solid outer shell. The exact purpose for such a large bill is not fully understood. The bird feeds primarily on fruits but will also feed on insects and lizards. Toucans are native to the tropical forests of Central and South America; the birds roost in holes in trees. Its tail has an unusual muscular and bony attachment that allows the tail to be folded over the bird's back. With the tail folded over the bird's back, the bird can fit more easily into roosting holes; toucans are usually very noisy. Prices for birds that are available may run more than $2,500.

The Toco Toucan

The Toco toucan is primarily black with a large white throat and chest patch. The bill is yellow with a black band around the bill where it joins the head; the bill may have bright-orange and blue markings. They have blue surrounding their large, black eyes. *See Figure 19–24.*

Swainson's Toucan

Swainson's toucan is also black but has a yellow chin and chest patch. The upper mandible of the bill is yellow, with brown on the lower part of the upper mandible; the lower mandible is brown, and the tip of the upper mandible is black. There is a green circle around the large black eyes.

The Channel-Billed Toucan

The channel-billed toucan is black with a white throat patch. There is a band of orange below the white; it has a blue band surrounding its black eyes. The upper part of the upper mandible is yellow; the lower part of the upper mandible and the lower mandible are brown. There is a white band around the bill where it joins the head. The upper tail coverts are yellow and orange.

Figure 19–24 The Toco toucan. *Photo by Isabelle Francais.*

CLASSIFICATION—PERCHING BIRDS

The order Passeriformes includes the perching birds. Of the approximately 9,000 species of birds, 5,100 are classified as perching birds; these are further classified into 65 families. Perching birds have three toes that point forward and one toe that points backward; the toes are not webbed. Perching birds are also noted for their pleasing songs. Of the 65 families, only 7 are discussed in this text.

MAJOR FAMILIES—PERCHING BIRDS

The Family Zosteropidae

The family Zosteropidae consists of 85 species classified into 16 genus groups and 381 species; the family is commonly referred to as white eyes and related species. They are native to Africa, Asia, New Guinea, and Australia. White eyes are small, only about 2½ inches long; the eyes have a ring around them. They feed primarily on nectar from flowers. In captivity, they feed on liquid nectars, fruits, peanut butter, cakes, mynah meal, or mynah pellets. Some species in this family are the gray-breasted white eye, *Zosterops lateralis;* the cinnamon white eye, *Hypocryptadius cinnamomeus*; and the Norfolk white-throated white eye, *Zosterops albogularis.*

The Family Emberizidae

The family Emberizidae is represented by one subfamily; the subfamily Emberizinae is commonly referred to as buntings. There are 281 species classified into 69 genus groups. Many buntings are native to the United States and cannot legally be captured or kept in cages. Two species that may be available, however, are the rainbow bunting and the red-headed bunting, *Emberiza melanocephala.*

The Rainbow Bunting

The rainbow bunting is native to Mexico. This bird is about 5 inches long. The upper parts are blue, and the underparts are yellow; some orange specks appear on the chest. The forehead and crown are bright green; there is a yellow ring around the dark eyes. Females are dull in comparison to males.

The Red-Headed Bunting

The red-headed bunting is native to India; it is about 7 inches long, and the primary color is brown. The head is a rich brown, and the color changes to a lighter brown down the nape of the neck and into the **mantle.** The wings are brown with lighter-brown edges to the feathers. The tail is brown with tinges of gray and black. The underparts of the bird are dull yellow. Females are lighter in color with tinges of green and black.

The Family Sturnidae

The family Sturnidae is commonly referred to as starlings. There are 106 species classified into 22 genus groups.

The Hill Mynah

One member of this family is the popular Hill mynah, also known as the talking mynah or Indian grackle, *Gracula religiosa.* The mynah bird is native to Asia and India. It is primarily black. There is a white patch on its wings; the bill is orange. A bare, yellow skin patch extends from below the eye back toward the nape of the neck; its legs are yellow. It feeds primarily on fruits. The mynah bird is noted for its ability to mimic the human voice and other sounds.

Keeping mynah birds can be more time consuming than keeping other species of birds because of their diet. They consume fruits, and soft-bill mynah pellets. Their cage needs to be cleaned at least once a day. Mynah birds command prices from about $300 to $500.

The Family Ploceidae

The family Ploceidae consists of three subfamilies, 18 genus groups, and 143 species. The subfamily Viduinae are referred to as whydahs. There are nine species of whydahs in the genus group *Vidua*. Whydahs are native to areas of Africa south of the Sahara Desert, where they are also called widow birds. Whydahs go through a period of about 6 to 8 months when they have bright colors, and another period when their colors fade to duller shades. The diet of these birds consists primarily of seeds.

Several species are rare and unavailable in the pet market. Species of whydah that may be available are the paradise whydah, *Vidua paradisaea,* and the pintailed whydah, *Vidua macroura.*

The Paradise Whydah

The paradise whydah is primarily black with two long, black central tail feathers that may reach 14 inches in length. A reddish-brown band encircles the neck, and the color extends down onto the chest. Two black bands extend from the crown back down the back of the head. Females are duller in appearance.

The Pintailed Whydah

The pintailed whydah is smaller than the paradise whydah. Its central tail feathers are only about 9 inches long and are very narrow. The primary color is white; the crown, wings, and tail are black. There is a white band across the flight feathers; the bill is red.

The Red-Collared Whydah

The red-collared whydah has a body that is about 4 inches long, and a tail that is 6 to 8 inches long. The primary body color is black. Brown shadings appear on the margins of the wing feathers, and a red-orange band encircles the neck; the bill is black. Several subspecies of the red-collared whydah may also be available in the pet market. Among these are the red-shouldered whydah, the yellow-backed whydah, the white-winged whydah, and the yellow-shouldered whydah.

The Orange Weaver

The weavers make up the subfamily Ploceinae. There are 94 species classified into seven genus groups; all are native to Africa. Male weavers go through color phases. The orange weaver, *Ploceus aurantius*, is bright orange. The head, chest, and lower abdomen are black; red tinges are found on the nape of the neck. The wings, tail, and rump have a brownish tinge.

Masked Weaver

Masked weavers are occasionally available in the pet market. Most are black with yellow markings on the face; their wings are brown with lighter margins. Among the masked weavers are the little-masked weaver, the half-masked or vitelline weaver, and the rofous-necked weaver.

The Red-Billed Weaver

The red-billed weaver is primarily light red with various shades of red on the head, neck, chest, and abdomen. A black patch covers the face, throat, and forehead; the beak, feet, and legs are red. They are very aggressive and should not be caged with other small birds. Two other species of weavers are the Baya weaver and the scaly crowned weaver.

The Family Pycnonotidae

The family Pycnonotidae consists of 118 species of bulbuls classified into 16 genus groups. These small birds are native to Africa, the Middle East, India, Southern Asia, the Far East, Java, and Borneo. They are soft-billed birds that feed on nectar, fruits, and soft-bill mynah pellets. Species that may be available are the black bulbul, *Hypsipetes madagascariensis;* the black-headed bulbul, *Pycnonotus xanthopygus;* the brown-eared bulbul, *Hypsipetesam aurotis;* the common bulbul, *Pycnonotus barbatus;* the finch-billed bulbul of the genus *Spizixos;* the pale olive-green bulbul, *Phyllastrephus fulviventris,* the red-vented bulbul, *Pycnonotus cafer;* and the red-whiskered bulbul, *Pycnonotus jocosus.*

The Family Fringillidae

The family Fringillidae consists of several canaries, finches, siskins, and grosbeaks. Family members have 9 instead of 10 large, primary feathers in each wing; they also have 12 large tail feathers. Females of the family build a cuplike nest to incubate the eggs. There are three subfamilies.

The subfamily Fringillinae consists of one genus group, *Fringilla,* and three species. The three species are the brambling, *F. montifringilla;* the Canary Islands chaffinch, *F. teydea;* and the chaffinch, *F. coelebs.*

The subfamily Drepanidinae consists of 15 genus groups and 28 species. All species are native to the Hawaiian iIslands and are either endangered, threatened, or thought to be extinct.

The subfamily Carduelinae consists of 122 species in 17 genus groups. Species in the subfamily are native to North and South America, Europe, Asia, and Africa. A few species are available in the pet market.

The Common Canary

One species that is very important as a pet is the common canary, *Serinus canarius domesticus.* The early history of the canary is not known, but it is believed they were native to the Canary Islands and other islands off the Northwestern Coast of North Africa. Spanish sailors and soldiers are believed to be responsible for bringing the first canaries to Europe as early as the 1500s. The Spanish controlled the sale and distribution of canaries for about 100 years. As the birds became available to the French, Dutch, German, and English birdkeepers, more research and development took place. Several new varieties were developed. Today, more than 12 different varieties in more than 50 different color combinations are available. Development in the breeding of canaries was concentrated into three areas: the physical appearance of the bird, the coloration, and the singing ability. *See Figure 19–25.*

The Gloster Fancy Canaries

A physical development that has become popular is the "crested" canary. The feathers on the top of the head fan out to form a crest or fringe of feathers that

Figure 19–25 A common canary. *Photo by Isabelle Francais.*

hangs out over the eyes and beak of the bird. These birds are known as the Gloster fancy canaries, and two types exist. One type is known as the "corona" and has the crest, and the other is known as a "consort" and has normal head feathering. When mating Gloster fancy canaries, it is important to mate a corona with a consort. If two corona types are mated, 25 percent of the offspring will die in the shell or shortly after hatching.

Red-Factor Canaries

The most important color varieties are those carrying the red color factor. Birds that carry red are usually more expensive; the more red, the more expensive they are. These birds are known as red-factor canaries.

The Roller Canary

Canaries are also popular because of their singing ability. The best known for its singing ability is the roller canary. Only the male canary sings, and the songs differ among birds.

Varieties of canaries that may be available include the lizard canary, Yorkshire canary, clear fife canary, Norwich canary, cinnamon canary, frilled canary, Belgian canary, Scotch fancy canary, and Lancashire canary.

The Family Estrildidae

The family Estrildidae consists of grass finches, waxbills, and mannikins. They are separated from the family Fringillinae because they have 10 large primary feathers. There are approximately 124 species in 27 genus groups. These are primarily ground-feeding birds, and their main food consists of seeds and insects.

The Zebra Finch

The zebra finch, *Poephila quttata,* is found wild over most of the continent of Australia, where it is as common as the sparrow is in the United States. They are active during most of the day and are very curious. Zebra finches are social, getting along well with other birds and people. They are considered very hardy and may live for several years. They are probably the most widely kept and bred finch in captivity.

The Wild (or Normal) Finch

The wild (or normal) finch has an orange beak and orange cheek patch. A white patch extends from the top front of the eye downward to the chin; a narrow black "tear" marking extends downward from the eye. The forehead, crown, top of the head, and nape of the neck are gray. The upper chest is black-and-white striped with zebra-like markings; there is a black chest patch. The underside is white, and the tail is black with white bars. The sides below the wings are reddish-brown with white spots. The females lack the zebra markings, cheek markings, and side marking; they are also lighter in color. *See Figure 19–26.*

Several other color patterns are recognized. Among them are the white, fawn, dominant silver or dilute normal, dominant cream or dilute fawn, pied,

Figure 19–26 Wild (or normal) finch. *Photo by Isabelle Francais.*

Figure 19–27 Several Bengalese (or society) finches. *Photo by Isabelle Francais.*

chestnut-flanked white, penguin, recessive silver or dilute normal, recessive cream or dilute fawn, and yellow-beaked varieties.

The Bengalese (or Society) Finch

The Bengalese (or society) finches, *Lonchura striata domestica,* are probably derived from the white-backed munia, *Lonchura striata.* These small birds are about 4 inches long. The white-backed munia is found in India and Sri Lanka. It is primarily brown with very dark, almost black, head and breast; the rump and underside are white. *See Figure 19–27.*

There are about eight different varieties of the domesticated Bengalese or society finch. The categories are self chocolate, pied chocolate, dilute chocolate and white, white with brown eyes, self fawn, pied fawn, dilute fawn and white, and white with ruby eyes.

The Owl Finch

The owl finch, *Poephila bichenovii,* is about 4 inches long. It is basically a gray-brown color with various shades of gray-brown on the body, wings, and tail. A dark ring surrounds the face, similar to that of a barn owl.

The Black-Throated Finch

The black-throated finch, *Poephila cincta cincta,* is primarily a cinnamon-brown color. The forehead, crown, cheeks, and nape of the neck are gray. The beak is black, and a black patch extends down from the lower beak to the chin and throat area; the wings are brown. The tail feathers are black, and the upper tail coverts are white. The eyes are brown, and the area in front of the eye (**lores**) is black; the legs and feet are flesh-colored. The black-throated finch can be aggressive to other birds and may need to be kept separated. They are about 4 inches long.

The Long-Tailed Finch

The long-tailed finch, *Poephila acuticauda acuticauda,* is very similar to the black-throated finch except it is more slender and is about 7 inches long. The beak is black, and there is a black chin and throat patch; the lores are also black.

The Double-Bar Finch

The double-bar finch, *Poephila bichenovii bichenovii,* is primarily dark brown. The lores, the area around the eyes, and the cheeks, chin, and throat are white with a black band bordering the white. The chest is white, and another black band runs across the lower chest; the abdomen is white tinged with yellow and brown. The wings are black with white spots, and the tail is black. The beak is blue-gray, the eyes are brown, and the legs and feet are gray.

The Shafttail Finch

The shafttail finch, *Poephila acuticaudata,* is primarily brownish-gray. The forehead, crown, head, and nape of the neck are gray. There is a large black chin and throat patch; the lores are black. The beak is orange, and the wing and tail feathers are brown. Some white and black may appear on the rump and tail. It is difficult to distinguish sex of males and females.

The Parson Finch

The Parson finch, *Poephila cincta,* is very similar to the shafttail finch except it has a black beak and the tail is shorter. In addition to their normal diet of seeds, they also feed on mealworms.

The Masked Finch

The masked finch, *Poephila personata personata,* is primarily a light cinnamon-brown color. The beak is yellow. A triangular-shaped black patch extends from the eyes forward to the beak. There is also a black chin and throat patch; a large black patch is also located on the flanks. The tail is black and the underside of the tail is white; the legs and feet are red. The masked finch is about 5 inches long.

The Diamond Firetail Finch

The diamond firetail finch, *Emblema guttata,* is primarily brown. The beak is red and the lores are black. The forehead, crown, top of the head, and nape of the neck are gray. There is a white chin and throat patch. A broad, black band extends across the upper chest and down the sides. The lower chest and abdomen are white. The wings are brown, and the long tail feathers are black; the upper tail coverts are red. The eyes are brown, and there is an orange ring around the eyes. The feet and legs are gray. The diamond firetail finch, sometimes referred to as a diamond sparrow, is about 5 inches long.

The Painted Finch

The painted finch, *Emblema picta,* is only about 4 inches long. The beak is long and pointed; the upper part of the beak is black with a red tip, and the lower beak is red with blue at the base. The forehead, lores, chin, throat, and area around the eyes are scarlet. An irregular scarlet patch runs down the center of the chest and abdomen. The crown, nape of the neck, and upper parts of the bird are brown. The wing feathers and tail feathers are brown; the underparts of the bird are black with numerous white spots. The feet and legs are flesh-colored.

The Red-Browed Finch

The red-browed finch, *Aegintha temporalis,* is about 5 inches long and has many different colors, although the primary color is gray. The beak is red with black. A red patch extends from the beak backward above and past the eye. The forehead, crown, nape of neck, cheeks, chin, throat, and lower portions of the bird are gray. The wings are an olive yellow, and the underside near the vent is white. The tail feathers near the base of the tail are red. The long tail feathers are gray. The feet and legs are flesh-colored.

The Plum-Headed Finch

The plum-headed finch, *Aidemosyne modesta,* is about 4 inches long. The primary color is brown. The feathers on the back, shoulders, wings, and tail are dark brown with white markings. The feathers on the underside of the bird are lighter brown with white stripes or bars. The long tail feathers are black, with the two outer tail feathers having white tips. The forehead, crown, and top of the head are red to violet (plum-colored). The ear coverts, cheek, and chin feathers are white with light-brown markings. The eyes are black with a white ring around them. The upper beak is black, and the lower beak is black with white at the base. The feet and legs are flesh-colored.

The Chestnut-Breasted Finch

The chestnut-breasted finch, also called a chestnut-breasted mannikin, *Lonchura castaneothorax,* is about 5 inches long. The colors of this finch are various shades of brown. The beak is blue-gray. The forehead, crown, and nape of the neck are gray-brown with darker brown markings; the lores, the area around the eyes, and the chin and throat area are dark brown. The upper chest is a chestnut color; a black band runs below this and up to the shoulders. The areas below the black band and the abdomen are straw yellow; the wings and tail feathers are light brown. The feet and legs are gray-brown. Other related species are the pictorella finch, *Lonchura pectoralis,* and the yellow-rumped finch, *Lonchura flaviprymna.*

Figure 19–28 Gouldian finch (red-headed). *Photo by Isabelle Francais.*

The Gouldian Finch

The Gouldian finch, *Chloebia gouldiae,* is recognized in three color forms. The three forms are distinguished by the black, red, or yellow coloration on the bird's head. Young finches are a dull green above and gray below. The young begin to molt at about 3 months of age and complete the process at 6 months. The adults are very brightly colored birds that reach a length of about 6 inches. *See Figure 19–28.*

The red-headed form of the Gouldian finch is primarily a grass-green color on the upper body, including the upper wings; the primary wing feathers and outer secondary feathers are brown. The rump and upper tail coverts are bright blue; the long tail feathers are black. The lores, forehead, crown, cheeks, and ear coverts are red. A narrow, black band borders the red and extends down and widens into a black chin and throat patch; a narrow, blue band borders the black. The chest area is light purple, and the lower chest and abdomen are yellow. The beak is gray-white with a red tip. The feet and legs are yellow. The black-headed form is very similar to the red-headed form, except that the lores, forehead, crown, cheeks, chin, and throat area are black. The yellow-headed form is very similar to the other two forms except the lores, forehead, crown, cheek, chin, and throat are yellow.

Some color variations in the Gouldian finch have been developed; some may have white chests, whereas others have a blue upper body and wings. All three forms breed indiscriminately. Female Gouldian finches usually have the same color patterns but are duller in appearance.

Temperature is very important in the survival of Gouldians. Gouldian finches should be kept at minimum temperature of 72°F.

The Crimson Finch

The crimson finch (or crimson waxbill), *Neochmia phaeton,* is primarily brown. The beak is red with white at the base of the lower mandible. The forehead, crown, back of neck, and shoulders are gray-brown, turning to a purple-red down the back; the upper tail coverts are crimson. The lower chest, abdomen, and under-tail coverts are black. The lores, cheeks, ear coverts, and chest are crimson with white spots; the feet and legs are yellow. Females have more brown on them.

The Star Finch

The star finch (or rufus finch), *Bathilda ruficauda,* is primarily olive brown on the upper parts with darker brown on the wings; the beak is red. The lores, forehead, and cheeks are crimson; the crown and nape of the neck are light yellow and white. The feathers from the throat down the chest and abdomen are yellow with white spots. The tail feathers are brown with a crimson tinge; the feet and legs are yellow. The star finch is about 4 inches long.

The Cordonbleu Finch

The cordonbleu, *Uraeginthus bengalus,* is native to Africa. This finch is about 4 inches long. The upper parts are brown, and the lores, chin, cheek, chest, and abdomen are blue. The male has red cheek patches. Other related species are the blue-breasted cordonbleu, *U. angolensis,* and the blue-capped cordonbleu, *U. cyano-cephala.* Both are very similar, except the male blue-breasted cordonbleu lacks the red cheek patch, and the blue-capped cordonbleu has more blue, especially on the crown.

The Blue-Faced Parrot Finch

Members of the *Erythrura* genus of finches are commonly referred to as parrot finches. All are native to countries and islands of the Southwest Pacific area. Among the species that sometimes become available is the blue-faced parrot finch, *Erythrura trichroa.* The primary color is green. The beak is black, and the lores, forehead, and cheeks are blue. The tail feathers are red, and the feet and legs are light brown. The blue-faced parrot finch is about 4 inches long.

The Red-Headed Parrot Finch

The red-headed parrot finch, *Erythrura psittacea,* is primarily green. The forehead, crown, cheeks, chin, and throat are red. The tail feathers are red, and the beak is black; the feet and legs are brown.

The Pintailed Nonpareil

The pintailed nonpareil, *Erythrura prasina,* is primarily green on the upper parts of the body. The forehead, chin, throat, and cheeks are blue. The rump and upper tail coverts are red; the center tail feathers are long and form a point. The chest is red, and the lower chest and abdomen are a duller shade of red. The beak is black, and the feet and legs are light brown. The pintailed nonpareil is about 5 inches long.

Other *Erythrura* finches that may be found in aviaries are the blue-green parrot finch, *E. tricolor;* the short-tailed parrot finch, *E. cyaneovirens;* the Mindanao (or many-colored) parrot finch, *E. coloria;* Kleinschmidt's (or the pink-billed) parrot finch, *E. kleinschmidti;* the green-faced parrot finch, *E. viridifacies;* the Papua parrot finch, *E. papuana;* and the green-tailed (or Bamboo) parrot finch, *E. hyperythra.*

The Red-Eared Waxbill

The genus group *Estrilda* is referred to as waxbills. They are small, nervous, alert, and very active birds; most are native to Africa. The red-eared waxbill, *E. troglodytes,* is one of the most popular. The primary color is light brown with a light-red tinge; the undersides are darker brown. The beak is red, and the red color extends back to and around the eyes. The tail and rump area are black. The red-eared waxbill is about 3½ inches long.

The Orange-Cheeked Waxbill

The orange-cheeked waxbill, *E. melpoda,* is primarily brown. The beak is red, and there is a large orange cheek patch. The rump is red, the tail is black, and there is orange on the lower abdomen. The orange-cheeked waxbill is about 3½ inches long.

The Lavender Waxbill

The lavender waxbill, *E. caerulescens,* is primarily a pearl-gray color except for a red rump and tail. White spots are scattered on the flanks and sides. The lavender waxbill is about 3½ inches long.

The Gold-Breasted Waxbill

The gold-breasted waxbill, *Amandava subflava,* is only about 3 inches long. The upper parts are primarily brown. The beak is red, the head and throat are red, the rump is orange, and the tail is black. The chest of the males is orange.

The Melba Finch

The melba finch (or waxbill), *Pytilia melba,* is larger than the other waxbills; it is about 5 inches long. The primary color is gray. The beak is red, and the head and throat are red. The lower crown is gray, changing to dark olive green down the shoulders, back, and rump. The upper tail coverts are red, and the wings are olive green. The throat and chest is an olive yellow with scattered white spots. The melba finch and other members of the *Pytilia* are insectivorous and should be supplied with a variety of small insects.

HOUSING AND EQUIPMENT

Various cages are available at pet shops and variety stores. Equipment used for the large parrot-type birds must be constructed of very heavy materials; wood and plastics cannot be used for parrot cages because they will shortly be destroyed. Heavily constructed metal cages must be used. The cage should be inspected for sharp points or edges that can cut or injure the bird's tongue or feet. Secure latches or locks should be used; these birds are very curious and ingenious and may soon learn to open their cages and escape.

If one chooses to construct a cage for macaws, the wire size should be ½-inch by 3-inch wire mesh screen in 12- to 14-gauge wire. Smaller mesh wire may be used for outside cages if mice, rats, or other birds are a problem. Finches should have rectangular cages to allow for long horizontal flight. Flying in a circular pattern is unnatural to the finch and may cause undue stress. *See Figures 19–29, 19–30,* and *19–31.*

Most new cages are equipped with plastic perches. These are hard and may be uncomfortable for many birds. If birds refuse to perch, one should suspect something is wrong with the perch. Size is also important. Wood makes the best perches. Parrots will soon destroy wood perches, and the perch will have to be replaced. Wood does give the parrot the opportunity to exercise its beak, which helps keep the beak trim and the bird busy. Limbs and tree branches can also be used for perches; one must make sure the limbs or branches are clean and free of molds or pesticides. If the perch tapers from end to end, this will give the bird the opportunity to find the most comfortable location. The ideal perch size for parrots is a 1-inch square perch. Budgerigars prefer an oval-shaped perch about ½ inch in diameter. Finches and canaries like round perches about ½ inch in diameter. *See Figure 19–32.*

Water containers should be made of glass, ceramic materials, or stainless steel, especially for the larger birds. These materials are hard and are easy to clean. Gravity-type waterers are excellent for birds; they hang on the outside of the cage, and a metal tube extends into the cage. The water is released on demand when the bird touches the end of the tube. Water intake should be closely monitored to make sure the bird learns how to use this device. Also available is the silo waterer, which consists of a tube or water reservoir that hangs on the outside of the cage and a small water bowl that fits through the bars of the cage. These work best for small birds. Another type would be the standard stainless steel or plastic water bowl that attaches to the inside of the cage. The type of waterer to use depends on the type and size of the bird.

The cage should be placed in an area of the room that is free from drafts, where constant temperature is maintained, not in direct sunlight, and away from poisonous plants or other hazards the bird can reach.

To prevent caged birds from becoming bored, toys may need to be provided for them. Many toys are available at pet stores. Large birds will like stainless steel chains with bells on them or the wooden blocks attached to string, which allows the bird to tug and chew on the toy. Chains with bells and ladders can be put in with smaller birds such as the budgerigars, canaries, and finches. *See Figure 19–33.*

When selecting cages, the following points should be kept in mind:

1. The cage should be big enough for the bird or birds. See Table 19–1.
2. The cage should be easy to clean.
3. The cage should be large enough to provide room for feeders, waterers, nest boxes, and other items.

Figure 19–29 Pictured here is a large rectangular cage for small birds. Finches need a cage like this that allows them more horizontal flight. *Photo by Isabelle Francais.*

Figure 19–30 This is a cage constructed of heavy materials, which is necessary to prevent larger birds, such as this yellow-napped Amazon, from destroying the cage. *Photo by Isabelle Francais.*

Figure 19–31 Pictured here are three tall, circular cages that provide plenty of room for small birds to move around. *Photo by Isabelle Francais.*

Parrot/2.5 cm
(1 in) square

Finch/1.25 cm
(0.5 in) round

Budgie/1.25 cm
(0.5 in) oval

Figure 19–32 Ideal
perch size and shape
for birds.

Figure 19–33 Birds need toys like
this chain with bells to help keep
them occupied and from getting
bored. *Photo by Isabelle Francais.*

	Caged Pair	**Single Bird**
Species	**Length × Width × Height**	**Length × Width × Height**
Canaries	18" × 10" × 10"	*
Cockatiels	4' × 2' × 3'	26" × 20" × 20"
Cockatoos	4' × 4' × 3.5'	3' × 2' × 3.5'
Conures	4' × 4' × 4'	4' × 3' × 4'
Finches	2' × 2' × 2'	12" × 12" × 12"
Lovebirds	4' × 4' × 4'	*
Macaws	6' × 6' × 6'	3' × 2' × 3.5'
Budgerigars	24" × 14" × 8"	*
Mynah birds	† 6' × 3' × 3'	6' × 3' × 3'
Amazon Parrots	4' × 3' × 4'	3' × 2' × 2'
African Gray	4' × 3' × 4'	3' × 2' × 2'

TABLE 19–1   Minimum Cage Size for Caged Birds.

*These birds prefer the company of other birds and should not be caged singly.

†The cage should be at least this size for a pair.

4. The cage should be affordable.

5. The cage should be made of stainless steel or anodized aluminum or be chrome-plated. These materials will not rust and are easy to design. Materials that rust present two hazards to birds: (a) the rust may be ingested and damage the linings of the digestive system and (b) the rust pieces may contain lead or other poisonous materials.

6. Galvanized wire contains zinc and should be avoided.

Feed containers for the large parrot-type birds must be constructed of sturdy materials. Glass, ceramic, or stainless steel are preferred; plastic material will be destroyed. Plastic feeders are fine for the smaller birds, however.

FEEDING

Many bird diets consist of seeds, except for the fruit-and nectar-consuming birds. Seeds consumed by birds are of two types: cereal seeds and oil seeds. Cereal seeds contain a higher proportion of carbohydrates compared with oil; examples of cereal seeds are canary seed, millet, corn, and dehusked oat kernels. Oil seeds are high in fat and low in carbohydrates; examples of oil seeds are sunflower seeds, peanuts, safflower, pine nuts, rape seeds, maw, niger, and linseed. Cereal seeds and oil seeds can be purchased separately and mixed together to provide a variety. Most people, however, probably prefer to buy premixed rations. When selecting seeds for a mix, one must make sure the seed is clean and free of dust and dirt. The seed should also be dry and free of molds; both shelled and unshelled peanuts should be examined for mold. Peanuts are attacked by a mold that produces strong toxins that can cause liver damage; fed over a period of time, they can cause the death of a bird.

Soaked seeds may be fed to young birds that are having trouble breaking hard seeds, and to birds during breeding and molting seasons. The seeds should be soaked in warm water for about 24 hours; this will stimulate the germination of the seeds. The chemical changes in the seeds during germination increase the protein content. When feeding soaked seeds, it is important that the water be poured off and that the seeds be washed thoroughly with tap water. Soaked seeds provide an ideal medium for the growth of molds and fungi; birds must not be allowed to consume seeds that have mold and fungi on them. Any seeds not consumed by the birds in a few hours should be removed and thrown away. The container that the seeds were in should be washed thoroughly before feeding more soaked seeds.

Birds can also be fed green plant materials. Chickweed, dandelion leaves, carrot tops, kale, and spinach are good choices that can be fed. One must be careful not to feed too much green plant material because it may cause diarrhea; this is especially true if the bird is not used to green plant material. Green material must be thoroughly washed to remove any residue of pesticides or other chemicals. Lettuce should be avoided because it is not as nutritious as other types of greens. If green foods are taken directly from the refrigerator, they should be allowed to warm to room temperature before being offered.

Birds cannot grind up their food. They can break the seeds open with their beaks, but the actual grinding of the food takes place in the ventriculus. For some species of birds, this is accomplished with the aid of **grit** that must be supplied in their diet. Grit is available from pet stores in two forms: soluble and insoluble. The soluble form is usually soluble oyster shell that breaks down and serves as a source of minerals. The insoluble form is usually crushed granite; the insoluble grit provides the primary base for which the food material is rubbed and worked against to grind the food up.

Cuttle bone should be provided as a source of calcium. Cuttle bone can be obtained from pet supply stores and clipped to the side of the bird's cage. Female birds need a supply of calcium for the making of eggshells. Some small birds may have trouble breaking or chewing off pieces of the cuttle bone; pieces may have to be cut or shaved off for them.

Mynah birds will not eat seeds and do not need grit or cuttlebone in their diet. Mynah birds are fed special soft-bill pellets or foods that are available from pet supply stores. Some of these foods can be fed directly from the package, and some need to be mixed with water before feeding. Mynah birds also need fruit in their diet; apple slices, grapes, orange slices, and banana slices are the most common. Dried fruits can also be used, provided they are soaked and rinsed off before feeding. Mynah birds also like live food such as mealworms.

Lories and lorikeets feed on nectar and pollen; nectar is usually available from pet supply stores. Nectar foods are usually powders that need to be mixed with water and provided to the birds in special feeders. Lories and lorikeets also eat fruit.

Hand-reared birds are in great demand because they are usually tamer and more easily handled. Birds that are hand-reared need to be provided a brooder or heating pad so that the birds can be kept warm. Hand-feeding is very time consuming; newly hatched birds need to be fed every hour and a half. As the birds approach weaning, they still need to be fed every 3 or 4 hours. Feeding is required from early morning to late evening. *See Figure 19–34.* Foods commonly used are dry baby cereals, fruits, and canned

Figure 19–34 In this photo a young sun conure is being hand-fed. *Photo by Isabelle Francais.*

baby food. Commercially formulated diets for hand-feeding some species are also available. These foods are mixed with water in a blender and then heated; the food must be warm and fairly runny. Supplements such as calcium and vitamins can be added. As the birds get older, they can be gradually weaned from the liquid diet to seeds.

TRAINING AND HANDLING

After bringing a new bird home, begin trying to offer the bird a treat through the bars of the cage. The bird may refuse the treat, but one should be patient and continue at regular intervals to offer a treat. One should talk softly to the bird, and soon the bird will take the treats. The next step is to open the door of the cage and offer the treat through the open door. One should move slowly and talk calmly to the bird. The bird may retreat, but again, one must be patient and the bird will soon be taking the treats. Next, a stick or perch should be introduced into the cage and the bird encouraged to step up onto this new perch. The perch should be pressed up against the bird's chest just above its legs. Parrots use their beaks to aid in climbing, so one should not be alarmed if the bird grasps the perch with its beak. The feet should step up on the perch next. After the bird becomes accustomed to climbing on a perch held into the cage, one can slide a hand or finger (depending on the size of the bird) in under the stick and attempt to get the bird to perch on the hand or finger. While attempting this, one should offer the bird a treat to distract it and continue to talk calmly to the bird. A leather glove may be needed when working with the large parrot-type birds. Once the bird feels comfortable perched on the hand, one can try to remove the bird from its cage while perched on the hand. The bird may jump from the hand at first, but one should keep trying; this process may take several days. Some birds are more adventurous than others. Patience is important. The next step is to try to get the bird to use a perch outside its cage or to perch on one's arm or shoulder.

When working with birds outside their cages, it is important to be aware of possible hazards. All windows must be closed, as should blinds or curtains. Aquariums and open containers of water should be covered. The bird should not be left unsupervised, especially where electrical wires and other hazards may exist. Other pets should be kept out of the room when the bird is outside its cage.

Budgerigars, cockatiels, parrots, macaws, cockatoos, and mynah birds can learn to talk. They actually mimic or imitate human words, the calls and sounds of other birds, and other sounds that they hear. Trying to teach a bird to talk should begin after it becomes tame and accustomed to its handler and surroundings. Toys, food, and any other distractions should be removed while working with the bird. Young birds are the easiest to teach, and males usually learn more words and learn more quickly than females, although this is not always true. One should develop a routine in the training. It is helpful to try to give the bird its lesson at the same time each day. Lessons should be only about 15 minutes in length; the bird may become bored if lessons are longer. Short words and phrases are easiest to mimic. One should slowly repeat several times the word or phrase he or she wants the bird to learn. Only one person in a family should be trying to teach the bird to talk. Birds may learn to talk for females and children more quickly. Patience is important.

DISEASES AND AILMENTS

Pet birds are usually very healthy because they do not come in contact with parasites or disease organisms that birds in the wild may encounter. However, pet birds are susceptible to diseases, and owners should be alert for the following signs that might indicate a problem:

1. Sick birds will sit on the perch with one or both wings drooped and the tail pointed downward.

2. A sick bird may stand unsteady on the perch, display restlessness, shift weight from one leg to the other, or favor one leg.

3. Runny feces could indicate a digestive ailment.

4. Fluffed feathers could indicate that the bird is trying to retain body heat; the bird may be very ill.

5. Lack of activity or failure to fly may indicate an ailment.

6. Eyes that are continually closed or have discharges are signs of ailments.

7. Irregular or difficulty breathing, wheezing, or noisy breathing could indicate a respiratory ailment.

8. Loss of appetite or failure to eat may indicate an ailment.

At the first sign of illness, a bird should be provided warmth, rest, food, water, and possibly medication. One should take the bird to a veterinarian for diagnosis and treatment.

Proper management is important in preventing diseases and ailments. This begins with selecting a healthy bird from a reliable source. Many ailments occur when the birds are first taken home. Make sure the bird is placed in a dry, warm, draft-free area. It is important that the bird be subjected to as little stress as possible; one should make sure there are no animals around to bother the bird. No new birds should be introduced in with other birds until they have gone through a quarantine and observation period. Sanitation is extremely important. The bird should receive fresh food and clean, fresh water. The cage and perches must be kept clean.

Internal Parasites

Internal parasites are rarely seen in pet birds because they are not exposed to wild birds or to the ground where exposure can occur. Occasionally, internal parasites may occur, especially in wild birds or in aviaries that do not maintain good sanitary practices and rodent control.

Roundworms

Infection occurs by ingestion of the worm eggs from contaminated feces, soil, or food. Infected birds show blockage of the intestine, poor plumage, loss of weight, and diarrhea. Several treatments can be used to control roundworms. A veterinarian should be consulted to identify the presence of roundworms and begin treatment.

Tapeworms

Tapeworms can be diagnosed by observing the small, rice-like, white segments in the feces of infected birds. Birds become infected by ingesting intermediate hosts, such as houseflies, fleas, ticks, or earthworms. Infection with tapeworms is slight if proper cleaning and sanitation practices are observed. A veterinarian will prescribe medication to kill the worms.

Trichomoniasis

Trichomoniasis is a disease caused by the flagellated protozoan *Trichomonas gallinae* and is spread through contaminated feces. The disease has primarily been observed in pigeons and in birds in outdoor aviaries. Birds indoors may become infected if infected birds are introduced into the cage. The protozoan lives in the head sinuses, mouth, throat, esophagus, and other organs. The first signs of the disease are small, yellowish lesions on the linings of the mouth and throat. These lesions, which are cheese-like in appearance, enlarge rapidly and may block the throat of the bird and prevent it from closing its mouth. Breathing may be difficult, and loss of weight may also occur. Watery discharge may occur from the eyes, and blindness can occur in advanced stages. If untreated, death usually occurs within 8 or 10 days. A veterinarian will prescribe medication.

External Parasites

Red Mites

External parasites such as red mites, *Dermanyssus gallinae,* feed on the blood of infected birds. The mites usually feed at night and then move to cracks in the cage and nest box. Heavy infestations can cause anemia and even death. Mites, although small, can be seen with the naked eye, and they appear as tiny red specks after feeding on the bird's blood. Birds infected with mites are restless and scratch and pick at their feathers. Mites can be spread to caged birds by introducing infected birds, or birds in outdoor aviaries can be infected from wild birds visiting the aviary. A veterinarian will prescribe a spray, powder, or other medication. Infected nest boxes should be destroyed or thoroughly disinfected. Cages and equipment should also be thoroughly cleaned and disinfected.

Feather Mite

The feather mite, *Knemidocoptes laevis,* should be considered in the case of feathers that look chewed, and where no other cause is known; severe infestation can cause a bird to lose its feathers. Feather mites bury into and feed on the feather quill just below the surface of the skin. Infestation usually occurs on the beak or neck first and spreads over the entire body.

Restlessness, severe scratching, feather picking, and skin irritation are signs of infection. The mites may be seen as small, gray-colored moving specks. Unlike red mites, they are on the bird both day and night. A veterinarian can identify the presence of these mites and prescribe proper treatment.

Scaly Face and Leg Mites

Scaly face and leg mites, *Knemidocoptes pilae,* and related species are mites that may infect budgerigars, lovebirds, and canaries. Scaly face mites tunnel into the soft tissues around the beak, ceres, and face. Scaly leg mites tunnel under the scales on the legs. These mites live their entire life cycle on the bird and can cause severe irritation. White, scaly deposits are the first signs; these deposits then become thickened, enlarged, and encrusted. Early diagnosis and treatment will prevent these unsightly deposits. A veterinarian can identify these mites and prescribe a spray, powder, or other medication.

Bacterial Diseases

Avian Psittacosis

Bacterial diseases such as psittacosis, also called parrot fever or chlamydiosis, is caused by the bacteria *Chlamydia psittaci.* The disease is also referred to as ornithosis when found in other birds such as turkeys and pigeons. Feces, nasal discharge, and contaminated feather dust are the primary source of infection; the bacteria can also be inhaled into the respiratory system. When infection occurs, the bacteria travel to the liver and spleen, where they multiply. Symptoms may include nasal discharges, closing of the eyes, listlessness, loss of appetite, loss of weight, greenish-colored diarrhea, and labored breathing. It is possible for birds to be carriers of the disease and show no signs. Pet birds should be purchased from reliable, disease-free sources. Psittacosis can also be transmitted to humans. Symptoms are flu-like and pneumonia-like with fever, headache, and coughing.

Colibacillosis or Sweating Disease

Colibacillosis is caused by the bacteria *Escherichia coli.* The disease primarily affects young birds and canaries. The bacteria normally exist as part of the intestinal flora of all animals. Infection from *E. coli* usually is associated with another disease. One form of colibacillosis is air sac disease, where infection produces severe, cheese-like inflammations over the heart, liver, and air sacs. Older birds may develop chronic infections of the oviduct as a result of bacteria entering the oviduct. Infection causes a swelling of the oviduct and a foul-smelling discharge. Many infections of the legs and wing joints, mouth, and respiratory tract are associated with *E. coli* bacteria. Prevention is probably more effective than treatment. Sanitation and cleanliness are of utmost importance.

Pasteurellosis

Pasteurellosisis is caused by the bacteria *Pasteurellosis multocida.* Symptoms may include joint, sinus, nasal, middle ear, or skull infections. Fever, depression, anorexia, ruffed feathers, mucous discharges, and difficulty breathing are more visible signs of the disease. Disease organisms may be spread by contaminated feed and equipment.

Bumblefoot

Bumblefoot is a painful ailment of the feet of birds and is usually associated with staphylococcal infections. The feet and joints become hot and swollen with a thick, grayish-white fluid. The bird may not want to walk or may have trouble bending its toes or clasping onto the perch.One should consult with a veterinarian for proper diagnosis and treatment. Suitable perches and sanitation of cages is important in preventing bumblefoot.

Tuberculosis

Tuberculosis is a disease caused by the bacteria *Mycobacterium avium*. The disease usually shows no outward signs; there is just a slow weight loss and deterioration of the bird's condition. Internal organs have hard, rounded gray to yellow pea-sized nodules. These nodules increase in size and number as the disease progresses. Regular tuberculin testing is recommended in breeding aviaries.

Viral Diseases

Psittacine Beak and Feather Disease

Psittacine beak and feather disease is a viral disease that attacks the immune system. Symptoms usually become evident at the first molt; new feathers may not emerge or may be deformed and soon break off. The beak and nails may be soft, overgrown, and lose their pigment. There is no cure for the disease. Treatment to control secondary diseases and supplements of vitamins and minerals may prolong the life of the bird. This disease is also referred to as French molt.

Newcastle Disease

"Newcastle disease, caused by a virus, is a significant threat to the poultry industry but can also infect pet birds. Transmission is by respiratory aerosols, fecal contamination of food or water, direct contact with infected birds, and contact with inanimate objects that harbor the virus. It is rarely seen in pet birds in the United States.

"Signs of infection include depression, loss of appetite and weight, sneezing, discharge from the eyes and nostrils, difficulty breathing, bright yellow-green diarrhea, lack of coordination, head bobbing, and spasms. In later stages, paralysis of the wings or legs, involuntary or jerky movements, twisting of the neck and unnatural position of the head, and dilated pupils may also be seen. Birds may die suddenly without signs.

"There is no cure, and treatment is not recommended. If suspected, Newcastle disease must be reported to the appropriate authorities due to its potentially devastating effects on domestic poultry. Vaccination is prohibited in birds entering the United States because it does not eliminate the carrier state and makes it difficult to detect the virus during quarantine."[1]

Pacheco's Parrot Disease

"Pacheco's disease is a contagious and often fatal viral hepatitis of parrots by a herpes virus. Many species of parrots such as African greys, Eclectus, Amazons, macaws, pionus, Quakers, Rosellas, lovebirds, and budgies are susceptible to Pacheco's. Nanday conures and Patagonian conures are thought to be resistant to the infection and may serve as sources of infection for other species of psittacines.

"Pacheco's virus is shed in oral secretions and feces. Some hens can pass the virus to their eggs, producing latently infected chicks. The virus may also be spread by people who come in contact with infected birds.

"Infected birds may die suddenly, without showing any signs of illness, or die several days later after a brief period of anorexia, depression, diarrhea, increased thirst and urination, lime green urates, ataxia (loss of balance), tremors, and convulsions. Any bird that recovers from Pacheco's disease or survives an outbreak should be considered a latent carrier.

"Latent carriers often appear normal but are able to shed large amounts of virus when they are stressed such as during the breeding season, when new birds are introduced, during changes in temperature, or when infected with other diseases.

"It is very difficult to determine which birds are asymptomatic carriers of this virus. An inactivated vaccination has been developed to protect high risk psitticines, such as those in pet shops, quarantine facilities, aviaries and bird shows. The vaccination is given subcutaneously to small psittacines and either subcutaneously or intramuscularly to large pssittacines. Young birds should be vaccinated twice—with the second vaccination given four to eight weeks after the first. Yearly booster vaccinations are recommended. Breeding birds should have their vaccinations completed a minimum of four weeks before breeding season."[2]

Nutritional Problems

Goiter

A deficiency of iodine in the diet of birds may lead to goiter. Symptoms of goiter are a swelling of the thyroid glands in the neck; the swelling can cause pressure on the windpipe and interfere with breathing. Budgerigars seem to have an especially high requirement for iodine; iodine blocks are available at pet supply stores.

Rickets

A deficiency of calcium or phosphorous, an imbalance of calcium and phosphorus, or a deficiency of vitamin D_3 can cause a deterioration or softening of bone. This condition is called osteomalacia, or rickets. Symptoms are lameness, a stiff-legged gait, or a constant resting in a squatting position. There is usually a decrease in growth. The beak can easily be bent from side to side, and leg and wing bones become soft, bowed, or twisted. Oyster shell or coarse limestone in the diet usually prevents rickets from occurring. Supplements of vitamin D_3 can also be provided.

Candidiasis

"Candidiasis is caused by infection with the yeast *Candida albicans*. It is the same organism that causes "thrush" in the mouths of human babies. This yeast is common in the environment and may be present in small numbers in a normal bird's digestive tract. It may, however, cause illness under certain conditions. Very young, unweaned birds, especially those on antibiotics, may develop candidiasis due to their immature immune systems. Adult birds on long-term antibiotics or suffering from malnutrition or other illnesses may also develop candidiasis. This occurs because some antibiotics can disturb digestion by killing off beneficial bacteria that normally live in the digestive tract, allowing other organisms, such as the *Candida* yeast, to overgrow.

"Candidasis most often affects the crop, although the stomach and intestines may be affected as well. It can also affect the skin, respiratory tract, and rarely, the central nervous system. The severity of infection often depends on the age of the bird and the state of its immune system. A very young or very ill bird may develop an infection that spreads to the blood, bone marrow, and other organs.

"Regurgitation of food, lack of appetite, and general signs of illness may be caused by delayed crop emptying caused by candidiasis. Some birds develop a swollen, mucus-filled crop. Adult birds can harbor low-grade candidiasis with few signs of illness. White spots may be present in the mouth if oral *Candida* is present. A veterinarian can determine whether these white spots are due to candidiasis or another disease.

"Good hygiene, including proper cleaning and disinfection of the cage, nest box, and feeding utensils, is critical to minimize the amount of *Candida* in the environment. In baby chicks affected with candidiasis, the crop must be emptied more often and smaller amounts fed until the crop begins to function normally again. Your veterinarian may prescribe medications to aid in effectively clearing the infection."[3]

Obesity

Obesity can be a problem with caged birds. Obese birds are usually lethargic and less active. Feeding too many sunflower seeds, which are high in fat content, can lead to obesity. A change of diet and more exercise can correct the problem.

Other Problems

Overgrown Claws

Overgrown claws may cause a bird to get caught on its cage and injure itself. The claws can easily be clipped with nail clippers designed for pets. One must be careful not to cut into the blood vessel that runs down the center of the claw; this vessel appears as a pinkish streak. One needs to be especially careful when trimming the claws of birds with black or dark-colored claws. The blood vessel is not visible and extra care should be taken.

Feather Plucking

Feather plucking may become a problem in some birds. The reason for the problem is not always apparent. The reason may be physical, such as diseases, parasites, or allergies; or the reason could be psychological, such as stress or boredom. Improper diet, lack of bathing facilities, or a frustrated desire to breed may be possible reasons for the feather plucking. One should consult with a veterinarian concerning the problem.

Birds that live indoors need to receive regular bathing or spraying; this encourages birds to **preen** themselves. Preening is the bird's natural cleaning using its beak. Regular preening reduces the chances of feather plucking. Regular bathing and spraying reduces feather dust and dirt. Small birds such as budgerigars, canaries, finches, mynahs, lories, and cockatiels should be provided with a container of water to bathe in. This container should be fairly heavy to prevent the bird from tipping or upsetting the container. The water container may be placed in the cage every few days for about 30 minutes at a time. Larger birds such as parrots and cockatoos can be sprayed with a plant sprayer that produces a fine mist. One should spray the mist above the bird so that the mist settles down. The bird may be startled at first but will soon spread its wings and enjoy it. The bird should not be completely saturated and should receive a gentle spray two or three times a week.

REPRODUCTION

The reproductive organs of a male bird are the testes, ductus deferens, and the cloaca. *See Figure 19–35.* The testes of the male, which produce the sperm, remain in the abdominal cavity and do not descend into a scrotum. The sperm is released from the testes and enters the ductus deferens, small tubes that lead to the cloaca. The male bird does not have a copulatory organ. The sperm is released from the ductus deferens into the cloaca. During mating, the cloaca of the male and female are joined together, and the sperm is transferred from the cloaca of the male to the female. The dorsal wall of the male's cloaca is lined with small finger-like projections called **papilla,** which aid in transferring the sperm to the female's cloaca. The sperm make their way up the oviduct of the female and are stored in the folds of the oviduct. The sperm of birds may be viable for up to 30 days, which is longer than the sperm of mammals; however, newer sperm is probably more active. When the egg yolk is released by the ovary, it drops down into the infundibulum. As the yolk moves into the oviduct, the walls are stretched, and the sperm are released to fertilize the yolk. After fertilization, the egg yolk moves down through the oviduct. The thick, white albumen is secreted around the yolk. Two shell membranes are added in the **isthmus.** The egg then enters the uterus, where the eggshell is formed. The eggshell takes about 20 hours to form; the egg moves to the vagina and is then laid. *See Figures 19–36* and *19–37.*

After the egg is laid, the embryo continues to grow inside the egg. This period of growth is referred to as the incubation period. The period varies with the species but averages around 21 days. Proper temperature and humidity are especially important in the incubation of eggs.

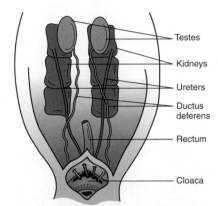

Figure 19–35 Urogenital system of a male bird.

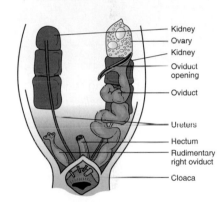

Figure 19–36 Urogenital system of a female bird. For more detail, see Figure 19–37.

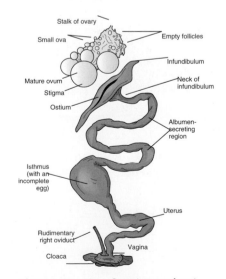

Figure 19–37 Ovaries and oviduct system of a female bird.

Some species of pet birds are easily bred in captivity, especially the smaller birds like budgerigars, finches, and canaries. Larger species may be more difficult to breed. The first problem that may be encountered in reproducing birds is getting both a male and female of a species. Determining the sex of some species is difficult if no dimorphism exists. Sex can be determined by DNA testing.

Canaries and finches are good choices for beginning bird fanciers who want to breed their birds, because these birds breed and reproduce easily in captivity. Most species show various degrees of sexual dimorphism, and pairs can more easily be selected. Breeding cages should be provided for these small birds and equipped with a wood or preferably a wire divider. The male is then placed on one side and the female on the other side. The divider should be removed when the male begins to call to the female and begins to feed her through the wire divider. A small wire, wood, plastic, or ceramic open-topped container should be provided for the bird to make a nest in. The container should be hung about halfway up the side of the cage. Straw, grass, cotton, feathers, or dry peat moss can be provided for nesting materials. The female usually lays one egg per day until a **clutch** of two to six eggs is laid; incubation lasts about 14 days. The female feeds the young. A soft food is needed so that she can feed them. Commercial foods are available. The young develop rapidly; their eyes open in about 7 days. At about 4 weeks, they can be separated from their parents. Soft foods can then be provided for the young; seeds should be cracked and soaked for 24 hours to soften them. *See Figure 19–38.*

It is fairly easy to distinguish sex of budgerigars. The males have blue ceres, and the females have pink or brown ceres. If a pair is introduced and they fight, they need to be separated for a couple of weeks and introduced again. Budgerigars may breed more successfully if two or more pairs are kept in a cage. A nest box should be placed in the upper third of the cage and should be about 7 inches square and 9 inches high. The nest box should have a 1½-inch hole placed about two-thirds the way up the side, and a wooden perch should be placed on the outside. Budgerigars usually do not use nesting materials. Rather, a concave depression should be made in the floor of the nesting box to hold the eggs together in a clutch. Budgerigars lay four to six eggs, one egg being laid every 2 days; the incubation period usually is 17 to 18 days. The young are fed a milk-like substance that is produced in the crop of the female; this substance is then regurgitated and fed to the young. After a few days, the mother mixes this substance with other soft foods. After 2 weeks, the male helps with the feeding duties. Soft foods like those fed to finches and canaries should also be provided. The young budgerigars' eyes open at 8 to 10 days. They can be separated from their parents at 5 to 6 weeks of age. *See Figure 19–39.*

It is difficult to determine the sex of lovebirds; surgically or DNA-sexed birds will prevent disappointment at trying to mate same-sexed birds. A pair of birds of the same sex will preen each other, and everything may appear to be going well. One bird may even try to mount the other.

A nest box similar to those used for budgerigars can be used for lovebirds. Straw, grass, small tree branches, and sticks can be provided for nesting materials. Some green materials may be provided to add moisture necessary

Figure 19–38 Young society finches. *Photo by Isabelle Francais.*

Figure 19–39 A young budgerigar. *Photo by Isabelle Francais.*

for incubation. The female lays four to seven eggs in a clutch; the incubation period varies from 18 to 24 days. The young begin to venture from the nest at about 5 weeks of age and are weaned at 7 weeks. Both parents assist in feeding the young.

Male cockatiels usually have brighter orange or yellow about their head and cheeks than the females. A nest box approximately 9 inches by 9 inches by 15 inches high should be provided. A hole 3 inches in diameter and a wooden perch should be provided. Wood chips and straw can be provided for nesting materials; however, some cockatiels may not use nesting materials. A concave depression should be made at the bottom of the nest box. The female lays three to six eggs in a clutch, with one egg laid every other day; the incubation period varies from 18 to 20 days. Both parents incubate the eggs and feed the young. At 4 to 5 weeks, the young leave the nest. They are ready to be separated from their parents at 7 to 8 weeks. *See Figure 19–40.*

The size of the nest box for the true parakeets will vary depending on the species. The smaller parakeets can use a nest box similar in size to those used for budgerigars; the larger parakeets such as the ringneck parakeets and rosettas need larger nest boxes. Boxes up to 13 inches square and 3 feet deep may be required; a hole 4 inches in diameter should also be provided. The box should then be filled up to the hole with pine shavings. The birds will remove the shavings until the desired level is reached. Those interested in breeding true parakeets should obtain more specific information for their species.

Most parrots use holes in the ground or in trees to make nests in the wild. In captivity, a wooden or metal nest box can be used. Metal might be preferred because parrots can destroy wooden nest boxes. Metal barrels or garbage cans serve as suitable nesting boxes. A wooden perch should be attached to the outside, and a hole 6 inches in diameter for an entrance should be made 3 to 12 inches from the top. Another hole with a lid should be added so that the process of incubation and hatching may be observed. The nest box should be hung high up in the corner of the bird's cage. A large cage or **aviary** is needed to provide plenty of room for breeding large birds. Wood shavings and peat moss make suitable nesting materials. Two to four eggs are usually laid per clutch, and the incubation period is 24 to 30 days. Both the male and female take turns incubating the eggs. When the young hatch, they are completely blind and helpless. At about 8 to 10 weeks of age, the young leave the nest; however, the parents continue to feed the young for another 3 to 7 weeks. The young are weaned and on their own at 11 to 17 weeks of age. *See Figure 19–41.*

Figure 19–40 A young cockatiel. *Photo by Isabelle Francais.*

Figure 19–41 Juvenile blue and gold macaw on the right. A juvenile military macaw on the left. *Photo by Isabelle Francais.*

SUMMARY

Birds are two-legged, egg-laying, warm-blooded animals with feathers and wings, and they are believed to have evolved from prehistoric reptiles. The first bird is believed to be *Archaeopteryx,* which lived approximately 150 million years ago during the Jurassic period. Modern birds appeared about 80 million years ago during the Cretaceous period, and by the Cenozic era some 70 million years ago, they were very similar to the birds of today.

Modern birds are classified into 28 orders containing 170 families and 9,000 species. Three orders are important as pets. The parrots, cockatoos, macaws, lovebirds, lories, lorikeets, budgerigars, and parakeets belong to the order Psittaciformes. Perching birds, which include the finches, waxbills, buntings, canaries, siskins, and grosbeaks, belong to the order Passeriformes. Toucans belong to the order Piciformes, which also include the woodpeckers and other related birds.

Cages for birds are available in a wide variety of sizes and styles. Cages should be selected depending on the species of bird, ease of cleaning, affordability, and the materials used in their construction.

Some birds' diet consists of seeds. Seeds consumed by birds are of two types: cereal seeds and oil seeds. Cereal seeds contain a higher proportion of carbohydrates compared with oil. Oil seeds are high in fat and low in carbohydrates. Nectar- and fruit-consuming birds are an exception because their diets consist of pollen, nectar, fruits, and other soft foods. Some green plant materials can be fed to birds, but the amount should be limited because overfeeding can cause digestive problems and diarrhea. Seed-eating birds must also receive grit in their diet. Grit is important in helping grind up the food material when it reaches the bird's gizzard. All female birds should receive ample amounts of calcium because it is important in the formation of egg shells.

Budgerigars, cockatiels, parrots, macaws, cockatoos, and mynah birds can learn to mimic the human voice and other sounds they hear. In teaching a bird to mimic the human voice or to learn other tricks, patience is very important; some birds may never learn to mimic sounds or learn tricks, but they still make excellent companions.

Pet birds are usually very healthy because they do not come in contact with parasites or disease organisms that birds in the wild may encounter. However, pet birds are susceptible to diseases, and owners should be alert for signs that might indicate a problem.

DISCUSSION QUESTIONS

1. What is a bird?
2. Why do many people believe that birds evolved from prehistoric reptiles?
3. What are the types of feathers found on birds, and what are their purposes?
4. What are the three orders of birds discussed, and what are the main characteristics of each?
5. Name three nutritional ailments of birds, and list the symptoms of each.
6. What are the guidelines to follow when placing a cage in a room?
7. What are some important factors to remember when selecting a cage?
8. Describe how a cage for large, parrot-type birds should be constructed.

9. What are two types of seeds consumed by birds, and what are some examples of each?

10. Why is grit important in the diet of some birds? Which birds do not need grit?

11. Describe the steps in getting a bird to feel comfortable with its handler and the steps in training.

12. What are some signs that might indicate poor health in a bird?

13. What are some external parasites of birds, and what are the symptoms that might indicate a bird has external parasites?

14. What are some internal parasites of birds, and what are the symptoms that might indicate a bird has internal parasites?

15. What are some bacterial diseases of birds, the symptoms, and the treatments for them?

16. What are some viral diseases of birds, the symptoms, and the treatments for them?

SUGGESTED ACTIVITIES

1. Visit a local pet shop and make a list of the different species of birds that are available. Using the local library, compile as much information as you can about the species.

2. Research the different orders, families, genus groups, and species of birds. Why are they classified as they are?

3. Contact government agencies and compile a current list of endangered, threatened, and extinct species of birds.

ADDITIONAL RESOURCES

1. Avian Web
   http://www.avianweb.com/index.html

2. Dr. Jungle's Animal-World
   http://www.animal-world.com/index.html

3. Nutty Birdwatcher
   http://www.birdnature.com/index.html

4. Birds Home
   http://www.birdshome.info

5. International Conure Association
   http://www.conure.org/species.htm

6. Zipcode Zoo
   http://zipcodezoo.com/default.asp

7. World Parrot Trust
   http://www.parrots.org

8. Animal Diversity Web
   http://animaldiversity.ummz.umich.edu/site/index.html

9. Birds of a Feather—Bird Links
   http://www.boaf.com/birdlinks.htm

END NOTES

1. *The Merck/Merial Manual for Pet Health.* Copyright © 2007. MERCK & CO., INC., Whitehouse Station, NJ, USA. All rights reserved. Used with permission.

2. Pesek, Linda, DVM. "Ask a Vet—Pacheco's Disease—What Is It?" January 1, 2008. http://theaviary.com/s0396-01.shtml.

3. *The Merck/Merial Manual for Pet Health.* Copyright © 2007. MERCK & CO., INC., Whitehouse Station, NJ, USA. All rights reserved. Used with permission.

CHAPTER 20

Fish

OBJECTIVES

After reading this chapter, one should be able to:

- 🐾 describe the characteristics of fish.
- 🐾 compare the three classes of fish.
- 🐾 describe and set up a freshwater and marine aquarium.
- 🐾 describe the feeding habits of fish.
- 🐾 describe the common diseases of fish.

TERMS TO KNOW

adipose fin

anal fin

anterior

aquarists

barbels

brackish

caudal fin

community
   aquariums

dorsal fin

gonopodium

labyrinthine
   chamber

neuromasts

pectoral fins

peduncle

pelvic fins

protrusive

shoals

spawning

species
   aquarium

symbiosis

vent

CHARACTERISTICS

Evidence of the early development of fish is found in fossils dating back to the Ordovician period 425 to 500 million years ago. These early fish were called Ostracoderms. They were slow, bottom-dwelling animals covered with thick, bony plates and scales. Ostracoderms had very poorly developed fins and did not have jaws. They are also believed to be the first animals to have a backbone. Ostracoderms became extinct about 250 million years ago.

From the Ostracoderms evolved two groups of fish with movable jaws. One group was called Placoderms and first appeared about 395 million years ago. They had thick, bony plates and paired fins. Their upper jaw was fused to the skull, but the lower jaw was hinged and movable. This group became extinct about 345 million years ago, but from them evolved the class Chondrichthyes. Today, Chondrichthyes are represented by the sharks, rays, skates, and other fish with cartilaginous skeletons.

The other group of jawed fish to evolve from Ostracoderms was called Acanthodians. It first appeared about 410 million years ago. From Acanthodians evolved the class Osteichthyes; these were fish with bony skeletons. Osteichthyes are divided into two subclasses: Sarcopterygii and Actinopterygii. Sarcopterygii are fish with fleshy fins or lobed fins, and their skeletons are made up of part cartilage and part bone. Today, only two orders exist: Dipteriformes, which is made up of lungfish, and Coelacanthiformes, which contains only one living species, *Latemeria chalumnae* (see Figure 1–2), Chapter 1. It is believed that one extinct order of Coelacanthiformes, the Phipidistia, crawled out of the water about 390 million years ago to become the first amphibians. The subclass Actinopterygii is represented today by more than 20,000 species in 34 different orders.

Fish are cold-blooded vertebrates that breathe with gills and move with the aid of fins. They are the most numerous vertebrates, with more than 30,000 species. The smallest fish is the Philippine island goby, *Pandaka pygmaea*, which is only ⅓ to ½ inch long. The largest is the whale shark, *Rhincodon typus*, which grows to about 50 feet and weighs several tons.

ANATOMY

Most fish are covered with scales, which are thin, bony plates that overlap each other and provide protection. *See Figures 20–1, 20–2, and 20–3.* The scales develop from and are embedded in a pocket of the dermis. The exposed part of the scale is covered with a thin layer of epidermis. The skin of fish

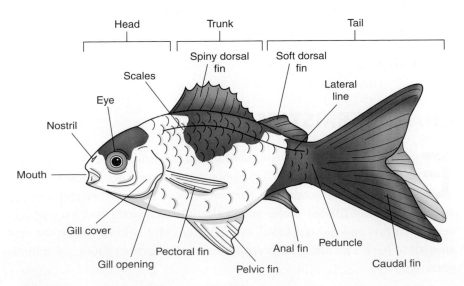

Figure 20–1 External anatomy of a fish.

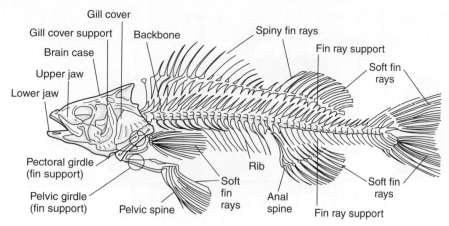

Figure 20-2   Skeletal structure of a fish.

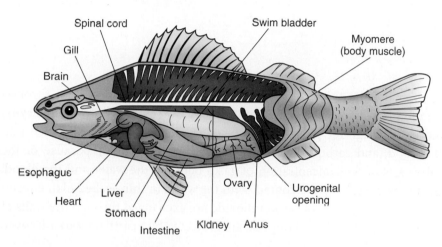

Figure 20-3   Internal anatomy of a fish.

contains glands that produce a slimy mucus, which makes the fish slippery and provides protection from bacteria. The skin contains chromatophores, pigment cells that give the fish its colors. The colors of a fish usually allow it to blend with its surroundings, and many fish are able to change their color if necessary. Sensory receptors are also contained in the skin.

The scales on modern fish are of four types: ctenoid, cycloid, ganoid, and placoid. Most aquarium fishes have either ctenoid or cycloid scales. Ctenoid scales have serrations on the edges and rough surfaces; fish that have these scales feel rough to the touch. Cycloid scales have smooth surfaces and edges that make the fish feel smooth and slick.

Fins are movable structures that aid the fish in swimming and maintaining its balance. Most modern bony fish have rayed fins. These fins consist of a web of skin supported by bone or cartilage rods called rays. Fish may have sharp, spiny, or soft rays. These fins are very flexible.

Most fish have at least one vertical **dorsal fin** along their back, a single **anal fin** on the underside near the tail, and a single **caudal fin** or tail fin. Some species of fish have an **adipose fin**—a small, fleshy fin located on the back between the dorsal and caudal fin. One set of identical paired fins, which is found on the side just behind the head, is called the **pectoral fins,** and one set called the **pelvic fins** is found below and just behind the pectoral fins.

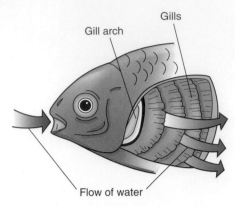

Figure 20–4 Fish breathe through organs called gills. Water is drawn in through the mouth and forced back into the pharynx and out through the gills by the constant opening and closing of the mouth.

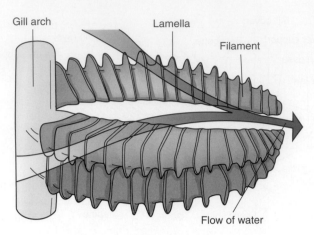

Figure 20–5 Each gill consists of many filaments.

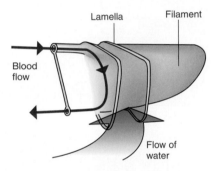

Figure 20–6 Each filament consists of lamella. Lamella are thin walled and contain the blood vessels through which oxygen and carbon dioxide are exchanged.

Fish breathe through organs called gills. Water is drawn in through the mouth by constant opening and closing, forced back into the pharynx, and out through the gills. The dissolved oxygen in the water is taken into the blood, and carbon dioxide is released into the water because of the differences in the concentration of gases between the water and the blood. The blood comes in close contact with the water in thin-walled filaments on the gills; the oxygen and carbon dioxide are exchanged through the walls of the filaments. The gills in bony fish are protected by a cartilaginous gill cover. *See Figures 20–4, 20–5,* and *20–6.*

All fish must maintain proper levels of salt and water in their bodies. Water flows from areas of weak salt solution to areas of strong solution in an effort to dilute the salt solution (osmosis). In freshwater fish, the body salts are a higher concentration than the salt in the water around them; therefore, water is drawn into their bodies. Freshwater fish have no need to drink extra water; their kidneys retain the salts and eliminate large amounts of water. Saltwater fish are surrounded by water that has a higher salt content than their bodies. Water is drawn from their bodies, so the fish must drink water to keep from dehydrating. The excess salts are eliminated in small amounts of very concentrated urine.

Most bony fish have a swim bladder in their abdominal cavity. The swim bladder is filled with gases produced by the blood, which enables the fish to maintain a particular depth. A few bony fish are able to breathe atmospheric air because their swim bladder is supplied with blood vessels, and this allows the swim bladder to function like a lung. This is an important adaptation for fish living in swamps and other waters that are poorly oxygenated.

A few species of fish are able to use atmospheric oxygen because part of their intestines is modified to allow for oxygen intake. These fish rise to the surface and gulp air into their mouths. The air is then swallowed into the digestive system, where it is taken into the blood. Anabantoid fish have special rosette-shaped plates in a **labyrinthine chamber** behind their gills. These rosette-shaped plates are supplied with numerous blood vessels that absorb oxygen from the atmospheric air they inhale through their mouth.

The fish's heart is a long, folded organ consisting of only two chambers: the atrium (or auricle) and the ventricle. The blood is pumped by the ventricle through the arteries to the gills. In the capillaries of the gills, the blood receives oxygen in exchange for carbon dioxide, and the blood then continues on through the arteries to other organs and parts of the body. As the blood flows through other organs and the other parts of the body, nutrients are picked up and later exchanged for waste products. Oxygen is exchanged for carbon dioxide, and waste products are eliminated. After these exchanges, the blood returns through veins to the atrium of the heart. From the atrium, the blood flows to the ventricle, where the circulation of the blood began, and the cycle is repeated. The blood makes a simple, single circuit in the fish's circulation system.

Most fish are able to react to changes in water pressure, temperature, currents, and sounds because of a lateral line running along their side from behind the gill cover to the base of the tail. A series of pressure-sensitive cells called **neuromasts** are contained in tubes along the lateral line. Pores opening into the tubes bring the cells in contact with water. *See Figure 20–1.*

The eyes of fish are similar to those of other vertebrates, but differ in a couple of ways. The fish's eye has a spherical lens that focuses by moving within the eyeball, not by changing the curvature of the lens. Fish do not have eyelids; the eye is kept moist by the flow of water. The size of the eye usually depends on the amount of light reaching the eye. Fish that live in shallow, brightly lit waters usually have small eyes, whereas fish living in dimly lit water have larger eyes.

All fish have inner ears. Species that have swim bladders have a more acute sense of hearing because the bladder acts as a resonator and amplifies the sound. Bony fish are sensitive to sounds in the range of 100 to 10,000 cycles per second.

Fish have taste buds in their mouth, on their lips, and on their body and fins. Their sense of smell and taste is highly developed. Some fish have "taste buds" on their **barbels,** which are whisker-like projections around their mouth.

Fish occupy different levels of an aquarium; some occupy the top, some the middle, and some the bottom levels. This usually gives an indication of the type of diet and feeding habits of the fish. The structure of their mouth also gives an indication of what level the fish will occupy. Fish that occupy the top level feed on insects and vegetable matter that fall into the water; their mouths are usually upturned to enable easier feeding at the surface. Some top-level fish feed on other small fish that also occupy the top level. Most top-level fish have long, streamlined bodies and can propel themselves through the water rapidly to catch their prey.

Some middle-water fish are found in flowing waters. These fish are active swimmers and have streamlined bodies. Some middle-water swimmers come from slower waters, live in shoals, and feed among the reeds and other vegetation. Fish that feed among the reeds usually have narrow bodies that enable them to easily move through the reeds. Fish species that feed on leafy vegetation usually have long, thin bodies with small fins that enable them to move among the leaves and to conceal themselves. Middle-water swimmers usually have small mouths that are positioned straight forward, neither upward nor downward turned.

Bottom-dwellers live on the bottom of streams, rivers, and lakes. Their mouths are downward turned or underslung, which enables them to pick up food from the bottom areas. Bottom-feeders are equipped with barbels to assist them in feeling their way along the bottom and to help them locate food.

Digestive System

The digestive system of fish varies depending on the type of food consumed. Fish that are predators have teeth on the front of the jaws, teeth on the roof of the mouth, and throat teeth just in front of the esophagus. Predator fish usually consume their prey whole, and the teeth function primarily to clasp and move their prey into place for easier swallowing, normally head first. Sharks and piranhas have cutting teeth for biting pieces from their prey. Some fish have rasp-like teeth to scrape plant and animal growth from rocks. Some species do not have teeth in their jaws but have teeth only in their pharynx to grind food material before it enters the esophagus. From the esophagus, food enters the stomach, which also varies considerably depending on the diet. The stomach is a tube-like organ, and food is digested here and then enters the intestine. Undigested material leaves the body through the anus. Lung fish, sharks, and rays have a cloaca through which the waste material passes.

GENERAL CLASSIFICATION

Kingdom	—	Animalia
Phylum	—	Chordata
Superclass	—	Pisces (all fish)
Class	—	Agnatha or Cyclostomata (jawless fish, possessing sucking or filter-feeding mouths)
Class	—	Chondrichthyes (fish with cartilaginous skeletons)
Class	—	Osteichthyes (fish with bony skeletons)
Subclass	—	Actinopterygii (ray-finned fish)
Order	—	Cypriniformes (characins, gymnotid eels, loaches, minnows, suckers)
Order	—	Siluriformes (catfish)
Order	—	Atheriniformes (flying fish, half beaks, killifish, needle fish)
Order	—	Perciformes (cardinalfish, glassfish, and related fish)
Order	—	Tetraodontiformes (box fish, ocean sunfish, puffers, triggerfish)
Order	—	Scorpaeniformes (scorpion fish, sculpins)
Order	—	Gasterosteiformes (pipefish, seahorses, sticklebacks, trumpetfish)
Order	—	Mormyriformes (mormyrids)
Order	—	Osteoglossiformes (bony tongues, freshwater butterfly fish, mooneyes)

The subclass Actinopterygil includes all ray-finned fish; they are divided into 34 orders. Nine of these orders have species that are common in the aquarium trade and are listed and discussed in this text.

CLASSIFICATION—CYPRINIFORMES

The order Cypriniformes is made up of three suborders, 19 families, and 3,500 species. Eight families are discussed here.

Suborder	—	Characoidei
Family	—	Characidae (tetras and characins)
Family	—	Gasteropelecidae (hatchet fish)
Family	—	Anostomidae (headstanders)
Family	—	Hemiodontidae (pencil fish)
Family	—	Citharinidae (moon fish)
Suborder	—	Cyprinoidei
Family	—	Cyprinidae (minnows and carps)
Family	—	Gyrinocheilidae (algae eaters)
Family	—	Cobitidae (loaches)

MAJOR FAMILIES—SUBORDER CHARACOIDEI

The Family Characidae

The family Characidae is a large family containing 1,300 species. About 1,000 species are found in Central America and South America; the remaining species are found in Africa. Most species are brightly colored and have a narrow dorsal fin and small adipose fin. They inhabit shallow, slow-moving rivers of the rainforest and live among sand and ground shoal areas. The waters are usually soft and slightly acidic. Most species are omnivorous and consume all types of foods; a few species are carnivorous. Most members of this family are sociable and do well in aquariums with other fish species, or **community aquariums.** Common members of the family include the following:

cardinal tetra, *Cheirodon axelrodi*

neon tetra, *Paracheirodon innesi (See Figure 20–7)*

silver-tipped tetra, *Hemigrammus nanus*

red phantom tetra, *Megalamphodus sweglesi (See Figure 20–8)*

rummy or red-nosed tetra, *Hemigrammus rhodustomus*

glow-light tetra, *Hemigrammus erythrozonus*

head and taillight tetra, *Hemigrammus ocellifer*

bloodfin, *Aphyocharax rubripinnis*

bleeding-heart tetra, *Hyphessobrycon erythrostigma (See Figure 20–9)*

black tetra, *Gymnocorymbus ternetzi*

lemon tetra, *Hyphessobrycon pulchripinnis*

rosy or rosaceous tetra, *Hyphessobrycon rosaceus*

jewel tetra, *Hyphessobrycon callistus*

Figure 20–7 Neon tetra. *Photo by Aaron Norman.*

Figure 20–8 Red phantom tetra. *Photo by Aaron Norman.*

Figure 20–9 Bleeding-heart tetra. *Photo by Aaron Norman.*

x-ray tetra or x-ray fish, *Pristella maxillaris*

emperor tetra, *Nematobrycon palmeri*

Congo tetra, *Phenacogrammus interruptus*

striped Pyrrhulina, *Pyrrhulina vittata*

penguin fish, *Thayeria obliquua*

All of these are small fish varying in length from 1½ to 3½ inches. Water temperature for these species should be kept at about 75°F, plus or minus 3°F. They consume all types of foods.

The Blind Cave Fish

The blind cave fish, *Astyanax mexicanus,* is native to the underground cave waters of Mexico; it grows to about 3½ inches long. Its body is primarily flesh-colored, and the fins are colorless. The young usually have eyes, but as they mature, skin grows over them and they become nonfunctional. This species does best in an aquarium with its own kind, which is referred to as a **species aquarium.** Although they are blind, they use their sense of smell to seek out and consume food. *See Figure 20–10.*

Figure 20–10 Blind cave fish. *Photo by Aaron Norman.*

The Spraying Characin

The spraying or splashing characin, *Copella arnoldi,* is unusual among aquarium fish because of its spawning habits; the female deposits her eggs on leaves, laying them 1 to 2½ inches above the water. She leaps out of the water and deposits her eggs, and the male immediately leaps out of the water to fertilize them. The male constantly splashes water on the eggs to keep them from drying out. When the young hatch, they drop into the water. In an aquarium, characins deposit their eggs on the cover glass or aquarium lid.

The Bucktoothed Tetra

The bucktoothed tetra (or Exodon), *Exodon paradoxus,* grows to about 5 inches in length. This species of tetra is very aggressive and should be kept in thickly planted aquariums with fish that are able to protect themselves. This species is gold-colored with reddish-orange fins. There are two large, black spots; one on the side and one on the **peduncle** of the tail (part in front of the caudal fin). *See Figure 20–11.*

Figure 20–11 Bucktoothed tetra. *Photo by Aaron Norman.*

Schreitmuller's Metynnis

Schreitmuller's Metynnis, *Metynnis schreitmulleri,* grows to about 5½ inches long; it has a disc-shaped silver body. There is a reddish tinge to the head, back, and fins. During breeding season, the females have a dark-red edge to the anal fin, and the male's is black. There are several species of *Metynnis,* and they are difficult to identify: All species have a broad-based adipose fin. They feed on vegetation and fruit; these species are native to South America.

The Glass (or Red-Eyed) Tetra

The glass (or red-eyed) tetra, *Moenkhausia oligolepis,* grows to slightly less than 5 inches in length. It is silver gray with dark edges to the scales on the upper body; the upper part of the eye is red. There is a yellow band around the peduncle and a black band at the base of the caudal fin.

The Red Piranha

The red piranha (or Natterer's piranha), *Serrasalmus nattereri,* is the most widespread of the piranha species found in South America; it grows to about 12 inches in length. The red piranha is a disc-shaped, muscular, and very powerful fish. Its diet consists of young fish, lean meat, meat-based flake foods, and insects. The primary color is steel gray; the back is a darker blue-gray color, and the underside is red. There are numerous black spots on the body. The most unusual feature of this species is its large, sharp teeth. With the exception of the bucktoothed tera, piranhas must be kept in an aquarium by themselves. Water temperature should be maintained at about 78°F. The red piranha is classified in some references in the separate family Serrasalmidae. There are 24 states that now prohibit the possession of piranha. Before purchasing piranha, one needs to check local and state laws to make sure they are legal in the area. *See Figure 20–12.*

Figure 20–12 Red piranha. *Photo by Aaron Norman.*

The Family Gasteropelecidae

Members of the family Gasteropelecidae are commonly referred to as hatchet fish. They have distinctive hatchet-shaped, deep, narrow bodies and are small fish, 1½ to 2½ inches long. Their pectoral fins are elongated and are used to fly short distances across the water's surface. Aquariums must be fitted with a cover lid to prevent hatchet fish from flying out. Water temperature must be maintained at about 79°F, plus or minus 6°F. Hatchet fish are native to South America. They are top-feeders and must be provided with floating foods, including insects and dried foods. Several common species of hatchet fish are listed as follows:

marbled hatchet fish, *Carnegiella strigata (See Figure 20–13)*

common hatchet fish, *Gasteropelecus sternicla*

silver hatchet fish, *Gasteropelecus levis*

black-winged hatchet fish, *Carnegiella marthae*

Figure 20–13 Marbled hatchet fish. *Photo by Aaron Norman.*

The Family Anostomidae

The family Anostomidae are commonly referred to as headstanders; they are native to South America. Headstanders reach a length of 3 to 5 inches. During rest, these fish position themselves vertically with their head down. They are found in shallow water among plants and rocks. Headstanders are middle- to bottom-dwelling fish and do well in community aquariums. The water should be soft, slightly acidic, filtered through peat, and maintained at 79°F. Members of the family are good jumpers, so the aquarium should be covered. They feed on all types of food, including lettuce, boiled spinach, and soaked oat flakes. Some common species of headstanders are listed as follows:

> marbled headstander, *Abramites hypselonotus*
>
> striped headstander, *Anostomus anostomus (See Figure 20–14)*
>
> spotted headstander, *Chilodus punctatus*
>
> banded Leporinus, *Leporinus fasciatus*
>
> nite tetra, flagtailed Prochilodus, *Prochilodus insignis*

Figure 20–14 Striped headstander. *Photo by Aaron Norman.*

The Family Hemiodontidae

The family Hemiodontidae (also called Lebiasinidae) is commonly referred to as pencil fish. These are 1½ to 7¾ inches in length and are slim, with a **protrusive** mouth. They are native to South and Central America. This family is distinguished from others in the suborder Characoidei by not having teeth in its lower jaw. Pencil fish have small, pointed mouths; some have an adipose fin, and some swim at an oblique angle. All pencil fish take on different coloration at night than during the day. They seek the upper and middle level of a community aquarium, where they feed on the surface. The water in the aquarium should be soft, slightly acidic, filtered through peat, and 76°F, plus or minus 6°F. Some common aquarium species are listed as follows:

> golden pencil fish or one-lined pencil fish, *Nannostomus unifasciatus (See Figure 20–15)*
>
> dwarf pencil fish, *Nannostomus marginatus*
>
> three-lined pencil fish, *Nannostomus trifasciatus*
>
> half-lined Hemiodus, silver Hemiodus, flying swallow, *Hemiodopsis* or *Hemiodus semitaeniatus (See Figure 20–16)*
>
> tube-mouthed pencil fish, *Nannostomus eques*
>
> common pencil fish, *Nannostomus espei*
>
> half-banded Pyrrhulina, *Pyrrhulina laeta*
>
> Rio Meta Pyrrhulina, *Pyrrhulina metae*

Figure 20–15 Golden or one-lined pencil fish. *Photo by Aaron Norman.*

Figure 20–16 Half-lined Hemiodus. *Photo by Aaron Norman.*

The Family Citharinidae

The family Citharinidae is sometimes referred to as moon fish and is native to Africa. There are about 70 different species, with a wide diversity of size and shape; four species are common in the aquarium trade.

The Six-Banded Distichodus

The six-banded distichodus, *Distichodus sexfasciatus,* gets its name from the six or seven vertical, dark bars on the side of the fish. The primary color is orange with silver or gold iridescence. The adipose fin is white with a dark edge, the other fins have a reddish tinge, and the caudal fin is bright red with black edges. The eyes are large and set in a rather small, pointed head. It does well in a community aquarium, but because of its large size, up to 10 inches, it may be aggressive to smaller fish. Plants cannot be maintained in an aquarium with distichodus because the fish eat the young, tender shoots. They feed on all types of food, including boiled lettuce, spinach leaves, and soaked oat flakes. Water should be maintained at 78°F, plus or minus 3°F. *See Figure 20–17.*

Figure 20–17   Six-banded distichodus. *Photo by Aaron Norman.*

The One-Striped African Characin

The one-striped African characin, *Nannaethiops unitaeniatus,* grows to about 2½ inches. The upper parts of the fish are brown and the lower parts yellow, with a white belly. A dark stripe runs from the mouth through the eye to the base of the caudal fin; above this is a golden iridescent stripe. This species does well in a community aquarium. Water temperature should be maintained at 78°F. They feed on all types of foods. *See Figure 20–18.*

Figure 20–18   One-striped African characin. *Photo by Aaron Norman.*

The Pike Characin

The pike characin, *Phago maculatus,* grows to about 6 inches. It is a very thin, narrow fish with long, beak-like jaws. Both upper and lower jaws have two rows of teeth. The caudal fin is large and deeply forked; other fins are small. The pike characin is light brown with a yellow tinge. A dark brown stripe runs from the eye back through the center of the caudal fin. There are dark, horizontal bars on the caudal fin. This very aggressive fish should be kept only in a species aquarium. Its diet consists of insects and other fish. Water temperature should be maintained at 81°F, plus or minus 2°F. *See Figure 20–19.*

Figure 20–19   Pike characin. *Photo by Aaron Norman.*

The African Redfin

The African redfin, *Neolebias ansorgii*, grows to about 1½ inches. This is a very colorful species; its primary color is dark blue with a white underside and red fins. There is an overall blue-gold iridescence to its color. This species occupies the lower levels of an aquarium and does best in **shoals** (groups of fish) of its own kind in a species aquarium. It feeds on all types of food. Water temperature should be maintained at 78°F, plus or minus 5°F.

MAJOR FAMILIES—SUBORDER CYPRINOIDEI

The Family Cyprinidae

The family Cyprinidae is a very large family consisting of more than 1,500 species; members are commonly referred to as carps and minnows. This family of fish does not have teeth in its jaws, but uses teeth in its throat (pharyngeal teeth) to grind and break up food. Some have one or two barbels at the corners of the mouth; none have an adipose fin. Most occupy the lower levels of an aquarium and do well in community aquariums. Water temperatures should be maintained at 75°F, plus or minus 2°F. These fish feed on all types of food. A few species of Cyprinidae are coldwater fish. Most of the tropical cyprinids are classified into three main genus groups: *Barbs, Rasbora,* and *Brachydanio,* and are commonly referred to as barbs, rasboras, and danios. All are colorful favorites of the aquarium trade. Most grow from 2 to 4 inches in length.

Some common species are listed as follows:

Arulius barb or longfin barb, *Barbus arulius*

rosy barb, *Barbus conchonius*

Cuming's barb, *Barbus cumingi*

striped barb or zebra barb, *Barbus fasciatus*

black ruby barb or purple-headed barb, *Barbus nigrofasciatus*

checker or island barb, *Barbus oligolepis*

dwarf or pygmy barb, *Barbus phutunio*

golden barb, *Barbus schuberti*

tiger barb, *Barbus tetrazona (See Figure 20–20)*

cherry barb, *Barbus titteya*

red-tailed rasbora, *Rasbora borapetensis*

Harlequin fish or red rasbora, *Rasbora heteromorpha*

pygmy or spotted rasbora, *Rasbora maculata*

red-striped or glow-light rasbora, *Rasbora pauciperforata (See Figure 20–21)*

Figure 20–20 Tiger barb. *Photo by Aaron Norman.*

Figure 20–21 Red-striped or glow-light rasbora. *Photo by Aaron Norman.*

scissor-tail or three-lined rasbora, *Rasbora trilineata*

pearl danio, *Brachydanio albolineatus*

leopard danio, *Brachydanio frankei*

blue or Kerr's danio, *Brachydanio kerri*

spotted or dwarf danio, *Brachydanio nigrofasciatus*

zebra danio or zebra fish, *Brachydanio rerio (See Figure 20–22)*

Figure 20–22   Zebra danio. *Photo by Aaron Norman.*

The Tinfoil Barb

The tinfoil barb, also called the goldfoil or Schwanenfeld's barb, *Barbus schwanenfeldi,* is a large member of this genus group that grows to 15 or more inches. It is native to Thailand, Malaysia, and Indonesia. The primary body color is silver with gold or blue iridescence. The fins are red, the dorsal fin has a black tip, and the caudal fin has black edges. Two pair of barbels are on the upper jaw. It does well in a community aquarium but should be kept only with other fish similar in size. Water temperature should be maintained at 72°F, plus or minus 5°F. Its diet consists of vegetation, and these fish should be fed lettuce leaves along with other types of food. *See Figure 20–23.*

Figure 20–23   Tinfoil barb. *Photo by Aaron Norman.*

The Spanner Barb

The spanner barb (or T-barb), *Barbus lateristriga,* is another large barb that grows to 7 inches. It is native to Thailand, Malaysia, and Indonesia. The primary color is greenish-gold, darker on the back and getting lighter down the sides, with silver on the underside. Two large, dark, vertical bars run from the back to the belly, and one dark, horizontal bar runs through the peduncle to the caudal fin. This community fish occupies all levels; however, it should be kept with fish similar in size. Water temperature should be maintained at 71°F, plus or minus 5°F. It consumes all types of food, including plant material.

The Clown Barb

The clown barb (or Everett's barb), *Barbus everetti,* is native to Singapore and Borneo; it grows to 6 inches. The primary color is reddish-brown with an orange tinge and silver or gold iridescence; the fins are light red. It does well in a community aquarium but should be kept with other fish similar in size. Water temperature should be maintained at 79°F, plus or minus 2°F. It consumes all types of food, including plant material.

The Giant Danio

The giant danio, *Danio malabaricus,* is found in Southwest India and Sri Lanka; it grows to 4 inches. The back color is bluish-green and the underside is light pink. The sides of the fish are blue, and there are three or four yellow, horizontal stripes running across the **anterior** half. The front part of the fish has irregular vertical stripes and spots. All the fins except the colorless pectoral fin have a reddish tinge. This species does well in a community aquarium but may show aggression toward smaller fish. Water temperature should be maintained at 72°F, plus or minus 3°F. It occupies the upper level of an aquarium and feeds on all types of foods. *See Figure 20–24.*

Figure 20–24 Giant danio. *Photo by Aaron Norman.*

The White Cloud Mountain Minnow

The white cloud mountain minnow, *Tanichthys albonubes,* is native to China and grows to 1¾ inches. This very hardy fish tolerates a wide range of aquarium temperatures from 61° to 72°F. The back is olive brown with green iridescence; the underside is white. A thin, dark-blue stripe runs the length of the body and is bordered above by a golden iridescent stripe and below by a reddish band. The fins are reddish with silver-blue tips. It does well in a community aquarium and feeds on all types of foods. *See Figure 20–25.*

Figure 20–25 White cloud mountain minnow. *Photo by Aaron Norman.*

The Dadio

The dadio, *Laubuca dadiburjori,* is native to Indonesia, Burma, Malaysia, Sri Lanka, and Thailand, where it is found in slow-moving waters. It grows to 1¾ inches. The primary color is golden-brown. A blue stripe runs the length of the body; the fins are yellow. The pectoral fins are well developed and positioned so that rapid movement of these fins propels the fish or helps it "fly" across the surface of the water to capture flying insects. Dadios spawn in pools of water on the upper surface of aquatic plants. This species does well in community tanks, where it occupies the upper levels. A cover glass or lid is needed. Water temperatures should be maintained at 75°F. It feeds on all types of food.

The Flying Fox

The flying fox, *Epalzeorhynchus kallopterus,* is native to Borneo, Indonesia, Sumatra, and Java. It is found in creeks with heavy vegetation. This species has an elongated body and grows to 5½ inches. The mouth faces downward and has two pair of barbels. The back is olive green, and the underside is white. A yellow stripe runs from the top of the snout, above the eye, and down to the peduncle. A larger black stripe runs below this from the snout, through the eye, the length of the body, and through the caudal fin. The fins are reddish-brown with yellow tips. This species does well in community aquariums and occupies the lower levels; it should be provided with plenty of vegetation and rocks to hide among. Water temperature should be maintained at 76°F, plus or minus 5°F. It feeds on all types of food, including plant material. *See Figure 20–26.*

Figure 20–26   Flying fox. *Photo by Aaron Norman.*

The Red-Tailed Shark

The red-tailed shark, *Epalzeorhynchus bicolor,* is native to Thailand. The dorsal fin is well developed, and gives this fish a shark-like appearance; it grows to 7 or 8 inches. It is completely black except for a red caudal fin. The mouth faces slightly downward, and there are two pairs of barbels. This species does better in a species aquarium. It can be introduced into a community aquarium but may be aggressive to smaller fish and to its own species. This fish prefers the middle and lower levels of an aquarium and should be provided with some vegetation and rocks. Water temperature should be maintained at 75°F, plus or minus 5°F. Water should be soft, slightly acidic, and filtered through peat. It feeds on all types of food, including plant material. *See Figure 20–27.*

Figure 20–27   Red-tailed shark. *Photo by Aaron Norman.*

The Goldfish

The cold-water Cyprinids that receive the most attention are the goldfish, *Carassius auratus.* Goldfish have been kept and developed by the Chinese for centuries; today, there are several varieties. Among them are the common goldfish, Bristal shubunkin, comet fantail, veiltail, moor, oranda, celestial, lionhead, bubble-eye, and pompon. The common goldfish, comet, and Bristal shubunkin have a single caudal fin. Fantails, veiltails, moor, and oranda have twin caudal fins and a single dorsal fin; the lionhead, celestial, pompom, and bubble-eye have twin caudal fins and no dorsal fin. Most goldfish grow from 3½ to 8 inches long. *See Figures 20–28 and 20–29.*

Figure 20–28   Bubble-eye variety of goldfish. *Photo by Aaron Norman.*

Goldfish occupy all levels of an aquarium. The single-tail varieties are the easiest to keep; the twin tails are more delicate and require more care. Water temperature should be maintained between 32° and 68°F for the single tails, and 46° to 68°F for the twin tails. They feed on all types of foods. The type of water is not critical, but it must be kept clean.

Figure 20–29 Oranda variety of goldfish. *Photo by Aaron Norman.*

Koi

Koi are a species of carp, *Cyprinus carpio.* They originated in Japan, where they have been kept and developed for centuries. There are three main varieties: single-colored, two-colored, and three-colored koi. They are also divided by the type of scales they have. Those that have a few large scales are called Doitsu, those with pinecone scales are called Matsuba, those with gold metallic speckled scales are called Kin-rin, and those with silver metallic speckled scales are called Gin-rin. The single-colored koi are yellow or orange. The two-colored koi are either white with red markings, called Kohaku, (*See Figure 20–30*), or gold and silver, called Hariwaki. The three-colored koi are light blue with orange and black markings, called Asagi; white with red and black markings, called Taisho Sanke; or black with red and white markings, called Showa Sante. Koi grow to 3 feet and can be worth thousands of dollars. Small koi are usually kept indoors and then moved to outdoor pools when they reach 5 inches; they occupy all levels. Water temperature should be maintained between 32° and 68°F. The water must be well filtered. They consume all types of foods.

Figure 20–30 Koi, Kohaku species. *Photo by Aaron Norman.*

The Family Gyrinocheilidae

The family Gyrinocheilidae contains a single species, the sucking loach or Chinese algae eater, *Gyrinocheilus aymonieri.* This fish is native to Thailand and grows to 10 inches. Its primary body color is gray-brown with darker blotches along its sides and back. There are smaller spots on the head and caudal fin. It has large, fleshy lips on the underside of the snout and can cling to vegetation and rocks and to the glass sides of an aquarium. It feeds primarily on algae and other vegetation but eats other foods. They do well in community aquariums when small but can become aggressive when they reach 4 to 5 inches in length. Water temperature should be maintained at 75°F, plus or minus 5°F. *See Figure 20–31.*

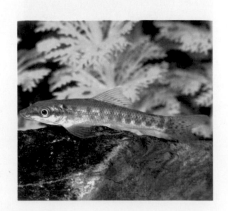

Figure 20–31 Chinese algae eater or sucking loach. *Photo by Aaron Norman.*

The Family Cobitidae

The family Cobitidae is commonly referred to as loaches. These are bottom-dwelling fish native to the waters of Indonesia. They are distinguished by their flat underside, mouths on the lower side of their snout, and three pairs of barbels. They are nocturnal and hide among vegetation and rocks during the day. There are several species, which grow from 2¼ to 24 inches. Most do best in aquariums with their own species. Loaches feed on all types of foods. Water temperature should be maintained at 80°F, plus or minus 5°F.

The clown loach, *Botia macracantha,* is the most colorful and popular; the primary color is bright orange. *See Figure 20–32.* Three large, black bands encircle the body. The first band covers the top of the head, the eyes, and the lower jaw; the second encircles the body in front of the dorsal fin; and the third encircles the peduncle and includes part of the dorsal and anal fins. The front portion of the dorsal and anal fins are orange and the back portion of these fins are black; the other fins are red. The clown loach does well in community aquariums when small.

Figure 20–32 Clown loach. *Photo by Aaron Norman.*

Some other species of loaches are listed as follows:

skunk Botia, Hora's loach, or mouse loach, *Botia horae*

Pakistani loach or reticulated loach, *Botia lohachata*

banded loach, *Botia hymenophysa*

orange finned loach, *Botia modesta*

chain or dwarf loach, *Botia sidthimunki*

zebra loach, *Botia striata*

Kuhli or Coolie loach, *Acanthophthalmus kuhli*

Nichol's loach, *Noemacheilus nicholsi*

CLASSIFICATION—SILURIFORMES

The order Siluriformes is made up of 24 families and more than 2,500 species; they are commonly referred to as catfish. Most species are covered with skin but lack scales. Some species are covered with bony plates; most species have an adipose fin. The pectoral and dorsal fins have stiff, sharp rays or spines. Their bodies are usually flattened on the underside, and they have barbels; these characteristics aid in their bottom-dwelling nocturnal habits. Six families are discussed here:

Family Siluridae (glass catfish)

Family Schilbeidae (three-striped glass catfish)

Family Mochokidae (upside-down catfish)

Family Pimelodidae (unarmored catfish)

Family Callichthyidae (armored catfish)

Family Loricariidae (sucker catfish)

MAJOR FAMILIES—CATFISH

The Family Siluridae

The family Siluridae contains one species commonly found in the aquarium trade, the glass catfish, *Kryptopterus bicirrhis*. This species is native to Indonesia. It is transparent, with the backbone and internal organs clearly visible. It has a very long anal fin, only one ray to the dorsal fin, and no adipose fin; the caudal fin is forked. There is one pair of barbels on the upper jaw that can be extended outward. The glass catfish grows to 4 inches in length and should be kept in small shoals; as a single specimen, it does not do well. This species does well in a community aquarium, where it occupies the upper level. The glass catfish swims at an oblique angle with its tail down. Water temperature should be maintained at 75°F, plus or minus 5°F. It feeds on all types of foods. *See Figure 20–33.*

Figure 20–33 Glass catfish. *Photo by Aaron Norman.*

The family Schilbeidae also contains one species of glass catfish that is common in the aquarium trade—the three-striped glass catfish, *Eutropiellus debauwi*. This species from Central Africa is blue-gray with three dark, horizontal stripes running behind the gill to cover the full length of the body. The center stripe runs through the center of the caudal fin. The upper and lower stripes end at the peduncle and appear as spots on the caudal fin. The anal fin is very long, and there is a small dorsal fin toward the front part of the back; it also has an adipose fin. This species grows to 3 inches and does well in a community aquarium, where it occupies the upper level and swims obliquely with its tail down. The three-striped glass catfish should be kept in shoals; a single specimen does not do well. It feeds on all types of foods. Water temperature should be maintained at 76°F, plus or minus 3°F. *See Figure 20–34.*

Figure 20–34 Three-striped glass catfish. *Photo by Aaron Norman.*

The Family Mochokidae

The family Mochokidae contains one genus *Synodantis* common in the aquarium trade—the upside-down catfish, *Synodontis nigriventris*. It is native to Central Africa and grows to 3 inches. The primary color is gold-brown with numerous dark spots. It has a very large adipose fin; the fins are colorless with dark spots. The upside-down catfish has three pair of barbels: one smooth pair on the upper jaw and two feathered pair on the lower jaw. Its upside-down habit enables it to feed on the underside of leaves and rocks. It feeds primarily on algae but takes other foods from the surface of the water. This species does well in a community aquarium and should be provided with vegetation and rocks. Water temperature should be maintained at 77°F, plus or minus 4°F. *See Figure 20–35.*

Figure 20–35 Upside-down catfish. *Photo by Aaron Norman*

The Family Pimelodidae

The family Pimelodidae is referred to as unarmored catfish. They are covered with skin but have no scales or bony plates. Two species, both from South America, are common in the aquarium trade: the spotted Pimelodella, *Pimelodella* or *Pimelodus pictus,* and the graceful Pimelodella, *Pimelodella gracilis.* The spotted Pimelodella grows to 4½ inches in length. It is silvery-white with numerous black spots along its body and caudal fin. It also has three pair of very long barbels. *See Figure 20-36.* The graceful Pimelodella grows to more than 6½ inches. Females are almost colorless with a dark stripe running horizontally from the gill cover to the caudal fin; the males are a blue-black irides-cent color. The graceful Pimelodella has three long pairs of barbels; the one pair on the upper jaw are extremely long. Both species do well in community aquariums, provided they have vegetation and rocks to hide among during the day; they become active at night. They occupy the middle and lower levels of the aquarium. Water temperature should be maintained at 72°F, plus or minus 5°F. They prefer soft, slightly acidic water. Both species are primarily carnivorous, feeding on insects, worms, and meat-based flaked foods.

Figure 20–36   Spotted pimelodella. *Photo by Aaron Norman.*

The Family Callichthyidae

The family Callichthyidae is commonly referred to as armored catfish and is native to Central and South America. They have two rows of overlapping bony plates, or scutes. They are small, ranging in size from 2½ inches to 4 inches. Their mouths are on the underside of the snout, and they have three pairs of barbels. They are primarily scavengers, feeding on the bottom of streams, usually at night. These species can be seen in an aquarium dashing to the surface, where they gulp in air that is later absorbed in their digestive tract. They do well in a community aquarium if provided with vegetation and rocks to hide among; single fish do not do well. Water temperature should be maintained at 73°F, plus or minus 6°F. *See Figure 20-37.*

Figure 20–37   Peppered or salt-and-pepper Corydoras catfish. *Photo by Aaron Norman.*

Some common species of armored catfish are listed as follows:

short-bodied catfish, *Brochis splendens*

bronze catfish, *Corydoras aeneus*

elegant catfish, *Corydoras elegans*

dwarf or pygmy Corydoras, *Corydoras hastatus*

leopard Corydoras, *Corydoras julii*

black-spotted Corydoras, *Corydoras melanistius*

peppered Corydoras, *Corydoras paleatus*

Swartz's Corydoras, *Corydoras schwartzi*

Myer's Corydoras, *Corydoras myersi*

reticulated Corydoras, *Corydoras reticulatus (See Figure 20–38)*

albino Corydoras, albino species of *C. paleatus* and *C. aeneus*

armored catfish, *Callichthys callichthys*

Port Hoplo or Atipa, *Hoplosternum thoracatum*

striped-tail catfish, *Dianema urostriata striata*

Figure 20–38   Reticulated Corydoras. *Photo by Aaron Norman.*

The Family Loricariidae

The family Loricariidae is commonly referred to as sucker catfish. These fish have three or four rows of bony plates, or scutes, and have underslung mouths designed for sucking or clinging onto vegetation and rocks found in the fast-moving mountain streams of South America. The whiptail catfish, *Loricaria filamentosa,* grows to about 10 inches in length. It has an elongated body with a very elongated peduncle; the upper lobe of the caudal tail is elongated into an appendage. The primary color is gray-brown to yellow-brown with several dark markings and spots. The front rays of the dorsal, pelvic, and pectoral fins form stiff, sharp spines; it does not have an adipose fin. It prefers water temperature of 75°F, plus or minus 3°F. *See Figure 20–39.*

Figure 20–39   Whiptail catfish. *Photo by Aaron Norman.*

The golden Otocinclus, *Otocinclus affinis,* is a small sucker-mouth catfish, growing only to a little over 1½ inches. *See Figure 20–40.* The primary color is golden-brown; a broad black stripe runs from the snout to the base of the tail. The fins are colorless except for black markings on the caudal fin; there is no adipose fin. The underside of the fish is silver-white. It prefers a water temperature of 73°F, plus or minus 6°F. The sucker catfish, *Plecostomus punctactus,* grows to more than 24 inches. The primary color is dark brown

Figure 20–40   Sucker catfish. *Photo by Aaron Norman.*

to black with dark spots. On the lighter fish, five oblique bars can be seen on the side. The fins are brown to black with dark spots, and there is an adipose fin. The fins are large, with the front rays forming stiff, sharp spines. It also prefers water temperatures of 75°F, plus or minus 6°F. All three of these sucker-mouth catfish do well in a community aquarium. They should be provided with vegetation and rocks to hide among during the day, although the plants may be rooted up. They feed on all types of food, including algae and other plant material.

CLASSIFICATION—ATHERINIFORMES

The order Atheriniformes is made up of three suborders and 15 families. Five families are discussed here.

Suborder	—	Exocoetoidei
Family	—	Exocoetidae (halfbeaks and flying fish)
Suborder	—	Cyprinodontoidei
Family	—	Cyprinodontidae (killifish or egg-laying tooth carps)
Family	—	Anablepidae (four-eyed fish)
Family	—	Poeciliidae (live-bearers or viviparous top minnows)
Suborder	—	Atherinoidei
Family	—	Atherinidae (rainbowfish or silversides)

MAJOR FAMILIES—SUBORDER EXOCOETOIDEI

The Family Exocoetidae

The family Exocoetidae or Hemiramphidae contains one species that is common in the aquarium trade, the halfbeak or wrestling halfbeak, *Dermogenys pusillus*. This fish is native to the waters of Thailand, Malaya, Sumatra, and Java. It is found in freshwater and **brackish** waters (waters where fresh- and saltwaters mix). The halfbeak grows to almost 3 inches; its distinguishing characteristic is that the lower jaw protrudes far beyond the upper jaw. This fish is greenish-brown, and the underside is greenish-yellow. The caudal fin is rounded, and the dorsal and anal fins are opposite each other, giving this fish a pike-like appearance. The male has a red spot on the lower front of the dorsal fin; the fins are greenish-yellow. The fish does well in a community aquarium and occupies the upper levels. It feeds on insects and floating foods. Water temperature should be maintained at 73°F, plus or minus 5°F. *See Figure 20–41.*

Figure 20–41 Celebes halfbeak. *Photo by Aaron Norman.*

Figure 20–42 Golden lyretail or golden lyretailed panchax. *Photo by Aaron Norman.*

Figure 20–43 Blue gularis. *Photo by Aaron Norman.*

Figure 20–44 American flag fish. *Photo by Aaron Norman.*

Figure 20–45 Clown killi or rocket panchax. *Photo by Aaron Norman.*

MAJOR FAMILIES—SUBORDER CYPRINODONTOIDEI

The Family Cyprinodontidae

The family Cyprinodontidae is referred to as killifish or egg-laying tooth carps. They are native to Africa, Asia, Europe, South America, Central America, and North America. They are small fish, reaching lengths of 1½ to 4 inches. They have elongated bodies, their heads are usually flattened, and their mouths are protrusive to pick up food. They are brightly colored; the males are usually more colorful than the females. The fins of the males are also larger, elongated, and more colorful. Some species lay their eggs on vegetation, whereas others lay their eggs in the mud on the bottom of stream beds. Those species that lay their eggs in the mud are referred to as *annual fish* because after their eggs are laid, the stream or pond dries up, and the parent fish die; the next year when the rains come, the streams and ponds are filled and the eggs hatch. The young grow quickly and become sexually mature in a single season. The eggs have very hard shells and are able to withstand the weeks or months that the dry period may last. Most species are very aggressive and should be raised as a pair in an aquarium without other fish. Water temperature should be maintained at about 75°F, plus or minus 5°F. The water should be soft, slightly acidic, and filtered through peat. These fish are usually not kept in large aquariums because of their reproductive habits and their short life spans.

Some examples of the killifishes are listed as follows:

lyretail or lyretailed panchax, *Aphyosemion australe (See Figure 20–42)*

red lyretail, *Aphyosemion bivittatum*

steel blue Aphyosemion, *Aphyosemion gardneri*

blue gularis, *Aphyosemion sjoestedti*

lampeyed panchax or lampeye, *Aplocheilichthys macrophthalmus*

plumed lyretail, *Aphyosemion filamentosum*

sabrefin or sicklefin killi, *Austrofundulus dolichopterus*

striped panchax, *Aplocheilus lineatus*

Ceylon killifish, *Aplocheilus dayi*

blue panchax, *Aplocheilus panchax (See Figure 20–43)*

black-finned pearl fish, *Cynolebias nigripinnis*

American flag fish, *Jordanella floridae (See Figure 20–44)*

Rachov's nothobranch, *Nothobranchius rachovi*

Medata or rice fish, *Oryzias latipes*

Playfair's panchax, *Pachypanchax playfairi*

clown killi or rocket panchax, *Pseudoepiplatys annulatus (See Figure 20–45)*

six-barred Epiplatys, *Epiplatys sexfasciatus*

featherfin panchax, *Pterolebius longipinnis*

red Aphyosemion, *Roloffia occidentalis*

green panchax or dwarf panchax, *Aplocheilus blocki (See Figure 20–46)*

Figure 20–46   Green panchax. *Photo by Aaron Norman.*

The Family Anablepidae

The family Anablepidae is a small family consisting of only one genus with two species. They are referred to as four-eyed fish, *Anableps anableps,* and inhabit waters of Mexico, Central America, and Northern South America. *See Figure 20–47.* This fish has an elongated body, a flat head, and large, protruding eyes. It swims just below the surface, and the large, protruding eyes are divided by the water's surface, which allows the fish to see above and below the water at the same time. It grows to 8 inches in length; its body is primarily blue-gray with a yellowish tinge to the underside. The dorsal fin is set far back on the body, and the caudal tail is rounded. Males can be distinguished from females by their prominent **gonopodium,** a modification of the anal fin into a tube-shaped organ through which sperm packets pass and enter the oviduct of the female. The gonopodium will extend either to the right or left side of the male's body. The female has a single scale that covers the cloaca and opens either to the left or to the right. Males and females have to be appropriately matched for fertilization to occur. The sperm packets break open and release sperm to fertilize the eggs. Four-eyed fish give birth to live young twice a year. They do well in community tanks but may be aggressive to smaller fish. Water temperature should be maintained at 75°F, plus or minus 2°F, and 5 teaspoons of salt should be added for every 2½ gallons of water.

Figure 20–47   Four-eyed fish. *Photo by Aaron Norman.*

The family Poeciliidae consists of live-bearers, viviparous top minnows, or live-bearing tooth carps. Members of this family of fish give birth to live young. The males are usually smaller than the females but have much brighter colors and striking patterns. The anal fin of the males is modified into a gonopodium. Sperm that is passed to the female may be stored in the folds of the oviduct, so further pregnancies can occur from one mating. The fertilized eggs develop within the female, and she gives birth about 20 to 60 days after fertilization, depending on the species. Live-bearers do well in community aquariums, but the young must be protected from the adults or they will become food. Water temperature varies with the species, but if maintained at 79°F, plus or minus 5°F, it should be sufficient for most species. Water should be medium-hard and slightly alkaline. Some species do best if salt is added to the water. Live-bearers consume all types of foods, including plant material. Beginning **aquarists** may want to check an additional reference before raising live-bearers. Some of the common live-bearers are listed as follows:

Piketop minnow, *Belonesox belizanus*

Eastern mosquito, *Gambusia affinis*

Girardinus, *Girardinus metallicus*

pseudo helleri, *Heterandria bimaculata*

merry widow, *Phallichthys amates*

guppy or millions fish, *Poecilia reticulata*

blue limia or black-bellied limia, *Poecilia melanogaster*

humpbacked limia, *Poecilia nigrofasciata*

Cuban limia, *Poecilia vittata*

sailfin molly, *Poecilia latipinna*

black molly, *Poecilia hybrid (See Figure 20–48)*

blue-eyed Priapella, *Priapella intermedia*

swordtail, *Xiphophorus helleri*

platy, *Xiphophorus maculatus*

Variatus platy, *Xiphophorus variatus*

Montezuma swordtail, *Xiphophorus montezumae*

Mexican sailfin molly, *Poecilia velifera (See Figure 20–49)*

Figure 20–48 Black sailfin molly. *Photo by Aaron Norman.*

Figure 20–49 Mexican sailfin molly. *Photo by Aaron Norman.*

The Family Atherinidae

The family Atherinidae is referred to as rainbowfish and silver-sides; they are found in the shallow coastal waters of Australia and Madagascar. *See Figure 20–50.* They are distinguished from other fish because they lack a lateral line; their pelvic fins are midway back on the underside, and they have two dorsal fins. The fins are positioned so that the top part of the fish is almost identical to the bottom half. Four species are found in the aquarium trade: the dwarf rainbowfish, *Melanotaenia maccullochi;* the red-tailed rainbow-fish, *Melanotaenia nigrans;* the Madagascar rainbowfish, *Bedotia geayi;* and the Celebes sailfish, *Telmatherina ladigesi.* These fish do well in community aquariums, although plenty of vegetation should be provided. Water should be medium-hard and maintained at about 75°F, plus or minus 3°F. Some salt may need to be added to the water for them to do best.

Figure 20–50 Madagascar rainbowfish. *Photo by Aaron Norman.*

CLASSIFICATION—PERCIFORMES

The order Perciformes consists of more than 6,000 species in 16 suborders and 160 families. Included in this order are both marine and freshwater species; 19 families are discussed here. Species in this order do not have their swim bladder connected to an open duct to the throat. The dorsal, anal, and pelvic fins usually have spines. The dorsal fin consists of an anterior (front) part that has spines and a posterior (rear) part that has soft rays.

Suborder	—	Percoidei
Family	—	Apogonidae (cardinalfish)
Family	—	Centropomidae (snooks, robalos, and glassfish)
Family	—	Grammidae (basslets)
Family	—	Nandidae (leaf fish)
Family	—	Cichlidae (cichlids)
Family	—	Pomacentridae (damsel fish, anemone fish)
Family	—	Monodactylidae (fingerfish)
Family	—	Toxotidae (archer fish)
Family	—	Platacidae (bat fish)
Family	—	Scatophagidae (scats)
Family	—	Chaetodontidae (butterfly fish)
Family	—	Labridae (wrasses)
Suborder	—	Blennioidei
Family	—	Acanthuridae (surgeons, tangs, and unicorn fish)
Family	—	Zanclidae (moorish idol)
Suborder	—	Mastacembeloidei
Family	—	Mastacembelidae (spiny eels)
Suborder	—	Anabantoidei
Family	—	Anabantidae (climbing perch)
Family	—	Helostomatidae (kissing gourami)
Family	—	Belontiidae (fighting fish and paradise fish)
Family	—	Osphronemidae (gourami)

MAJOR FAMILIES—SUBORDER PERCOIDEI

The Family Apogonidae

The family Apogonidae is a group of small, narrow fish less than 4 inches in length. They have short, compact bodies, and the peduncle is unusually large. These species are mouth-brooders, and the male incubates the eggs in his mouth. Most species are nocturnal and hide among the vegetation and rocks during the day. Water temperature should be maintained at 77°F, plus or minus 2°F.

The Pajama cardinalfish, *Sphaeramia nematoptera,* has an interesting color pattern. The part of the fish in front of the dorsal and pectoral fins is yellow and white; next there is a large, black band from top to bottom, and the back half of the fish is white with brown spots. The anterior dorsal fin and pelvic fins are yellow with red on the front edges. The posterior dorsal fin, pectoral fins, and caudal fin are colorless. *See Figure 20–51.*

Figure 20–51 Pajama cardinalfish. *Photo by Aaron Norman.*

Two other species are the barred cardinalfish, *Apogon binotatus,* and the sawcheek cardinalfish, *Apogon quadrisquametus.*

The Family Centropomidae

The family Centropomidae is a group of marine fish found in the waters near India, Thailand, and Burma. *See Figure 20–52.* One species found in the aquarium trade is the Indian glassfish, *Chanda ranga.* This narrow-bodied species grows to about 3 inches long. It is transparent, and the skeleton and internal organs are visible; in light, it shows a gold, yellow, and green iridescence. It is a peaceful fish and does well in a community tank. Water temperature should be maintained at 71°F, plus or minus 6°F. The water should be hard and slightly alkaline. Three to six teaspoons of salt should be added for every 2½ gallons of water.

Figure 20–52   Indian glassfish. *Photo by Aaron Norman.*

The Family Grammidae

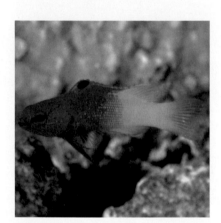

The family Grammidae consists of small fish found around the coral reefs of the Indo-Pacific and Caribbean. *See Figure 20–53.* One species found in the aquarium trade is the royal Gramma or fairy basslet, *Gramma loreto,* which is native to the Caribbean. It grows to about 3 inches. The head and front half of the fish are bright pinkish-purple, and the back half is bright yellow. There is a black spot on the dorsal fin and an oblique, narrow yellow stripe running through the eye from the mouth. This species adapts well to a community aquarium and should have some rocks and coral to hide around. It occupies all levels of the aquarium and feeds on chopped foods and live foods. Water temperature should be maintained at 77°F, plus or minus 2°F.

Figure 20–53   Royal Gramma or fairy basslet. *Photo by Aaron Norman.*

The Family Nandidae

The family Nandidae is a small family made up of very aggressive freshwater predators. The South American leaf fish, *Monocirrhus polyacanthus,* is a very narrow fish, brown and yellow in color, and resembles a leaf. Its head is pointed, the forehead slightly concave, and it has a large protrusive mouth. The lower jaw is elongated and appears to have a worm-like growth. It moves slowly among vegetation in search of prey. The leaf fish is able to change its coloration so it blends in with changes in the vegetation; it must be kept in a species aquarium. Water temperature should be maintained at 79°F, plus or minus 4°F. The Schomburgk's leaf fish, *Polycentrus punctatus,* is a narrow, short-bodied fish with a short peduncle. *See Figure 20–54.* It is another predator that slowly stalks its prey. It is usually gray-blue with numerous dark markings but can change its coloration to blend into its surroundings. It must be kept in a species tank. Water temperature should be maintained at 76°F, plus or minus 3°F.

Figure 20–54   Schomburgk's leaf fish. *Photo by Aaron Norman.*

The Badis, *Badis badis,* is another fish in this family, but it is more peaceful and can be kept in a community aquarium. *See Figure 20–55.* Its general color is reddish-brown with blue and red spots; the dorsal fin has several dark markings. It prefers the middle and lower levels of an aquarium. Water temperature should be maintained at 81°F, plus or minus 2°F.

Figure 20–55   Badis. *Photo by Aaron Norman.*

The Family Cichlidae

The family Cichlidae, commonly referred to as cichlids, is a large family of heavy-bodied fish. Most cichlids are native to South America and Africa; there are more than 700 species. They have one pair of nostrils instead of two like most other fish. The dorsal fin has a long base, and the first three rays form spines. The anal fin is short, and the first three rays are also spined; the caudal fin is usually rounded. There are two main patterns of spawning. Some species lay their eggs on rocks, leaves, or logs, or in the holes dug by the male. Other species, known as shelter breeders, lay their eggs in spawning areas out in the open that have been cleared. Spawning sites are usually defended by both parents. They tend the nest until the young are hatched, and the female may stay with them for a short time; a few species are mouth-brooders.

The Oscar, *Astronotus ocellatus,* is one of the most popular and easily recognized cichlids. In an adequate aquarium, it can grow to 20 or more inches in length. The coloration varies considerably, but it is usually gray with black, vertical blotches along the sides; some have reddish-orange scales along the lower sides. The one constant in coloration is a black spot encircled with a reddish-orange ring on the peduncle. Oscars are aggressive and should be kept in a species aquarium; they occupy all levels. Rooted plants are not recommended because they will dig them up. Water temperature should be maintained at 75°F, plus or minus 3°F. They consume all types of food, including chopped meat, live minnows, and small fish. *See Figure 20–56.*

Figure 20–56   Oscar. *Photo by Aaron Norman.*

The angelfish, *Pterophyllum scalare,* is another popular aquarium fish from South America. *See Figure 20–57.* It is a narrow, disc-shaped cichlid with large fins. The dorsal, pelvic, and anal fins are usually elongated. It grows to about 6 inches. Angels are usually silver with brown to black markings; numerous patterns are available. It is very difficult to determine the sex of angels. They are peaceful fish that do well in community aquariums. Water temperature should be maintained at 79°F, plus or minus 7°F. They consume all types of foods.

Figure 20–57   Angelfish. *Photo by Aaron Norman.*

Figure 20–58   Keyhole cichlid. *Photo by Aaron Norman.*

Other cichlids available in the aquarium trade are listed as follows:

Port acara, *Aequidens portalegrensis*

keyhole cichlid, *Cleithacara maronii (See Figure 20–58)*

blue acara, *Aequidens pulcher*

Agassiz's dwarf cichlid, *Apistogramma agassizi*

yellow dwarf cichlid, *Apistogramma reitzigi*

Jack Demsey, *Cichlasoma biocellatum*

Festivuum, *Cichlasoma festivum*

Rio Grande perch, *Cichlasoma cyanoguttatum*

firemouth cichlid, *Cichlasoma meeki*

zebra cichlid, *Cichlasoma nigrofasciatum*

pike cichlid, *Grenicichla lepidota*

orange chromide, *Etroplus maculatus*

green chromide, *Etroplus suratensis*

earth-eaters, *Geophagus jurupari*

jewel cichlid, *Hemichromis bimaculatus*

golden Julie, *Julidochromis ornatus*

Egyptian mouth-brooder, *Hemihaplochromis multicolor*

Fuelleborn's cichlid, *Labeotropheus fuelleborni*

Trewavas's cichlid, *Labeotropheus trewavasae*

Golden-eyed dwarf cichlid, *Nannacara anomala*

Kribensis, *Pelvicachromis pulcher*

Malawi blue cichlid, *Pseudotropheus zebra*

Mozambique mouth-brooder, *Sarotherdon mossambicus*

brown discus, *Symphysodon aequifasciata (See Figure 20–59)*

blue discus, *Symphysodon aequifasciata haraldi*

discus, *Symphysodon discus*

waroo, *Uaru amphiacanthoides*

Figure 20–59   Brown discus. *Photo by Aaron Norman.*

The Family Pomacentridae

The family Pomacentridae includes the damsel fish, anemone fish, and sergeant majors. These are marine fish of the coral reefs of tropical areas. They are brightly colored active fish, and most are territorial and aggressive.

Clown Fish

One group, the clown fish, is also called anemone fish because they live in and around the tentacles of the sea anemone. *See Figure 20–60.* Most fish would be fatally stung by the tentacles, but clown fish are immune and enjoy the safety of the sea anemone. Clown fish provide bits of food to the sea anemone in exchange for safety. These small fish do best if sea anemones are also kept in the aquarium; this mutually beneficial relationship is known as **symbiosis.** Common species are the tomato or fire clown fish, *Amphiprion ephippium frenatus,* and the common clown fish, *Amphiprion ocellaris.* The tomato or fire clown fish is tomato red in color with a black blotch on its side. A white vertical stripe runs down the head just behind the eye; a narrow black stripe runs along the edge of the white stripe. The common clown fish is orange with three white bands encircling the body; the white bands are edged with black. The fins are orange, edged in black. If more than one clown fish is kept in an aquarium, plenty of space should be provided, and each should have its own sea anemone.

Figure 20–60   Common clown fish. *Photo by Aaron Norman.*

The damsel fish and sergeant majors are related to the clown fish but do not live among the sea anemone; they prefer more open areas and seek protection among the coral branches. The yellow-backed damsel or black-footed sergeant major, *Abudefduf melanopus,* grows to about 2¾ inches in length. The primary body color is gray-turquoise, but the back is yellow from the upper lip through the eye to the back of the dorsal fin. The top and bottom of the caudal fin are yellow, and the pelvic and anal fins have black edges.

The Yellow-Tailed Blue Damsel

The yellow-tailed blue damsel, *Chrysiptera parasema,* is about 4 inches long. The body is blue, and the peduncle and caudal fin are yellow. *See Figure 20–61.*

Figure 20–61   Yellow-tailed blue damsel. *Photo by Aaron Norman.*

The Sergeant Major

The sergeant major, *Abudefduf saxatitis* grows to 7 inches. *See Figure 20–62.* It is silver-blue with a yellow tinge to the upper body. There are six dark, vertical bands on the side.

Figure 20–62   Sergeant major. *Photo by Aaron Norman.*

The Humbug Damsel

The humbug damsel, *Dascyllus aruanus,* is silver-white with three black bands encircling the body. It grows to 4½ inches.

The Domino Damsel

The domino damsel, *Dascyllus trimaculatus,* is all black except for a white spot on each side by the middle of the dorsal fin and one on the forehead. *See Figure 20–63.* It grows to 4¾ inches in length. The blue puller, *Chromis coeruleus,* is blue-green. It grows to 5 inches. All members of this family prefer water temperatures of 77°F, plus or minus 2°F.

Figure 20–63 Domino damsel. *Photo by Aaron Norman.*

The Family Monodactylidae

The family Monodactylidae is referred to as fingerfish. One species, the mono, *Mondactylus argenteus,* is native to the brackish coastal waters of Africa and India. This tall, disc-shaped fish grows to about 9 inches. The dorsal and anal fins are long-based, large, and appear almost identical in shape; the fish is silver. The dorsal fin is tinged with yellow, and the anal fin has a black front edge. Young can be kept in hard freshwater, but as they grow, salt should be added and the amount gradually increased. Water temperature should be maintained at 75°F, plus or minus 3°F. They feed on all types of food. *See Figure 20–64.*

Figure 20–64 The Mono. *Photo by Aaron Norman.*

The Family Toxotidae

The family Toxotidae contains one genus group and six species. They are called archerfish and are found in the brackish coastal waters of Southeast Asia and the Western Pacific. *See Figure 20–65.* Archerfish shoot drops of water at insects to knock them into the water. The head is pointed; the tongue and roof of the mouth form a tube through which water can be forced out in a straight stream. One species of archerfish, *Toxotes jaculator,* grows to about 10 inches. The fish is silver with black marks and blotches on its side; the back is greenish-brown. The dorsal fin is black with yellow markings, and the anal fin is black. The edges of the caudal fin are tinged with greenish yellow. Salt should be added to the aquarium water, and the temperature should be maintained at 80°F, plus or minus 3°F. The archerfish feeds on insects and worms.

Figure 20–65 Archerfish. *Photo by Aaron Norman.*

The Family Platacidae

The family Platacidae is referred to as bat fish and is native to the marine waters of India and the Western Pacific. *See Figure 20–66.* These large, disc-shaped fish have tall, wide-based dorsal and anal fins. They grow to 20 inches in length. The round bat fish, *Platax orbicularis,* is easy to keep in marine aquariums. It prefers tall aquariums, and because it grows quickly, a large aquarium should be made available. The round bat fish is light brown with two dark, vertical bars crossing the body. One bar runs from the top of the head down through the eye to the bottom of the lower jaw. The second bar runs from just in front of the dorsal fin down to just behind the pelvic fins. The pelvic fins are dark brown; the leading and trailing edges of the brown dorsal fin are black. The base of the anal fin is brown, and the rest is black. The base of the caudal fin is brown; the rest is colorless. Bat fish are peaceful fish but are easily frightened. Coral and rocks should be provided for them to hide among. Water temperature should be maintained at 77°F, plus or minus 2°F.

Figure 20–66 Round bat fish. *Photo by Aaron Norman.*

The Family Scatophagidae

The family Scatophagidae consists of a small number of species referred to as scats. *See Figure 20–67.* They are native to the coastal waters of the Indo-Pacific. They are narrow, almost square-shaped fish, and their heads are short. The Argus fish, *Scatophagus argus,* grows to about 12 inches. The young live in freshwater or brackish water, but the adults live in coastal sea areas. They are silver-brown with dark spots on the body. The fins are colorless on the young; on older individuals, the fins become yellow-brown with dark spots. They feed on all types of foods. Water temperature should be maintained at 76°F, plus or minus 6°F.

Figure 20–67 Argus fish. *Photo by Aaron Norman.*

The Family Chaetodontidae

The family Chaetodontidae contains the butterfly fish and angelfish that live in the marine waters of the Indo-Pacific region. They are brightly colored, attractive fish that live among the coral reefs. Most of them have small, elongated snouts that they use to pick and scavenge among the crevices of the coral for food. They are territorial, and fish of the same species will fight. Angelfish can be distinguished from butterfly fish because they have a backward-pointing spine on the gill cover. Butterfly fish and angelfish can be fed a variety of fresh meat, frozen brine shrimp, flake foods, and live foods. Water temperature

Figure 20–68 Long-nosed coral fish or forceps fish. *Photo by Aaron Norman.*

should be maintained at 77°F, plus or minus 2°F. It is impossible to describe the coloration of these fish.

Common species are listed as follows:

threadfin butterfly fish, *Chaetodon auriga*

Pakistani butterfly fish or collared coral fish, *Chaetodon collare*

copperbanded butterfly fish, beaked coral fish, or long-nosed butterfly fish, *Chelmon rostratus*

long-nosed coral fish or forceps fish, *Forcipiger longirostris See Figure 20–68)*

wimple fish, pennant coral fish, poor man's moorish idol, or featherfin bullfish, *Heniochus acumineatus*

black and gold angelfish, oriole angelfish, two-colored angelfish, or bicolor cherub, *Centropyge bicolor (See Figure 20–69)*

yellow-faced angelfish, blue-faced angelfish, *Euxiphipops xanthometapon*

rock beauty, *Holacanthus tricolor*

blue ring angelfish, *Pomacanthus annularis*

black or gray angelfish, *Pomacanthus arcuatus*

imperial or emperor angelfish, *Pomacanthus imperator*

Figure 20–69 Black and gold angelfish, Oriole angelfish, two-colored angelfish, bicolored angelfish, or bicolored cherub. *Photo by Aaron Norman.*

The Family Labridae

The family Labridae is referred to as wrasses, hogfish, or cleaner fish. *See Figure 20–70.* They are found in temperate and tropical marine areas. Wrasses are found in a wide range of colors and patterns; most species change color as they mature. They live among the coral and feed on parasites they pick off the skin of other fish. They lie on the bottom or bury themselves into the sand at night. Common species are listed as follows:

Cuban hogfish, *Bodianus pulchellus*

clown labrid, *Coris angulata*

bird wrasse, *Gomphosus coeruleus*

cleaner wrasse, *Labroides dimidiatus*

green wrasse, *Thalassoma lunare*

Figure 20–70 Cuban hogfish. *Photo by Aaron Norman.*

MAJOR FAMILIES—SUBORDER BLENNIOIDEI

The Family Acanthuridae

The family Acanthuridae is made up of several species of brightly colored fish referred to as tangs, surgeons, or unicorn fish. *See Figure 20–71.* They are oval-shaped fish with bony, scalpel-like projections on the peduncle. These projections can give the unwary prey a painful wound. Three of the more common species are the powder-blue surgeon, *Acanthurus leucosternon;* the sailfin tang, *Zebrasoma veliferum;* and the smooth-headed unicorn fish, *Naso lituratus.* All of these fish are native to the Indo-Pacific oceans. It is recommended that they receive vegetative-type foods in their diet.

Figure 20–71   Smooth-headed unicorn fish. *Photo by Aaron Norman.*

The Family Zanclidae

The family Zanclidae contains only one species—the moorish idol, *Zanclus canescens* or *Zanclus cornutus. See Figure 20–72.* The scientific name can be found both ways in different reference books. It is closely related to the tangs and surgeons. The coloration is a series of yellow, white, and black vertical bands on the body and includes the fins. The dorsal fin is elongated into a long, trailing fin. This species is not very hardy and is difficult for inexperienced marine aquarists to keep. It grows to 9 inches in length.

Figure 20–72   Moorish idol. *Photo by Aaron Norman.*

MAJOR FAMILIES—SUBORDER MASTACEMBELOIDEI

The Family Mastacembelidae

The family Mastacembelidae consists of the spiny eels. There are about 50 species native to the freshwaters from Africa to Southeast Asia. They are eel-like in appearance, with the dorsal, caudal, and anal fins merging into one continuous fin. There are several spiny rays at the front part of the dorsal fin. The upper jaw and snout are elongated and extend past the lower jaw.

The spotted spiny eel, *Mastacembelus dayi,* grows to about 14 inches. It is mainly brown. Four or five dark-brown spots with light-brown lines encircling the spots are located on the dorsal fin. The lesser spiny eel, *Macrognathus aculeatus* grows to about 10 inches. It is dark brown with numerous light-brown spots and markings. Both species should be kept in species aquariums. Spiny eels occupy the middle to the lower levels and may bury themselves into the sand or gravel on the bottom, with only the heads sticking out. Rocks or other hiding places should be provided. Water temperature should be maintained at 77°F, plus or minus 2°F. They feed primarily on worms and insects. *See Figure 20–73.*

Figure 20–73   Spotted spiny eel. *Photo by Aaron Norman.*

MAJOR FAMILIES—SUBORDER ANABANTOIDEI

The suborder Anabantoidei, order Perciformes, is a group of small- to medium-sized freshwater fish referred to as labyrinth fish. They are native to Africa, Southeast Asia, the Philippines, and Indonesia. All members of this suborder have a labyrinth organ that lies in the gill cavity and allows the fish to breathe atmospheric air. This is an advantage for the fish that live in the poorly oxygenated tropical waters. In most species, the male builds a floating nest made from saliva and air bubbles. As the eggs are released from the female, the male fertilizes them, and the eggs float up under the bubble nest. The male then guards the nest and may even guard the young fry. There are four families in this suborder: Anabantidae, Helostomatidae, Belontidae, and Osphronemidae.

The Family Anabantidae

The family Anabantidae consists of about 20 species. Included in the family is the climbing perch, *Anabas testudines*. This species has an elongated body with large scales and a large mouth. The pectoral fins are well developed, and the fish can climb out of shallow pools and crawl across the ground to another body of water. Its general color is grayish-brown, and the fins are tinged with yellow. It grows to 10 inches in length and is easy to keep. Water temperature should be maintained at 80°F, plus or minus 5°F. *See Figure 20–74.*

Figure 20–74   Climbing perch. *Photo by Aaron Norman.*

The Family Helostomatidae

The family Helostomatidae contains only one species—the kissing gourami, *Helostoma temmincki. See Figure 20–75.* This narrow, oval fish grows to 12 inches in length. The main characteristic of the species is its habit of extending its lips and "kissing." This "kissing" occurs between two males and is believed to be a contest of strength. No harm is done in this contest. They are silver with a greenish or yellowish tinge to the fins. Kissing gourami consume all types of food material but are primarily vegetarians. They do well in a community aquarium. Water temperature should be maintained at 79°F, plus or minus 4°F.

Figure 20–75   Kissing gourami. *Photo by Aaron Norman.*

The Family Belontidae

The family Belontidae consists of about 50 species, including the fighting fish, and a few gourami and paradise fish. *See Figure 20–76.* The Siamese fighting fish, *Betta splendens,* is a popular freshwater labyrinth species. It has an elongated body with long, flowing dorsal, caudal, anal, and pelvic fins. In the wild, they are reddish-brown with blue-green iridescence. There are numerous red, green, and blue dots along the sides. By selective breeding, they are

Figure 20–76   Male Siamese fighting fish. *Photo by Aaron Norman.*

now available in green, blue, red, and violet. The males are very aggressive and must never be put together. One or two females and one male can exist in a community aquarium; however, they will probably do better and display more intense coloration in a species-only aquarium. They feed on all types of food. Water temperature should be maintained at 80°F, plus or minus 3°F.

Several species of gourami are in this family, including the honey gourami, *Colisa chuna;* dwarf gourami, *Colisa lalia (See Figure 20–77);* banded gourami, *Colisa fasciata;* pearl gourami, *Trichogaster leeri;* and the three-spot gourami, *Trichogaster trichopterus. See Figure 20–78.* These gourami have pelvic fins that are elongated into long filaments. They do well in community aquariums. Water temperatures should be maintained at 77°F, plus or minus 5°F. They feed on all types of foods.

Figure 20–77   Dwarf gourami. *Photo by Aaron Norman.*

Two species of paradise fish are commonly found in the aquarium trade: the paradise fish, *Macropodus opercularis,* and the brown-spike-tailed paradise fish, *Macropodus cupanus dayi.* The coloration on the paradise fish, *M. opercularis,* varies considerably; the color is normally reddish-brown with blue-green vertical bars. Coloration on some individuals may be primarily blue with reddish-brown markings and stripes on the caudal fins. *M. opercularis* is aggressive and should be kept in a species aquarium. *M. cupanus dayi* is primarily reddish-brown, with two dark, horizontal bars on the sides. The dorsal and anal fins are reddish-brown with blue edges. The caudal tail has red elongated center rays, and the outer rays are reddish-brown with blue edges. *M. cupanus dayi* can be kept in a community aquarium but may be aggressive to smaller fish. Water temperature for *M. opercularis* should be maintained at 67°F, plus or minus 8°F; *M. cupanus dayi* should be kept at 71°F, plus or minus 7°F. Both species consume all types of foods.

Figure 20–78   Three-spot gourami. *Photo by Aaron Norman.*

The Family Osphronemidae

The family Osphronemidae consists of one species—the gourami, *Osphrone-mus goramy.* It grows to 24 inches in length and thus is only suited for the aquarist that has an aquarium larger than 1,500 gallons. It is brownish-gray with silver iridescence. It has a small head, and the lower lip projects forward in older fish. The pelvic fins are elongated and thread-like. The gourami consumes all types of food, but it is primarily a vegetarian. Lettuce and soaked oat flakes can provide the needed plant material to their diet. Water temperature should be maintained at 73°F, plus or minus 6°F. *See Figure 20–79*

Figure 20–79   Gourami. *Photo by Aaron Norman.*

CLASSIFICATION— TETRAODONTIFORMES

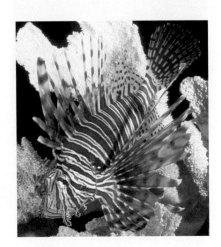

The family Balistidae, order Tetraodontiformes, is made up of a group of saltwater species referred to as triggerfish. *See Figure 20–80.* They have three dorsal spines, with the second locking the first into an upright position. They are found throughout the world. As with other species of colorful marine fish, it is almost impossible to describe all the various colorations, patterns, and shapes of these unique fish. Common species are the undulate or orange-green triggerfish, *Balistapus undulatus,* and the Picasso or white-barred triggerfish, *Rhinecanthus aculeatus.* Both species have strong jaws and sharp teeth. They can be kept in community aquariums but may show aggression to smaller fish. Water temperature should be maintained at 75°F, plus or minus 2°F. They feed on chopped meat, shellfish, and meat-based flake foods.

Figure 20–80   Picasso or white-barred triggerfish. *Photo by Aaron Norman.*

CLASSIFICATION—SCORPAENIFORMES

The family Scorpaenidae, order Scorpaeniformes, is made up of the lion fish, also called the dragon fish or turkey fish, *Pterois volitans. See Figure 20–81.* This species grows to about 14 inches in length. It is native to the Indo-Pacific areas and the Red Sea. Its primary color is reddish-brown with numerous white bands encircling the body. The fin rays are long, and the spiny ray is poisonous. It feeds on live fish or chopped meats and has a voracious appetite. It can be kept in a community aquarium with other fish that are equal in size. Water temperature should be maintained at 77°F, plus or minus 2°F.

Figure 20–81   Lion fish, dragon fish, or turkey fish. *Photo by Aaron Norman.*

CLASSIFICATION—GASTEROSTEOIDEI

The family Syngnathidae, order Gasterosteoidei, contains the seahorses. *See Figure 20–82.* The yellow or oceanic seahorse, *Hippocampus kuda,* grows to about 2 inches in length, and it is native to the Indo-Pacific marine waters. Seahorses may vary in color from black to yellow. They have a hard body covering and no caudal or anal fins; they swim in a vertical position. Seahorses have a prehensile tail that they can use to anchor themselves to plants and coral. They are fairly difficult to keep and should be kept in species aquariums. Water temperature should be maintained at 77°F. They feed on live foods.

Figure 20–82   Golden or yellow seahorse. *Photo by Aaron Norman.*

CLASSIFICATION—MORMYRIFORMES

The family Mormyridae, order Mormyriformes, contains 130 species of fish referred to as mormyrs and elephant-snout fish. The long-nosed elephant-snout fish, *Gnathonemus petersi,* is occasionally seen in the aquarium trade. *See Figure 20–83.* It is native to the dark, turbid freshwaters of Africa. It grows to 9 inches in length. Elephant-snout fish are dark brown-black, with a violet iridescence. There are two yellowish-white, irregular, vertical stripes running between the single dorsal and anal fins. The edges of the dorsal, anal, and caudal fins are edged with white. The lower lip is elongated into a tool that is used to dig for food on the bottom of streams. They also emit an electrical field around their bodies that enables them to move around in the darkness. They feed on insects, insect larvae, and worms. They will adapt to a community aquarium provided it has some dark and secluded places to hide. Water temperature should be maintained at 75°F.

Figure 20–83 Long-nosed elephant-snout fish. *Photo by Aaron*

CLASSIFICATION— OSTEOGLOSSIFORMES

The family Notopteridae, order Osteoglossiformes, has six species consisting of fish with long anal fins and reduced or no dorsal fins. *See Figure 20–84.* Several species of knife fish exist, and there are some discrepancies in their scientific name. They move about by the undulating motion of their long anal fin. Members of the family are commonly referred to as knife fish. The African knife fish, *Xenomystus nigri,* grows to about 8 inches in length. It is primarily light brown with a lighter underside. It occasionally rises to the surface to swallow atmospheric air. The African knife fish is peaceful and can be kept in a community aquarium with other peaceful fish. Darkened areas and plenty of vegetation should be provided. Water temperature should be maintained at 79°F, plus or minus 4°F.

Figure 20–84 Knife fish. *Photo by Aaron Norman.*

AQUARIUM TANKS AND EQUIPMENT

Modern aquarium tanks are available in all sizes and shapes. The glass used in tank construction can be bonded together with a sealant to form a strong, leak-proof container. These all-glass aquariums are the cheapest to purchase. Framed tanks are available for additional cost; the frames are usually for decoration and add very little to the structural strength of the aquarium. Plexiglas tanks are available on the market but must be cleaned carefully to prevent scratching the surfaces; these tanks are rather inexpensive. Acrylic aquariums are becoming very popular. These tanks are lighter, stronger, have less distortion, are clearer and insulate better than glass. If they become scratched, they can be polished to remove the scratches.

The size of tank will depend on the amount of money a person wants to invest in an aquarium and the species he or she plans to purchase. The larger the aquarium the more money it will cost for the tank and accessory equipment. Most people will want to start small and decide later whether they want to get more involved and invest more money.

To determine the number of fish that can be put in a tank, one should multiply the length by the width to determine the square inches of surface area. A rule of thumb is to have no more than 1 inch of fish for every 10 square inches of surface area in a tropical freshwater aquarium; 1 inch of fish for every 30 square inches of surface area in a cold freshwater aquarium; and 1 inch of fish for every 48 square inches in a marine aquarium. An aquarium tank 30 inches long by 12 inches wide has 360 square inches of surface area. Dividing the 360 square inches by 10 indicates that the tank will hold 36 inches of tropical freshwater fish. That could be three 12-inch fish, eighteen 2-inch fish, or twelve 3-inch fish. One should also allow for rocks, coral, or branches that are in the aquarium.

The aquarium should be situated in a draft-free area and out of direct sunlight. Aquariums placed in direct sunlight have an excessive growth of algae, and the tank may overheat during the summer and cause an oxygen deficiency. Aquariums situated in drafty areas may have fluctuations in the water temperature. The surface should be level to prevent cracking the glass. A thin piece of styrofoam can be placed under the aquarium to compensate for an uneven surface.

The water in an aquarium must be kept clean and contain adequate amounts of dissolved oxygen; aquariums use filters and air pumps to accomplish this. Air pumps introduce compressed air (oxygen) into the water and at the same time disperse carbon dioxide (aeration). As the air pump introduces air into the water, air bubbles rise to the surface. Air is usually forced through airstones (fused, porous glass) to break up the flow into minute bubbles. As the bubbles rise, they pull the water with them and cause the water in the tank to circulate. This circulation brings water from the bottom of the tank to the surface, where aeration (the exchange of oxygen and carbon dioxide) can take place. The constant circulation of the water also helps equalize the temperature throughout the aquarium. *See Figure 20–85.*

Figure 20–85 A modern glass aquarium system.

The purpose of the filter is to remove solid waste and uneaten food materials from the water. Filter systems work in one of three ways: through the mechanical removal of waste and food materials, by chemical removal of dissolved materials, and by biological filtration or conversion of harmful substances into harmless ones.

Older aquariums used internal mechanical filters driven by airstones. *See Figure 20–86.* As the air rose, this system pulled water through the media in the box. The media separated the waste and food materials out of the water, and the water was returned to the tank. Newer models still used the air-driven uplift principle, but instead use external box filters or internal sponge filters. *See Figure 20–87.*

Simple internal filters

Figure 20–86 Internal filters. An internal filter can consist of a sponge or plastic box containing wool and carbon media. Dirty water enters the hole in the sponge or holes in the plastic box, where suspended materials are trapped in the filter media. The air pulls the dirty water into the filter and forces the clean water out through the tube.

External box filters

Figure 20–87 An external box filter hangs on the back of the aquarium. The dirty water is fed into the filter by a siphon tube. Air from the pump enters the bottom of the filter. As the air rises, it pulls the dirty water up the siphon tube and through the filter media. The filtered, clean water flows back into the aquarium.

Return clean water
to tank

Dirty water
drawn into filter

Wool filter

Magnet on impeller

Magnet

Motor box

Carbon

To power source

Figure 20–88 A motorized or power filter operates the same as an external filter except that a motorized pump is used to draw the dirty water through the filter media, and the filtered, clean water is pumped back into the aquarium.

Most aquariums today use power filters with an electric motor connected to an impeller by a magnetic device through the wall of the filter box. By not housing a drive shaft through the filter box, there is less chance for leaks in the system. Power filters increase water flow through the medium and remove more waste materials. It is important that the connection of the return hose to the aquarium is secure. If the hose is loose, large amounts of water can be pumped onto the floor, and the entire tank could possibly be emptied. Undergravel filters and internal and external canister filters are now available with the use of electric power pumps. *See Figure 20–88.*

There are at least a half-dozen different types of chemical filtration units on the market today. One method of chemical filtration is the use of activated charcoal; the charcoal soaks up dissolved minerals and chemicals. Activated charcoal can be placed in box filters, or return water can flow through the charcoal. The charcoal must be replaced periodically because it reaches a point at which it can no longer absorb dissolved material.

Aeration is achieved by the use of electrically driven air pumps; pumps are either a vibrator-diaphragm type or a rotary-vane type. Pumps are available in several sizes; the size needed depends on the size of the aquarium, the number of airstone filters, and other equipment used.

Biological filters neutralize toxic substances, especially ammonia excreted from fish during respiration and produced from decaying waste and food materials. In this system, a slotted plastic plate is placed in the bottom of the aquarium. *See Figure 20–89.* A 2- or 3-inch layer of gravel is placed over the slotted plate; the gravel should be 0.125 inch in diameter. Gravel that contains large particles allows food to fall into the spaces around the gravel and decay; particles that are too small block the action of the air pump.

Aerobic bacteria in the gravel break down
ammonia wastes into less harmful nitrates

Air supplied
from an electric
pump

Water is
drawn down
through the
filter bed

Rising air bubbles cause
water to flow upward

Water flows under filter
plate toward uplift tube

Figure 20–89 An undergravel filter system. A slotted
plastic filter is placed at the bottom of the aquarium, and
a layer of gravel is placed on the filter. Air is pumped
through an airstone. As the air rises in the tube, it lifts
water upward inside the tube. As the water rises, it pulls
the dirty water through the gravel, where suspended
materials are filtered and aerobic bacteria convert the
harmful ammonia wastes into less harmful nitrites and
then nitrates.

Marine aquariums should have coral sand placed above the slotted plate. In
some aquariums, peat should be placed over the slotted plate. The slotted
plate has an uplift tube on the corner; this tube should be placed at the back
of the aquarium. When the tank is filled, an air line with an aerator attached
is placed down the uplift tube; this line is connected to an air pump. Air
is pumped down the line and through the airstone. As the air bubbles rise
in the uplift tube, water is drawn through the gravel, up through the uplift
tubes, and back into the aquarium at the surface. The gravel layer acts as a
filter for suspended particles in the water. After several hours of aeration,
colonies of aerobic bacteria (bacteria that use oxygen) begin to grow in the
gravel; this process is called nitrification. *Nitrosomonas* bacteria convert
ammonia into nitrites, and then *Nitrobacter* bacteria convert the nitrites
into nitrates. Nitrates may inhibit the growth of fish, but they are not nearly
as toxic as ammonia. Nitrites and nitrates may be utilized by plants in the
aquarium.

Several factors contribute to an increase in ammonia and nitrites:

- Increase in waste material and uneaten food on the bottom of the
aquarium
- Dirty filters
- Failure to change water at regular intervals (One-fourth to one-fifth
of the water should be changed every 3 to 4 weeks.)
- Overfeeding that leads to an accumulation of uneaten food
- Overpopulation of fish

Nitrite levels can be measured with test kits. It is very important to remove nitrites from tapwater used in marine aquariums; many marine species are sensitive to nitrites.

In order for the aerobic bacteria to flourish, the correct pH and oxygen levels must be maintained; pH is a measure of acidity or alkalinity. The nitrification process is most efficient at a pH of 9. The pH scale reads from 0, which is very acidic, to 14, which is very alkaline; 7.0 is neutral. Most freshwater tropical fish prefer slightly acid conditions. The pH of an aquarium is easily measured with a pH test kit.

A protein skimmer can be added to marine systems to eliminate dissolved proteins and other waste products that pollute and discolor the water; they operate on the airlift water principle. As the air bubbles reach the surface of the airlift tube, the foaming residue that contains the proteins and other wastes is collected in a chamber; these residues settle into a yellow liquid that can be poured out.

Tropical aquariums must have water heaters to heat the water and thermostats to maintain water temperature. Heaters are of two types. One type is a heater that is placed in the aquarium and controlled by a separate thermostat; the electrical connection between the heater and the thermostat is on the outside of the aquarium. The second type is a condensed heater and thermostat; the heater and the thermostat are connected together inside a glass tube. This tube is placed into the aquarium, and the temperature is controlled by an adjustment knob at the top of the tube. Heaters are available in several standard sizes. Approximately 10 watts of power for each 1½ gallons of water is required. A 30-gallon aquarium would need a heater of approximately 250 watts. Large aquariums may have better distribution of heat if two heaters are used. It is important to use a heater that is of a recommended size. If a large heater is used and the thermostat unit fails while the heater is on, the water could become overheated. If a small heater is used and the room temperature should drop, the heater might not produce enough heat to maintain the water temperature.

Another internal type of heater uses a heating cable buried under the gravel. The advantage is that fish do not come in contact with the heater and cannot be burned by or dislodge it.

Several types of thermometers are available for monitoring water temperature. Liquid crystal thermometers in the form of an adhesive strip stuck to the outside of the aquarium are very popular. Mechanical dial thermometers can also be used and attached to the outside of the aquarium. Conventional float-type thermometers that float or are attached to the inside of the aquarium with a suction cup are also available.

Water hardness is another factor that needs to be considered. Hardness is caused by dissolved magnesium and calcium salts. Hardness is of two types: temporary and permanent. Temporary hardness can be removed by boiling the water. Permanent hardness needs to be removed by chemical means or distillation. The hardness can also be diluted by removing some of the hard water and adding soft water.

Hardness can be measured in several different ways. A German scale measures as degrees of hardness, °DH; and the Clark or English scale measures as degrees of hardness, °H. Hardness can also be measured as parts per million (ppm) of either $CaCo_3$ or CaO; and as grains per gallon (gpg). Conversion factors are as follows:

1 °DH on the German scale is 17.9 ppm of $CaCO_3$

1 °H on the English scale is 14.3 ppm of CaO

gpg $CaCo_3$ = ppm $CaCo_3$ × 0.058 ppm

Test kits are available to measure hardness as parts per million of calcium carbonate (ppm of $CaCo_3$). To avoid confusion with the various scales, we use the ppm scale: 17.9 to 89.5 ppm of $CaCo_3$ is considered very soft, 89.5 to 179 soft, 179 to 358 medium-hard, 358 to 537 hard, and above 537 ppm is considered very hard water.

Chlorine in tapwater may be a concern to aquarists; however, it is easily removed by aerating the water for 12 to 24 hours. Commercial dechlorinators or activated charcoal filters are also effective.

The water in marine aquariums needs to be measured for the amount of salt (density); this is done with a hydrometer. This instrument floats higher or lower in the water depending on the density of the water. The density or specific gravity of the water can be read on a scale located on the glass tube; the recommended specific gravity is 1.020 to 1.026. Water should be tested at the recommended operating temperature. Marine mixes can be added to freshwater to provide a marine water environment.

Lighting in an aquarium is necessary for visual display and for plant growth. Plants need light for photosynthesis, which in turn uses carbon dioxide and releases oxygen. The amount of light depends on the kind and number of plants, the height of the aquarium, the type of water, and the thickness of the glass cover. Lighting can be provided with tungsten filament bulbs or fluorescent bulbs; fluorescent bulbs are available in different colors and can add to the attractiveness of an aquarium. Fluorescent bulbs operate at cooler temperatures and do not increase the water temperature. A standard level of fluorescent lighting is approximately 0.10 to 0.14 watts per square inch of surface area. Lighting should be provided for 10 to 15 hours per day.

Decorative items may be added to an aquarium to make it more attractive; these also provide shelter and hiding places for aquarium inhabitants. Decorations added should make the aquarium look as natural as the original environment of the fish. In freshwater aquariums, rocks and wood pieces are often added; in marine aquariums, coral, shells, and rocks may be used. Typically wood is not used because there are too many problems and it is not good for the environment. Wood pieces should be boiled several times to kill bacteria and resins that may pollute the aquarium. Many manufactured look-alike pieces are now available for aquarium use as well. Petrified woods, bamboo canes, and pieces of slate can also be used.

Plants can be added for attractiveness and to provide a means of using carbon dioxide while releasing oxygen. Some common floating plants for tropical aquariums include water lettuce, *Pistia stratiotes,* and duckweed, *Lemna minora.* Rooted plants include Japanese dwarf rush, *Acorus gramineus;* Madagascar lace plant, *Aponogeton madagascariensis;* baby's tears, *Bacopa monnieria;* fanwort, *Cabomba aquatica;* hornwart, *Ceratophyllum demersum;* water trumpet, *Cryptocoryne blassii;* Amazon sword plant, *Echinodorus grandiflorus;* water weed, *Egeria densa;* hairgrass, *Eleocharis acicularis;* water thyme or Canadian pondweed, *Elodea canadensis;* water star, *Hygrophila polysperma;* water wisteria, *Hygrophila difformis;* Ambulia, *Limnophilia aquatica;* Ludwigia, *Ludwigia repens;* Java fern, *Micosorium pteropus;* water milfoil, *Myriophyllum aquaticum;* giant Indian water star, *Nomaphila stricta;* European curly pondweed, *Potamogeton crispus;* arrowhead, *Sagittaria graminea;* giant eel grass, *Vallisneria gigantea;* and Java moss, *Vesicularia dubyana.*

Japanese dwarf rush, baby's tears, hornwort, water weed, hairgrass, water thyme, duckweed, Ludwigia, water milfoil, European curly pondweed, arrowhead, giant eel grass, and Java moss can be used in cold-water aquariums in addition to willow moss, *Fontinalis antipyretica.* These plants should be planted to a depth where the crown of the plant is level to or just above the gravel. Plants may be fertilized with liquid or tablet plant foods specially formulated for aquarium use. Plastic plants are also available and add the same attractiveness but do not filtrate the aquarium.

FEEDING

Fish consume all types of foods and should be given a variety so that they do not become bored. Fish should be fed two or three times a day and should receive only what they can consume in 3 to 5 minutes.

Flake foods are ideal for small fish up to 4 or 5 inches long, if appropriate. These foods are produced from the meat of fish, fish eggs, wheat, and vegetables. They usually contain additional vitamins and minerals. These food materials are also processed into pellet form for larger fish; floating fish sticks are good for large top-feeding fish, and small pellets that sink slowly are best for middle-feeders, and sinking tablets are appropriate for bottom-feeders.

Several live foods are also available; among them are small crustaceans called water fleas, *Daphnia.* Brine shrimp, *Artemia,* are small shrimp that live in salt lakes or brackish waters. These shrimp should be rinsed off and may be fed to both small and large fish. River shrimp and bloodworms are available for still larger fish. Earthworms, flies, maggots, wood lice, caterpillars, crickets, and grasshoppers are other live foods that can be used. One problem with feeding live foods is the parasites and diseases that they may carry can be harmful to fish.

Freeze-dried and frozen foods are basically the same as live foods except they are in a safe form. Some freeze-dried or frozen foods available are *Mysis* shrimp, Pacific shrimp, tubiflex worms, krill, and plankton.

Carnivorous fish can be fed minced or chopped meat; common foods include beef heart, liver, raw fish meat, and shellfish meat. Pieces of turkey and chicken can also be used. Meat pieces can also be thinly sliced and frozen; these can then be broken and fed. Some of the larger carnivorous fish may need to be fed other fish; small goldfish may be used for this purpose.

Many fish also require vegetative material in their diets. Chopped or shredded lettuce, chopped spinach leaves, canned peas, wheat germ, and oat flakes can be used. Any vegetables not consumed within 8 hours should be removed.

Marine species of invertebrates are of four types:

1. Those that feed on plankton that is filtered from the water, including stone and horny coral, tubeworms, bivalves, some species of sea cucumbers, and crustaceans. Commercially prepared plankton foods and frozen foods are available for the plankton feeders.

2. Those that feed on plant material, including sea urchins, mollusks, and sea slugs. A diet of lettuce and spinach will hopefully prevent them from feeding on aquarium plants.

3. Those that are carnivorous, including crabs, starfish, sea anemones, shrimp, and lobsters. They can be fed small pieces of crab, fish, shrimp, and flake foods. Sea anemones should be fed only when their tentacles are out (in bloom). Drop the food onto their tentacles.

4. Those that are scavengers, including sea cucumbers. They feed on debris and uneaten foods on the bottom of the aquarium.

DISEASES AND AILMENTS

Aquarists need to be able to recognize symptoms of diseases in their fish as soon as they begin to appear. Early recognition and treatment increase the chances for recovery. Some signs that may indicate a disease problem include the following:

1. Unusual behavior, such as unusual swimming movements
2. A fish leaning to the side, floating to the surface, or sinking to the bottom
3. Fins not extended and laying flat against the body
4. Caudal fins closed or rolled
5. A fish shying away from the shoal
6. A fish slower in its movements than others
7. Respiration faster and deeper than normal
8. Fish up on the surface gasping for air
9. Fish that are rubbing against decorations on the bottom
10. Fish that are not eating
11. Fish that are thin with sunken sides
12. Abnormally large or swollen belly
13. Abnormal color
14. Frayed fins
15. Malformation of the back and spinal column
16. Cloudiness of the eyes
17. Scales sticking out away from the body instead of lying flat

Parasites

White Spot

White spot, or "Ich," is caused by the parasitic organisms *Ichthyophthirius multifiliis*. This is one of the most common diseases of aquarium fish; the organism is often present in fish but usually does not cause any problems. When the fish is stressed, the parasite multiplies rapidly, causing numerous white spots on the body and fins; the parasite lives under the epidermis. When the parasite matures, it breaks through the wall of the epidermis and swims about. The mature parasite eventually settles on a stone or plant and forms a cyst (resting stage). During this resting stage, the cell divides into 1,000 to 2,000 ciliospores (reproductive spores). These break from the cyst and swim freely in the water. When they make contact with a fish, they burrow into the epidermis and mature in about 4 days. The parasites living under the skin and the ciliospores in the cysts cannot be destroyed by chemicals; however, the free-swimming adult parasites and ciliospores can be destroyed with chemicals added to the water. Chemicals are readily available at most pet stores with aquarium products. The organism *Cryptocaryon irritans* causes the same type of problem in marine fish.

Slime Disease

Slime disease is caused by parasites that attack the skin of fish. The fish's body produces large amounts of mucus or slime in reaction to the irritation caused by the parasites. The heavy mucus produced dulls the colors of the fish, weakens the fish, and may cause death if the parasites attack the gills. Parasites of the genus *Chilodonella* may cause slime disease in freshwater; *Cyclochaeta (Trichodina)* and *Costia (Ichthybodo)* in all types of water; and *Brooklynella* in marine water. Chemical water treatments are available for the treatment of slime disease; salt baths may also be effective. Infected fish should be removed from the aquarium and put in a quarantine tank for treatment.

Hole-in-the-Head

Hole-in-the-head disease is caused by the parasitic organism *Hexamita*. These organisms live under the skin and in the muscle tissues of the host fish. As the organisms multiply, the tissues break down, the skin breaks open, and the decayed tissue breaks away, leaving "holes" in the body. These large open sores or ulcers are most common around the head, along the lateral line, and at the base of the dorsal fin. The disease may be linked to vitamin deficiencies, and supplements may aid in healing. Metronidazole or dimetridazole added to the water is the recommended treatment. One should consult with a veterinarian.

Velvet Disease

The organism *Oodinium* causes a condition called velvet disease in freshwater fish and coral disease in marine fish. *Oodinium* is a member of a group of parasites known as dinoflagellates, which are related to algae and contain chlorophyll. The organisms penetrate the skin cells and feed on them. Both diseases appear as a velvet-like gold coating of dust on the sides of the fish.

The gills are usually affected, and the fish will have rapid movement of the gills. Copper-based treatments can be added to the water. Some marine invertebrates cannot tolerate copper-based treatments, however, and the affected fish should be removed and treated in a hospital aquarium.

White Fungus Growth

A fungus organism called *Saprolegnia* causes white fungus growths on the fins, mouth, eyes, and gills. These fungus growths sometimes appear as secondary infections in fish already infected with spot or bacterial diseases. Fish should be removed to a hospital aquarium where the water has been treated. Treated fish should be kept in a hospital aquarium until the fungus disappears. Some commercial treatments are available at pet stores with fish supplies.

Flukes

Two species of flukes that can cause problems are *Dactylogyrus* and *Gyrodactylus*. Resulting damage from both can be fatal depending on the degree of infestation. Flukes are tiny worm-like parasites that attach themselves to the gills and body of fish. They have numerous hooks on the rear part of their body and a sucking mouth. A single fluke has both male and female reproductive organs (hermaphroditism). *Dactylogyrus* permanently attach themselves to the gills. Fish affected with *Dactylogyrus* will have rapidly moving gills, swim at the surface, and pant heavily while trying to obtain more oxygen. Their gills will be covered with slime, and parts of the gills may be eaten away. Fish may also scrape their bodies against objects, trying to relieve the irritation. *Gyrodactylus* primarily attach themselves to the body of the fish. The infected fish will also scrape against objects; their color will fade; the fish will produce more slime in response to the irritation; and the fins may also become ragged and eaten away. Commercial treatments are available at pet stores with fish supplies. Infected fish should be removed and treated in a hospital aquarium quarantine. The aquarium that the fish were removed from should be thoroughly cleaned before returning fish to it.

Anchor Worms

Anchor worms are members of the class Crustacea, phylum Arthropoda. Two species, *Lernaea cyprinacea* and *Lernaea carassii,* are found regularly on goldfish; other species of tropical fish are rarely affected. Only the females of *Lernaea* species are parasitic. They have a chitinous adhesive organ around their mouth and will burrow under the scales and attach themselves to the muscle tissue. Two whitish-colored egg sacs can be seen trailing from the parasite. Commercial water treatments are available. The aquarist may need to remove the fish from the aquarium, remove the parasite with tweezers, and treat the wound.

Fish Lice

Argulus foliaceus is a species of fish lice that fish can be infected with from feeding on infected plankton. These flat, oval-shaped crustaceans attach themselves to the body of the fish with sucking mouth parts (proboscis); they pierce the skin with a spine and discharge a poison. This poison can be fatal to young fish. Fish should be removed, and the lice should be removed with tweezers. The fish should then be placed in a hospital tank to which a commercially available treatment has been added. The original tank should be thoroughly cleaned before returning fish to it.

Bacterial Disease

Finrot

Finrot is a condition especially prevalent in brightly colored fish or those with long, trailing fins like the black mollies. The edges of the fins start to lose their color, the tissue between the rays begins to break down, and the fins become ragged. The fins become shorter as the condition persists. The disease is believed to be caused by bacteria, along with a contributing factor such as a vitamin-deficient diet, poor water conditions, or fin nipping by other fish. Commercially prepared treatments are available; treatment of marine species may be more difficult.

Mouth Fungus

Mouth fungus is a misleading name because the disease is actually caused by bacteria, *Chondrococcus columnaris.* Symptoms appear as white, tufty material around the mouth and white patches on the skin. Severely affected fish may be unable to close their mouths and will have respiratory trouble. Commercially prepared treatments are available. If the condition persists, the infected fish may need to be removed, placed in a hospital aquarium, and treated. One may want to consult a veterinarian.

Neon Disease

Neon disease is a disease that primarily affects the neon tetra but may also affect other species of tropical fish. The disease is caused by the parasite *Plistophora hyphessobryconis.* The organism lives in the tissue of the fish's body, where it multiplies and produces spores. These spores are released when the fish dies or when parts of the fish slough off. The spores later hatch, and the newly hatched organisms can infect other fish. The first symptoms of the disease are gray to white discolorations on the fish's body. These areas usually enlarge and bulge outward. After a period of time, the fish dies; no treatment or cure is available.

Tuberculosis

Tuberculosis of fish is caused by the bacterium *Mycobacterium;* cold-water species are usually the most frequently affected. Once the bacteria have invaded the fish, they move to various tissues and organs; the symptoms can be varied. The fish usually become slow, move off to an area of the aquarium by themselves, and die. Usually, there are discolored areas or bulges. The back may be disfigured because of deformities to the spinal cord. All weak or diseased fish should be removed from the aquarium and destroyed; other fish should be moved to a hospital aquarium. The aquarium, gravel, decorations, and equipment should be thoroughly cleaned and disinfected before fish are returned to the aquarium. The causal bacteria in tuberculosis of fish is different then the causal bacterium of tuberculosis in humans. The risk for obtaining tuberculosis from fish or fish aquariums is low; however, one should be careful when working with diseased fish and equipment if he or she has cuts or abrasions.

Pseudomonas and Aeromonas

Bacterial infections from *Pseudomonas* and *Aeromonas* species sometimes occur in cold-water fish. These bacteria infect only fish that have already been weakened by other disease conditions or old age. Infected fish have skin lesions or ulcers, and their bellies may be swollen as a result of an accumulation of fluid. This swelling causes the scales to stand out from the body; swelling behind the eyes may also cause the eyes to protrude.

Environmental Control

Fish living in an aquarium that has a lack of oxygen and an excess supply of carbon dioxide may show signs of disease. If single fish are seen swimming near the surface, this is normal; if almost all of the fish are swimming at the surface, it may be a sign of insufficient oxygen. One-fourth to half of the water in the aquarium should be replaced immediately, and the cause, the lack of oxygen, should be corrected.

Aquariums with an incorrect water pH can cause respiratory distress and excess mucus production in fish. This excess mucus clouds the color of the fish. Partial replacement of the water and the addition of chemicals can correct the pH; any changes should be gradual.

After a period of time, the chemical filtration abilities of an aquarium are reduced and a buildup of ammonia occurs. The first sign of this buildup is a cloudiness or murky appearance of the water; there is also a buildup of anerobic bacteria. Anerobic bacteria produce foul-smelling poisonous gases that are harmful to fish. A characteristic sign of gravel filtration system pollution is a black color to the gravel, which is caused by the buildup of iron sulphite produced by the interaction of hydrogen sulphite and iron compounds in the gravel. The hydrogen sulphite is produced by the anaerobic bacteria; at this point, the aquarium should be cleaned. Fish living in a polluted aquarium show signs of respiratory distress and eventually die.

REPRODUCTION

Fish reproduce by means of fertilization of the female's eggs by the sperm (milt) from the male. This may take place outside the female's body as the eggs are being expelled, or as the milt is being deposited in the **vent** (the urogenital opening) of the female and fertilization takes place internally. When fertilization takes place internally, the eggs are retained in the body of the female, and she gives birth to live, free-swimming young. The young may be nourished within her body (viviparous), or the eggs may hatch within her body (ovoviviparous), with the young then expelled.

Egg-Layers

Most aquarium fish are egg-laying species. The eggs are expelled from the female and then fertilized by the male during a reproduction ritual called **spawning.** Egg-laying fish fall into one of five groups.

Egg-Scatterers

Egg-scatterers are fish that lay their eggs in a haphazard manner on the floor of an aquarium. Some species lay adhesive eggs that stick to the gravel on the aquarium floor or on decorations and plant materials. Other species lay non-adhesive eggs. There is no parental care of the eggs or young; once the eggs are laid, they are forgotten. In many cases, steps must be taken to prevent the parents from eating the eggs and the newly hatched fry. To reduce the number of eggs eaten by the parents, several steps can be taken:

1. Cut down on the amount of light; this encourages spawning and increases the number of eggs produced.

2. Have aquarium plants so dense that adult fish will have difficulty swimming among the plants to find eggs and newly hatched fry.

3. Place one or two layers of glass marbles on the floor of the aquarium; the eggs will settle among them, and the adults will not be able to get to the eggs.

4. Place spawning grass, Spanish moss, or manmade grass on the aquarium floor; this material also can be placed over layers of marbles.

5. Drape a nylon curtain in the tank; the adults can swim above the curtain, and as the eggs are laid, they will fall through the curtain to the floor of the aquarium.

6. Use a large spawning tank that allows the eggs to go undiscovered and the newly-hatched fry to hide or evade the adults.

Examples of egg-scatterers are the barbs, danios, some species of rasbora, some species of characins, and goldfish.

Egg-Buriers

Egg buriers are fish that lay their eggs in the mud of rivers and ponds; these are the killifish and other annual fish that live in areas where the rivers and ponds dry up. The adults lay their eggs in the mud and die when the rivers and ponds dry up. The eggs survive in the mud, and when the next rains come, the young hatch. In aquariums, annual fish lay their eggs in the material on the floor of the aquarium. Two or three inches of peat is the ideal material. Because the water does not dry up, they live for more than 1 year. These fish may also lay eggs in dense plant growth or spawning grass. The eggs can be removed with tweezers and stored in moist peat. One should take the peat containing the eggs, squeeze out the water, place the peat in plastic bags, and seal them. After 2 or 3 months, the peat containing the eggs should be placed in water so that the eggs will hatch.

Egg-Depositors

Egg-depositors have complex spawning routines and are excellent parents; these fish usually select their own mate. They clean off a nesting site where the female deposits her eggs and the male fertilizes them. The parents usually take turns guarding the nest; they constantly fan the nest and eggs with their fins to keep the eggs clean from dirt, silt, and dust that may settle on them. When the young fry hatch, the parents keep watch for several days to protect them from predators. Most of the cichlids fit into this group, as well as the corydoras, catfish, sunfish, and some rasboras.

Mouth-Brooders

Mouth-brooders are fish that carry their eggs in their mouth until they hatch, after which the young may continue to be carried in the parent's mouth until they are ready to fend for themselves. In some species, the young, after leaving the parent's mouth, may return if threatened. After the eggs deposited by the female are fertilized by the male, the female, and in some species the male, goes around and picks up the eggs by mouth. During this incubation period of 2 to 3 weeks, the parent does not eat. Some species of cichlids and some species of *Betta* are mouth-brooders.

Nest-Builders

The nest-builders construct a nest in which the eggs are deposited; it is usually constructed by the male. This nest may be a bubble nest on the surface made from saliva-blown bubbles or may be prepared from materials found on the floor of the aquarium. After the eggs are fertilized, the female should be removed because the male will become aggressive toward her; the male then guards the nest. Gouramis, some sunfish, some cichlids, a few catfish, and *Betta splendens* are nest-builders.

Live-Bearers

In live-bearing species, the male's anal fin is modified so that he can deposit milt into the vent of the female. The modification is called a gonopodium and is a characteristic that helps identify the male of the species. Live-bearers may give birth to 20 or as many as 200 young fry. Females store the milt in their oviducts for several months; thereby several broods can be produced from one mating. In viviparous species, the young are nourished from the female's bloodstream. In ovoviviparous species, the young are nourished by the yolk sac. Guppies, platies, swordtails, mollies, and half beaks are common live-bearing species.

In preparing to breed fish, several conditions should be observed. The temperature of the water should be increased 2° or 3°F; this stimulates the fish into breeding. Filtration and healthy water conditions are vital to successful breeding and reproduction of fish; however, a system that provides too much uplift or suction may draw the young fry into the gravel on the bottom, and if the aeration is too strong, the young can be injured. An air-operated sponge filter and a gentle flow of air from an airstone provide adequate conditions.

Young fry of live-bearers can begin taking food immediately; fry of egg-laying fish survive the first few days from the nutrients in the egg sac. Commercially prepared foods are available in liquid and powdered form for both live-bearing and egg-laying fry. Young fry should also receive brine shrimp. These are available from aquarium supply stores and also can be hatched from eggs. After a few days, the young can be fed dried foods that have been mashed into small pieces.

SUMMARY

Evidence of the early development of fish is found in fossils dating back to the Ordovician period 425 to 500 million years ago. These early fish were called Ostracoderms. From the Ostracoderms evolved two groups of fish: the Placoderms and Acanthodians. From Placoderms evolved our present-day sharks, rays, skates, and other fish with cartilaginous skeletons. From the Acanthodians evolved the Osteichthyes, represented today by fish with bony skeletons.

Fish are cold-blooded vertebrates that breathe with gills and move with the aid of fins. They are the most numerous vertebrates, with more than 30,000 species; these are divided into three main classes and 34 orders. Most fish are covered with scales; these are thin, bony plates that overlap each other and provide protection. Fins are movable structures that aid the fish in swimming and maintaining balance. Gills are organs through which fish obtain oxygen from the water.

Fish have a long, folded heart that consists of only two chambers: the atrium and the ventricle. Most species have a lateral line that is composed of a series of pressure-sensitive cells, called neuromasts, which enable the fish to react to changes in water pressure, temperature, currents, and sounds.

The teeth and digestive systems of fish vary depending on the type of food consumed. Some fish are predators and consume other fish or animals; others feed primarily on plant material. The eyes of fish are similar to those of other vertebrates but have spherical lenses that focus by moving within the eyeball, not by changing the curvature of the lens. Fish do not have eyelids; the eye is kept moist by the flow of water.

The senses of smell and taste in fish are highly developed; they not only have taste buds in their mouth and on their lips, but also on their body and fins. Some fish also have taste buds on barbels around the mouth.

All fish have inner ears. Species that have swim bladders have a more acute sense of hearing because the bladder acts as a resonator and amplifies the sound.

Modern aquarium tanks are available in many sizes and shapes. The size of the aquarium depends primarily on the species kept and the amount of money the aquarist is able to spend. The size of the aquarium one buys will influence the amount of money that will be needed for supplies and accessories.

The water in an aquarium must be kept clean and contain adequate amounts of dissolved oxygen; filters and air pumps are used to accomplish this. Filter systems work in one of three ways: through the mechanical removal of waste and food materials, by chemical removal of dissolved materials, and by the biological filtration or convection of harmful substances into harmless ones.

Fish consume all types of foods and should be given a variety so that they do not become bored. Fish should be fed two or three times per day and should receive what they can consume in 3 to 5 minutes. Common foods for aquarium fish are commercially prepared flake foods and pellets; live foods such as earthworms, flies, maggots, wood lice, caterpillars, small crustaceans, and brine shrimp; and freeze-dried and frozen foods. Carnivorous fish may be fed minced or chopped beef hearts, liver, raw fish, and shellfish. Many species of fish also eat vegetable and plant material.

Aquarists need to be able to recognize symptoms of diseases in their fish as soon as they appear. Some common diseases of fish are white spot, or "ich"; slime disease; hole-in-the-head; and velvet disease of freshwater fish; and coral disease of marine fish. Parasitic fungus, flukes, worms, and lice may also cause ailments in fish; bacterial infections are also common. Maintaining water cleanliness and quality is important in controlling many of these diseases and ailments.

Fish reproduce by means of fertilization of the female's eggs by the milt from the male. This fertilization can take place after the eggs are expelled from the female's body, or the milt can be deposited in the vent of the female and fertilization takes place internally. When fertilization takes place internally, the eggs are retained in the body of the female, and she gives birth to live, free-swimming young. Egg-laying fish fall into one of five groups: egg-scatterers, egg-buriers, egg-depositors, mouth-brooders, and nest-builders.

In preparing to breed fish, several conditions should be observed; water cleanliness, quality, and temperature are extremely important in successful breeding.

DISCUSSION QUESTIONS

1. Describe the characteristics of fish.

2. What are the three main classes of fish, and what are their important characteristics?

3. Describe the general characteristics of the 12 orders that are commonly found in the aquarium trade.

4. What are the three types of filter systems, and how does each one work?

5. Describe the nitrogen cycle that takes place in an aquarium.

6. What is a power filter, and what is the advantage of using one?

7. What are some common foods that fish consume?

8. What are the five types of egg-laying fish, and what are their habits?

9. What are signs that might indicate a fish has a disease or other ailment?

10. What are common diseases caused by parasites, and what are the symptoms and treatments for each?

11. What are two species of flukes that attack aquarium fish, and what are the symptoms and treatments for each?

12. What causes finrot, and how can it be controlled?

SUGGESTED ACTIVITIES

1. Visit the local pet shop and make a list of the different species of fish available. Using the local library, compile as much information as you can about the different species.

2. Set up an aquarium and put some common species of aquarium fish in it. Observe and record their habits.

3. Set up an aquarium with live-bearers and try raising some young fish. Observe and record their daily habits.

4. Species of fish such as guppies can be used to conduct color and genetic studies. Raise young guppies, and select and mate them for differences in color and fins. Are some characteristics dominant?

5. Obtain test kits and sample the aquarium water for dissolved oxygen, ammonia, nitrites, pH, hardness, and carbon dioxide. Keep a daily or weekly record. What changes occur and why?

6. Determine the species, number, and size that can be raised in an aquarium 30 inches long, 19 inches high, and 12 inches wide. What is the ideal temperature for the species you selected?

ADDITIONAL RESOURCES

1. Fish Link Central
   http://www.fishlinkcentral.com

2. Tropical Fish
   http://www.tropicalfishcentre.co.uk

3. Discus Breeder' Web site
   http://www.aquaworldnet.com/dbws

4. Fins: Fish Information Service
   http://fins.actwin.com

5. Pet Education.com
   http://www.peteducation.com/default.cfm

6. Dr. Jungle's Animal-World
   http://animal-world.com/index.html

7. Badman's Tropical Fish: Starting a Brackish Aquarium
   http://badmantropicalfish.com/brackish/brackish.html

CHAPTER **21**

Hedgehogs

OBJECTIVES

After reading this chapter, one should be able to:

 recognize and describe the common types of hedgehogs.

 list common management practices for maintaining the health of hedgehogs.

 list common diseases and ailments of hedgehogs.

TERMS TO KNOW

anointing

induced ovulator

keratin

parturition

polyestrous

quick

CLASSIFICATION

Kingdom — Animalia
Phylum — Chordata
Class — Mammalia
Order — Insectivora
Family — Erinaceidae
Subfamily — Erinaceinae
Genus — 4 genus classes
Species — 14 species

HISTORY

For people living in Britain, Europe, North Africa, and Asia, the hedgehog is a very common animal. The hedgehog is not an indigenous species to the United States. The first hedgehog brought to the United States was a female obtained by the New York Zoological Society in 1900. The National Zoological Park in Washington and the New York Zoo also brought hedgehogs into the United States. However, the number brought into the United States was few, and the animals were little known to the general public. Several individuals obtained animals from zoos and other sources. By 1990 several small herds were established. A group of interested breeders and hobbyists met in 1993. This meeting in 1993 was the beginning of the North American Hedgehog Association (NAHA). The NAHA is no longer in existence. The International Hedgehog Association, which was founded in 1999, appears to be the major organization formed to educate the public in the care, knowledge, and promotion of hedgehogs.

The hedgehog probably received its name from its habit of slowly waddling along hedge rows in search of food and because it often gives off low grunts as it searches for food. The grunts are similar to the grunts of pigs or hogs.

MAJOR GENUS

There are four genus groups and 14 species of hedgehogs. The genus *Atelerix* consists of four species. Of all the 14 species of hedgehogs, *A. albiventris*, commonly referred to as the white-bellied hedgehog, is the most common of all the species of hedgehogs found as pets in the United States. They are also referred to as dwarf hedgehogs, pygmy hedgehogs, and African, or four-toed hedgehogs. *See Figure 21–1.* This species is smaller than the other species. Its size ranges from 7 to 10 inches, and it weighs 9 to 25 ounces. A unique characteristic of the four-toed hedgehog is the absence of a large toe on the back foot. If the large toe is present, it is usually smaller than what would be normal. This species is native to Central Africa.

GENERAL CHARACTERISTICS

The white-bellied hedgehogs have long snouts, four digits on each foot, and 36 teeth. They are primarily nocturnal, beginning to move about at twilight, and may remain active till dawn. *See Figure 21–2.*

The most notable characteristic of hedgehogs are the spines that the animal has developed as a means of defense. The spines cover the back, sides, and up over the crown of the head. The spines are modified hairs made of **keratin.** Keratin is the protein that makes up hair, nails, and horns. A hedgehog has from 5,000 to 7,000 spines. The outside of the spine is solid and the inside is filled with soft air pockets. The face, forehead, chest, and underside are covered with hair. *See Figure 21–3.* Although the hedgehog has spines, hedgehogs are not related to other animals that have spines. Porcupines, for example, are rodents and members of the order Rodentia. Hedgehogs are in the order Insectivora. There are 390 species of Insectivora, and 14 of those speciesare hedgehogs. Ancestors of today's insectivores lived around 100 million years ago during the age of the dinosaurs. Hedgehogs have large eyes and

Figure 21–1   A typical pygmy hedgehog. *Photo by Isabelle Francais.*

Figure 21–2   Hedgehogs are very active and curious. This hedgehog has been exploring a cloth napkin. *Photo by Isabelle Francais.*

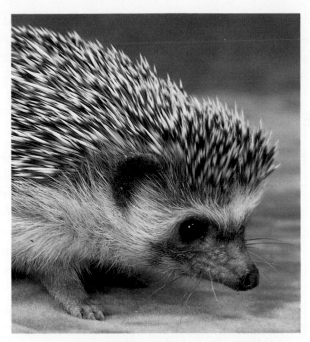

Figure 21–3 The spines on the hedgehog cover the back, sides, and up over the crown of the head. The face, forehead, chest, and undersides are covered with hair. *Photo by Isabelle Francais.*

ears. Their tails are very short and most species have short legs. The color of the hedgehog is determined by skin color and not spine color. The primary color is a mixture of brown, black, and white. *See Figure 21–4.* Some are very light while others may be very dark. There are also albino and cream-colored individuals. *See Figure 21–5.* Some may be light colored but have a dark mask across the face. The International Hedgehog Association now recognizes 92 colors in the white-bellied and Algerian hedgehogs.

Figure 21–4 This is a common colored hedgehog, a mixture of brown, black, and white. A hedgehog being held in the hand will often lick the hand of the one holding it, as this hedgehog appears to be doing. *Photo by Isabelle Francais.*

Figure 21–5 There are also albino-colored hedgehogs. This photo shows an albino female with two young albino-colored hoglets. *Photo by Isabelle Francais.*

The hedgehog is not a very long-lived animal; the typical life span is about 5 years. Like many animals, their life span in captivity is usually longer than in the wild. Hedgehogs have very little odor. They make very little noise outside of some low grunts and clicks.

HEDGEHOGS AS PETS

Hedgehogs have received attention only since 1993, when they became a "fad" pet. With proper care and management they can become somewhat tame. *See Figure 21–6.* As with any other pet, one must consider the amount of time available and one's ability to care for a hedgehog. It is probably best to start with a single animal before considering having several as pets. A hedgehog not familiar with you or your scent will roll up into a ball when you get near it. This is the animal's natural reaction to danger. The spines on the hedgehog are very sharp. *See Figure 21–7.* When one attempts to pick up an animal that has rolled up into a ball, it is possible to get some painful pricks from the spines. Gloves would be recommended for handling hedgehogs that are rolled up in a ball. *See Figures 21–8* and *21–9.* As an animal gets familiar with its handler and his or her scent, it will not sense any danger and will not roll up into a ball. Hedgehogs can then be picked up with one hand on each side of the body and under the belly. A hedgehogs that is familiar with its handler will many times walk up into his or her cupped hands.

When considering purchasing a hedgehog for a pet, look for many of the same signs that you would look for in selecting any other animal. Check to see that the animal's cage is clean. Make sure the animal is clean and check for mites. The eyes should be clear with no liquid running from them. The nose

Figure 21–6 Hedgehogs can become used to being handled; however, as with any pet, small children should be carefully supervised when handling hedgehogs. *Photo by Isabelle Francais.*

Figure 21–7 When scared or frightened, a hedgehog will roll up into a ball to protect itself. *Photo by Isabelle Francais.*

Figure 21–8 Hedgehogs that are not used to being handled may need to be handled with gloves to prevent painful pricks from the spines. *Photo by Isabelle Francais.*

Figure 21–9 The spines on a hedgehog will usually lay flat against the body. The animal an be more easily be handled when the spines are flat. *Photo by Isabelle Francais.*

should be moist but not runny. Check the anal region for signs of diarrhea. If there is any doubt about the health of an animal, it would be best not to select it, and instead look for one you are sure is healthy. Before trying to handle a new pet hedgehog, let it get used to its surroundings for a day or two. Once it is accustomed to its surroundings, you can start to try to handle the animal. Again, it might be best to use a glove unless you are sure the animal is familiar with you and can be picked up. *See Figure 21–10.* Hedgehogs have teeth and will bite if unaccustomed to you or if they are frightened. *See Figure 21–11.*

Figure 21–10 These hedgehogs are not used to being handled; they need to be handled with gloves. *Photo by Isabelle Francais.*

Figure 21–11 Hedgehogs do have teeth. This animal is used to being handled, or the handler would not want to attempt to check its teeth without using gloves. *Photo by Isabelle Francais.*

The animal may also make a clicking sound when rolled up in a ball. This is a warning. The animal may also hiss and attempt to bite when one tries to pick it up. It will raise its spines in an attempt to protect itself. When touched, the animal may jump upward in a further effort to protect itself. Hedgehogs are solitary animals. Males and females will not tolerate each other except for mating.

HOUSING AND EQUIPMENT

Hedgehogs in the wild are very active animals and will cover considerable distances in search of food. Pet hedgehogs also need room to move about. Aquariums, large plastic tubs, or plastic cat or dog carriers are appropriate for hedgehogs. Minimum floor space should be 2 square feet; however, a larger cage that would allow room for a litter pan, nest box, and exercise wheel would be more suitable. *See Figure 21–12.*

Housing should be of sufficient size to allow for the animal to exercise and have room for a nest box. If the housing is large enough, an exercise wheel can be put in to allow for the animal to get exercise. *See Figure 21–13.* Small housing areas for hedgehogs that don't allow for exercise may lead to stress and illness. Plastic, fiberglass, and glass houses have the advantage over wood of being easily cleaned. Mites are less likely to be able to hide and breed on these surfaces than on the wood surface.

A nest box should be provided for the hedgehog. It should be a little longer, wider, and taller than the animal. The animal must be able to easily enter, turn around, and exit the box. An entrance hole to the nest box should be made that will allow room for the hedgehog to easily enter and exit. The nest box should be attached to the cage or be heavy enough so the animal can't push it around the cage. Aspen shavings are recommended as a floor covering. Sawdust is not recommended because it may cause respiratory problems and also may cling to the anal and genital areas of the animal. The cage should have a solid floor. Hedgehogs can get their legs and feet caught in wire floors and cause injury to themselves.

Figure 21–12 Aquariums like the ones used for other small animals can be used for hedgehogs. Also, note the gravity-type water bottle in the aquarium. *Photo by Isabelle Francais.*

Figure 21–13 If a hedgehog is confined to a small cage, an exercise wheel similar to the one in the photo should be provided for the animal to exercise. *Photo by Isabelle Francais.*

Figure 21–14 A hedgehog should be provided with a heavy and sturdy food bowl that it can't upset or push around the cage. *Photo by Isabelle Francais.*

Hedgehogs are very clean and prefer to urinate and defecate in the same area. A litter tray should be provided. The tray should be approximately 12 inches square. Plastic is the preferred material because it is easily cleaned. Litter trays are available in pet stores. Cat litter can be used in the litter tray, but should be the type that isn't dusty or clumping. Natural plant litter is probably the best because it is biodegradable. The litter tray should be placed as far as possible from the animal's food container. The litter tray should be cleaned daily of soiled litter and all the litter should be changed at least once a week.

The hedgehog will also need a food bowl. It should be heavy and sturdy so the animal can't push it around the cage or upset the bowl. *See Figure 21–14.* Water is provided by means of the typical vacuum-type water bottles or ceramic bowl.

FEEDING

Hedgehogs are insectivores and are nocturnal in the wild. They will come out at night to feed not only on insects, but worms, snails, bird eggs, and small vertebrates such as frogs and snakes. Feeding hedgehogs in captivity is usually very simple. One needs to make sure the animal gets a diet that includes proteins, fats, minerals, vitamins, and fiber. It is important to find out what the animal has been eating. Try to maintain the original diet and gradually make changes if necessary. A rapid change in diet may cause diarrhea or put the animal off its feed and lead to stress and illness.

Foods formulated for hedgehogs are now available in some localities. These would be the preferred diet. Other foods that one can try feeding are dry kitten formulas and boiled or scrambled eggs. The commercial foods can be supplemented with some small pieces of fruits and vegetables. Hedgehogs love earthworms, wax worms, and crickets. Feed the worms and crickets primarily as treats, two or three at a time, every 2 or 3 days. Experiment with different foods to see which foods the hedgehog likes and dislikes.

Hedgehogs are nocturnal and will feed at night. To determine how much to feed, one needs to monitor carefully how much the pet eats and the pet's weight and adjust the amount accordingly. One should feed an amount that will be cleaned up. Large amounts of food should not be left in the bowl because the pet may overeat and become obese.

DISEASES AND AILMENTS

A change in eating habits such as reduced appetite or drinking excessive amounts of water or a lack of thirst may be the first signs of an illness. Dark colored, torpedo-shaped fecal material is the sign of a healthy animal. A soft, green-colored or liquid fecal material may indicate a problem. Spending time with one's pet is important in being able to recognize changes or signs that indicate illness.

Hedgehogs are hardy animals; however, like most animals, they are subject to diseases and ailments—many of which are caused by poor hygienic conditions. It is important to keep the housing area clean. Fresh bedding should be provided at least once per week. Food bowls should be cleaned daily. Never leave spoiled food in the bowls.

Mites may occasionally infest hedgehogs. Mites crawl on the host and cause irritation by biting the skin and sucking blood. Severe infestations in hedgehogs can cause loss of the spines. One should consult a veterinarian on proper treatment and control of the mites. The bedding must be changed and the housing unit and nest box treated as needed.

Intestinal threadworms (*Capillaria* species), intestinal worms (Brachylaemus species), and coccidia (*Isospora* species) can be present in the intestines of hedgehogs. Diarrhea, weight loss, and decreased activity level are signs of infestation. Fecal samples examined by a veterinarian can confirm infestation and the veterinarian can prescribe proper treatment, as well as steps for prevention.

Eye infections are possible when sand, dirt, or other foreign material gets in the eyes. If the animal is holding an eye closed or is seen pawing at an eye, it may indicate a problem. A veterinarian should be consulted to determine the cause and proper treatment.

Diarrhea may be the result of a change in the diet of an animal. A veterinarian needs to be consulted to diagnose the problem and recommend treatment.

Hedgehogs will develop overgrown toenails. Overgrown nails will hamper the animal's ability to move about. The nails can be clipped using small pet clippers. A second person may be needed to hold the animal. A trip to the veterinarian's office may be needed. The animal may need general anesthesia to get it to relax and keep it from curling into a ball. As in the case of other animals, one must be careful not to clip into the **quick,** because this will cause bleeding.

REPRODUCTION

Female hedgehogs (sows) may become sexually mature at 8 weeks of age; however, it is not recommended that they be bred until they are more physically mature, at about 6 months of age. Males (boars) may also be sexually mature at 8 weeks of age, but may not be able to successfully impregnate a female until they are more mature.

Female hedgehogs are **polyestrous,** meaning they are capable of having several litters during a year. In the wild, the female's estrous cycle is controlled by the temperature, daylight hours, and food supply. In captivity, where the conditions are more constant, the female can be bred throughout the year.

When preparing to have a sow bred, she should be taken to the boar's cage and put in with him. Careful observation of the pair is important to determine if the pair is compatible and so that no injury occurs to either animal. The female hedgehog is an **induced ovulator,** which means the act of mating will induce the female's eggs to be released into the oviduct for fertilization. If the female is not ready to breed, she will reject the male in a forceful manner, and she will make puffing, hissing, or snorting sounds. She should be removed from the cage and the process tried again the next night. Once the female is ready to mate, she will lie flat on her stomach and lower the spines on her back so that they lie flat. The male will then mount her. He may grab hold of some of the spines on the back of the female's neck. The actual mating period is short but may be repeated several times. After the mating is completed, the female should be removed from the boar's cage and returned to her cage.

The gestation period is usually 35 days. Litter size will range from one to seven hoglets, with three being more typical. During the gestation period, the sow's food intake will gradually increase. The pregnant female will become larger as the fetuses develop. A nest box should be provided in her cage. Make sure the entrance is large enough so she can easily enter and exit the box. About 1 week before the expected birth, the cage should be thoroughly cleaned and fresh bedding placed in the cage. Do not clean the cage or disturb the female any more than necessary for the next 3 weeks. She will need to be fed and supplied with water. Any build-up of fecal material can be removed. The sow should not be disturbed during the birth or for several days after, even though one may want to check the litter. Squeaks from the litter will alert you to the birth of the litter. Disturbing the female at this time may cause her to kill the litter.

The young are born deaf, blind, and helpless at birth. They are completely dependent on their mother. Spines are present but they are covered by a thin membrane that protects the mother from injury during **parturition.** This membrane dries and shrinks shortly after birth and allows the spines to protrude. The eyes and ears open at about 10 to 12 days. The weaning process should begin at about 3 weeks. The length of time it takes to wean the litter depends on the mother and the size of the litter. One can begin handling the young at this time, so they become used to being handled. *See Figure 21–15.*

Sexing and separating the young should begin shortly after they are weaned. The young become sexually mature at an early age, and to avoid inbreeding and unwanted pregnancies, they need to be separated. *See Figure 21–16.* The young hoglets are fairly easy to separate by sex. The females have five pair of mammary glands and the vulva or urogenital area

Figure 21–15 Young hedgehogs can be handled beginning at about 3 weeks of age. *Photo by Isabelle Francais.*

Figure 21–16 Young hedgehogs will need to be separated shortly after they are weaned to avoid inbreeding and unwanted pregnancies. *Photo by Isabelle Francais.*

Figure 21–17 Young hoglets can easily be separated by sex. In the male, on the left above, the penis is located near the mid-belly region. The vulva and urogenital area of the female, pictured on the right, is close to the anus.

is close to the anus. The male's penis is located more near the mid-belly region. *See Figure 21–17.*

ANOINTING

One of the most peculiar and least understood behaviors of the hedgehog is the process of **anointing.** When the animal comes across a new smell or object, it licks at the smell or object, producing a flow of foamy saliva. The animal will then stiffen its front legs and swing its head from side to side. Using its tongue it spits and smears its spines with the saliva. The reason for this behavior is not known. It may be a form of protection from predators, or a method of attracting a mate. *See Figure 21–18.*

Figure 21–18 The hedgehog will spit saliva on its spines and then use its tongue to smear the saliva around on the spines. This process is called anointing.

HIBERNATION AND ESTIVATION

Hedgehogs in the wild that are found in cold climates survive the winter's cold temperatures by going into a state of hibernation. During hibernation the animal's heart beat drops by 90 percent, its body temperature declines, and respiration is reduced. This is a tremendous savings in energy requirements. The animal is sustained during hibernation by fat reserves that it built up in its body during warmer weather. Hibernation may last from a few weeks to 6 months depending on the severity of the winter. As the weather warms, the animal comes out of hibernation.

Estivation is similar to hibernation but not as intense. Hedgehogs that are found in hot, dry climates with reduced food supplies will go into a state of estivation. The animals seek out a cool hiding place. An animal in a state of estivation will have a reduced metabolic rate but not to the low levels that occur with animals in hibernation. They may wake and exit their hiding places on a regular basis. Animals that have been kept in captivity where conditions are more constant will not need to hibernate or go into estivation.

SUMMARY

Hedgehogs belong to the order of insect-eating mammals, or Insectivora. Although they are in the order Insectivora, hedgehogs are omnivorous, eating both plant and animal material. To people living in Britain, Europe, North Africa, and Asia, the hedgehog is a very common animal. The hedgehog is not indigenous to the United Stares but was brought into the United States in 1900.

The most notable characteristic of the hedgehog is the spines that the animal has developed as a means of self-defense. They have long snouts, four digits on each foot, and 36 teeth. In the wild they are nocturnal, beginning to move about at twilight, and may remain active till dawn.

Hedgehogs will roll up into a ball when they are confronted by a predator or when they fear they are in danger. When the hedgehog is rolled into a ball, the spines are extended and protect the animal.

Housing should have a solid floor and be of sufficient size to allow for the animal to exercise and have room for a nest box.

Formulated foods are available for hedgehogs, and this should be the basic diet for the hedgehog. Dry kitten food can be used to supplement the diet, along with boiled or scrambled eggs. Pieces of fruit and vegetables can also be fed. For treats, earthworms, wax worms, and crickets may be fed.

Hedgehogs become sexually mature at about 8 weeks of age. Young hoglets should be separated shortly after weaning to prevent inbreeding and unwanted pregnancies. The gestation period for hedgehogs is about 35 days. Litter size will range from one to seven hoglets, with three being typical.

One of the most peculiar and least understood behaviors of the hedgehog is the process of anointing. The reason for this process of smearing saliva over their spines is not known.

Hedgehogs in the wild that are found in cold climates survive the winter's cold temperature by going into a state of hibernation. Hedgehogs that are found in hot, dry climates with reduced food supplies will go into a state of estivation.

DISCUSSION QUESTIONS

1. What is the genus name of the common hedgehog?

2. What things should be considered when buying a hedgehog for a pet?

3. What is the major identifying characteristic of the hedgehog? What are some other characteristics of the hedgehog?

4. What is meant by the term *anointing*? What are the possible reasons for this behavior?

5. What are the common foods for hedgehog?

6. How should a hedgehog be handled?

7. What are some common diseases and ailments of hedgehogs?

8. Define the terms *hibernation* and *estivation*. How are they similar and how do they differ?

SUGGESTED ACTIVITIES

1. Hedgehogs belong to the family Erinaceidae, subfamily Erinaceinae. What is the other subfamily of the Erinaceidae family? Use the Internet and research to try to find the other subfamily and its common members.

2. Using the Internet, research the many different color possibilities for hedgehogs.

3. With the approval of the instructor, invite a local veterinarian into the classroom to discuss the diseases and ailments of hedgehogs and other small animals.

ADDITIONAL RESOURCES

1. Dr. Jungle's Animal World
   http://animal-world.com/index.html

2. Hill Top Animal Hospital
   http://www.hilltopanimalhospital.com/pocketpets.htm

3. Pet Education.com
   http://www.peteducation.com/index.cfm?cls=18

4. Exotic Nutrition Pet Company
   http://www.exoticnutrition.com/index.html

5. Animal Hospitals—USA
   http://animalhospitals-usa.com/home.html

CHAPTER 22

Sugar Gliders

CLASSIFICATION

Kingdom — Animalia
Phylum — Chordata
Class — Mammalia
Subclass — Marsupialia
Order — Diprotodontia
Suborder — Phalangeriforms
Super-
family — Petauroidea
Family — Petauridae
Subfamily — Petaurinae
Genus — *Petaurus*
Species — *breviceps*

OBJECTIVES

After reading this chapter, one should be able to:

- recognize and describe the sugar glider.
- compare and contrast marsupial mammals and placental mammals.
- list common management practices for maintaining the health of sugar gliders.
- list common diseases and ailments of sugar gliders.

TERMS TO KNOW

bulbous	marsupium	syndactyl
joeys	patagium	
marsupial	priapism	

HISTORY

The sugar glider, *Petaurus breviceps,* is one of six species of small arboreal **marsupials** native to northeastern regions of Australia, Tasmania, Indonesia, and Papua New Guinea. In the wild they live in groups of one or two males, several females, and out-of-the-pouch youngsters called **joeys.** These groups consist of from 12 to 40 depending on the number of hollows in trees needed for nesting and the amount of food available in their territory. Dominant males mark their territory and members of their group with scent gland secretions. Young adults leave the group as they mature and seek out other young adults for mating and starting their own groups.

Sugar gliders are nocturnal. They are omnivores, coming out at night to hunt for insects, small spiders, worms, eggs of nesting birds, and to feed on the sap and gum from eucalyptus and acacia trees, nectar, and pollen.

Sugar gliders are marsupials. The females have a pouch called a **marsupium** in which the young are raised during early infancy. The young are born at 15 to 17 days into the gestation period. They then crawl up their mother's belly and into the pouch. Once inside the pouch the young will attach them-selves to a nipple, where they are nourished. The young remain in the pouch for 70 to 74 days. They will then leave the pouch but return for warmth and nourishment.

CHARACTERISTICS

Sugar gliders have a body length of 5 to 6 inches with a tail of near equal length, giving the animal a total length of approximately 10 to 12 inches. They weigh 3 to 5 ounces, with the males being slightly larger than the females. The tail is used for balance, stability, and as a rudder as the animal glides among the tree tops. *See Figure 22–1.*

The head shape of the sugar glider is somewhat triangular. The muzzle is short and the nose is rounded. The sugar glider is primarily nocturnal. Their eyes are large and round, which is common for animals that move about at night. Sugar gliders have 40 to 46 teeth, including the two incisors that are large and protrude forward. These two large incisors are used for chewing into the bark of trees. *See Figure 22–2.*

Figure 22–2 This is a close-up showing the head of a sugar glider. One can see the short, rounded muzzle and the large, round eyes. Also, note the rather large, furless ears that move independently of each other. *Photo by Isabelle Francais.*

Figure 22–1 This picture shows a typical sugar glider. *Photo by Isabelle Francais.*

Sugar gliders normally have soft, gray-colored fur with cream-colored undersides. Some may be more brown in color. They have a black stripe that runs from between the eyes, over the head, down the center of the back, to the base of the tail. Adult males have a scent gland that appears as a bald spot on top of the head, and this bald spot separates the black stripe as it crosses over the head. Sugar gliders have black rings around the eyes and black stripes on the front of each leg. Sugar gliders have a bushy tail, which is also gray in color except for the last couple of inches, which is black. Their ears are furless and fairly large. The ears move independently of each other and are constantly in motion, trying to pick up sounds.

Sugar gliders have five toes on each forefoot, which end with short-curved claws. The inner opposable toe on the hindfoot is clawless but is enlarged or **bulbous,** which aids in grasping and holding on to branches. The second and third toes or digits on the rear feet are **syndactyl** or fused together and have two claws. This acts like a comb and aids in grooming by helping to remove dirt and other debris clinging to the fur.

One of the most distinguishing characteristics of the sugar glider is the **patagium,** or thin, furred membrane that stretches from the wrists to the ankles. *See Figure 22–3.* When not gliding, the membrane appears as excess skin on the side of the animal. When gliding, the sugar glider extends all four legs, which stretches the membrane out, and this allows the animal to glide. Sugar gliders are capable of gliding distances of up to 150 feet. Their sharp claws aid in grasping on to branches when they land.

Figure 22–3   In this picture, with the animal's legs extended, one can see the patagium, the thin, furred membrane that stretches from the wrists to the ankles. This extended membrane enables the sugar glider to glide among the tree tops. *Photo by Isabelle Francais.*

The female sugar glider has a pouch. This pouch can be seen as a small slit at the lower abdominal area. Male sugar gliders have scent glands on the top of their head, on their chest, in the anal region, on the surfaces of their feet, at the corners of the mouth, and on the inside surfaces of their ears. Dominant males use the scent glands on the top of the head and on the chest to mark members of their group. Gliders live in family units and are territorial. *See Figure 22–4.* The adult males in the group also mark the limits of their territory. Members of outside groups wandering into the territory of another group will be subject to attack. The female sugar glider has scent glands in her pouch, in the anal region, on the surfaces of her feet, at the corners of her mouth, and on the inside surfaces of the ears.

Sugar gliders will normally live 5 to 8 years.

Figure 22–4   Sugar gliders are very sociable animals. This picture shows a family unit that includes a female, male, and young joey. *Photo by Isabelle Francais.*

SUGAR GLIDERS AS PETS

Before purchasing sugar gliders, check with state and local laws to make sure they can be kept legally. Also check with landlords to make sure they have no objections to sugar gliders as pets and that those pets can be kept in their buildings.

Sugar gliders are very sociable animals. In the wild they live in groups, and if one is planning to obtain a sugar glider for a pet, then one should plan on getting more than a single glider. *See Figures 22–5 and 22–6.*

Figure 22–5   Being very sociable animals, sugar gliders will need human companions willing to provide hours of companionship. This is especially true if one only has a single glider. *Photo by Isabelle Francais.*

Figure 22–6   If one doesn't have time to spend providing companionship to the pet, a second glider should be purchased. *Photo by Isabelle Francais.*

They do produce a skunk-type scent, but it is not as strong and doesn't last as long as the scent of a skunk. Remember, the male will mark his territory, so the glider's cage and housing area will need to be cleaned so the scent doesn't build up.

If possible, obtain sugar gliders from local sources, because shipping and handling can be stressful to the animal. Shipping can also be costly. Buying locally also gives one a chance to see where the animals were raised, how and what they were fed, and how they were handled and cared for. One may want to ask for and check references from other buyers.

Again, it is best to buy more than one sugar glider. Two males or two females will get along as well as a male and female. If one isn't interested in breeding sugar gliders, two animals of the same sex may be the best purchase.

When one is selecting individual animals, choose recently weaned gliders. These will become used to their human companion more easily. Select sugar gliders that have their eyes open and are eating on their own. A young glider that has been out of its mother's pouch for less than 3 weeks will have short, smooth hair on its tail and not a fluffy tail like an adult. A young glider out of its mother's pouch less than 3 weeks is too young and isn't ready to be weaned.

Other things to look for when selecting individual gliders are:

1. The eyes should be bright, clean, and clear.

2. The animal should show signs that it has been eating well.

3. There should be no running or discharges from the nose.

4. The fur should be soft, thick, and clean.

5. There should be no signs of diarrhea.

6. The animal should be active and alert.

7. The animal should be friendly and curious.

HOUSING AND EQUIPMENT

Sugar gliders, being arboreal in the wild, are very active and very curious. They like to climb and explore. When one is considering the cage for these animals, it needs to be large enough to allow them to explore, climb, exercise, and glide. The larger the cage, the better suited it is for gliders. The minimum recommended size is 20 × 20 × 30 inches tall. *See Figure 22–7.* This will accommodate one or two gliders; however, a larger size would probably be more suitable. Other factors one may need to consider in selection of a cage for pet sugar gliders include the amount of time one has to spend, the time needed to clean a cage, the appearance of the cage in the home, and the cost. Large bird cages or aviaries may also be considered for housing sugar gliders. The openings in the wire mesh should be no more than 1 inch square for adults. If baby gliders are present, than smaller-size openings such as ½ by 1 inch would be preferred. Some commercial cages have a plastic pull-out drawer that can aid in cleaning. One may want to purchase a plastic storage box or a cat litter box to place under the cage so that fecal material, food, and debris don't get kicked out of the cage onto the floor or table top. The door to the cage should be large enough to allow for easy cleaning and for putting food bowls, nest boxes, and other furnishings into and taking them out of the cage.

Figure 22–7 Cages for sugar gliders should be at least 20 × 20 × 30 inches tall. The larger the cage the better, so the gliders can climb, explore, and exercise. The cage pictured here probably is too small for sugar gliders, especially if more than one glider will be housed in it. *Photo by Isabelle Francais.*

Tree limbs or branches should be added to the cage to give the cage a more natural environment. The gliders will climb around the branches, foraging for food and providing exercise. Branches from non-toxic woods such as apple, pear, plum, or willow are recommended. Do not use branches from conifer trees such as cedar, pine, or fir, because these have harmful sticky saps and resins. Gliders will often busy themselves by stripping the bark from branches. They may even take the bark back to their nest box to use as bedding. If the bark from the branches is stripped off or the branches become soiled with urine and feces, then they should be replaced. *See Figure 22–8.*

Figure 22–8 Sugar gliders like to climb and explore. Branches like the one pictured here should be put in their cage to provide exercise. *Photo by Isabelle Francais.*

Food can be placed in the cage in heavy bowls that the gliders can't tip over or in food dishes used by birds that can be fastened to the side of the cage. *See Figure 22–9.* Water can be supplied in heavy bowls or in bird dishes fastened to the side of the cage. Bowls used for food and water should be cleaned daily to

Figure 22–9 Food should be provided in heavy bowls that the glider can't tip over. Food can also be provided in bird feeders that fasten to the side of the cage. *Photo by Isabelle Francais.*

prevent bacterial growth. Food and water containers should be placed high up in the cage where the sugar gliders will feel more safe and secure, and this will prevent feces and urine from falling into them and contaminating the contents.

Sugar gliders should be supplied with toys to keep them occupied. *See Figure 22–10.* Toys used for birds such as swings, perches, ladders, and bells that are well made will be enjoyed by gliders. Gliders can also be supplied with pieces of PVC pipe or cardboard paper towel rolls. The gliders will explore these or use them to curl up in for a nest box. Nest boxes made of wood or plastic can also be used. These should be placed high in the cage where the gliders will feel more safe and secure. Cloth pouches attached to the top of the cage can be used by gliders as a place to curl up and rest.

Figure 22–10 Hamster wheels can provide enjoyment and exercise for gliders; however, the one pictured here is not recommended. The gliders can get their feet or tail caught in this wire-type wheel, causing injuries. The plastic wheel that is enclosed at the back side is recommended. *Photo by Isabelle Francais.*

Shredded paper, sheets of non-printed paper, and hardwood shavings can be put in the cage to help soak up urine, feces, and spilled food and water. Do not use pine or cedar for the nest box or for bedding. Cedar and pine contain oils that are toxic to small animals. Do not use sawdust, because it can lead to respiratory problems and can get in the eyes of the gliders and cause eye irritation.

Cleaning the cage is very important. Fecal material should be removed daily, and material used on the floor of the cage should be removed and replaced at least once a week. The cage wiring, perches, and other cage furnishings should also be cleaned weekly.

FEEDING

Sugar gliders are omnivorous, feeding on both plant and animal material. The diet of pet gliders will consist of fresh fruits and vegetables, grains, meats, eggs, and insects. Commercial diets for sugar gliders are available and should be included as part of their diet. The diet should consist of approximately 75 percent fruits and vegetables and 25 percent protein such as meats and insects.

A list of suggested fresh fruits includes apples, apricots, bananas, melons, cherries, dates, figs, grapes, oranges, peaches, pears, pineapple, plums, and strawberries. Some of the larger fruits can be cut into quarters or into small pieces for the gliders. Feed three or four different types of fruit in the evening. One may need to experiment with different fruits to see what the sugar gliders prefer. Remove any uneaten fruit in the morning and clean the food bowls. A list of vegetables that one may feed includes asparagus, beans, beets, broccoli, carrots, cauliflower, peas, sweet potato, and sweet corn. Never feed lettuce, celery, and other leafy vegetables, because these are high in water content and don't have as high a nutritional value as other vegetables.

Lean, cooked beef and poultry are good sources of protein. Don't feed chicken skin, because it is high in fat, and don't feed chicken bones, because these may splinter and cause injury to one's pet. Hard boiled eggs, chopped or mashed, and scrambled eggs are good protein sources and may be fed occasionally.

The sugar glider will enjoy crickets, grasshoppers, meal worms, and wax worms. These can be fed as treats. Feed live food sparingly because some insects are high in fat and low in calcium.

As one can see, wide variety of foods can be fed to sugar gliders. The key is to provide a balanced diet that provides all the nutrients the glider needs. In feeding sugar gliders, it is also very important that they have a plentiful supply of clean, fresh water. One may also want to consult with a veterinarian to make sure the glider is being fed a balanced diet.

Sugar gliders are nocturnal animals and will start to become active at dark. Foods should be available for them at this time. Start with ¼ cup per animal. See how much they eat in an hour. If they are consuming all of the food in the first hour, one may want to provide more. One should also check the next morning to see if any food is left. Adjust the amount fed based on the observations.

REPRODUCTION

Adult male sugar gliders are easily recognized by the bald scent spot on the top of their head. Females have a pouch in the area where the navel would be on placental animals. *See Figure 22–11.*

Figure 22–11 This drawing shows the underside of a female sugar glider. The number 1 indicates the location of the marsupium, or pouch, where the young will be raised during early infancy. The number 2 indicates the cloacal opening, or common opening for the reproductive system, urinary ducts, and alimentary tract.

Figure 22–12 This is a drawing of the underside of a male sugar glider. The number 1 indicates the location of the scent gland on mature gliders. The number 2 indicates the location of the scrotum. The scrotum on a young male appears as a wart-like swelling on the lower abdomen. On the adult male the scrotum is covered with fur. The number 3 indicates the cloacal opening.

Sugar gliders have a common opening that serves for the reproductive system, urinary ducts, and the alimentary tract. This opening is called the cloaca. The male's penis can sometimes be seen protruding from the cloacal opening. This is especially true when he is cleaning and grooming himself. *See Figure 22–12.*

Sugar gliders are usually bred in pairs, with a male and female being housed together, or in a colony, where one male is housed with two or more females. Colony mating involves larger cages and requires more experience on the part of the pet breeder.

During mating, the male will grasp the fur on the female's neck to hold her. Sugar gliders usually mate at night and will mate several times. The estrous cycle in female sugar gliders is 29 days.

The gestation period is about 16 days. The young gliders, called joeys, are only about two-tenths of an inch long. After being born, the young gliders will crawl their way into the mother's pouch, where they take hold of one of four nipples. The nipple swells in the young glider's mouth so that it is securely attached. It will be about 2 weeks before the young are large enough to be detected in the pouch. There are usually one or two in the litter. The young gliders will remain attached to the nipples for approximately 30 to 40 days and remain in the pouch for another 30 days. At about 60 to 70 days (86 days after conception), the young gliders will leave the pouch, returning occasionally for warmth, security, and to nurse. At this point, they are fully furred and their eyes are open. Weaning takes place at about 105 to 125 days (121 days after conception). When the young gliders leave the pouch, one can start to handle them on a regular basis so they can get used to being around their human companions and make the young gliders more desirable pets. *See Figure 22–13.* After the young gliders have emerged from the pouch, they ride around clinging to the back or stomach of one of their parents. *See Figure 22–14.* Young gliders will become sexually mature at about 8 months of age. To avoid unwanted pregnancies, the young should be separated prior to 8 months of age.

Figure 22–13 Tame gliders that have been handled regularly make desirable pets. A sugar glider might like to climb up its human companion and ride around on his shoulders, as pictured in this photo. *Photo by Isabelle Francais.*

Figure 22–14 Young joeys that have emerged from their mother's pouch will cling to the back or stomach of one of their parents. *Photo by Isabelle Francais.*

DISEASES AND AILMENTS

Sugar gliders have very few health problems. This is especially true if the following guidelines are followed:

1. Feed a balanced nutritious diet.

2. Keep the animals clean.

3. Keep their cage, toys, feed bowls, nest boxes, and other furnishings clean.

4. Provide them with adequate space for exercise.

5. Keep them in a group so they will be happier and healthier.

"Because sugar gliders are nocturnal, they should be kept in a fairly quiet area and allowed to sleep during the day. They can be easily stressed if awakened and taken out of their cages in daytime hours. This can increase the risk of illness. Sugar gliders are most active and playful in the evenings and at night. This is also when they are most vocal."[1]

"The sharp claws of sugar gliders sometimes get caught in the fabric of clothing or other objects. Care must be taken when freeing them from the cloth or object; their toes, wrists, or ankles could easily be broken."[2]

"Behavior disorders can occur in sugar gliders housed alone, with incompatible mates, or in inappropriate cages. It is very important to provide sugar gliders with a secure nest box or pouch. Anxiety may lead to over-grooming and fur loss, particularly at the base of the tail. Deliberately causing injury to themselves, over- or under eating, abnormally excessive thirst, eating their own droppings, cannibalism, and pacing are also associated with stress. **Priapism** (persistent erection of the penis) has also been reported in adult male sugar gliders. This may result in trauma to the penis requiring surgical removal."[3]

"The sugar glider's overall appearance and behavior should be watched for signs of illness. Generally, sugar gliders should have bright eyes, a moist nose, pink nose and gums, the ability to grip with all four feet, a smooth hair coat, and good elasticity of their gliding membranes.

"Wounds and signs of illness are similar to those seen in other animals, including depression, inactivity, and loss of appetite or weight. Other signs that your sugar glider is not well may include watery eyes, lack of energy, red and scaly skin, sores, abnormal droppings, excessive shedding or bald patches, labored breathing, and dragging the hind legs. If you notice any of these signs, you should bring your pet to a veterinarian immediately. Sugar gliders can very quickly pass the point of recovery if they do not receive prompt medical attention.

"Sugar gliders can easily become dehydrated either from a lack of drinking water or because of a medical condition such as vomiting or diarrhea. This can be deadly if not addressed promptly. Signs of dehydration include sunken eyes, loose skin (the skin on the back will stay up after it is pinched), dry mouth and nose, lack of energy, abnormal breathing, and seizures."[4]

"Sugar gliders can be infected by several common bacteria, including *Pasteurella*, staphylococci, streptococci, mycobacteria, and clostridia. Several microorganisms that can be transmitted to and cause illness in humans also affect sugar gliders, including *Salmonella, Giardia, Crytosporidium, Leptospira,* and *Toxoplasma* species. It is, therefore, very important to wash your hands thoroughly after cleaning your pet's enclosure and items within its cage, or after handling the sugar glider itself." [5]

Aflatoxicosis

Aflatoxicosis is a liver disease caused by toxins produced by a certain fungi in or on foods and feeds. Corn, peanuts, and cottonseed are the most likely to be contaminated with aflatoxins. Sugar gliders can get aflatoxicosis by eating contaminated peanuts or by eating crickets that have been fed contaminated corn. Therefore, it is important not to feed your sugar gliders peanuts and to know what kind of feed your insect supplier feeds its insects. Signs of aflatoxicosis are loss of appetite, jaundice, lack of energy, and diarrhea. Sudden death may occur. If diagnosed in time, aflatoxicosis is reversible. Take your sugar glider to a veterinarian immediately, and change its feed.

Constipation

"Small, hard, dry animal droppings, or none at all, may be a sign of constipation. Possible causes of constipation are not enough liquids or fiber in the diet, poor overall diet, stress, lack of exercise, or digestive system problems. Some medications can also cause this problem. A medical examination by a veterinarian to determine the cause should be performed as soon as constipation is noticed.

Diarrhea and Vomiting

"Several microscopic parasites, including *Crypto-sporidium, Giardia,* and *Trichomonas* can cause diarrhea and occasional vomiting in sugar gliders, along with abdominal cramps, weight loss, and dehydration. Your veterinarian can perform tests to determine the cause of these signs and prescribe the appropriate medication. Because several of these parasites can also infect humans, it is important to use care when handling sick gliders and to wash your hands thoroughly after cleaning the cage and before touching any food or any item that will be near your mouth. Until a follow up test shows that the parasite has been eliminated, the sugar glider should be quarantined from other pets. The cage should also be thoroughly cleaned to reduce the possibility of reinfection." [6]

Leptospirosis

Sugar gliders may become infected with this disease and pass it on to humans if they come into contact with water or food that has been contaminated with *Leptospira* bacteria. Signs include fever and kidney and liver problems. A veterinarian can test for the presence of these bacteria. Strict attention to thorough cage cleaning, food dish and water supply sanitation, and careful hand washing following contact with the pet or its environment are important to prevent both reinfection and transmission to you or other humans.

Lumpy Jaw (Actinomycosis)

Lumpy jaw is a condition in which the bacteria *Actinomyces israelii* infect the face and neck, creating a slowly enlarging, hard lump. The bacteria can also infect the lungs, intestinal tract, and other parts of the body. Discharge from the eyes and weight loss are other signs of infection. The facial tissues can become infected with the bacteria after surgery, trauma, or another infection. A common cause in sugar gliders is an abscess in the mouth. Lumpy jaw is deadly if left untreated; therefore, you should contact your veterinarian immediately if a hard lump appears on the face or neck of your pet or if you notice eye discharges or sudden weight loss. Treatment to get rid of the infection requires a prescribed medication. Because several disease-causing organisms may be involved, your veterinarian will perform tests to determine the most effective medication.

Hypocalcemia

"Hypocalcemia is an abnormally low level of calcium in the blood and is due mainly to an imbalance of dietary calcium, phosphorus, and vitamin D."[7] Signs of the disease are lameness, difficulty moving about, and pelvic deformity. Calcium deficiency can lead to weak or broken bones. Hypocalcemia is most common among females that are raising young, but males may also develop calcium deficiency. A veterinarian should be consulted for diagnosis and treatment. Feeding a balanced diet is important in preventing the disease.

Obesity

Sugar gliders that are fed diets that are high in fat content will become obese. Obesity may lead to several health problems, such as infertility, listless behavior, and a limited life span. If the pet sugar glider is becoming overweight, the amount of fat in its diet should be reduced.

SUMMARY

Sugar gliders belong to the subclass of mammals called marsupials. The female has a pouch called a marsupium in which the young are raised during early infancy. The young are born 15 to 17 days into the gestation period. They then crawl up their mother's belly and into the pouch. Once inside the pouch the young will attach themselves to a nipple, where they are nourished. The young remain in the pouch for 70 to 74 days. They will leave the pouch and return for warmth and nourishment.

The sugar glider, *Petaurus breviceps,* is a small marsupial native to the northeastern regions of Australia, Tasmania, Indonesia, and Papua New Guinea. They are arboreal, living among the tree tops, where they move from tree to tree by gliding. Sugar gliders have a body length of approximately 5 to 6 inches with a tail of near equal length, giving the animal a total length of approximately 10 to 12 inches. They weigh 3 to 5 ounces, with the males being slightly larger than the females.

Sugar gliders normally have a soft, gray-colored fur with cream-colored undersides. They have a black stripe that runs from between the eyes, over the head, down the center of the back, to the base of the tail. Sugar gliders have black rings around the eyes and black stripes on the front of each leg. They have a bushy tail that is also gray, except for the last couple of inches of the tail, which is black. Sugar gliders are nocturnal animals and have large eyes, which is typical of animals moving about in the dark.

In the wild, sugar gliders live in family units and are territorial. The adult males in the group mark the limits of their territory. Members of outside groups wandering into the territory of another group will be subject to attack.

Sugar gliders, being arboreal in the wild, are very active and curious. They like to climb and explore. When one is considering the cage for these animals, it needs to be large enough to allow them to explore, climb, glide, and exercise.

Sugar gliders are omnivorous, feeding on both plant and animal material. The diet of pet gliders will consist of fresh fruits and vegetables, meats, eggs, and insects. The diet should consist of approximately 75 percent fruits and vegetables and 25 percent protein such as meats and insects.

Adult male sugar gliders are easily recognized by the bald scent spot on the top of their head. Females have a pouch in the area where the navel would be on placental animals. Sugar gliders are usually bred in pairs, with a male and female being housed together, or in a colony, where one male is housed with two or more females.

Sugar gliders are subject to stress and several diseases and ailments. This is especially true if guidelines for their care and management are not followed.

DISCUSSION QUESTIONS

1. What are the countries of origin for sugar gliders?

2. What is a marsupial?

3. What are the differences between marsupial mammals and placental mammals?

4. Describe the stages in the birth and growth of young sugar gliders.

5. Describe the identifying characteristics of sugar gliders.

6. What should one look for when selecting a sugar glider for a pet?

7. Describe the proper environment for a sugar glider.

8. List some of the suggested foods that can be fed to sugar gliders.

9. What is stress, and what are conditions that can lead to stress?

SUGGESTED ACTIVITIES

1. Feeding of dry cat and dog foods is not recommended for sugar gliders. Research the reasons for this recommendation.

2. Search the Internet for more information on calcium deficiency in sugar gliders.

3. There are five other species of winged gliders, in the genus *Petaurus*. Using the Internet, search for the other five species. What are their characteristics and where are they found?

ADDITIONAL RESOURCES

1. Tasmania—Primary Industries and Water
   http://www.dpiw.tas.gov.au/inter.nsf/WebPages/BHAN-53J8XS?open

2. Exotic Nutrition Pet Company®
   http://www.exoticnutrition.com/Sugar-Glider-Information.htm

3. AmeyZoo Exotic Pets—Caresheets
   http://www.ameyzoo.co.uk/Caresheets/caresheets.htm

4. Ruth's Sugar Glider Page
   http://www.sugarglider.com/archives/ruth

5. Animals Exotique Sugar Glider Diets
   http://www.animalsexotique.com/sugar_glider_diet.html

END NOTES

1. *The Merck/Merial Manual for Pet Health.* Copyright © 2007 by Merck & Co., Inc. Whitehouse Station, NJ, USA. All rights reserved. Used with permission.

2. Ibid.

3. Ibid.

4. Ibid.

5. Ibid.

6. Ibid.

7. Ibid.

GLOSSARY

A

abomasum: the fourth compartment of a ruminant's stomach, sometimes referred to as the "true stomach" because it functions very similarly to the stomach in monogastric animals.

abscess: a swelling caused by the accumulation of pus from an infection.

absorption: the taking in of fluids.

adipose fin: a small, fleshy fin located on the back and between the dorsal and caudal fins on certain species of fish.

adrenal glands: glands located near the front of the kidneys that produce hormones involved in sexual activity and metabolism.

aeration: the exchange of carbon dioxide in water for oxygen.

aerobic bacteria: bacteria that live in the presence of oxygen.

agile: graceful, quick, and easy movement.

agouti: color pattern where one color is streaked or intermingled with another color.

airstones: fused, porous glass devices used in aquariums to oxygenate the water.

albino: an animal that lacks color pigmentation. Albino animals are white with red eyes.

albumen: the thick, white matter that surrounds the yolk of an egg.

algae: a group of single-celled aquatic plants that have chlorophyll.

allergy: a hypersensitivity to animal hair, dust, and other allergens.

aloof: one who holds his or her head high in a proud, confident manner.

amiable: mild, easygoing.

amino acids: a compound that contains both the amino (NH_2) group and the carboxyl (COOH) group; considered the building blocks of proteins.

amniotic sac: a sac or thin membrane that surrounds the embryo and contains a watery amniotic fluid.

amphibians: members of the class Amphibia, which includes frogs, toads, and newts. They live part of their lives in water and part on land.

amphibious: being able to live both on land and in the water.

amphiumas: a form of amphibian commonly known as Congo eels. They have long, slender bodies with tapered heads; have two pairs of tiny weak limbs with one, two, or three toes on each limb; and are found in muddy, swampy areas of the Southern United States.

anaerobic bacteria: bacteria that live in the absence of oxygen.

anal fin: a fin located on the underside of a fish between the vent and the caudal fin.

anal pores: openings located just forward of the vent on certain species of geckos. They can be used to identify the male of the species.

anal sacs: glands located at the four o'clock and eight o'clock positions on either side of the dog's anus. These glands produce a substance that allows the dog to mark its territory when it defecates. These glands may sometimes become infected or impacted.

analgesics: substances used to make an animal insensitive to pain without losing consciousness.

anatomist: person who studies the structure and form of plants and animals.

anemia: a condition in which the blood is lacking in red blood cells; may result from a lack of iron and copper in the blood.

anesthetics: substances used to produce a lack of awareness or sensitivity.

angora: a type of long fur on cats and rabbits.

animal breeder: also known as animal technician, a person concerned with the propagation, feeding, housing, health, production, and marketing of animals.

animal caretaker: one hired to take care of animals. The job ranges from day-to-day normal activities for a healthy animal to caring for sick, injured, or aged animals.

animal curators: persons responsible for the day-to-day management of a specific portion of a zoo or aquarium's animal collection.

animal groomer: person who bathes pets and keeps them free of ticks, fleas, and other pests. Groomers brush the pets, trim their hair, and may cut and trim their nails.

animal handler: anybody who works directly with the training, breaking in, or managing of animals. Handlers may be caretakers at a local park's petting zoo, or the activist who reintroduces wild animals to national parks.

animal health technician: see *veterinary technician*.

animal husbandry: the management and care of domestic animals.

animal husbandry technician: worker primarily trained in the breeding and feeding of animals.

animal hygiene: the management of animals so that good health is maintained.

animal psychology: understanding the basic principles and processes that form the basis of the behavior of animals.

animal rights: the position that animals should not be exploited— that is, not used for food, clothing, entertainment, medical research, or product testing. This also includes the use of animals in rodeos, zoos, circuses, and even as pets.

animal shelters: the places where abandoned and unwanted animals are taken for care or to be euthanized.

animal technician: see *animal breeder*

animal trainer: person who teaches animals to obey commands, to compete in shows or races, or to perform tricks to entertain audiences. They also teach dogs to protect lives or property, or act as guides for visual- or audio-impaired individuals, and to hunt or track.

animal welfare: the position that animals should be treated humanely—that is, properly cared for with adequate housing, good nutrition, disease prevention and treatment, responsible care, humane handling, and humane euthanasia or slaughter.

Animalia: the animal kingdom, the division of life forms that includes all living and extinct animals. The animal kingdom includes the invertebrates and vertebrates.

annual fish: fish that lay their eggs in the mud of streams and ponds that are drying and then die. When the rains come, the eggs hatch and the young grow quickly to adults.

anointing: the process where a hedgehog will lick at a new smell or object producing a flow of foamy saliva. The animal then spits and smears its spines with the saliva.

anterior: the front of an animal.

antibodies: substances produced by an animal's body in response to antigens. Antibodies neutralize the effects of microorganisms and the toxins produced by microorganisms.

anticoagulant infusion: the process of slowly introducing anti-blood-clotting solution into the veins of patients having heart surgery.

anticonvulsant: a substance that controls or prevents convulsions.

antifungal agent: a substance that kills fungi.

antiseptic: substance applied to a wound to prevent the growth of microorganisms.

antivivisectionists: those who are against the use of living animals for experimentation and study.

apathetic: showing no interest.

aplastic anemia: a disease common to female ferrets that occurs as a result of not being bred. The female ferret stays in heat if not bred, and the continued production of the hormone estrogen causes a decline in the production of red and white blood cells and the blood clotting factor in the bone marrow.

aquarist: a person responsible for the maintenance of the aquatic exhibits in aquariums, oceanariums, and marine research institutes.

aquarium: a water tank or container used to raise fish or other aquatic animals.

aquarium curators: person responsible for the day to day care of an aquarium.

aquarium directors: administrators who coordinate the business affairs of an aquarium.

aquatic: living in a water habitat.

aquatic biologist: person who studies animals and plants that live in water and how they are affected by their environmental conditions, such as salt, acid, and oxygen content of the water, and temperature, light, and other factors.

arachnids: eight-legged arthropods of the class Arachnida, which includes spiders, scorpions, mites, and ticks.

arboreal: living in trees.

aristocratic: holding the head high and having a certain air of nobility.

arrogant: excessively proud, to the point of being overbearing and hard to handle.

arthritis-like: symptoms similar to arthritis; inflammation of the joints, stiffness, and difficulty in moving.

arthropods: invertebrate animals such as insects, arachnids, and crustaceans that have jointed bodies and legs and an external skeletal structure.

ascarids: intestinal parasitic worms; the common roundworm is an ascarid.

ascorbic acid: vitamin C.

assimilation: conversion of food material into a form that can be absorbed.

assistant laboratory animal technician: a person who assists laboratory animal technicians in feeding and caring for animals used in laboratories; the lowest level of certification for laboratory animal care workers.

atria: part of an animal's heart that receives blood from the veins and pumps blood to the ventricle.

atrophy: when a part of the body decreases in size or wastes away.

Aves: the class for birds; warm-blooded vertebrate animals having two legs, wings, and feathers and that lay eggs.

aviary: a place where large numbers of birds are kept.

avidin: a protein in egg white that combines with biotin in the intestinal tract to make the biotin unavailable to the animal.

awn fluff: a soft, wavy wool with a guard-hair tip found between the underwool and the awn hair of Giant Angora rabbits.

awn hair: strong, straight hair (guard hair) protruding above the awn fluff of Giant Angora rabbits.

B

bacteria: single-celled organisms that lack chlorophyll and reproduce by fission. They live in colonies and can cause diseases in animals.

barbels: the whisker-like projections on the mouth of several species of fish. These projections help the fish feel its way along the bottom.

bias: a view or belief that is slanted or prejudice.

bib: distinct marking under the chin and on the upper chest.

binomial nomenclature: Linnaeus's system of naming species, which used the Latin language. Each species has a name composed of two words: first, the genus that is written with the first letter capitalized and second, the species, which is written with a small first letter.

biochemical reaction: chemical reactions that take place in the cells of plants and animals and include respiration, digestion, and assimilation.

biochemist: person who studies the chemical processes of living organisms.

biological technician: another name for a biotechnician.

biologist: person who studies the origin, development, anatomy, function, distribution, and other basic principles of living organisms. The biologist may also be called a biological scientist or life scientist.

biology: the study of living organisms.

biophysicist: person who studies the atomic structure and the electrical and mechanical energy of cells and organisms.

biotechnician: trained to assist the biological scientist in his or her endeavors.

biotechnology: industries that combine the sciences of biology and engineering to help or aid humans.

Black Plague: disease that killed more than one-fourth of the population during the 1300s.

black tongue disease: a disease in dogs resulting from niacin deficiency.

black tongue factor: also called pellagra. Black tongue factor is a disease in humans caused by a deficiency in niacin. Not to be confused with black tongue disease in dogs.

bold: showing no hesitation or fear.

botanist: person who studies plants.

brackish: a mixture of freshwater and saltwater found where rivers and streams enter the saltwaters and also found in some inland swamp areas that receive some saltwater.

brawny: very muscular and strong.

breeder: one who breeds animals.

brille: a transparent layer that covers the eye of a snake.

broad-spectrum antibiotics: antibiotics that are effective against a large range of diseases.

broken: white-furred animal with a spotted or colored pattern.

brood: the incubation of eggs where the female keeps them warm. The term brood also describes a group of young produced by a female.

buck: male rabbit.

bulbous: the bulb-like enlargement at the tips of the inner toes of the hindfeet on sugar gliders, which are adapted for grasping and hanging onto branches.

butterfly nose: a dark nose marking on some breeds of rabbits; covers the nose, mouth, and whisker bed.

C

caecilians: class Amphibia, order Gymnophiona; wormlike amphibians that do not have limbs and are usually without sight.

caiman: class Reptilia, order Crocodilia, family Alligatoridae. Reptiles found in Central and South America that grow up to 15 feet; smaller species may be found in the pet trade.

calico: also known as tortoiseshell and is made up of three colors: black, orange, and white.

camouflage: the body coloring of an animal that allows it to hide or conceal itself in its environment.

canines: sharp, pointed teeth on either side of the jaws.

cannibalism: practice of eating one's own kind.

caped: a type of rat that is white with a darker color about the head, ears, and neck.

capybara: *Hydrochoerus hydrochaeris,* found in South America, is the largest of the rodents, growing to 4 feet in length and weighing up to 120 pounds.

carapace: the hard, protective upper shell on a turtle, tortoise, or terrapin; it also forms part of the vertebrae and ribs.

carbohydrates: compounds such as sugars and starches composed of carbon, hydrogen, and oxygen that are needed by an animal for energy.

Carboniferous period: geologic time period from 280 to 345 million years ago that saw the growth of trees, ferns, and conifers as well as an increase in amphibians and the appearance of the first reptiles.

carboxylation: the combining of organic acids with carbon dioxide to form carboxylic acid.

carnivorous: a meat-eating animal.

carrier: an animal that appears healthy but carries disease-causing organisms that can infect other animals.

cartilaginous: species of fish that has a skeleton wholly or partly composed of cartilage.

casque: a structure on the top of the head that resembles a helmet.

catheter: a tube that is placed internally so that fluids can be injected or withdrawn.

cation: a positively charged ion in chemical reactions.

caudal fin: the tail fin of a fish.

cavy: another name for a guinea pig.

cecum: where the small intestine joins the large intestine. It serves no purpose in most animals, but in horses and rabbits, bacterial digestion of roughages takes place here.

Cenozoic era: the geologic period of time from the present to 70 million years ago during which placental animals, carnivores, large mammals, and humans appeared.

central scutes: the scutes located down the center of the shell on turtles, tortoises, and terrapins.

cere: the fleshy portion of the beak that surrounds the nostrils of birds.

chatty: very talkative; a cat that makes a considerable amount of vocal noise.

chemistry: the study of various substances and how they react with each other.

cherts: sections of rocks that show fossilized remains of blue-green algae.

chimaera: class Chondrichthyes, order Chimaerae. A group of fish with cartilaginous skeletons and having their upper jaw fused to the skull.

chitinous: material that forms the hard outer covering on insects and crustaceans.

Chordata: a phylum with four distinctive characteristics: embryonic notochord, which is usually replaced by the vertebra; a dorsal tubular nerve cord running down the back side; pharyngeal or throat-area gill slits; and a rear area tail.

chromatophores: specialized pigment cells that lie under the outer layer of skin of chameleons.

ciliospores: reproductive spores of certain species of parasites.

civet-like animal: an animal that has a long body and short legs similar to the African civet cat.

clan: colony structure.

class: a division of the classification of life forms beneath phylum and above order.

classical LD50 test: determines the dosage required to kill 50 percent of a population of test animals within a specified period.

climax: period of greatest abundance of a particular group of wild animals.

cloaca: a common chamber where the digestive and urinary tracts empty, and the reproductive tracts of birds, reptiles, amphibians, and fish are located.

clutch: a nest of eggs or a group of young birds.

coagulation: clotting of the blood.

coastal scutes: scutes located between the central and marginal scutes on turtles, tortoises, and terrapins. See also *scutes*.

cochlea: the organ within the ear that enables an animal to detect sounds.

cold-blooded: animals that cannot regulate their internal body temperature. They take on the temperature of their surrounding environment.

collar: a band or stripe around the neck of an animal.

colony mating: placing several females in with one male for mating. This method is practiced in hamsters, mice, rats, and guinea pigs.

colony structure: living in groups for protection against enemies, finding and storing food, and sharing the duties of raising the young.

colorpoint: the color of the nose, ears, feet, and tail of the animal. See also *points*.

colostrum: first milk produced by the mother. It is rich in proteins and minerals and contains antibodies.

community aquarium: an aquarium that has more than one species of fish.

composed: does not become overexcited, anxious, or hurried.

conception: the point when the sperm fertilizes the egg.

confinement systems: animals are confined to cages or pens in partially enclosed or totally enclosed buildings so that production improves through closer control of the environment.

conformation: the size, shape, and form based on what an ideal animal of the species should look like.

congenial: easygoing and cooperative.

congenital: present at birth.

conjunctiva: the mucous membrane lining the inner surface of the eyelids and covering the front part of the eyeball.

conjunctivitis: inflammation of the membrane lining the inner surface of the eyelids.

consignee: the person to whom animals are being delivered.

consignor: the person who ships animals to be delivered.

constitution: the physical makeup of an animal.

consumer: the person who buys products or goods for his or her own use, not for a business.

contagious: a disease that can be spread from one animal to another by direct or indirect contact.

contour feathers: the feathers that give the bird its outward form.

coprophagy: the consuming of fecal material.

copulation: the act of mating.

cornea: the transparent covering over the eyeball.

corneal ulceration: a tear or break in the cornea of the eye that becomes infected.

coronavirus: a disease that causes severe vomiting and diarrhea in dogs and cats; is usually spread through contaminated feces.

cough suppressant: a substance that checks, restrains, or subdues a cough.

coverts: the smaller feathers that cover the base of the quills of the wing and tail feathers.

crepuscular: animals that are active during the twilight hours.

crest: the tuft of hair or feathers on the top of the head.

Cretaceous period: geologic period 135 to 70 million years ago during which the dinosaurs and marine reptiles reached their greatest abundance followed by their extinction.

crown: the top of the head.

crude fiber: substance composed mostly of nondigestible bulk or roughage.

ctenoid scales: scales on certain fish that have serrations on the edges that makes the fish feel rough.

cud: regurgitated food not fully digested in the rumen. The animal chews on this and swallows it again for further digestion.

cycloid scales: scales on certain fish that have smooth edges that make the fish feel smooth and slick.

cyst: a capsule formed around an organism that is in a dormant or spore stage. Also a sac-like structure or pocket on an animal that is filled with fluid or pus.

cytochromes: iron-containing enzymes.

cytologist: sometimes called cell biologist, examines the cells of plants and animals.

D

decarboxylation: the breakdown of carboxylic acid.

defecate: the movement of solid waste from the anus.

degenerate: to decrease in size or waste away.

demodectic mites: class Arachnida, order Acari. Mites found on the skin of many animals can cause demodicosis if present in large numbers.

demodicosis: skin disease caused by large numbers of demodectic mites; also called demodectic mange.

deoxyribonucleic acid: see *DNA*.

dermal: relating to the dermis.

dermis: the sensitive layer of skin that lies below the epidermis.

devilish: over playful; tending to get into trouble.

Devonian period: geologic period about 370 million years ago, during which the first amphibians appeared and freshwater fish became abundant.

dew-drop valve: a water valve fitted on bottles to provide water on demand to small animals.

dewlap: loose fold of skin under the chin or hanging from the throat.

dexterous: skillful in use of hands or body.

digestion: the breakdown of food materials and the absorption of nutrients.

dignified: an excellent-appearing animal that shows beauty of form and grace.

dilate: to widen, expand, or increase in size.

dimorphism: the male and female of a species having different physical characteristics so that they can be identified.

disc-shaped: circular or round in form.

disposition: how an animal acts under various circumstances.

dissection: cutting an animal into pieces or parts so that it can be studied or examined.

diurnal: animals that are active during the daylight hours.

DNA (deoxyribonucleic acid): genetic material that carries the hereditary codes.

docile: tame, easygoing, and obedient.

doe: female rabbit.

dog trainer: person who teaches dogs to hunt or track, to obey signals or commands, to guard lives or property, to run races, or to lead the blind.

domesticated: animals that have adapted to a life around humans.

dominant gene: in a pair of genes, the one that controls the particular genetic characteristic.

dorsal: the topside or back of an animal.

dorsal fin: fin located on the back of a fish.

down feathers: small, soft feathers located under the contour feathers that conserve body heat.

Draize eye test: measures the effects of products such as cosmetics on the unprotected eyes of restrained rabbits. Rabbits do not have tear ducts to wash the products from their eyes.

ductus deferens: small tubes in the male bird that lead from the testes to the cloaca.

dull: the lack of usual shine or luster.

duodenum: the first section of the small intestine.

E

ecologist: person who studies the relationship of plants and animals to the environment.

ecoterrorism: terrorism or crimes committed under the disguise of saving nature.

ectoparasite: an external parasite that attacks the outside of an animal's body.

ectotherm: an animal that is cold-blooded.

egg-buriers: fish that lay their eggs in the mud of rivers and ponds.

egg-depositors: fish that prepare a nest, deposit their eggs, fertilize them, and guard the eggs until they hatch.

egg-scatterers: fish that lay their eggs in a haphazard manner on the bottom of streams, rivers, and lakes.

ELISA (enzyme-linked immunosorbent assay): test that can be performed in a veterinarian's office to detect both primary and secondary stages of viremia.

embryo: the early developing stages of a vertebrate animal before it is born or hatched.

embryologist: person who studies the development of animals from egg to birth.

encephalitis: an inflammation of the brain.

endoparasite: an internal parasite that attacks the inside of an animal's body.

enteritis: an inflammation of the intestinal tract.

enterprising: hardworking, determined, and aggressive in its work.

entomologist: person who studies insects and their relationship to other life forms.

entropion: a condition in which the eyeball sinks into its socket or in which there is a spasm of the eyelid due to discomfort. Entropion can be a congenital defect or a result of an injury.

enzyme: complex protein substance produced by the body and necessary for biochemical reactions to occur.

Eocene epoch: geologic period 40 to 60 million years ago, during which placental mammals spread and increased.

epidermal: relating to the epidermis.

epidermis: the layer of skin that lies above the dermis.

epinephrine: a substance made from the adrenal glands of cattle used to relieve symptoms of hay fever, asthma, and some allergies in humans. It is also used as a heart stimulant, as a muscle relaxant, and to constrict blood vessels.

epithelial cells: the tissue that covers the inside of the intestines.

epoch: geologic division of time less than a period.

era: refers to one of the five major geologic divisions of time.

estivation: a deep sleep-like state triggered by excessively warm temperatures.

estrogen: a female sex hormone that promotes estrus and is responsible for the development of secondary sex characteristics.

estrus: female's heat period or season.

ethical: belief or ideal that is basically good or bad; dealing with what is good and bad.

euthanasia: to induce the death of an animal quickly and painlessly.

evulsion: a tear or a pulled-out wound.

exoskeleton: the hard, external protective shell of a turtle, tortoise, or terrapin.

expectoration: to expel material from the throat or lungs by coughing or spitting.

external parasite: an ectoparasite.

extroverted: not shy; forward and aggressive.

F

factory farming: practice of keeping chickens in cages and veal calves in small crates.

fad: a passing craze; interest for a short time.

family: a division of the classification of life forms beneath order and before genus.

fats: nutrient made up of the same chemical elements as carbohydrates but in different combinations. Fats contain 2.25 times as much energy as an equivalent amount of carbohydrates and proteins.

fatty acids: the saturated and unsaturated monocarboxylic acids in the form of glycerides in fats and oils.

feces: solid waste material discharged through an animal's anus.

feisty: nervous, excitable, and easily agitated or irritated.

femoral pores: openings located on the inside of the rear legs of agamas and some species of iguana. They can be used to determine sex.

fennecs: a small, fox-like animal native to Africa.

feral: wild.

ferreting: the practice of using ferrets to drive rabbits and rats from burrows.

fetal development: the natural and normal growth of the fetus within the uterus.

fetus: the unborn or unhatched young of a vertebrate.

filoplume feathers: short, hair-like feathers.

fish: used when referring to several species of fish.

flight feathers: the longer contour feathers that extend beyond the body and that are used in flight.

fly-back fur: animal fur that quickly returns to its normal position after the animal is stroked.

formalin: a solution of water, formaldehyde, and a small amount of methanol.

fossils: hardened or petrified remains of a plant or animal from previous periods of history.

freshwater: water habitat that is neither brackish nor marine.

Fungi: plant kingdom including the molds, yeasts, and fungi that obtain their food by absorption and that lack chlorophyll.

fungi: a group of plants that lack chlorophyll and obtain their nourishment from other plants and animals. Fungi include molds, mildews, smuts, mushrooms, and certain bacteria.

fungus disease: disease that is caused by a fungus.

G

gaiety: being cheerful, happy, or merry.

gait: the way an animal moves.

gaming dog business: the business of raising and racing dogs.

ganoid scales: scales on gar, paddlefish, reed fish, sturgeons, rope fish, and bichirs. These scales are very thick, heavy, and similar to scales on ancient species of fish.

gastrointestinal tract: the stomach and intestines.

general curators: persons in charge of the day-to-day management of a specific portion of a zoo or aquarium's animal collection.

geneticist: person who specializes in the study of heredity and how characteristics vary in all forms of life.

genetics: the study of heredity and how and what characteristics are passed from generation to generation.

genus: a division of the classification of life forms beneath family and before species.

gestation: period of time from conception until birth of young.

gizzard: a common name for the ventriculus. The ventriculus is the largest organ of the bird's digestive system.

gloves: white paws.

glucagon: a substance produced from cattle pancreas that increases the sugar content in blood and is used to counteract insulin shock in humans.

goiter: a swelling of the thyroid glands in the neck area. The ailment is caused by a lack of iodine in the diet.

gonopodium: a modification to the anal fin of live-bearing male fish that aids in transferring sperm to the vent and cloaca of the female.

gram-negative bacteria: bacteria that will not hold the purple dye used in Gram's method of identifying bacteria.

gram-positive bacteria: bacteria that will hold the purple dye used in Gram's method of identifying bacteria.

gregarious: associates in groups with others of their own kind; not solitary.

grit: crushed granite, oyster shell, or other insoluble material that aids in the grinding and breaking down of coarse food material in the gizzard or ventriculus of a bird. Gravel-like material fed to birds that accumulates in the gizzard and aids in the grinding of food material.

grotzen: the center back strip of a chinchilla pelt.

guard hairs: the longer, coarse hairs that form a protective coat above the shorter underfur of an animal. The guard hair protects the underfur from rain and cold.

H

handler: person who holds and commands trained animals during a show, sporting match, or hunt.

hardness: the soluble minerals in water.

haw: another term for the nictitating membrane.

heat period: the estrus of females.

heel: a command for the dog to walk close to the side of the handler, usually the left side.

hemipenes: the modified cloaca in male reptiles of certain species used to transfer sperm to the cloaca of the female.

hemoglobin: red-colored, iron-containing protein component of blood that is responsible for carrying oxygen to the body tissues.

hemorrhage: uncontrolled bleeding.

herbicide: a chemical substance used to destroy plants.

herbivorous: animals that feed on plant material.

hertz: a unit of frequency; one cycle per second.

hibernation: going into a resting or dormant state during winter.

histopathologist: investigates diseased tissue in humans and animals.

hob: a male ferret.

hookworms: bloodsucking parasitic worms that have hooks near their mouth so they can attach themselves to the intestinal lining of animals.

hormones: substances that are produced in one part of the body, but influence cells in another part of the body. Hormones may influence growth.

host: an animal from which a parasitic organism derives nourishment.

humane: kind, gentle, merciful, sympathetic.

humane shelters: places where unwanted or abandoned animals are taken for care or to be euthanized.

humanize: to believe that animals have the same rights as humans and to treat them as humans.

husbandry: care and management of domestic animals.

hutch: house used to keep rabbits.

hutch burn: a chapped or burned condition around the external genital area of female rabbits caused by urine-soaked bedding.

hypomagnesemia: a condition that is caused by low blood magnesium.

I

ichthyologist: scientist who studies fish.

IFA (indirect immunofluorscent antibody assay): test that must be sent out to a diagnostic laboratory. IFA tests for secondary viremia only.

ileum: the last section of the small intestine. It is located between the jejunum and the large intestine.

immune gamma globulin: antibodies.

immune system: the body system that prevents the development of disease organisms and counteracts their effects.

immunize: to give an animal protection against disease-causing organisms.

immunoglobulins: protein substances that are produced in response to antigens and act as antibodies.

immunologist: scientist who studies the manner in which the body resists disease.

immunosuppressive therapy: therapy used to correct an immune system problem.

impaction: a condition in which the small intestines, cecum, or large intestines become tightly packed with food material or feces.

impetuous: abrupt, hasty, and rushing headlong into work without much efficiency.

in vitro: refers to in glass or in laboratory containers. Cultured skin can be grown in vitro in laboratories.

inbreeding: breeding closely related animals.

incisor: front teeth between the canines in either jaw.

incorrigible: unable to be controlled or corrected.

incubation: the period of time in which the embryo is developing within the egg. Also the act of sitting on or brooding an egg.

independent: wanting to have its own way and not wanting to be controlled.

indigenous: native to a particular region.

indirect ovulator: the female's eggs are induced by the presence of the male into being released into the oviduct for fertilization.

indiscriminately: the mating habits of animals where different species, types, or varieties mate and produce young.

indolent: being lazy or inactive.

induced ovulation: the act of mating will induce the female's eggs to be released into the oviduct for fertilation.

infectious diseases: diseases that are caused by microorganisms such as viruses, bacteria, protozoa, and fungi that are capable of invading and growing in living tissues.

infundibulum: the funnel-shaped opening to the oviduct in which the egg yolk drops after being released from the ovary.

inherent: belief or ideal that is by nature right or wrong; a belief or ideal that we are born with; a deep-seated belief or ideal.

inherited: passed on to offspring from parents or ancestors.

inherited defects: problems, ailments, or deformities that are passed on to offspring from parents or ancestors.

inhumane: cruel and savage.

inquisitive: wanting to check things out or wanting to investigate things.

insecticide: a chemical substance that is used to destroy insects.

insectivorous: animals that feed primarily on insects.

insulin: a substance produced from cattle pancreas and used to treat people with diabetes.

intensive operations: farming operations in which the farmer or manager tries to increase output through better breeding, feeding, and management.

intermediate host: a host that a parasitic organism lives on or in during an immature stage.

internal parasite: an endoparasite.

intradermal: within the skin.

intramuscular: within the muscle.

intravenous: entering by way of the vein.

invertebrates: members of the animal kingdom without backbones, such as arthropods and worms.

inviolate sanctuaries: a place where animals are protected and no hunting is allowed.

iridescence: colors that have a brilliant shine or glitter.

isotopes: atomic forms of elements.

isthmus: the area in a bird's oviduct where the two shell membranes are added to surround the yolk and albumen.

J

Jacobson's organ: smell and taste receptors located in the roof of the mouth of cats and reptiles.

jaundice: abnormal yellowing of the skin, tissues, eyes, and urine due to bile pigments building up in the blood as a result of liver disease or excessive breakdown of red blood cells.

jejunum: the second section of the small intestine. It is located between the duodenum and ileum.

jill: a female ferret.

joey: a young sugar glider.

jovial: good-natured, happy, easygoing.

Jurassic period: geologic period 135 to 180 million years ago, during which the first birds and mammals appeared and the dinosaurs became abundant.

K

keen: very sensitive, alert, and intense.

kennel attendant: person who feeds and waters animals, keeps their quarters clean, exercises animals, and keeps records on the animal's health, feeding, and breeding.

keratin: protein that makes up hair, nails, and horns of animals and the spines on hedgehogs.

keratitis: an inflammation of the cornea.

kilohertz: a unit of frequency; 1,000 cycles per second.

kindling: giving birth to a litter of rabbits.

kingdom: one of the five major areas in which all life forms are classified—Monera, Protista, Plantae, Fungi, and Animalia.

kinked: twisted or curled fur.

kits: the name given to young ferrets.

L

laboratory animal care workers: persons who provide food and water, clean cages, check for illness, and keep records on each animal.

laboratory animal technicians: the second level of certification of laboratory animal care workers. They work with scientists and carry out their instructions.

laboratory animal technologists: the highest level of certification of laboratory animal care workers. They work with scientists and carry out their instructions.

labyrinthine chamber: a chamber behind the gills of certain species of fish that consists of rosette-shaped plates supplied with numerous blood vessels that absorb oxygen from the atmosphere.

labyrinthine organ: rosette-shaped plates within the labyrinthine chamber.

laceration: a wound, tear, or cut that has rough edges.

lactating: producing milk.

lamellae: special adaptations on the underside of the toes of most geckos that allow them to grip onto smooth surfaces.

lateral line: a line running horizontally on the side of most fish comprised of pressure-sensitive cells called neuromasts. See also *neuromasts.*

lethal gene: a recessive gene that, if carried by both parents, causes death of the offspring.

lethargy: lack of energy, sluggish.

lice: class Insecta, phylum Arthropoda. Lice are small wingless insects belonging to the order Mallophaga (biting lice) and Anaplura (sucking lice). They can cause great irritation and loss of blood.

life scientist: another name given to a biologist.

limber: very agile and flexible.

limnologists: Scientists who study freshwater organisms and their environment.

litter: multiple offspring produced at one birth.

live-bearers: fish that give birth to live young.

locomotor problems: problems involving the coordination of movement.

longy: a Manx cat with a complete tail.

lores: the area between the eye and the bill of a bird.

luster: shine by reflected light.

lutinos: birds, especially cockatiels, that are primarily yellow in color.

M

macrominerals: minerals needed in the largest quantity and most likely to be lacking in the feed supply; includes calcium, phosphorus, potassium, sodium, sulfur, chlorine, and magnesium. Macrominerals are also called major minerals.

malocclusion: a condition in which the upper and lower teeth grow to the extent that the animal cannot eat. The condition results because the teeth do not wear down properly, or it may be an inherited condition in which the teeth do not come together properly.

Mammalia: vertebrates that possess mammary glands, have a body that is more or less covered with hair, and possess a well-developed brain.

mandibles: both the lower and upper segments of a bird's bill; the lower jaws of mammals.

mane: the hair from the head down the back of the neck.

mantle: the feathers on the back of a bird.

marginal scutes: the scutes located around the outer edge on the shell of turtles, tortoises, and terrapins.

marine: saltwater habitats.

marine biologists: biologists who specialize in the study of marine species and their environment.

marsupials: mammals that lack a placenta and have an abdominal pouch (marsupium) in which the young develops.

marsupium: the abdominal pouch of marsupial animals that contains teats to nourish the young as they develop.

maxilla: the upper jaws of mammals.

meek: showing a lack of aggression and weakness.

melanin: a chemical in the cells of chameleons that causes the cells to darken depending on changes in light, temperature, and their emotional state.

meningitis: inflammation of the membranes covering the brain and spinal cord.

metabolism: the physical and chemical processes that take place in an animal's body that produce the energy necessary for the body's activities.

metamorphosis: a series of physical changes that an animal goes through from birth to adult.

metritis: condition where the fetuses are retained and leads to bacterial contamination and inflammation of the uterus.

microbiologist: biologists who study bacteria, viruses, molds, algae, yeasts, and other organisms of microscopic or submicroscopic size.

microfilariae: the very small larvae form of the filament nematodes, *Filaria,* that causes heartworm disease in dogs.

microminerals: minerals that are needed in trace amounts; includes iron, iodine, copper, cobalt, fluorine, manganese, zinc, molybdenum, and selenium. Also called trace minerals.

milt: the sperm-containing substance of male fish.

mimic: ability to imitate sounds or words.

minerals: inorganic compounds needed in small amounts by an animal's body for the growth of bones, teeth, and tissues and to control various chemical reactions. They are components of the ash or noncombustible part of the ration that is left after the material has been burned.

mites: class Arachnida, order Acari, phylum Arthropoda; small animals that have four pairs of legs. They can infest larger animals by burrowing into the skin and feeding on the blood.

mitt: the feet or paws of a ferret.

molds: a fungus that produces a soft, woolly growth usually on damp, decaying materials.

molecular biologists: biologists who apply their research on animal and bacteria systems toward the goal of improving and better understanding human health.

molt: the shedding of feathers, skin, or shells to allow for the internal growth of an animal.

Monera: the kingdom including the bacterial organisms that lack a true nucleus in the cell and reproduce by fission.

monk: a man of a religious order who lives in a monastery.

monogamous: pair bond may last throughout the life of the animals; having only one mate at a time.

monogastric: having a single stomach.

moral: a belief or ideal of what is basically right or wrong.

mouth-brooders: fish that carry their eggs in their mouth until they hatch.

mutant: an individual or species of animal produced by a change in the genetic structure or hereditary material. See also *sport*.

mutation: a sudden change in an inherited trait; a departure from the parent type. Mutant animals are also called sports.

mycologist: biologists who study edible, poisonous, and parasitic fungi, such as mushrooms, molds, yeasts, and mildews, to determine which are useful to medicine, agriculture, and industry.

N

nape: the back of the neck.

naturalists: workers whose primary role is to educate the public about the environment and maintain the natural environment on land specifically dedicated to wilderness populations.

nature: the mental and emotional qualities of an animal.

nematologists: biologists who study nematodes, which are parasitic in animals and plants.

nest box: a box that animals use to give birth, lay eggs in, and raise their young.

nest-builders: fish that construct a nest in which their eggs are deposited.

neuromasts: a series of pressure-sensitive cells along the lateral line of fish.

neuter: the surgical procedure performed on male animals to prevent reproduction by the removal of the testicles.

neutered: castrated or spayed animal.

newts: class Amphibia, order Caudata; small semiaquatic amphibians similar to salamanders.

nictitating membrane: the third eyelid of certain animals that moves diagonally under the eyelid to help lubricate the cornea. It is also referred to as the haw.

night feces: the fecal material consumed by rabbits. See also *coprophagy*.

nitrification: nitrogen cycle.

nitrogen cycle: the process by which ammonia is converted into nitrites and nitrates.

nitrogen-free extracts (NFEs): the easily and completely digested sugars and starches.

nocturnal: sleeping during the day and being active at night.

no-kill shelters: shelters where all adoptable and treatable animals are saved and where only unadoptable or non-rehabilitative animals are euthanized.

noncontagious: a disease or ailment that cannot be spread to another animal by contact.

noninfectious diseases: diseases caused by physical injuries and genetic defects; diseases that are not contagious.

nonruminant animals: single-stomached, or monogastric, animals. All of the small animals discussed in this book are nonruminants.

normal: fur type that is regular or the most common.

nose break: the curve the nose makes where it connects with the skull.

notochord: a flexible cord of cells in the embryo stage of growth that eventually becomes the vertebra.

nutrient: a single class of foods or group of food materials of the same general chemical composition, which aids in the support of animal life.

nutrition: receiving a proper and balanced ration so that an animal can grow, maintain its body, reproduce, and produce what we expect from it.

nutritionist: person who studies how food is used and changed into energy.

O

obesity: excessive accumulation of fat in the body.

oblique: running at approximately a 45-degree angle.

observant: quick to see things and attentive.

obstinate: determined to have its own way; stubborn.

occlusion: teeth on both jaws coming together correctly.

olfactory mucosa: an area of the nasal lining that picks up smells from the air as the animal breathes.

olm: class Amphibia, order Caudata (*Proteus anguinus*). An olm is a dull, white animal with red, feathery, external gills. It is about 1 foot long, blind, and found in underground lakes, streams, and rivers in parts of Europe.

omasum: the third compartment of a ruminant's stomach. It removes large amounts of water from food as it moves from the rumen.

omnivorous: animals that feed on both animal and plant material.

oocyst: an egg that has not fully matured.

ophthalmic ointments: a salve or cream used in the eyes to relieve pain or irritation.

oral mucosa: the mucous membrane of the mouth.

order: a division of the classification of life forms beneath class and above family.

Ordovician period: geologic period about 420 million years ago, during which the first land plants were believed to have become established. During this time invertebrates, such as arthropods and worms, appeared on land.

osmosis: the flow of a substance from an area of high concentration to an area of lower concentration through a semipermeable membrane.

Osteichthyes: the bony fish, a large group of vertebrate animals that live in the water and have permanent gills for breathing, fins, and a body covered with scales.

osteomalacia: a disease in which the bones of the body become soft and weak.

osteoporosis: a condition in which the bones become porous and brittle.

otitis: an inflammation or infection of the ear.

oviparous: animals that lay eggs.

ovoviviparous: animals that retain eggs within their bodies until they hatch and then give birth to live young.

ovulation: release of egg cells from the ovaries into the oviduct.

ozone: a gas found in a layer of the atmosphere. It filters out ultraviolet radiation from the sun and prevents heat loss from the earth.

P

palatable: a food material that will be eaten and digested by an animal.

palate bone: the bone of the roof of the mouth.

Paleocene epoch: geologic period 60 to 70 million years ago, during which placental mammals appeared.

paleontologists: a branch of geology in which scientists study prehistoric life through the study of fossils.

palpating: a method of using the hands to externally examine internal structures for physical diagnosis, for example, pregnancy or tumors.

papilla: small projections on the dorsal wall of a male's cloaca that aid in transferring sperm to the female's cloaca.

papillae: knobs on the tongue that help hold food or prey; some carry taste buds.

paralytic: a partial or complete loss of motion and sensation in a part of the body.

parasite: organism that lives on or within another organism and derives nourishment from the host.

parasitologists: biologists who study animal parasites and their effects on humans and other animals.

pariah dogs: semi-feral dogs that live in close approximation to towns, where they survive by scavenging on garbage.

parturition: the birthing process.

patagium: the fold of skin between the front and hind legs of some animals. When the animal extends its legs, it enables them to glide from branch to branch or tree to tree.

pathogenic organisms: organisms capable of causing disease.

pathologist: a person who studies the effects of diseases on plants and animals.

peccaries: small nocturnal mammals native to Central and South America that are related to pigs.

pectoral fins: fins located on each side of a fish behind the gill opening; they correspond to the forelimbs on other animals.

pedigree: a record of an animal's ancestry.

peduncle: the section of a fish's tail from the main body to the caudal fin.

pellagra: Also called the black tongue factor. A disease in humans related to niacin deficiency.

pelvic fins: a pair of fins located toward the anterior underside of a fish; they correspond to the rear limbs on other animals.

penicillin: an antibiotic produced from molds.

periodontal disease: a disease affecting the tissues that surround the teeth.

peripheral vision: able to see in a wide circle or a large surrounding area.

Permian period: the geologic time period from 230 to 280 million years ago. During the Permian period there was an increase and spread of reptiles and the extinction of many marine invertebrates.

perosis: an ailment of birds caused by malnutrition. Symptoms include the shortening and thickening of the bones in the limbs and enlargement of the hock.

pesticide: a chemical substance used to destroy pests.

pet care counselor: a person employed by pet shops to counsel, guide, and recommend the correct pet for the customer.

pet care workers: general workers who perform several tasks in animal hospitals, boarding kennels, animal shelters, pet grooming parlors, pet training schools, and pet shops.

pet groomer: person who combs, cuts, trims, and shapes the fur of all types of dogs and cats.

pet shops: specialized stores in which consumers can purchase a pet and a full line of pet products.

pH: the acidity or alkalinity of a substance. A pH of less than 7 is acidic, and a pH of greater than 7 is alkaline.

pharmaceutical: medicinal drug industry.

pharmacologist: person employed as a researcher by pharmaceutical companies to study the effects of drugs and other substances such as poisons and dusts on living organisms.

pharyngeal: the area of the pharynx.

pharyngeal teeth: teeth located in the pharynx of certain fish species that are used to grind food material.

phenol: a natural occurring chemical in cedar and pine shavings that may be toxic to small animals.

philosophers: persons who attempt to understand the values, reality, and truths about life.

photosynthesizing: the formation of carbohydrates from carbon dioxide and water in the chlorophyll-containing cells of plants.

phylum: a division of the classification of life forms directly beneath the kingdom.

physiologist: biologists who specialize in studying all the life stages of plants and animals.

piebald: mice with spots, patches, or broken patterns.

pied markings: markings of two or more colors that appear as blotches of color all over the body.

pika: a small, short-eared animal related to rabbits. Order Lagomorpha, family Ochotonidae. They are found in Asia and Western North America.

pinworms: small parasitic worms of the family Oxyuridae that infect the intestines of animals.

placenta: sac that contains a newborn; must be removed before the baby animal can breathe.

placental: having a vascular organ that unites the fetus to the uterus. The developing fetus receives nourishment through this organ.

placental mammals: mammals having a placenta through which the embryo and fetus are nourished while in the uterus.

placental membrane: same as placental sac.

placid: calm and quiet.

placoid scales: scales found on sharks and rays that are pointed and similar to teeth.

Plantae: plant kingdom, including multicellular photosynthesizing organisms, higher plants, and multicellular algae.

plastron: lower part of the shell of turtles, tortoises, and terrapins.

Pleistocene epoch: the geologic period of time that occurred from 500,000 years to 3 million years ago; during this time period, early humans appeared.

pocket pets: refers to small animals that are kept as pets and could fit in one's pocket.

Pointer: a breed of dog that gets its name from the stance it takes in the presence of game.

points: nose, ears, tail, and feet of an animal.

poliomyelitis: a disease caused by a virus that results in paralysis of the body and a decrease in muscle size, often with permanent disability and deformity.

polyestrous: capable of breeding and producing several litters during a year.

polygamous: having more than one mate at a time.

popping: method used to determine the sex of juvenile snakes by trying to turn the hemipenes of a male outward using the thumb and finger.

posterior: the rear part of an animal.

pound seizures: dogs and cats from animal pounds and shelters are sold to laboratory animal dealers and to research and educational facilities for use in animal experimentation.

powder-down feathers: feathers that have tips that break down as they mature and then release a talc-like powder that provides waterproofing and luster to the feathers.

preanal pores: openings located just forward of the vent on certain species of geckos. They can be used to identify the male of the species. Also called anal pores.

precocious: showing premature development.

predator: an animal that seeks out smaller or weaker animals for food.

preen: the process a bird uses to trim and clean its feathers with its bill.

preferential: have a greater liking for or use of an element.

preferred optimal temperature zone (POTZ): a zone of temperature that reptiles try to maintain during daily activities.

prehensile: an animal's tail that is adapted for grasping or wrapping around objects.

preplacenta sac: sac through which some oxygen and nutrient exchange takes place, but not to the extent of a placenta in mammals.

priapism: persistent erection of the penis. This condition may result in trauma to the penis requiring surgical removal.

primary viremia: initial spread of virus in the blood from the first site of infection.

prime fur: a pelt from a chinchilla that has completely shed and been replaced by new fur.

priming line: new hair that comes up through old hair as a chinchilla sheds. This new hair forms a distinct line called the priming line.

probing: a method used in determining the sex of adult snakes where a probe is inserted into the vent. The probe will travel deeper into a male then into a female.

proboscis: the long, tubular, sucking mouthparts on certain species of fish.

proestrus: period preceding a dog's estrus.

progeny: offspring.

prolapse: an organ of an animal's body that falls out of its normal position. The anus, vagina, and uterus, for example, may turn inside out and extend out from the body opening.

proteins: compounds made up of amino acids that supply materials needed for an animal to produce muscle tissue, hair, hooves, horns, skin, and various internal organs.

Protista: kingdom including the single-celled or microscopic animals and single-celled algae.

protozoa: microscopic, single-celled animals.

protrusive: species of fish that can protrude their mouth outward to aid them in capturing prey.

proventriculus: the true stomach of a bird where gastric acids and enzymes are secreted to begin the chemical digestive process.

provoked: to arouse or incite an animal to bite or attack.

public-health microbiologist: conducts experiments on water, food, and the general environment of a community to detect the presence of harmful bacteria so pollution and contagious diseases can be controlled or eliminated.

puppy mills: places where dogs are bred without regard to the pedigrees, defective traits, and abnormalities. The primary purpose of the breeder is to produce puppies for the pet trade. Also called puppy factories.

purebred breeder: a person who breeds animals from parent stock eligible for registration.

pyometra: condition where pus accumulates in the uterus and usually follows a case of metritis.

pyrethrin: natural extracts made from flowers of chrysanthemum plants contained in pet shampoo to control parasites such as lice, mites, fleas and ticks.

Q

quarantine: to isolate animals while treating them for diseases or while determining if they have a disease or ailment.

queen: female cat.

quick: the inner pink center of the nail that contains blood vessels.

R

rabbitry: place where domestic rabbits are kept.

rabies: a viral disease that affects the nervous system of warm-blooded animals.

random source: obtaining animals from animal shelters or humane shelters for use in laboratories.

rays: the hard, spiny structures that support the fins on fish.

recessive gene: in a pair of genes, the one that does not control the particular genetic characteristic; see also *dominant gene*.

rectal prolapse: the protrusion of the rectum from the anus.

red nose: a condition in gerbils, usually caused by bacteria, that causes hair loss and red, swollen areas of the skin around the nose and muzzle.

regurgitated: not fully digested food material is brought back to the mouth from the stomach or rumen.

reptile: class of cold-blooded vertebrates having lungs, an entirely bony skeleton, a body covered with scales or horny plates, and a heart with two atria and usually a single ventricle.

Reptilia: the class that includes turtles, tortoises, terrapins, snakes, pythons, boas, iguanas, lizards, crocodiles, alligators, caiman, and gharials.

reserved: being shy, cautious, or restrained.

reservoir: an immune host.

respiration: the process by which oxygen is supplied to the cells and tissues in exchange for carbon dioxide; breathing.

reticulum: the second stomach compartment of ruminants often referred to as the "hardware stomach" because foreign bodies such as nails or pieces of wire can be held there for long periods of time without causing harm; also called the honeycomb because of its beehive-like interior appearance.

retractile: the claws of a cat that can be drawn back into the toes.

rex: the type of fur on the rex breed of rabbit. The guard hairs and the underfur are the same length, giving the fur a very soft, plush feel.

robust: a strong, rough, healthy-looking individual.

rodenticide: chemical substance used to kill rats and mice.

rodents: order Rodentia. Rodents are small, gnawing mammals that have a single pair of continuously growing incisors in both the upper and lower jaw.

roll-back fur: fur that gently returns to the original position after the animal is stroked.

Roman nose: a nose that is shorter and more rounded than the nose of similar species.

rosettes: the circles or swirls of fur found on Abyssinian guinea pigs.

roundworms: round, unsegmented worms that occupy the intestinal tract of animals.

ruddy: color of a cat breed; an orange-brown color, ticked with dark brown or black.

rumen: the largest compartment of ruminants that makes up about 80 percent of the total capacity of the stomach.

ruminant animals: forage-consuming or multistomached animals; includes farm animals such as cattle, sheep, and goats, and zoo animals such as elk, deer, giraffes, buffalo, camels, and antelope.

rumpy: a Manx cat that has a dimple where the tail should be.

rustic looking: a rough, coarse-looking animal.

S

salamander: class Amphibia, order Caudata. Salamanders are small, lizard-like amphibians lacking scales and breathing through gills during their larval stage.

sanitation: proper cleaning or sterilizing to prevent the growth of microorganisms.

saprophyte: a plant that lives on and receives its nourishment from dead or decaying organic material.

satin: fur with a very shiny appearance due to its transparent outer shell.

scalloped feathers: feathers that are black, with the outer margins or edges being of a different color.

scapulars: the small feathers that cover the base area on a bird's wing.

scurvy: a disease caused by the lack of vitamin C; a condition characterized by swollen, painful joints and bleeding gums.

scutes: the hard, tough scales or plates on the shell of turtles, tortoises, and terrapins.

secondary viremia: occurs when primary viremia has resulted in infection of additional tissues via the bloodstream, in which the virus has replicated and once more entered the circulation.

selective breeding: selecting animals to breed that have certain traits or qualities.

self: one-color fur.

self-propagating cells: cells that reproduce when transferred to an artificial medium as in tissue culture.

semiaquatic: animals that live both in water and on land.

serene: calm and quiet.

serrations: the notches found on the rear edge of the carapace on certain species of turtles.

serum: the liquid portion of blood and other body fluids left after the solids have coagulated.

Setter: breed of bird dogs that take a half-sitting stance when on point.

sheen: brightness, shininess.

shoal: a group or school of fish.

sirens: class Amphibia, order Caudata, family Sirenidae. Sirens are eel-like amphibians that live in freshwater areas of the Southern and Central United States. They have feathery, external gill structures; small, weak front legs; and no rear legs.

skin irritancy test: measures the effects of products such as cosmetics on the skin of restrained rabbits. A patch of fur is shaved from the animal. Products are applied to the area and then observed for any reactions.

small animal breeder: a person who specializes in raising small animals for the pet trade or for use in laboratories.

solitary: prefers being alone; by itself.

solubility: capable of being dissolved. Vitamins are classified as fat soluble or water soluble. Fat-soluble vitamins can be stored by the body; water-soluble vitamins are stored in limited amounts.

sore hock: an ulceration to the foot pad caused by the animal's body weight pressing down on the rough wire of cages.

spawning: the act of depositing and the fertilization of eggs by certain species of fish.

spayed: a female animal that has had her ovaries removed by surgical procedure to prevent her from reproducing.

specieism: belief that any use of animals by humans reflects a bias or that humans are superior to animals.

species: a division of genus groups into individuals that are related.

species aquarium: an aquarium used to keep only one species of fish.

spermatophore: a capsule containing sperm deposited by the male.

spiracles: openings behind the eyes of sharks and rays through which water is drawn in and forced out the gills.

spirit: showing a lot of pep, vigor, and energy.

spirochete: a type of bacteria that is spiral shaped, is flexible, and does not have rigid cell walls. Bacteria of this type cause leptospirosis and syphilis in animals.

sport: an offspring that shows an unusual characteristic; a mutation. See also *throwbacks*.

spraying: the spraying of urine on objects by male cats to mark territory.

status animal: an animal regarded as a symbol of high social status.

status symbol: something that would be an indication of a person's wealth, position, or importance.

staunchness: strong, steady, and steadfast while on point.

steady: staying on task.

steatitis: a deficiency disease, also called yellow fat disease, caused by feeding too much oily fish such as tuna.

sternum: the breastbone of an animal.

streptococcal bacteria: round-shaped bacteria that forms rows of cells in a chain-like arrangement.

stress: any condition that is not normal. Stress can be caused by putting an animal in a situation or under conditions that it is not used to.

stromatolites: layered structures found in rock created through the activity of primitive algae or bacteria.

stumpy: a Manx cat with a short tail.

styptic: a substance used to stop bleeding by causing the blood to clot.

subclass: a division of the classification of life forms beneath class and above order. Life forms placed in certain classes are sometimes further divided into subclasses.

subcutaneous: beneath the skin; between the skin and muscle.

suborder: a division of the classification of life forms beneath order and above family. Life forms placed in certain orders are sometimes further divided into suborders.

subordinate: being submissive to. In a colony structure, there is one lead or dominant male, the others are subordinate or submissive to him.

subphylum: a division of the classification of life forms beneath phylum and above class. Life forms placed in certain phylums are sometimes further divided into subphylums.

substratum: the layer of gravel, peat, or soil that is placed in the bottom of an aquarium or vivarium.

suppressants: a drug that is used to keep under control or in check various conditions, ailments, or diseases.

sustenance: food or nourishment necessary to maintain life.

symbiosis: two different species living together in a mutually beneficial relationship.

syndactyl: an animal with two or more toes connected or fused together.

synthesis: making or producing of a compound.

systemic: a substance that moves throughout all body systems.

systemic antibiotics: antibiotics such as tetracycline, gentamicin, tylosin, and neomycin.

T

tabby: cat with striped markings.

tapeworm: a parasitic segmented worm that lives in the intestines of a host animal. Segments are called proglottids.

tapirs: small, hoofed animals found in South America. They have four toes on each front foot and three toes on each rear foot. They have a short, movable trunk.

taxa: the seven categories of Linnaeus's classification system that include kingdom, phylum, class, order, family, genus, and species.

taxidermist: a person who skins, prepares, and mounts the skins of animals for display.

taxonomy: the science that is concerned with the naming and classification of organisms.

teased: masses of hair are untangled and straightened with a comb.

temperament: the emotional and mental qualities of an individual animal.

temperate zone: the mild climate zone between the Tropic of Cancer and the Arctic Circle and the zone between the Tropic of Capricorn and the Antarctic Circle.

terrapin: class Reptilia, order Testudines, family Emydidae, genus *Malaclemys*. Turtles found in freshwater and brackish water that are considered excellent eating.

terrarium: a cage used for keeping land animals.

terrestrial: an animal that lives on land.

Tertiary period: the geologic time period from 3 million to 70 million years ago, during which the first placental animals appeared, spread, and increased in numbers. Monkeys, apes, whales, large carnivores, and grazing animals also appeared during this period.

testosterone: a male sex hormone that is responsible for the development of secondary sex characteristics.

theologians: persons who study the theory and teachings of religion.

therapy: methods used in the treatment of a disease, disorder, or condition. Also methods used to maintain good health.

thorough: describes a dog that works every bit of ground and cover.

thrombin: a substance made from cattle blood that aids in the clotting of blood and is used in skin grafting.

throwbacks: an offspring that shows an ancestor's characteristics that have not appeared in recent generations.

ticked coat: darker colors are found on the tips of each hair of the animal's coat.

ticking: a fur coat that has hairs with darker tips mixed in.

timid: showing lack of confidence, somewhat afraid.

tom: male cat.

tortoise shell: two colors producing a spotted or blotched pattern.

toxemia: a condition in which there is a buildup of toxins in the blood.

toxicologist: scientist who studies the effects of toxic substances on humans, animals, and plants.

toxicology: the science that deals with poisons and their effects on an organism.

toxin: a poisonous substance produced by an organism.

trace minerals: see *microminerals*.

traits: characteristics or qualities of an animal that are usually inherited.

tranquil: quiet and peaceful.

tranquilizers: substances used to reduce mental anxiety or tension.

Triassic period: geologic time period from 180 to 230 million years ago, during which the dinosaurs appeared along with mammal-like reptiles.

trinomial nomenclature: adds a third word to the binomial nomenclature for an organism that has a subspecies name.

trophozoites: the growing stage of a protozoan parasite distinguished from the dormant or reproductive stage.

tympanum: a membrane that picks up vibrations and transfers them to the inner ear.

Tyzzer's disease: a disease primarily of rabbits caused by the bacterium *Bacillus piliformis*. The disease causes severe diarrhea and death among young rabbits.

U

udder development: the natural and normal growth of the mammary system of female animals.

ultrasonic: frequencies above what can be detected by the human ear; usually above 20,000 cycles per second.

ultraviolet: a portion of light rays that are beyond the visible violet spectrum because the wavelengths are shorter.

umbilical cord: cord that connects the fetus with the placenta to carry food to and remove waste from the fetus.

underfur: the very dense, soft, short fur that is beneath the longer, coarser guard hairs.

underwool: the short, dense, soft fur underneath the longer fur of Giant Angora rabbits.

undulating: moving in a wavy form or in a series of waves similar to the way snakes move.

unethical: a belief that something is going against one's moral principles and values.

unprovoked: an animal bite or attack in which the animal was not aroused or incited.

unthrifty: not thriving, flourishing, or growing vigorously; showing impaired performance.

uremic poisoning: a condition that occurs with severe kidney disease. Toxic substances that are normally removed from the blood by the kidneys build up and cause a toxic condition.

uterus: the organ in the female mammal in which the young is nourished and develops.

V

vain: appearing conceited, overconfident, or showing excessive pride.

vegetarian: person who lives on a diet primarily of vegetables, fruits, grains, and nuts.

veil: refers to the jet-black tips of the fur on black chinchilla mutations called Black Velvets.

vent: the external opening of the cloaca.

ventral: the underside or belly of an animal.

ventricle: part of the heart that receives blood from the atria and pumps blood into the arteries.

ventriculus: commonly called the gizzard, the largest organ of a bird's digestive system; it grinds and crushes food, which then enters the small intestine.

vermin: small, destructive animals such as wild mice and rats.

versatile: having a wide range of skills.

vertebrate: animals that have a segmented spinal column.

veterinarian: Doctors of Veterinary Medicine who treat and control animal injuries and diseases. They immunize healthy animals, inspect animals and meat used as food, perform surgery, set broken bones, establish diet and exercise routines, and prescribe medicines.

veterinary technician: a person who works under the supervision of the veterinarian to soothe and quiet animals under treatment, draw blood, insert catheters, and conduct laboratory tests.

veterinary technologist: the next step in the advancement of a veterinary technician taken by completing 2 more years of college and receiving a bachelor's degree.

vicious: overly aggressive, ferocious, or violent.

vigilant: always alert and watchful.

vigor: healthy growth; vitality; physical or mental strength.

viruses: infective agents that can cause diseases. They are only capable of growth and multiplying in living cells.

vitamins: organic compounds needed in small amounts by an animal body for various chemical reactions to occur.

vivariums: cages in which the environment is duplicated as close as possible to the natural environment of the species.

viviparous: animals that retain the young within their bodies in a preplacenta sac.

vivisection: research consisting of surgical operations and experiments to study the structure and function of organs.

vogue: something that is popular, fashionable, or in style at a given period of time.

W

waltzing mice: mice with a genetic defect that causes them to lose their sense of hearing and balance; they are unable to walk in a straight line and dance in circles.

wary: showing caution, being somewhat afraid.

wet tail: hamster enteritis; symptoms include wetness around the tail and rear area of the animal caused by runny diarrhea.

whelping: the act of dogs giving birth.

wholesale: selling animals in large numbers, usually for resale.

wildlife biologist: biologists who study the habitats and the conditions necessary for the survival of birds and other wildlife.

witty: intelligent, clever, or quick to learn.

Y

yeast: a type of fungus that causes fermentation and is used in the making of alcoholic beverages and baking bread.

Z

zoo administrators: persons who are employed at zoos throughout the country. There are zoo directors, animal curators, veterinarians, and resident zoologists. Their job is to provide services for the entertainment and education of the public.

zoo curators: the chief employees responsible for the day-to-day care of a zoo.

zoo directors: administrators who coordinate the business affairs of a zoo.

zookeepers: persons who provide the day-to-day care for animals in zoological parks.

zoologists: biologists who study animals.

zoology: the study of animals.

zoonoses: diseases that can be transmitted from animals to humans; rabies is an example.

APPENDIX A

COLLEGES OF VETERINARY MEDICINE

U.S. Colleges of Veterinary Medicine accredited by the American Veterinary Medical Association

Alabama:

Auburn University
College of Veterinary Medicine
104 J. E. Greene Hall
Auburn University, AL 36849-5517
(334) 844-4546
Status: Full Accreditation
Last Evaluation: 2005
Next Evaluation: 2012
SAVMA Chapter: Yes
http://www.vetmed.auburn.edu

Tuskegee University
School of Veterinary Medicine
Tuskegee, AL 36088
(334) 727-8174
Status: Limited Accreditation
Last Evaluation: 2006
Next Evaluation: 2008
SAVMA Chapter: Yes
http://tuskegee.edu

California:

University of California
School of Veterinary Medicine
Davis, CA 95616-8734
(530) 752-1360
Status: Full Accreditation
Last Evaluation: 2004
Next Evaluation: 2011
SAVMA Chapter: Yes
http://www.vetmed.ucdavis.edu

Western University of Health Sciences
College of Veterinary Medicine
309 E Second Street - College Plaza
Pomona, CA, 91766-1854
(909) 469-5628
Status: Limited Accreditation
Last Evaluation: 2008
Next Evaluation: 2010
http://www.westernu.edu/cvm.html

Colorado:	Colorado State University
College of Veterinary Medicine
and Biomedical Sciences
Fort Collins, CO 80523-1601
(970) 491-7051
Status: Full Accreditation
Last Evaluation: 2001
Next Evaluation: 2008
SAVMA Chapter: Yes
http://www.cvmbs.colostate.edu

Florida:	University of Florida
College of Veterinary Medicine
PO Box 100125
Gainesville, FL 32610-0125
(352) 392-2213
Status: Full Accreditation
Last Evaluation: 2001
Next Evaluation: 2008
SAVMA Chapter: Yes
http://www.vetmed.ufl.edu

Georgia:	University of Georgia
College of Veterinary Medicine
Athens, GA 30602
(706) 542-3461
Status: Full Accreditation
Last Evaluation: 2006
Next Evaluation: 2013
SAVMA Chapter: Yes
http://www.vet.uga.edu

Illinois:	University of Illinois
College of Veterinary Medicine
2001 South Lincoln Avenue
Urbana, IL 61802
(217) 333-2760
Status: Full Accreditation
Last Evaluation: 2006
Next Evaluation: 2013
SAVMA Chapter: Yes
http://www.cvm.uiuc.edu

Indiana:	Purdue University
School of Veterinary Medicine
1240 Lynn Hall
West Lafayette, IN 47907-1240
(765) 494-7607
Status: Full Accreditation
Last Evaluation: 2004
Next Evaluation: 2011
SAVMA Chapter: Yes
http://www.vet.purdue.edu

Iowa:	Iowa State University
College of Veterinary Medicine
Ames, IA 50011

(515) 294-1242
Status: Full Accreditation
Last Evaluation: 2006
Next Evaluation: 2010
SAVMA Chapter: Yes
http://www.vetmed.iastate.edu

Kansas:	Kansas State University
College of Veterinary Medicine
Manhattan, KS 66506
(785) 532-5660
Status: Full Accreditation
Last Evaluation: 2003
Next Evaluation: 2010
SAVMA Chapter: Yes
http://www.vet.ksu.edu

Louisiana:	Louisiana State University
School of Veterinary Medicine
Baton Rouge, LA 70803-8402
(225) 578-9900
Status: Limited Accreditation
Last Evaluation: 2005
Next Evaluation: 2012
SAVMA Chapter: Yes
http://www.vetmed.lsu.edu

Massachusetts:	Tufts University
School of Veterinary Medicine
200 Westboro Road
North Grafton, MA 01536
(508) 839-5302
Status: Full Accreditation
Last Evaluation: 2004
Next Evaluation: 2011
SAVMA Chapter: Yes
http://www.tufts.edu/vet

Michigan:	Michigan State University
College of Veterinary Medicine
G-100 Veterinary Medical Center
East Lansing, MI 48824-1314
(517) 355-6509
Status: Full Accreditation
Last Evaluation: 2005
Next Evaluation: 2012
SAVMA Chapter: Yes
http://cvm.msu.edu

Minnesota:	The University of Minnesota
College of Veterinary Medicine
1365 Gortner Avenue
St. Paul, MN 55108
(612) 624-9227
Status: Full Accreditation
Last Evaluation: 2007
Next Evaluation: 2014

SAVMA Chapter: Yes
http://www.cvm.umn.edu

Mississippi: Mississippi State University
College of Veterinary Medicine
Mississippi State, MS 39762
(662) 325-3432
Status: Full Accreditation
Last Evaluation: 2007
Next Evaluation: 2014
SAVMA Chapter: Yes
http://www.cvm.msstate.edu

Missouri: University of Missouri-Columbia
College of Veterinary Medicine
Columbia, MO 65211
(573) 882-3877
Status: Full Accreditation
Last Evaluation: 2006
Next Evaluation: 2013
SAVMA Chapter: Yes
http://www.cvm.missouri.edu

New York: Cornell University
College of Veterinary Medicine
Ithaca, NY 14853-6401
(607) 253-3700
Status: Full Accreditation
Last Evaluation: 2003
Next Evaluation: 2010
SAVMA Chapter: Yes
http://www.vet.cornell.edu

North Carolina: North Carolina State University
College of Veterinary Medicine
4700 Hillsborough Street
Raleigh, NC 27606
(919) 513-6210
Status: Full Accreditation
Last Evaluation: 2007
Next Evaluation: 2014
SAVMA Chapter: Yes
http://www.cvm.ncsu.edu/

Ohio: The Ohio State University
College of Veterinary Medicine
1900 Coffey Road
Columbus, OH 43210-1092
(614) 292-1171
Status: Full Accreditation
Last Evaluation: 2006
Next Evaluation: 2013
SAVMA Chapter: Yes
http://www.vet.ohio-state.edu

Oklahoma: Oklahoma State University
Center for Veterinary Health Sciences

Stillwater, OK 74078
(405) 744-6595
Status: Full Accreditation
Last Evaluation: 2003
Next Evaluation: 2010
SAVMA Chapter: Yes
http://www.cvm.okstate.edu

Oregon: Oregon State University
College of Veterinary Medicine
Corvallis, OR 97331
(541) 737-2098
Status: Full Accreditation
Last Evaluation: 2007
Next Evaluation: 2014
SAVMA Chapter: Yes
http://www.vet.orst.edu

Pennsylvania: University of Pennsylvania
School of Veterinary Medicine
3800 Spruce Street
Philadelphia, PA 19104-6044
(215) 898-5438
Status: Full Accreditation
Last Evaluation: 2002
Next Evaluation: 2009
SAVMA Chapter: Yes
http://www.vet.upenn.edu

Tennessee: University of Tennessee
College of Veterinary Medicine
2407 River Drive
Knoxville, TN 37996
(865) 974-7262
Status: Full Accreditation
Last Evaluation: 2001
Next Evaluation: 2008
SAVMA Chapter: Yes
http://www.vet.utk.edu

Texas: Texas A&M University
College of Veterinary Medicine &
Biomedical Sciences
College Station, TX 77843-4461
(979) 845-5051
Status: Full Accreditation
Last Evaluation: 2001
Next Evaluation: 2008
SAVMA Chapter: Yes
http://www.cvm.tamu.edu

Virginia-
Maryland:
Virginia Tech
Virginia-Maryland Regional
College of Veterinary Medicine
Blacksburg, VA 24061-0442
(540) 231-7666
Status: Full Accreditation

Last Evaluation: 2007
Next Evaluation: 2014
SAVMA Chapter: Yes
http://www.vetmed.vt.edu

Washington: Washington State University
College of Veterinary Medicine
Pullman, WA 99164-7010
(509) 335-9515
Status: Full Accreditation
Last Evaluation: 2003
Next Evaluation: 2010
SAVMA Chapter: Yes
http://www.vetmed.wsu.edu

Wisconsin: University of Wisconsin-Madison
School of Veterinary Medicine
2015 Linden Drive West
Madison, WI 53706-1102
(608) 263-6716
Status: Full Accreditation
Last Evaluation: 2001
Next Evaluation: 2008
SAVMA Chapter: Yes
http://www.vetmed.wisc.edu

Canada:
Calgary, Alberta: University of Calgary
Faculty of Veterinary Medicine
Calgary, Alberta Canada T2N 4N1
Phone: 403-210-3961
Status: Letter of Reasonable Assurance
Last Evaluation: 2007
Next Evaluation: to be determined
http://www.vet.ucalgary.ca

Ontario: University of Guelph
Ontario Veterinary College
Guelph, ON N1G 2W1
CANADA
Phone: 519-824-4120
Status: Full Accreditation
Last Evaluation: 2002
Next Evaluation: 2009
SAVMA Chapter: No
http://www.ovc.uoguelph.ca/

Prince Edward   University of Prince Edward Island
Island: Atlantic Veterinary College
550 University Avenue
Charlottetown, PE C1A 4P3
CANADA
Phone: 902-566-0882
Status: Full Accreditation
Last Evaluation: 2004
Next Evaluation: 2011
SAVMA Chapter: Yes
http://www.upei.ca/~avc

Quebec: Université de Montréal
Faculté de médecine vétérinaire
C.P. 5000
Saint Hyacinthe, PQ J2S 7C6
CANADA
Phone: 450-773-8521
Status: Full Accreditation
Last Evaluation: 2006
Next Evaluation: 2012
SAVMA Chapter: No
www.medvet.umontreal.ca

Saskatchewan: University of Saskatchewan
Western College of Veterinary Medicine
52 Campus Drive
Saskatoon, SK S7N 5B4
CANADA
Phone: 306-966-7447
Status: Full Accreditation
Last Evaluation: 2003
Next Evaluation: 2010
SAVMA Chapter: No
http://www.usask.ca/wcvm

Foreign:
Australia: Murdoch University
Division of Veterinary and Biomedical
Sciences
Murdoch
WESTERN AUSTRALIA
Phone: 61 8 9360 2566
Status: Full Accreditation
Last Evaluation: 2002
Next Evaluation: 2009
http://www.vetbiomed.murdoch.edu.au/

The University of Melbourne
Faculty of Veterinary Science
Werribee, VIC Australia
Phone: 61 3 9731 2261
Status: Full Accreditation
Last Evaluation: 2006
Next Evaluation: 2013
http://www.vet.unimelb.edu.au

The University of Sydney
Faculty of Veterinary Science
NSW
AUSTRALIA
Phone: 61 2 9351 6936
Status: Full Accreditation
Last Evaluation: 2005
Next Evaluation: 2012
http://www.vetsci.usyd.edu.au/

England: University of London
The Royal Veterinary College
Royal College Street

London NW1 OTU
ENGLAND
Phone: 44 (0)20 7468-5000
Status: Full Accreditation
Last Evaluation: 2005
Next Evaluation: 2012
http://www.rvc.ac.uk

Ireland:
University College Dublin
School of Agriculture, Food Science &
   Veterinary Science
Belfield, Dublin 4 Ireland
Phone: 353 1 716 6100
Status: Full Accreditation
Last Evaluation: 2007
Next Evaluation: 2014
http://www.ucd.ie/agfoodvet

The Netherlands:
State University of Utrecht
Faculty of Veterinary Medicine
PO Box 80.163
3508 TD Utrecht
THE NETHERLANDS
Phone: 31 30 253-4851
Status: Full Accreditation
Last Evaluation: 2007
Next Evaluation: 2014
http://www.vet.uu.nl

New Zealand:
Massey University College of Sciences
Institute of Veterinary, Animal, and
   Biomedical Sciences

Palmerston North
NEW ZEALAND
Phone: 64 6 350-5714
Status: Full Accreditation
Last Evaluation: 2007
Next Evaluation: 2014
http://ivabs.massey.ac.nz

Scotland:
University of Glasgow
Faculty of Veterinary Medicine
Glasgow, Scotland G61 1QH
Phone: 44 (0)141 330-5700
Status: Full Accreditation
Last Evaluation: 2006
Next Evaluation: 2013
http://www.gla.ac.uk/faculties/vet

The University of Edinburgh
Royal (Dick) School of Veterinary Studies
Summerhall
Edinburgh EH9 1QH
SCOTLAND
Phone: 44 131 650-1000
Status: Full Accreditation
Last Evaluation: 2001
Next Evaluation: 2008
http://www.vet.ed.ac.uk

Reprinted with permission of the American Veterinary Medical Association 1931 North Meacham Road, Suite 100, Schaumberg, Illinois 60173-4360.

APPENDIX B

VETERINARY TECHNOLOGY PROGRAMS

American Veterinary Medical Association Programs in Veterinary Technology

Programs Accredited by the AVMA Committee on Veterinary Technician Education and Activities (CVTEA)
1931 North Meacham Road, Suite 100, Schaumburg, Illinois 60173-4360
Phone (708) 925-8070

Alabama:
Jefferson State Community College
Veterinary Technology Distance Learning Program
2601 Carson Road
Birmingham, AL 35215-3098
205-856-8519
www.jeffstateonline.com/vet_tech/
(Christie Wallace, DVM Director)
Initial Accreditation-March 16, 2007
PROVISIONAL ACCREDITATION

Arizona:
Anthem College
1515 E. Indian School Road
Phoenix, AZ 85014
602-279-9700
www.anthem.edu
(Jeremy Prupas, DVM-Director)
Associate in Science
Initial Accreditation-March 3, 2008
PROVISIONAL ACCREDITATION

Kaplan College
Veterinary Technology Program
13610 N. Black Canyon Hwy., Suite 104
Phoenix, AZ 85029
602-548-1955
http://getinfo.kaplancollege.com/KaplanCollegePortal/
KaplanCollegeCampuses/Arizona/PhoenixEast/
(James Flom, DVM, Program Director)
Associate of Occupational Studies
Initial Accreditation-August 19, 2005
PROVISIONAL ACCREDITATION

Mesa Community College
Veterinary Technology/Animal Health Program
1833 W. Southern Avenue
Mesa, AZ 85202
480-461-7488
www.mc.maricopa.edu
(Jill Sheport, DVM Director)
2 year Associate in Applied Science

Initial Accreditation February 3, 2006
PROVISIONAL ACCREDITATION

Penn Foster College
Veterinary Technician Distance Education
    Program
14624 N. Scottsdale Road, Suite 310
Scottsdale, AZ 85254
800-275-4410
www.pennfostercollege.edu
(Margi Sirois, Ed.D. Director)
Associate of Science
Initial Accreditation January 5, 2006
PROVISIONAL ACCREDITATION

Pima County Community College
Veterinary Technology Program
8181 E. Irvington Road
Tucson, AZ 85709-4000
520-206-7414
FAX 520-206-7803
www.pima.edu
(Daniel DeNoon, DVM-Director)
2 years-Associate in Applied Science
Initial Accreditation-January 29, 2003
FULL ACCREDITATION

California: California State Polytechnic University-Pomona
College of Agriculture
Animal Health Technology Program
3801 W. Temple Ave.
Pomona, CA 91768
909-869-2136
www.csupomona.edu
(James Alderson, DVM-Coordinator)
4 years-Bachelor of Science
Initial Accreditation-April, 1996
FULL ACCREDITATION

Cosumnes River College
Veterinary Technology Program
8401 Center Pkwy.
Sacramento, CA 95823
916-691-7355
http://www.crc.losrios.edu
(Christopher Impinna, DVM-Director)
2 years-Associate in Science
Initial Accreditation-April, 1975
FULL ACCREDITATION

Foothill College
Veterinary Technology Program
12345 El Monte Rd.
Los Altos Hills, CA 94022
650-949-7203
FAX 650-949-7375
www.foothill.fhda.edu

(Karl Peter, DVM-Director)
2 years-Associate in Science
Initial Accreditation-April, 1977
FULL ACCREDITATION

Hartnell College
Animal Health Technology Program
156 Homestead Ave.
Salinas, CA 93901
831-755-6855
www.hartnell.cc.ca.us
(Sharon McKinney-Radel, RVT-Director)
2 years-Associate in Science
Initial Accreditation-April, 1981
FULL ACCREDITATION

Los Angeles Pierce College
Veterinary Technology Program
6201 Winnetka Ave.
Woodland Hills, CA 91371
818-347-0551
www.macrohead.com/rvt
(Elizabeth White, RVT-Director)
2 years-Associate in Science
Initial Accreditation-December, 1975;
    April, 1993
FULL ACCREDITATION

Mt. San Antonio College
Animal Health Technology Program
1100 N. Grand Ave.
Walnut, CA 91789
909-594-5611
www.mtsac.edu
(Jean Hoffman, RVT-Director)
2 years-Associate in Science
Initial Accreditation-April, 1977
PROBATIONAL ACCREDITATION

Western Career College-Citrus Heights Campus
7301 Greenback Lane, Suite A
Citrus Heights, CA 95621
916-722-8200
www.westerncollege.edu
Ms. Patti Hebert, RVT, Program Director
pattih@westemcollege.com
Associate in Science
Initial Accreditation-June 16, 2006
FULL ACCREDITATION

Western Career College-Pleasant Hill
Veterinary Technician Program
380 Civic Drive, #300
Pleasant Hill, CA 94523
925-609-6650
www.westerncollege.com/ph_campus.asp
(Andrea Monroe, DVM-Director)

2 years-Associate in Science
Initial Accreditation-March 31, 2004
PROVISIONAL ACCREDITATION

Western Career College-Sacramento
Veterinary Technology Program
8909 Folsom Blvd.
Sacramento, CA 95826
916-361-1660
www.westerncollege.com/ph_campus.asp
(Janelle Emmett, DVM-Program Director)
18 months-Associate in Science
Initial Accreditation-August 20, 2004
PROVISIONAL ACCREDITATION

Western Career College-San Jose
Veterinary Technician Education Program
6201 San Ignacio Ave.
San Jose, CA 95119
408-360-0840
www.westerncollege.edu
(Barbara McManus, RVT Interim Director)
18 months-Associate in Science
Initial Accreditation-December 1, 2006
PROVISIONAL ACCREDITATION

Western Career College-San Leandro
Veterinary Technology Program
170 Bayfair Mall
San Leandro, CA 94578
510-276-3888
www.westerncollege.com/ph_campus.asp
(Laura Jamison, RVT-Program Director)
18 months-Associate in Science
Initial Accreditation-August 18, 2004
PROBATIONAL ACCREDITATION

Western Career College-Stockton Campus
Veterinary Technician Education Program
1313 West Robinhood Drive, Suite B
Stockton, CA 95207
209-956-1240 × 44116
www.westerncollege.edu
(Jamie Larson, RVT, Interim Program Director)
tyeatts@westerncollege.com
18 months-Associate in Science
Initial Accrediation-June 14, 2006
PROBATIONAL ACCREDITATION

Yuba College
Veterinary Technology Program
2088 N. Beale Rd.
Marysville, CA 95901
530-741-6962
www.yccd.edu/yuba/vettech/index.html
(Scott Haskell, DVM-Director)
2 Years-Associate in Science

Initial Accreditation-April, 1978
FULL ACCREDITATION

Canada:

Ridgetown College, University of Guelph
Veterinary Technology Program
Main Street East
Ridgetown, Ontario N0P 2C0
519-674-1666
www.ridgetownc.on.ca
(Dr. Irene Moore, Program Director)
Diploma
Initial Accreditation-February 4, 2005
FULL ACCREDITATION

AHT/VT Programs in Canada

Canadian recognition

At its June 2006 meeting, the AVMA Executive Board approved a recommendation that the AVMA recommends that veterinary technician credentialing (i.e., licensing, registration, or certification) entities in the US recognize graduates of Canadian Veterinary Medical Association (CVMA)-accredited veterinary technology programs as eligible for credentialing. In turn, the CVMA recommends that Canadian provincial licensing bodies recognize graduates of AVMA-accredited veterinary technology programs as being eligible for licensure. As always, eligibility for licensure/registration/certification of veterinary technicians is the purview of each state and provincial credentialing agency.

Colorado:

Bel-Rea Institute of Animal Technology
1681 S. Dayton St.
Denver, CO 80231
800-950-8001
www.bel-rea.com
(Nolan Rucker, DVM-President)
2 years-Associate of Applied Science
Initial Accreditation-April, 1975
FULL ACCREDITATION

Colorado Mountain College
Veterinary Technology Program
Spring Valley Campus
3000 County Rd. 114
Glenwood Springs, CO 81601
970-945-8691
www.coloradomtn.edu
(Jeff Myers, DVM-Coordinator)
2 years-Associate of Applied Science
Initial Accreditation-December, 1975
FULL ACCREDITATION

Community College of Denver
Veterinary Technology Program

1070 Alton Way, Bldg. 849
Denver, CO 80230
303-365-8300
www.ccd.rightchoice.org
(Shannon Burkhalter, CVT-Director)
2 years-Associate of Applied Science
Initial Accreditation-March, 2002
FULL ACCREDITATION

Front Range Community College
Veterinary Research Technology Program
4616 S. Shields
Ft. Collins, CO 80526
970-226-2500
www.frcc.cc.co.us
(Elizabeth Bauer, DVM-Director)
2 years-Associate of Applied Science
Initial Accreditation-April, 1996
FULL ACCREDITATION

Connecticut: NW Connecticut Community College
Veterinary Technology Program
Park Place East
Winsted, CT 06098
860-738-6490
www.nwctc.commnet.edu/vettech/
(Sheryl Keeley, CVT or Nancy Marchetti,
  DVM-Coordinator)
2 years-Associate in Science
Initial Accreditation-March, 2000
FULL ACCREDITATION

Quinnipiac University
Veterinary Technology Program
Mt. Carmel Ave.
Hamden, CT 06518
203-582-8958
www.quinnipiac.edu
(Steve Carleton, DVM-Director)
4 years-Bachelor of Science
Initial Accreditation-April, 1980
FULL ACCREDITATION

Delaware: Delaware Technical and Community College
Veterinary Technology Program
PO Box 610, Route 18
Georgetown, DE 19947
302-855-5918
FAX 302-858-5460
www.dtcc.edu
(Valerie Quillen, DVM-Department Chair)
2 years-Associate in Science
Initial Accreditation-May 1, 2002
FULL ACCREDITATION

Florida: Brevard Community College
Veterinary Technology Program

1519 Clearlake Rd.
Cocoa, FL 32922
321-433-7594
FAX 321-634-4565
www.brevard.cc.fl.us
(Laura Earle-lmre, DVM-Director)
2 years-Associate in Science
Initial Accreditation-November, 1999
FULL ACCREDITATION

Miami-Dade College
Veterinary Technology Program
Medical Center Campus
950 NW 20th Street
Miami, FL 33127
305-237-4473
FAX 305-237-4278
www.mdc.edu/medical/AHT/Vet/
  default.asp
(Lois Sargent, DVM, Coordinator)
2 years-Associate in Science
Initial Accreditation-February 21, 2003
PROVISIONAL ACCREDITATION

St. Petersburg College
Veterinary Technology Program
Box 13489
St. Petersburg, FL 33733
727-341-3652
www.spcollege.edu/hec/vt/
(Richard Flora, DVM-Director)
On-Campus Program
2 years-Associate in Science
Initial Accreditation-April, 1978
FULL ACCREDITATION

Distance Learning Program
Associate in Science
Bachelor of Science
www.spcollege.edu/bachelors/vtech.php?
  program=vtech
Initial Accreditation-November, 1995
FULL ACCREDITATION

Georgia: Athens Technical College
Veterinary Technology Program
800 US Highway 29N
Athens, GA 30601
706-355-5107
www.athenstech.edu
(Carole Miller, DVM-Director)
7 Quarters-Associate of Applied Science
Initial Accreditation-February 28, 2003
FULL ACCREDITATION

Fort Valley State University
Veterinary Technology Program

1005 State University Drive
Fort Valley, GA 31030
478-825-6353
FAX 478-827-3023
www.fvsu.edu/
(Seyedmehdi Mobini, DVM-Interim
    Director)
2 years-Associate of Applied Science
4 years-Bachelor of Science
Initial Accreditation-November, 1978; (4 yr)
    September, 2001
FULL ACCREDITATION

Gwinnett Technical College
Veterinary Technology Program
5150 Sugarloaf Pkwy.
Lawrenceville, GA 30043
770-962-7580
FAX 770-962-7985
www.gwinnetttechnicalcollege.com
(Bonnie Ballard, DVM-Director)
2 years-Associate of Applied Science
Initial Accreditation-March, 2000
FULL ACCREDITATION

Ogeechee Technical College
Veterinary Technology Program
1 Joe Kennedy Blvd.
Statesboro, GA 30458
912-688-6037
(Dr. Janice Grumbles, Program Director)
Initial Accreditation-August 12, 2005
Associate of Applied Science
PROVISIONAL ACCREDITATION

Idaho: College of Southern Idaho
Veterinary Technology Program
315 Falls Ave.
Twin Falls, ID 83303-1238
208-733-9554 ext. 2408
FAX 208-736-2136
www.csi.cc.id.us
(Jody Rockett, DVM-Director)
2 years-Associate of Applied Science
Initial Accreditation-March, 1999
FULL ACCREDITATION

Illinois: Parkland College
Veterinary Technology Program
2400 W. Bradley Ave.
Champaign, IL 61821
217-351-2224
www.parkland.edu
(Paul Cook, DVM-Director)
2 years-Associate of Applied Science
Initial Accreditation-November, 1976
FULL ACCREDITATION

Joliet Junior College
Agriculture Sciences Department
1215 Houbolt Road
Joliet, IL 60431
815-280-2746
FAX 815-280-2741
www.jjc.cc.il.us
(R. Scott Keller, DVM-Coordinator)
2 years-Associate of Applied Science
Initial Accreditation-March 2001
FULL ACCREDITATION

Vet Tech Institute at Fox College
18020 Oak Park Ave.
Tinley Park, IL 60477
(Dr. Jordan Siegel, Program Director)
708-636-7700
www.foxcollege.edu
16 months-Associate of Applied Science
Initial Accreditation-October 22, 2007
PROVISIONAL ACCREDITATION

Indiana: Purdue University
School of Veterinary Medicine
Veterinary Technology Program
West Lafayette, IN 47907
765-494-7619
www.vet.purdue.edu/vettech/index.html
(Pete Bill, DVM-Director)
On-Campus Program
2 years-Associate of Science
4 years-Bachelor of Science
Initial Accreditation-November, 1976
FULL ACCREDITATION

Distance Learning Program
Associate of Science
Initial Accreditation-July 31, 2002
FULL ACCREDITATION

The Vet Tech Institute at International Business
    College at Fort Wayne
Veterinary Technology Program
5699 Coventry Lane
Fort Wayne, IN 46804
260-459-4500
www.vettechinstitute.edu/camp_wayne.php
(Stephanie Coates, DVM-Director)
16 months Associate of Applied Science
Initial Accreditation-February 23, 2007
PROVISIONAL ACCREDITATION

The Vet Tech Institute at International Business
    College-Indianapolis
7205 Shadeland Station
Indianapolis, IN 46804
317-813-2300

www.ibcindianapolis.edu
(Peg Villanueva, DVM-Director)
Associate of Applied Science
Initial Accreditation-February 29, 2008
PROVISIONAL ACCREDITATION

Iowa: Des Moines Area Community College
Veterinary Technology Program
2805 SW Snyder Dr., Suite 505
Ankeny, IA 50023
800-362-2127
www.dmacc.edu
(Frank Cerfogli, DVM-Program Director)
2 years-Associate of Applied Science
Initial Accreditation-March 14, 2005
PROBATIONAL ACCREDITATION

Iowa Western Community College
Veterinary Technology Program
2700 College Road, Box 4-C
Council Bluffs, IA 51502
712-325-3431
http://iwcc.cc.ia.us/programs/departments/
    veterinary-tech.asp
(Dan DeWitt, DVM-Program Director)
2 years-Associate of Applied Science
Initial Accreditation-March 2, 2007
PROVISIONAL ACCREDITATION

Kirkwood Community College
Animal Health Technology Program
6301 Kirkwood Blvd., SW
Cedar Rapids, IA 52406
319-398-4978
www.kirkwood.edu
(Anne Duffy, RVT-Coordinator)
2 years-Associate of Applied Science
Initial Accreditation-April, 1989
FULL ACCREDITATION

Kansas: Colby Community College
Veterinary Technology Program
1255 S. Range
Colby, KS 67701
785-460-5466
http://www.colbycc.edu/?m=4&s=80&l=163
(Jennifer Martin, DVM-Director)
2 years-Associate of Applied Science
Initial Accreditation-November, 1974
FULL ACCREDITATION

Kentucky: Morehead State University
Veterinary Technology Program
25 MSU Farm Dr.
Morehead, KY 40351
606-783-2326
www.morehead-st.edu

(Scott Rundell, DVM-Coordinator)
2 years-Associate of Applied Science
Initial Accreditation-November, 1977
FULL ACCREDITATION

Murray State University
Animal Health Technology Program
Department of Agriculture
100 AHT Center
Murray, KY 42071
270-762-7001
breathitt.murraystate.edu/bvc/
(Terry Canerdy, DVM-Director)
4 years-Bachelor in Science
Initial Accreditation-November, 1986
FULL ACCREDITATION

Louisiana: Delgado Community College
615 City Park Avenue
Building 4, Room 301
New Orleans, LA 70119-4399
504-671-6234
http://www.dcc.edu/campus/cp/ahealth/vet_tech
Dr. James Hurrell, Program Director
Associate of Applied Science
PROVISIONAL ACCREDITATION effective
    September 22, 2006

Northwestern State University of Louisiana
Veterinary Technology Program
Department of Life Sciences
225 Bienvenu Hall
Natchitoches, LA 71497
318-357-5323
www.nsula.edu
(Brenda Woodard, DVM-Director)
2 years-Associate in Science
Initial Accreditation-April, 1981
FULL ACCREDITATION

Maine: University College of Bangor
Veterinary Technology Program
85 Texas Avenue, 217 Belfast Hall
Bangor, ME 04401-4367
207-262-7852
www.uma.maine.edu/bangor
(Robert Gholson, DVM, Director)
2 years-Associate in Science
Initial Accreditation-March, 2002
FULL ACCREDITATION

Maryland: Essex Campus of the Community College
Of Baltimore County
Veterinary Technology Program
7201 Rossville Blvd.
Baltimore, MD 21237
410-780-6306

FAX 410-780-6145
www.ccbcmd.edu
(Jack Stewart, RVT-Coordinator)
2 years-Associate of Applied Science
Initial Accreditation-April, 1978
FULL ACCREDITATION

Massachusetts: Becker College
Veterinary Technology Program
964 Main Street
Leicester, MA 01524
508-791-9241
FAX 508-892-8155
www.beckercollege.com
(James Q. Knight, DVM-Director)
2 years-Associate in Science
4 years-Bachelor of Science
Initial Accreditation-November, 1978
FULL ACCREDITATION

Holyoke Community College
Veterinary Technician Program
303 Homestead Ave.
Holyoke, MA 01040-1099
413-538-7000
www.hcc.mass.edu
(Walter Jaworski, DVM-Coordinator)
2 years-Associate in Science
Initial Accreditation-April, 1989
FULL ACCREDITATION

Mount Ida College
Veterinary Technology Program
777 Dedham St.
Newton, MA 02459
617-928-4545
FAX 617-928-4072
www.mountida.edu
(Tracy Blais, CVT-Director)
3 years-Associate in Arts
4 years-Bachelor of Science
Initial Accreditation-November, 1985
FULL ACCREDITATION

North Shore Community College
Veterinary Technology Program
1 Femcroft Road
Danvers, MA 01923
978-762-4000
www.northshore.edu
(Karen Komisar, DVM-Director)
2 years-Associate of Applied Science
Initial Accreditation-March 28, 2007
FULL ACCREDITATION

Michigan: Baker College of Cadillac
Veterinary Technology Program

9600 East 13th Street
Cadillac, Ml 49601
231-775-8458
www.baker.edu
(Tina Burke, DVM-Coordinator)
2 years-Associate of Applied Science
Initial Accreditation-May 9, 2003
PROVISIONAL ACCREDITATION

Baker College of Flint
Veterinary Technology Program
1050 W. Bristol Road
Flint, Ml 48507
800-964-4299 or 810-766-4153
FAX 810-766-2055
www.baker.edu
(Aaron Walton, DVM Director)
2 years-Associate of Applied Science
Initial Accreditation-November 20, 2002
FULL ACCREDITATION

Baker College of Jackson
Veterinary Technology Program
2800 Springport Road
Jackson, MI 49202
517-789-6123
www.baker.edu
(Katrina Bowers, LVT Director)
3 years-Associate of Applied Science
Initial Accreditation March 21, 2006
PROVISIONAL ACCREDITATION

Baker College of Muskegon
Veterinary Technology Program
1903 Marquette Avenue
Muskegon, MI 49442
800-937-0337 or 231-777-5275
FAX 231-777-5201
www.baker.edu
(Marcy O'Rourke, DVM-Department Chair)
2 years-Associate of Applied Science
Initial Accreditation-October 18, 2002
FULL ACCREDITATION

Baker College of Port Huron
Veterinary Technology Program
3403 Lapeer Road
Port Huron, MI 48060-2597
810-985-7000
www.baker.edu
(Norman Bayne, DVM-Director)
2 years-Associate of Applied Science
Initial Accreditation-February 2, 2007
PROVISIONAL ACCREDITATION

Macomb Community College
Veterinary Technician Program

Center Campus
44575 Garfield Rd.
Clinton Township, MI 48044
586-286-2096
FAX 586-286-2098
www.macomb.cc.mi.us
(Lori Renda-Francis, LVT-Director)
2 years-Associate of Applied Science
Initial Accreditation-April, 1979
FULL ACCREDITATION

Michigan State University
College of Veterinary Medicine
Veterinary Technology Program
A-10 Veterinary Medical Center
East Lansing, MI 48824
517-353-7267
www.cvm.msu.edu/vettech
(Helene Pazak, DVM-Director)
2 years-Certificate
4 years-Bachelor of Science
Initial Accreditation-August, 1973
FULL ACCREDITATION

Wayne County Community College District
Veterinary Technology Program
c/o Wayne State University
Div. of Laboratory Animal Resources
540 E. Canfield
Detroit, MI 48201
313-577-1156
FAX 313-577-5890
www.dlar.wayne.edu/vtp
(Karen Hrapkiewicz, DVM-Director)
2 years-Associate of Applied Science
Initial Accreditation-November, 1976
FULL ACCREDITATION

Minnesota: Argosy University
Twin Cities
Veterinary Technician Program
1515 Central Parkway
Eagan, MN 55121
888-844-2004
www.argosy.edu
(Paula Lind, CVT-Interim Director)
19 months-Associate of Applied Science
Initial Accreditation-November, 1981
FULL ACCREDITATION

Duluth Business University
Veterinary Technology Program
4724 Mike Colalillo Dr.
Duluth, MN 55807
800-777-8406
www.dbumn.edu
(Deserie Bremer, DVM-Program Director)

2 years-Associate of Applied Science
Initial Accreditation-August 6, 2004
PROVISIONAL ACCREDITATION

Globe University
Veterinary Technology Program
8089 Globe Drive
Woodbury, MN 55125
651-714-7360 or 800-231-0660
FAX 651-730-5151
http://www.msbcollege.edu/campus-locations/
    woodbury-mn/
(Brian Hoefs, DVM-Interim Program Director)
7 Quarters-Associate of Applied Science
Bachelor of Science
Initial Accreditation-February 14, 2003
PROVISIONAL ACCREDITATION

Minnesota School of Business-Blaine
Veterinary Technology Program
3680 Pheasant Ridge Dr. NE
Blaine, Minnesota 55449
763-225-8000
www.msbcollege.edu
(Mary Mortenson, CVT, Program Director)
llewandowski@msbcollege.edu
Associate of Applied Science
Bachelor of Science
PROVISIONAL ACCREDITATION effective
    June 2, 2006

Minnesota School of Business-Plymouth
Veterinary Technology Program
1455 County Road 101 North
Plymouth, MN 55447
763-476-2000
www.msbcollege.edu
(Carrie Padget, CVT-Interim Director)
2 years-Associate in Applied Science
Initial Accreditation-February 4, 2004
FULL ACCREDITATION

Minnesota School of Business-Rochester
2521 Pennington Dr. NW
Rochester, MN 55901
507-536-9500
www.msbcollege.edu
(Julie Legred, CVT-Director)
Associate of Applied Science
Initial Accreditation-February 15, 2008
PROVISIONAL ACCREDITATION

Minnesota School of Business-St. Cloud
Veterinary Technology Program
1201 2nd Street South
Waite Park, MN 56387
320-257-2000

www.msbcollege.edu
(Sarah Bjorstrom, CVT Director)
2 years-Associate of Applied Science
Initial Accreditation-January 12, 2007
PROVISIONAL ACCREDITATION

Minnesota School of Business-Shakopee
Veterinary Technology Program
1200 Shakopee Town Square
Shakopee, MN 55379
952-345-1200
Associate of Applied Science
(Ms. Teresa Reuteler, CVT; Program Director)
Initial Accreditation-June 8, 2005
PROVISIONAL ACCREDITATION

Ridgewater College
Veterinary Technology Dept.
2101 15th Ave., NW
Willmar, MN 56201
320-235-5114
FAX 320-222-8260
www.ridgewater.mnscu.edu
(Allen Balay, DVM-Director)
2 years-Associate of Applied Science
Initial Accreditation-November, 1993
FULL ACCREDITATION

Rochester Community and Technical College
Animal Health Technology Program
851 30th Avenue SE
Rochester, MN 55904-4999
800-247-1296
www.rctc.edu
(Dr. Kimberly Rowley, Program Director)
2 years-Associate of Applied Science
Initial Accreditation-January 28, 2005
PROVISIONAL ACCREDITATION

Mississippi: Hinds Community College
Veterinary Technology Program
1100 PMB 11160
Raymond, MS 39154
601-857-3456
FAX 601-857-3577
www.hindscc.edu/Departments/Agriculture/
   VeterinaryTech.aspx
(Bobby Glenn, DVM-Director)
2 years-Associate of Applied Science
Initial Accreditation-April, 1990
FULL ACCREDITATION

Missouri: Crowder College
601 LaClede Avenue
Neosho, MO 64850
417-455-5772
www.crowder.edu

Stephanie Watson, DVM, Program Director
swatson@crowder.edu
Associate of Applied Science
PROVISIONAL ACCREDITATION effective
   October 20, 2006

Jefferson College
Veterinary Technology Program
1000 Viking Dr.
Hillsboro, MO 63050
636-942-3000
FAX 636-789-4012
www.jeffco.edu
(Dana Nevois, RVT-Director)
2 years-Associate of Applied Science
Initial Accreditation-April, 1978
FULL ACCREDITATION

Maple Woods Community College
Veterinary Technology Program
2601 NE Barry Rd.
Kansas City, MO 64156
816-437-3235
FAX 816-437-3239
www.mcckc.edu/vettech
(Chris Morrow, DVM-Director)
2 years-Associate of Applied Science
Initial Accreditation-April, 1975
FULL ACCREDITATION

Sanford Brown College-St. Peters
Veterinary Technology Program
100 Richmond Center Blvd.
St. Peters, MO 63376
(Dr. Amy Wolff, Program Director)
888-769-2433
www.sanford-brown.edu
2 years-Associate of Applied Science
Initial Accreditation-August 31, 2007
PROVISIONAL ACCREDITATION

Sanford Brown College-Fenton
Veterinary Technology Program
1345 Smizer Mill Road
Fenton, MO 63026
(Ms. Angela Hutchinson, RVT)
888-769-2433
www.sanford-brown.edu
2 years-Associate of Applied Science
Initial Accreditation-October 11, 2007
PROVISIONAL ACCREDITATION

Nebraska: Nebraska College of Technical Agriculture
Veterinary Technology Program
RR3, Box 23A
Curtis, NE 69025
308-367-4124

www.ncta.unl.edu
(Barbara Berg, LVT-Department Chair)
2 years-Associate of Applied Science
Initial Accreditation-August 1973
FULL ACCREDITATION

Northeast Community College
Veterinary Technician Program
801 E. Benjamin Ave.
Norfolk, NE 68702-0469
402-371-2020
FAX-402-844-7400
www.northeastcollege.com
(Mike Cooper, DVM-Director)
2 years-Associate of Applied Science
Initial Accreditation-November, 1997
FULL ACCREDITATION

Vatterott College
Veterinary Technician Program
11818 I Street
Omaha, NE 68137-1237
402-392-1300
www.vatterott-college.com
(Joni Brunssen, DVM-Director)
23 months-Associate of Applied Science
Initial Accreditation-November, 1978
FULL ACCREDITATION

Nevada.   The College of Southern Nevada
Veterinary Technology Program
6375 W. Charleston Blvd.
Las Vegas, NV 89146-1164
702-651-5852
www.ccsn.nevada.edu
(Dennis Olsen, DVM Director)
2 years-Associate of Applied Science
Initial Accreditation March 8, 2006
FULL ACCREDITATION

Pima Medical Institute
Veterinary Technician Program
3333 E. Flamingo Road
Las Vegas, NV 89121
702-458-9650
www.pmi.edu
(Dennis Lopez, LVT Director)
15–18 month Occupational Associate
Initial Accreditation March 10, 2006
PROVISIONAL ACCREDITATION

Truckee Meadows Community College
Veterinary Technology Program
7000 Dandini Blvd.
Reno, NV 89512
775-850-4005
(Ms. Wendi Ford, LVT; Program Director)

Associate of Applied Science
Initial Accreditation-September 16, 2005
FULL ACCREDITATION

New Hampshire: Great Bay Community College (formerly
New Hampshire Community Technical
College)
Veterinary Technology Program
277 Portsmouth Ave.
Stratham, NH 03885-2297
603-772-1194
FAX 603-772-1198
www.greatbay.edu
(Margaret MacGregor, VMD-Director)
2 years-Associate in Science
Initial Accreditation-April, 1998
FULL ACCREDITATION

New Jersey: Camden County College
Animal Science Technology Program
P.O. Box 200
Blackwood, NJ 08012
856-227-7200
www.camdencc.edu
(Margaret Dorsey, VT-Coordinator)
2 years-Associate in Animal Technology
Initial Accreditation-November, 1978
FULL ACCREDITATION

Northern New Jersey Consortium for
Veterinary Technician Education
(Includes: Bergen Community College, County
College of Morris,
Sussex County Community College)
400 Paramus Road
Paramus, NJ 07652
201-612-5389
FAX 201-612-3876
www.bergen.cc.nj.us
(Harriet Terodemos, CVT-Coordinator)
2 years-Associate of Applied Science
Initial Accreditation-March, 2000
PROBATIONAL ACCREDITATION

New Mexico: Central New Mexico Community College
Veterinary Technology Program
525 Buena Vista SE
Albuquerque, NM 87106
505-224-5043
(Dr. Bonnie Snyder, Program Director)
Associate of Applied Science
Initial Accreditation-June 17, 2005
FULL ACCREDITATION

San Juan College
Veterinary Technology Distance Learning
Program

4601 College Blvd.
Farmington, NM 87402
505-566-3182
www.sjc.cc.nm.us/pages/1.asp
(David Wright, DVM Director)
Associate of Applied Science
Initial Accreditation-November 8,
  2006
PROVISIONAL ACCREDITATION

New York: Alfred State College
Veterinary Technology Program
Agriculture Science Building
Alfred, NY 14801
607-578-3009
FAX 607-587-4721
www.alfredstate.edu
(Melvin Chambliss, DVM-Director)
2 years-Associate of Applied Science
Initial Accreditation-September, 2001
FULL ACCREDITATION

La Guardia Community College
The City University of New York
Veterinary Technology Program
31-10 Thomson Ave.
Long Island City, NY 11101
718-482-5470
www.lagcc.cuny.edu
(Robin Sturtz, DVM-Director)
2 years-Associate of Applied Science
Initial Accreditation-April, 1983
FULL ACCREDITATION

Medaille College
Veterinary Technology Program
18 Agassiz Cr.
Buffalo, NY 14214
716-884-3281
FAX 716-884-0291
www.medaille.edu
(Joseph Savarese, DVM-Director)
2 years-Associate in Sciencev
Initial Accreditation-November, 1996
FULL ACCREDITATION

Mercy College
Veterinary Technology Program
555 Broadway
Dobbs Ferry, NY 10522
914-674-7530
www.mercy.edu/acadivisions/natscivettech/
  vettech.cfm
(Jack Burke, DVM-Director)
4 years-Bachelor of Science
Initial Accreditation-April, 1987
FULL ACCREDITATION

State University of New York-Canton
Agricultural & Technical College
Health Sciences & Medical Technologies
Veterinary Science Technology Program
34 Cornell Drive
Canton, NY 13617
315-386-7410
www.canton.edu
(Mary O'Horo-Loomis, DVM-Director)
2 years-Associate of Applied Science
Initial Accreditation-November, 1979
FULL ACCREDITATION

State University of New York-Delhi
College of Technology
Veterinary Science Technology Program
156 Farnsworth Hall
Delhi, NY 13753
607-746-4306
www.delhi.edu
(Dawn Dutka, DVM, Interim Director)
2 years-Associate of Applied Science
4 years-Bachelor in Business Administration
Initial Accreditation(2 yr) Dec., 1975 (4 yr)
  March, 2002
FULL ACCREDITATION

State University of New York-Ulster (Ulster
  County Community College)
Veterinary Technology Program
Cottekill Road
Stone Ridge, NY 12484
800-724-0833, ext 5233
www.sunyulster.edu
(Dr. Beth Alden, Program Director)
2 years-Associate of Applied Science
Initial Accreditaion-December 3, 2004
PROVISIONAL ACCREDITATION

Suffolk Community College
Veterinary Science Technology Program
Western Campus
Crooked Hill Rd.
Brentwood, NY 11717
631-851-6289
www.sunysuffolk.edu
(Elia Colon-Mallah, DVM-Director)
2 years-Associate of Applied Science
Initial Accreditation-April, 1996
FULL ACCREDITATION

North Carolina: Asheville-Buncombe Technical Community
  College
Veterinary Medical Technology Program
340 Victoria Road
Asheville, NC 28801
828-254-1921 ext. 273

www.abtech.edu/ah/vet/
(Dianne Cotter, LVT Director)
2 years-Associate of Applied Science
Initial Accreditation March 3, 2006
FULL ACCREDITATION

Central Carolina Community College
Veterinary Medical Technology Program
1105 Kelly Dr.
Sanford, NC 27330
919-775-5401
FAX 919-775-1221
www.ccarolina.cc.nc.us
(Paul Porterfield, DVM-Director)
2 years-Associate of Applied Science
Initial Accreditation-November, 1974
FULL ACCREDITATION

Gaston College
Veterinary Medical Technology Program
201 Hwy. 321 South
Dallas, NC 28034-1499
704-922-6200
FAX 704-922-6440
www.gaston.cc.nc.us
(Kristine Blankenship, DVM-Director)
2 years-Associate of Applied Science
Initial Accreditation-November, 1996
FULL ACCREDITATION

North Dakota: North Dakota State University
Veterinary Technology Program
Van Es Laboratories
Fargo, ND 58105
701-231-7511
FAX 701-231-7514
vettech.ndsu.nodak.edu
(Tom Colville, DVM-Director)
4 years-Bachelor of Science
Initial Accreditation-May, 1978
FULL ACCREDITATION

Ohio: Columbus State Community College
Veterinary Technology Program
550 E. Spring St.
Columbus, OH 43216
614-287-3685
www.cscc.edu
(Brenda Johnson, DVM-Director)
2 years-Associate of Applied Science
Initial Accreditation-May, 1974
FULL ACCREDITATION

Cuyahoga Community College
Veterinary Technology Program
11000 Pleasant Valley Rd.
Parma, OH 44130

216-987-5450
www.tri-c.cc.oh.us
(Kathy Corcoran, DVM Director)
2 years-Associate of Applied Science
Initial Accreditation-April, 1998
FULL ACCREDITATION

UC Raymond Walters College
Veterinary Technology Program
9555 Plainfield Road
Blue Ash, OH 45236
513-936-7173
FAX 513-558-0734
www.rwc.uc.edu
(Jennifer Wells, DVM-Director)
2 years-Associate of Applied Science
Initial Accreditation April, 1975
FULL ACCREDITATION

Stautzenberger College
Veterinary Technology Program
1796 Indian Wood Circle
Maumee, OH 43537
419-866-0261
FAX 419-866-0261 x237
www.stautzen.com
(Laurie Pfaff-Cherry, RVT Director)
2 years-Associate of Applied Science
Initial Accreditation-April, 1996
FULL ACCREDITATION

Stautzenberger College-Strongsville
Veterinary Technology Program
12925 Pearl Road
Strongsvilie, OH 44136
440-846-1999
www.learnwhatyoulove.com
(Jeanette Kwiatkowski, RVT Director)
2 years-Associate in Applied Science
Initial Accreditation-December 8, 2006
PROVISIONAL ACCREDITATION

Vet Tech Institute at Bradford School
2469 Stelzer Road
Columbus, OH 43219
614-416-6200
www.bradfordschoolcolumbus.edu
Angela Beal, DVM Program Director
abeal@bradfordschoolcolumbus.edu
Associate of Applied Science
PROVISIONAL ACCREDITATION effective
    August 11, 2006

Oklahoma: Murray State College
Veterinary Technology Program
One Murray Campus
Tishomingo, OK 73460

580-371-2371
FAX 580-371-9844
www.msc.cc.ok.us
(Carey Floyd, DVM-Director)
2 years-Associate of Applied Science
Initial Accreditation-November, 1980
FULL ACCREDITATION

Oklahoma State University-Oklahoma City
Veterinary Technology Program
900 N. Portland Ave.
Oklahoma City, OK 73107
405-945-9112
www.osuokc.edu
(Sally Henderson, DVM-Coordinator)
2 years-Associate of Applied Science
Initial Accreditation-March, 2000
FULL ACCREDITATION

Tulsa Community College
Veterinary Technology Program
7505 W. 41st St.
Tulsa, OK 74107
918-595-8212
FAX 918-595-8216
www.tulsacc.edu
(Jan Weaver, DVM-Coordinator)
2 years-Associate of Applied Science
Initial Accreditation-March, 2000
FULL ACCREDITATION

Oregon: Portland Community College
Veterinary Technology Program
P.O. Box 19000
Portland, OR 97219
503-244-6111
www.pcc.edu
(Brad Krohn, DVM-Division Chair)
2 years-Associate of Applied Science
Initial Accreditation-April, 1989
FULL ACCREDITATION

Pennsylvania: Harcum College
Veterinary Technology Program
750 Montgomery Ave.
Bryn Mawr, PA 19010-3476
610-526-6055
FAX 610-526-6031
www.harcum.edu
(Nadine Hackman, VMD-Director)
6 semesters-Associate in Science
Initial Accreditation-April, 1976
FULL ACCREDITATION

Johnson College
Veterinary Science Technology Program
3427 N. Main Ave.

Scranton, PA 18508
1-800-2WE-WORK -or- 570-342-6404
FAX: 570-348-2181
www.johnsoncollege.com
(Rosemary Cook, CVT-Chair)
2 years-Associate in Science
Initial Accreditation-November, 1995
FULL ACCREDITATION

Lehigh Carbon & Northampton
    Community Colleges
Veterinary Technology Program
3835 Green Pond Rd.
Bethlehem, PA 18020
610-861-5548
FAX 610-861-4132
www.lccc.edu
(Susan Stadler, DVM-Director)
2 years-Associate of Applied Science
Initial Accreditation-March, 1999
FULL ACCREDITATION

Manor College
Veterinary Technology Program
700 Fox Chase Road
Jenkintown, PA 19046
215-885-2360
www.manorvettech.com
(Joanna Bassert, VMD-Director)
2 years-Associate in Science
Initial Accreditation-April, 1992
FULL ACCREDITATION

The Vet Tech Institute
Veterinary Technician Program
125 Seventh Street
Pittsburgh, PA 15222
412-391-7021 or 800-570-0693
FAX 412-232-4345 x218
www.vettechinstitute.com
(Barbara Karner, DVM-Director)
16 months-Associate in Specialized
    Technology
Initial Accreditation-June 14, 2002
FULL ACCREDITATION

Sanford-Brown Institute (formerly Western
    School of Health and Business Careers)
Veterinary Technology Program
421 7th Avenue
Pittsburgh, PA 15219
412-281-2600
www.westernschool.com
(Cynthia Watkins, CVT-Program Director)
2 years-Associate of Specialized Technology
Initial Accreditation-September 22, 2004
PROVISIONAL ACCREDITATION

Wilson College
Veterinary Medical Technology Program
1015 Philadelphia Ave.
Chambersburg, PA 17201
717-264-4141
FAX 717-264-1578
www.wilson.edu
(Freya S. Burnett, CVT-Director)
4 years-Bachelor of Science-College for
    Women
Initial Accreditation-May, 1984
FULL ACCREDITATION

Puerto Rico: University of Puerto Rico
Veterinary Technology Program
Medical Sciences Campus
P.O. Box 365067
San Juan, PR 00936-5067
787-758-2525, ext. 1051 or 1052
www.cprsweb.rcm.upr.edu
(Elizabeth Rivera, DVM-Director)
4 years-Bachelor of Science
Initial Accreditation-April, 1996
FULL ACCREDITATION

South Carolina: Tri-County Technical College
Veterinary Technology Program
P.O. Box 587
Pendleton, SC 29670
864-646-8361
www.tctc.edu
(Peggy Champion, DVM-Director)
2 years-Associate in Applied Health
    Science
Initial Accreditation-April, 1979
FULL ACCREDITATION

Trident Technical College
Veterinary Technology Program
1001 South Live Oak Drive
Moncks Comer, SC 29461
843-899-8011
www.tridenttech.edu
(Paul Kerwin, DVM-Director)
2 years-Associate in Allied Health Sciences
Initial Accreditation-March, 2002
FULL ACCREDITATION

South Dakota: National American University
Allied Health Division
Veterinary Technology Program
321 Kansas City St.
Rapid City, SD 57701
800-843-8892
www.national.edu/veterinary tech.html
(Margaret Behrens, DVM-Director)
8 quarters-Associate in Applied Science

Initial Accreditation-April, 1981
FULL ACCREDITATION

Tennessee: Chattanooga State Technical Community
    College
4501 Amnicola Highway
Chattanooga, TN 37406-1097
423-697-4400
www.chattanoogastate.edu
(Susie Matthews, DVM-Director)
Associate of Applied Science
Initial Accreditation-March 28, 2008
PROVISIONAL ACCREDITATION

Columbia State Community College
Veterinary Technology Program
P.O. Box 1315, Health Sciences 105
Columbia, TN 38401
931-540-2722
www.coscc.cc.tn.us
(Boyce Wanamaker, DVM-Director)
2 years-Associate of Applied Science
Initial Accreditation-April, 1979
FULL ACCREDITATION

Lincoln Memorial University
Veterinary Technology Program
Cumberland Gap Pkwy.
LMU Box 1659
Harrogate, TN 37752
423-869-6278
FAX 423-869-7151
www.lmunet.edu
(Mary Hatfield, DVM-Director)
2 years-Associate in Science
4 years-Bachelor of Science
Initial Accreditation-November, 1987
FULL ACCREDITATION

Texas: Cedar Valley College
Veterinary Technology Program
3030 N. Dallas Ave.
Lancaster, TX 75134
972-860-8127
FAX 972-860-8057
ollie.dcccd.edu/vettech
(Steven Grubbs, DVM-Director)
On-Campus Program
2 years-Associate of Applied Science
Initial Accreditation-November, 1978
FULL ACCREDITATION

Distance Learning Program
Associate of Applied Science
Initial Accreditation-September, 2001
FULL ACCREDITATION
(Kelly Alan Black, DVM-Director)

McLennan Community College
Veterinary Technology Program
1400 College Drive
Waco, TX 76708
254-299-8750
www.mclennan.edu/departments/workforce/
   vtech/
(Jennifer Garretson, DVM-Director)
2 years-Associate of Applied Science
Initial Accreditation-January 23, 2004
FULL ACCREDITATION

Midland College
Veterinary Technology Program
3600 N. Garfield
Midland, TX 79705
432-685-4619
FAX 432-685-6431
www.midland.edu
(Kerry Coombs, DVM-Director)
2 years-Associate of Applied Science
Initial Accreditation-November 1992,
   April 1995
FULL ACCREDITATION

Palo Alto College
Veterinary Technology Program
1400 W. Villaret Blvd.
San Antonio, TX 78224-2499
210-486-3355
FAX 210-486-3356
www.accd.edu
(Fonzie Quance-Fitch, DVM-Director)
2 years-Associate of Applied Science
Initial Accreditation-April, 1998
FULL ACCREDITATION

Sul Ross State University
School of Agriculture & Natural Resource
   Sciences
Veterinary Technology Program
P.O. Box C-114
Alpine, TX 79830
432-837-8205
FAX-432-837-8409
www.sulross.edu
(Darwin R. Yoder, DVM-Director)
2 years-Associate Degree
Initial Accreditation-November, 1977
PROBATIONAL ACCREDITATION

Lone Star College-Tomball
Veterinary Technology Program
30555 Tomball Pkwy.
Tomball, TX 77375-4036
281-351-3357
FAX-281-351-3384

wwwtc.nhmccd.cc.tx.us/
(George W. Younger, DVM-Director)
2 years-Associate of Applied Science
Initial Accreditation-April, 1990
FULL ACCREDITATION

Utah: Utah Career College
               Veterinary Technician Program
               1902 West 7800 South
               West Jordan, UT 84088
               801-304-4224
               FAX 801-676-2220
               www.utahcollege.com/ushs/veterinary_techni-
                  cian.html
               (Alex Arangua, DVM-Program Director)
               2 years-Associate of Applied Science
               Initial Accreditation-August 22, 2003
               PROVISIONAL ACCREDITATION

Vermont: Vermont Technical College
               Veterinary Technology Program
               Randolph Center, VT 05061
               802-728-3391
               www.vtc.vsc.edu
               (Amy Woodbury-St. Denis, DVM-Director)
               2 years-Associate of Applied Science
               Initial Accreditation-April, 1991
               FULL ACCREDITATION

Virginia: Blue Ridge Community College
               Veterinary Technology Program
               Box 80
               Weyers Cave, VA 24486
               540-234-9261
               FAX 540-234-9066
               www.br.cc.va.us
               (Stuart Porter, VMD-Director)
               2 years-Associate of Applied Science
               On-Campus Initial Accreditation-
                  November, 1976
               Distance Learning Initial Accreditation
                  (VA residents only)-November, 1999
               FULL ACCREDITATION

               Northern Virginia Community College
               Veterinary Technology Program
               Loudoun Campus
               1000 Harry Flood Byrd Hwy.
               Sterling, VA 20164-8699
               703-450-2525
               FAX 703-450-2536
               www.nv.cc.va.us
               On Campus Program
               (Leslie Sinn, DVM-Head)
               2 years-Associate of Applied Science
               Initial Accreditation-April, 1980
               FULL ACCREDITATION

Distance Learning Program
Associate of Applied Science
(Terri Champney, DVM-Director)
Initial Accreditation-February 20, 2004
FULL ACCREDITATION

Washington: Pierce College Ft. Steilacoom
Veterinary Technology Program
9401 Farwest Dr., SW
Lakewood, WA 98498
253-964-6668
www.pierce.ctc.edu
(Salvador Hurtado, DVM-Director)
2 years-Associate in Veterinary Technology
Initial Accreditation-May, 1974
FULL ACCREDITATION

Pima Medical Institute-Seattle
Veterinary Technology Program
9709 Third Ave. NE, Suite 400
Seattle, WA 98115
(Dr. Carol Mayer, Program Director)
800-477-PIMA
www.pmi.edu
15-18 months-Occupational Associate Degree
Initial Accreditation-October 2, 2007
PROVISIONAL ACCREDITATION

Yakima Valley Community College
Veterinary Technology Program
P.O. Box 22520
Yakima, WA 98907-1647
509-574-4759
FAX 509-547-4751
www.yvcc.edu
(Susan Wedam, DVM-Coordinator)
2 years-Associate of Applied Science
Initial Accreditation-March, 1999
FULL ACCREDITATION

West Virginia: Pierpont Community & Technical College
Veterinary Technology Program
1201 Locust Ave.
Fairmont, WV 26554
304-367-4589
www.fscwv.edu
(Anna Romano, RVT-Coordinator)
2 years-Associate of Applied Science
Initial Accreditation-April, 1981
PROBATIONAL ACCREDITATION

Wisconsin: Madison Area Technical College
Veterinary Technician Program
3550 Anderson
Madison, WI 53704
608-246-6100
www.madison.tec.wi.us/matc

(Jane Clark, DVM-Director)
2 years-Associate in Applied Science
Initial Accreditation-November, 1974
FULL ACCREDITATION

Moraine Park Technical College
Veterinary Technology Distance Learning
Program
235 N. National Ave.
Fond du Lac, WI 54936
920-922-8611
www.morainepark.edu
(Laura Lien, BS, CVT-Program Director)
2 years-Associate of Applied Science
Initial Accreditation-September 7, 2007
PROVISIONAL ACCREDITATION

Wyoming: Eastern Wyoming College
Veterinary Technology Program
3200 W. "C" St.
Torrington, WY 82240
800-658-3195, ext. 8268
www.ewc.wy.edu
(Patti Sue Peterson, AHT-Director)
2 years-Associate in Applied Science
Initial Accreditation-April, 1976
FULL ACCREDITATION

Classification of Accreditation

The final appraisal of each program is based on review of the total educational program by an evaluation site team visit and accreditation decision by CVTEA, with consideration given to each stated objective and the Standards as listed. Classification of accreditation of a program in veterinary technology shall be in effect from the last day of the site visit until acted upon by CVTEA. The accreditation status of each program will be published semi-annually and made available upon request.

Full Accreditation—Those programs that meet or exceed all Standards of Accreditation.

Provisional Accreditation—Provisional accreditation is granted to new programs in veterinary technology where students have not completed the entire curriculum and/or the programs have not produced sufficient numbers of graduates to adequately assess outcomes. Programs may remain on provisional accreditation for a period not to exceed five years. A classification of provisional accreditation is not punitive. Graduates of a provisionally accredited program are graduates of an AVMA-accredited program.

Probational Accreditation—Probationally accredited programs do not fully meet all accreditation standards, but are still accredited. Probational accreditation is not an adverse accreditation action and should be viewed as remedial, not punitive. Programs may remain on probational accreditation for a maximum of two years. Graduates of a probationally accredited program are graduates of an AVMA-accredited program.

Terminal Accreditation—Terminal accreditation is granted to programs that are voluntarily discontinuing. Students enrolled at the time that a program is terminally accredited will graduate from an AVMA-accredited program if they complete the program within the time limit of the accreditation status.

Accreditation Withdrawn—Previously accredited programs that have had their accreditation withdrawn will be classified as Accreditation Withdrawn. Graduates of the program at the time of withdrawn accreditation will not be considered graduates of an AVMA accredited program.

Graduates of a program classified as Provisional, Probational, or Terminal Accreditation are considered to be graduates of an AVMA-accredited program in veterinary technology.

General Statistics

Total Number of Programs: 145
Number offering 4yr degree: 16
Number offering distance learning: 9

The following states do not have AVMA-accredited veterinary technology programs: Alaska, Arkansas, District of Columbia, Hawaii, Montana, Rhode Island.

Reprinted with permission of the American Veterinary Medical Association 1931 North Meacham Road, Suite 100, Schaumberg, Illinois 60173-4360.

APPENDIX C

RECOGNIZED VETERINARY SPECIALTY ORGANIZATIONS

American Board of Veterinary Practitioners
Mr. Jeff Allen, Executive Director
618 Church St, Suite 220
Nashville, TN 37219
Phone: 615-254-3687
Fax: 615-254-7047
E-mail: Jallen@XMI-AMC.com
Web site: http://www.abvp.com

American Board of Veterinary Toxicology
Dr. Birgit Puschner, Secretary-Treasurer
CAHFS Toxicology Laboratory
School of Veterinary Medicine
University of California
West Health Sciences Drive
Davis, CA 95616
Phone: 530-752-6322
Fax: 530-752-3361
E-mail: bpuschner@ucdavis.edu
Web site: http://www.abvt.org

American College of Laboratory Animal Medicine
Dr. Melvin W. Balk, Executive Director
96 Chester Street
Chester, NH 03036
Phone: 603-887-2467
Fax: 603-887-0097
E-mail: mwbaclam@gsinet.net
Web site: http://www.aclam.org

American College of Poultry Veterinarians
Dr. Sherrill Davison, Secretary/Treasurer
University of Pennsylvania
382 West Street Road
Kennett Square, PA 19348
Phone: 610-444-5800, Ext 2710
Fax: 610-925-8106
E-mail: acpv@vet.upenn.edu
Web site: http://www.acpv.info/

American College of Theriogenologists
Dr. Charles Franz, Executive Director
P.O. Box 3065
Montgomery, AL 36109
Phone: 334-395-4666
Fax: 334-270-3399
E-mail: charles@franzmgt.com
Web site: http://www.theriogenology.org/

American College of Veterinary Anesthesiologists
Dr. Lydia Donaldson, Executive Secretary
P.O. Box 1100
Middleburg, VA 20118
Phone: 540-687-5270
Fax: 540-687-6615
E-mail: ldonldsn@earthlink.net
Web site: http://www.acva.org/

American College of Veterinary Behaviorists
Dr. Bonnie V. Beaver, Executive Director
Texas A&M University
Department of Small Animal Medicine & Surgery
4474 TAMU
College Station, TX 77843-4474
Phone: 979-845-2351
Fax: 979-845-6978
E-mail: mail@dacvb.org
Web site: http://www.dacvb.org/

American College of Veterinary Clinical Pharmacology
Dr. Albert Boeckh, Secretary-Treasurer
Merial Limited
Pharmaceutical R&D
3239 Satellite Blvd., Building 500 - #258
Duluth, GA 30096
Phone: 678-638-3634
Fax: 678-638-3636
E-mail: albert.boeckh@merial.com
Web site: http://www.acvcp.org

American College of Veterinary Dermatology
Ms. Alexis Borich, Executive Secretary
5610 Kearney Mesa Road, Suite B1
San Diego, CA 92111
Phone: 858-560-9393
Fax: 858-560-0206
E-mail: itchypet@aol.com
Web site: http://www.acvd.org

American College of Veterinary Emergency and Critical Care
Dr. Armelle de Laforcade, Executive Secretary
Tufts Cummings School of Veterinary Medicine
200 Westboro Road
North Grafton, MA 01536
Phone: 508-839-5395 (ext. 84415)
Fax: 508-839-7922
E-mail: Armelle.delaforcade@tufts.edu
Web site: www.acvecc.org

American College of Veterinary Internal Medicine
Ms. June Pooley, Executive Director
1997 Wadsworth, Suite A
Lakewood, CO 80214
Phone: 303-231-9933 or 800-245-9081
Fax: 303-231-0880
E-mail: acvim@acvim.org
Web site: http://www.acvim.org

American College of Veterinary Microbiologists
Dr. Chris Hayhow, Secretary-Treasurer
30705 W. 84th Circle
DeSoto, KS 66018
Phone: 913-894-0230
Fax: 913-894-0236
E-mail: chayhow@biomunecompany.com
Web site: http://www.acvm.us

American College of Veterinary Nutrition
Dr. Wilbur Amand, Executive Director
6 North Pennell Road
Media, PA 19063-5520
Phone: 610-892-4812
Fax: 610-892-4813
E-mail: Wbamand@aol.com
Web site: www.acvn.org

American College of Veterinary Ophthalmologists
Ms. Stacee Daniel, Executive Director
P.O. Box 1311
Meridian, ID 83680
Phone: 208-466-7624
Fax: 208-466-7693
E-mail: office06@acvo.org
Web site: http://www.acvo.org

American College of Veterinary Pathologists
Ms. Wendy Coe, Executive Director
ACVP Executive Offices
2810 Crossroads Drive, Suite 3800
Madison, WI 53718
Phone: 608-443-2466 Ext 149
Fax: 608-443-2478
E-mail: wcoe@acvp.org
Web site: http://www.acvp.org/

American College of Veterinary Preventive Medicine
Dr. Russell W. Currier
Executive Vice President
P.O. Box 22219
Clive, IA 50325
Phone: 515-331-4439
Fax: 515-331-4947
E-mail: evp@acvpm.org
Web site: http://www.acvpm.org

American College of Veterinary Radiology
Dr. Robert Pechman, Executive Director
5205 W. 4th Avenue
Stillwater, OK 74074
Phone: 405-372-5331
E-mail: Pechman@mail.vetmed.lsu.edu
Web site: http://www.acvr.org

American College of Veterinary Surgeons
Dr. Alan J. Lipowitz, Executive Secretary
Ann T. Loew, EdM, Executive Director
11 N. Washington St., Suite 720
Rockville, MD 20850
Phone: 301-610-2000
Fax: 301-610-0371
E-mail: acvs@acvs.org
Web site: http://www.acvs.org

American College of Zoological Medicine
Dr. Craig A. Harms, Secretary
North Carolina State University
College of Veterinary Medicine
Department of Clinical Sciences
Center for Marine Sciences and Technology
303 College Circle
Morehead City, NC 28557
Phone: 252-222-6339
Fax: 252-222-6311
E-mail: craig_harms@ncsu.edu
Web site: http://www.aczm.org/aczmmain.html

American Veterinary Dental College
Dr. Colin E. Harvey, Secretary
University of Pennsylvania
VHUP 3113
3900 Delancey Street
Philadelphia, PA 19104-6043
Phone: 215-898-5903
Fax: 215-898-9937
E-mail: ceh@vet.upenn.edu
Web site: http://www.AVDC.org

Reprinted with permission of the American Veterinary Medical Association 1931 North Meacham Road, Suite 100, Schaumberg, Illinois 60173-4360.

APPENDIX D

REPORTED CASES OF LYME DISEASE

Reported Lyme Disease Cases by State, 1993–2006

State	1993	1994	1995	1996	1997	1998	1999	2000	2001	2002	2003	2004	2005	2006	Incidence 2006*
Alabama	4	6	12	9	11	24	20	6	10	11	8	6	3	11	0.2
Alaska	0	0	0	0	2	1	0	2	2	3	3	3	4	3	0.4
Arizona	0	0	1	0	4	1	3	2	3	4	4	13	10	10	0.2
Arkansas	8	15	11	27	27	8	7	7	4	3	0	0	0	0	0.0
California	134	68	84	64	154	135	139	95	95	97	86	48	95	85	0.2
Colorado	0	1	0	0	0	0	3	0	0	1	0	0	0	0	0.0
Connecticut	1350	2030	1548	3104	2297	3434	3215	3773	3597	4631	1403	1348	1810	1788	51.0
Delaware	143	106	56	173	109	77	167	167	152	194	212	339	646	482	56.5
DC	2	9	3	3	10	8	6	11	17	25	14	16	10	62	10.7
Florida	30	28	17	55	56	71	59	54	43	79	43	46	47	34	0.2
Georgia	44	127	14	1	9	5	0	0	0	2	10	12	6	8	0.1
Hawaii	1	0	0	1	0	0	0	0	0	0	0	0	0	0	0.0
Idaho	2	3	2	2	4	7	3	4	5	4	3	6	2	7	0.5
Illnois	19	24	18	10	13	14	17	35	32	47	71	87	127	110	0.9
Indiana	32	19	19	32	33	39	21	23	26	21	25	32	33	26	0.4
Iowa	8	17	16	19	8	27	24	34	36	42	58	49	89	97	3.3
Kansas	54	17	23	36	4	13	16	17	2	7	4	3	3	4	0.1
Kentucky	16	24	16	26	20	27	19	13	23	25	17	15	5	7	0.2
Louisiana	3	4	9	9	13	15	9	8	8	5	7	2	3	1	0.0
Maine	18	33	45	63	34	78	41	71	108	219	175	225	247	338	25.6
Maryland	180	341	454	447	494	659	899	688	608	738	691	891	1235	1248	22.2
Massachusetts	148	247	189	321	291	699	787	1158	1164	1807	1532	1532	2336	1432	22.2
Michigan	23	33	5	28	27	17	11	23	21	26	12	27	62	55	0.5
Minnesota	141	208	208	251	256	261	283	465	461	867	474	1023	917	914	17.7
Mississippi	0	0	17	24	27	17	4	3	8	12	21	0	0	3	0.1
Missouri	108	102	53	52	28	12	72	47	37	41	70	25	15	5	0.1
Montana	0	0	0	0	0	0	0	0	0	0	0	0	0	1	0.1
Nebraska	6	3	6	5	2	4	11	5	4	6	2	2	2	11	0.6
Nevada	5	1	6	2	2	6	2	4	4	2	3	1	3	4	0.2
New Hampshire	15	30	28	47	39	45	27	84	129	261	190	226	265	617	46.9
New Jersey	786	1533	1703	2190	2041	1911	1719	2459	2020	2349	2887	2698	3363	2432	27.9
New Mexico	2	5	1	1	1	4	1	0	1	1	1	1	3	3	0.2

(continued)

Reported Lyme Disease Cases by State, 1993–2006 (continued)

State	1993	1994	1995	1996	1997	1998	1999	2000	2001	2002	2003	2004	2005	2006	Incidence 2006*
New York	2818	5200	4438	5301	3327	4640	4402	4329	4083	5535	5399	5100	5565	4460	23.1
North Carolina	86	77	84	66	34	63	74	47	41	137	156	122	49	31	0.4
North Dakota	2	0	0	2	0	0	1	2	0	1	0	0	3	7	1.1
Ohio	30	45	30	32	40	47	47	61	44	82	66	50	58	43	0.4
Oklahoma	19	99	63	42	45	13	8	1	0	0	0	3	0	0	0.0
Oregon	8	6	20	19	20	21	15	13	15	12	16	11	3	7	0.2
Pennsylvania	1085	1438	1562	2814	2188	2760	2781	2343	2806	3989	5730	3985	4287	3242	26.1
Rhode Island	272	471	345	534	442	789	546	675	510	852	736	249	39	308	28.8
South Carolina	9	7	17	9	3	8	6	25	6	26	18	22	15	20	0.5
South Dakota	0	0	0	0	1	0	0	0	0	2	1	1	2	1	0.1
Tennessee	20	13	28	24	45	47	59	28	31	28	20	20	8	15	0.2
Texas	48	56	77	97	60	32	72	77	75	139	85	98	69	29	0.1
Utah	2	3	1	1	1	0	2	3	1	5	2	1	2	5	0.2
Vermont	12	16	9	26	8	11	26	40	18	37	43	50	54	105	16.8
Virginia	95	131	55	57	67	73	122	149	156	259	195	216	274	357	4.7
Washington	9	4	10	18	11	7	14	9	9	11	7	14	13	8	0.1
West Virginia	50	29	26	12	10	13	20	35	16	26	31	38	61	28	1.5
Wisconsin	401	409	369	396	480	657	490	631	597	1090	740	1144	1459	1466	26.4
Wyoming	9	5	4	3	3	1	3	3	1	2	2	4	3	1	0.2
U.S. TOTAL	8,257	13,043	11,700	16,455	12,801	16,801	16,273	17,730	17,029	23,763	21,273	19,804	23,305	19,931	8.24

*cases per 100,000 population

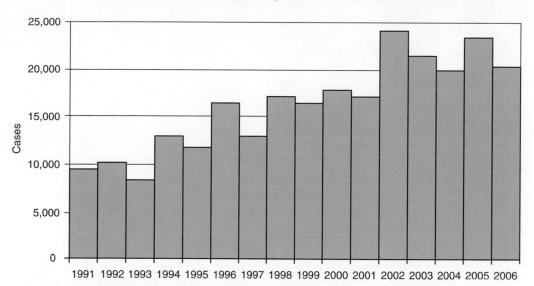

In 2006, 19,931 cases of Lyme disease were reported, yielding a national average of 8.2 cases per 100,000 persons. In the 10 states where Lyme disease is most common, the average was 30.2 cases per 100,000 persons. *Courtesy of the Center For Disease Control and Prevention (CDC).*

INDEX

A

C

T